Sustainable Biomanufacturing

This book discusses the role of sustainable biomanufacturing in advancing Industry 5.0, achieving carbon neutrality, and fostering innovation in diverse sectors. It emphasizes the integration of cutting-edge technologies, such as artificial intelligence and machine learning, with sustainable practices to address global challenges in manufacturing, environmental conservation, and resource management. The initial chapters explore the foundational concepts, historical evolution, and applications of sustainable biomanufacturing, alongside its contributions to the UN Sustainable Development Goals and the circular economy. The subsequent chapters present real-world case studies highlighting biomanufacturing innovations across industries, including bioenergy, biomedical engineering, agriculture, food production, personal care, and cosmetics. It further examines the role of advanced tools and techniques in creating sustainable enzymes, biocatalysts, and biochemicals while discussing industry-driven solutions for CO_2 sequestration and carbon neutrality.

Key Features:

- Explores the foundations of sustainable biomanufacturing and its role in addressing global sustainability challenges
- Highlights real-world applications across industries, including bioenergy, healthcare, agriculture, and cosmetics
- Discusses innovative tools, such as artificial intelligence (AI) and machine learning, to enhance sustainable production systems
- Provides insights into governance policies, life cycle assessment, and research trends for sustainable manufacturing
- Examines future opportunities, bridging the gap between industry needs and sustainable innovation

This book is useful for professionals in biotechnology and environmental sciences.

Sustainable Industrial and Environmental Bioprocesses

This book series aims to provide a comprehensive collection of books focusing on bioprocesses in industrial and environmental biotechnology. The multidisciplinary content encompasses the chemical and biochemical engineering, industrial microbiology, and energy biosciences, all with a central theme of sustainable development and circular economy principles. The books align with the Sustainable Development Goals (SDGs) and offer state-of-the-art information and in-depth knowledge on the subject matter. The book in these series also emphasizes on the application of emerging tools, such as machine learning and artificial intelligence, for the advancement of bioprocesses. While primarily targeting academicians and researchers, the series is valuable for policy planners and industry professionals, with carefully tailored contents to cater to their specific needs and interests.

Series Editor: Dr Ashok Pandey

Professor Ashok Pandey is currently Distinguished Scientist at the Centre for Innovation and Translational Research, CSIR-Indian Institute of Toxicology Research, Lucknow, India. His major research and technological development interests are industrial & environmental biotechnology and energy biosciences, focusing on biomass to biofuels & chemicals, waste to wealth & energy, etc. He has 16 patents, 120 books, >1,000 papers and book chapters, etc. with h index of 131 and >74,000 citations. Professor Pandey is the recipient of many national and international awards and honours, which include Fellow, The World Academy of Sciences (TWAS), Highly Cited Researcher (Top 1% in the world), Clarivate Analytics (since 2018- till date), Rank #1 in India in Biology in India and Biochemistry and #417 in the world, Research.Com (2023), Rank #1 in India in Microbiology under Enabling and Strategic Technologies sector, Elsevier (2021), Rank #1 in India in Biotechnology and #8 in the world, Stanford University Report (2020-2021-2022), etc.

Biodegradation of Toxic and Hazardous Chemicals: Detection and Mineralization
Kashyap K. Dubey, Kamal. K. Pant, Ashok Pandey and Maria Angeles Sanromán

Biodegradation of Toxic and Hazardous Chemicals: Remediation and Resource Recovery
Kashyap K. Dubey, Kamal K. Pant, Ashok Pandey and Maria Angeles Sanromán

Waste Management in Climate Change and Sustainability- Lignocellulosic Waste
Sunita Varjani, Izharul Haq, Ashok Pandey, Vijai Kumar Gupta, and Xuan-Thanh Bui

Ashwagandha for Quality of Life: Scientific Evidence
Sunil C Kaul & Renu Wadhwa

Microfluidics in Food Processing:Technologies and Applications
Ayon Tarafdar, Ranjna Sirohi, Barjinder Pal Kaur, Ashok Pandey, and Claude-Gilles Dussap

Sustainable technologies in Food waste management
Ranjna Sirohi, Ayon Tarafdar, Luciana Porto de Souza Vandenberghe, Mohammad J. Taherzadeh, and Ashok Pandey

Microplastic Pollution: Occurrence, Health Risk and Challenges
Navish Kataria, Vinod Kumar Garg, Changseok Han and Eldon Rene

Diatoms and Sustainable Agriculture
Archana Tiwari and Ashok Pandey

Biochar and the Circular Economy: Innovations in Production and Applications
Hrudayanath Thatoi, Shibani Mohapatra, Ashok Pandey, Lam Su Shiung

Sustainable Technologies for Value Addition and Management of Plastic Waste
Abha Kumari, Ashok Pandey, Rohidas Bhoi, Ahn Tuan Hoang

Sustainable Biomanufacturing: Innovations for a Circular Bioeconomy and Industry 5.0
Navish Kataria, Vinod Kumar Garg, and Ashok Pandey

For more information about this series, please visit: https://www.routledge.com/Sustainable-Industrial-and-Environmental-Bioprocesses/book-series/SIEB

Sustainable Biomanufacturing

Innovations for a Circular Bioeconomy and Industry 5.0

Edited by
Navish Kataria
Vinod Kumar Garg
Ashok Pandey

CRC Press
Taylor & Francis Group
Boca Raton London New York

CRC Press is an imprint of the
Taylor & Francis Group, an **informa** business

Designed cover image: Shutterstock

First edition published 2027
by CRC Press
2385 NW Executive Center Drive, Suite 320, Boca Raton FL 33431

and by CRC Press
4 Park Square, Milton Park, Abingdon, Oxon, OX14 4RN

CRC Press is an imprint of Taylor & Francis Group, LLC

ISBN: 978-1-041-06799-3 (hbk)
ISBN: 978-1-041-06800-6 (pbk)
ISBN: 978-1-003-63715-8 (ebk)

DOI: 10.1201/9781003637158

Typeset in Times
by SPi Technologies India Pvt Ltd (Straive)

Contents

Author the Editors

Navish Kataria is an Assistant Professor in the Department of Environmental Sciences at J.C. Bose University of Science and Technology, YMCA, Faridabad, Haryana, India. He earned his Ph.D. in Environmental Science and Engineering from Guru Jambheshwar University of Science and Technology, Hisar, Haryana, India. He was awarded the Dr. D.S Kothari Postdoctoral Fellowship by the University Grants Commission (UGC), New Delhi, India. He has published more than 60 research articles (with more than 3,500 citations) in high-impact international journals and 11 book chapters. Dr. Kataria has published four books (three with CRC Press and one with Springer). He has also served as managing guest editor for the reputed journal ESPR (Springer journal) and has published an editorial. He has also served as a reviewer for reputed publishers, including some by Elsevier, Springer, Taylor & Francis, MDPI, and De Gruyter. His major research interests are environmental pollution management, water and wastewater treatment technologies, green/sustainable technologies for pollution, and biowaste management.

Vinod Kumar Garg is a Senior Professor in the Department of Environmental Science and Technology at the Central University of Punjab, Bathinda, India. He previously served at Guru Jambheshwar University of Science and Technology and Chaudhary Charan Singh Haryana Agricultural University, Hisar, India, in various capacities. Professor Garg has published over 200 research and review articles, as well as two books. He was honored with the "Thomson Reuters Research Excellence – India Citation Award 2012." He is an active member of various scientific societies and organizations, including the Biotech Research Society of India and the Indian Nuclear Society. In 2011, he was elected a Fellow of the Biotech Research Society of India.

Ashok Pandey is the Executive Director of the Centre for Energy and Environmental Sustainability, Lucknow, Uttar Pradesh, India. He previously served as a Distinguished Scientist at the Centre for Innovation and Translational Research, Council of Scientific & Industrial Research (CSIR)-Indian Institute of Toxicology Research, Lucknow; Eminent Scientist at the Centre of Innovative and Applied Bioprocessing, Mohali; and Chief Scientist and Head of the Centre for Biofuels at CSIR-National Institute for Interdisciplinary Science and Technology, Trivandrum, Kerala, India. His major research and technological development interests include industrial and environmental biotechnology and energy biosciences, with a focus on biomass-to-biofuels and chemicals, as well as waste-to-wealth and energy conversion. Professor Pandey has received numerous national and international awards and honors. He is a Fellow of five international and four national academies, including the World Academy of Sciences and the National Academy of Sciences, India. He is also a Distinguished Fellow of the Biotech Research Society of India and an Academician of the European Academy of Sciences and Arts, among others.

1 Sustainable Biomanufacturing
Concept, Historical Perspectives, and Applications

Kavita Sharma and Amit Singh Kshatri
Department of Fisheries Resource Management, College of Fisheries Science, CCS HAU, Hisar, Haryana, India

Vinod Kumar Garg
Department of Environmental Science and Technology, Central University of Punjab, Bathinda, Punjab, India

Navish Kataria
Department of Environmental Sciences, J.C. Bose University of Science and Technology, YMCA, Faridabad, Haryana, India

1.1 INTRODUCTION

Biomanufacturing is a manufacturing technique that uses engineered biological systems, such as microorganisms, cultured animal or plant cells, tissues, enzymes, or synthetic enzymatic systems, to produce high-value, commercially significant biomolecules (Shih et al., 2013; Oluyemi et al., 2024). It spans applications across pharmaceuticals (e.g., vaccines, insulin, and monoclonal antibodies [MABs]), food and nutrition (e.g., enzymes and amino acids), materials (e.g., bioplastics and textiles), agriculture, and energy. Conventional biomanufacturing has not automatically ensured sustainability despite its technological maturity in several sectors, since many of its processes are still resource dependent, energy-intensive, and economically constrained at scale. Sustainable biomanufacturing refers to biomanufacturing that delivers value-added products with minimal environmental footprint, strong resource circularity, economic viability, and positive social impact (Rasor et al., 2021; Xu et al., 2021). It aims to produce chemicals, materials, and other products through biological processes to enhance resource efficiency and reduce environmental impact. This approach utilizes renewable resources, optimizes processes to maximize energy efficiency, and minimizes waste generation. Nevertheless, accomplishing these objectives concurrently remains a significant scientific and engineering challenge, particularly when bringing an invention from the laboratory to industrial scale. Its products may be derived from natural sources, blood, microbial cultures, or plant or animal cells grown under controlled conditions or using specialized equipment. These processes can use naturally occurring cells, tissues, enzymes, or they can be altered via synthetic biology, genetic engineering, metabolic engineering, or protein engineering (Del Hierro et al., 2024).

Biomanufacturing covers several important areas, including biopharmaceuticals, such as vaccines, therapeutic proteins, and MABs. Bioproducts, such as biofuels and sustainable biomaterials, are applicable across sectors such as healthcare, agriculture, and energy. Additionally, it extended to advanced fields such as cell and gene therapy, synthetic biology, and microbial fermentation, which contribute to the development of innovative and sustainable biosynthesis. It facilitates the development of climate-resilient crop varieties and promotes environmentally friendly agricultural practices (Iragavarapu et al., 2023). It supports the general shift to a more sustainable and circular bioeconomy by enhancing access to clean water and sanitation, promoting sustainable industrial growth, and mitigating climate change (Bhatia et al., 2024). However, there are still significant limitations in terms of integration with current industrial infrastructure, techno-economic viability, process scalability, and feedstock sustainability.

By leveraging biological processes such as those involving microorganisms, biomanufacturing supports the development of bioproducts that are typically more sustainable and resource-efficient than those produced through conventional manufacturing methods. However, supply-chain inefficiencies, unpredictable biomass availability, and energy-intensive downstream processing frequently limit the circular potential of biomanufacturing. As the world

DOI: 10.1201/9781003637158-1

moves toward a circular economy and the next phase of industrial development, Industry 5.0, there is a stronger focus on making manufacturing processes more eco-friendly, socially fair, and economically practical. In India, the Union Cabinet approved the Biotechnology for Economy, Environment, and Employment (BioE3) policy on August 24, 2024, in response to the need for sustainability. The policy emphasizes six thematic areas: bio-based chemicals, smart proteins, precision therapeutics, climate-resilient agriculture, carbon capture and utilization and marine and space biotechnology to enhance domestic biomanufacturing capacity, foster innovation, and develop a skilled workforce.

The aim should be to create a regenerative manufacturing model aligned with the national goals of net-zero carbon growth, green innovation, and job creation through creating biomanufacturing and bio-artificial intelligence (Bio-AI) hubs and biofoundries across the country. Although the policy offers a solid institutional framework, its success will rely on resolving issues with sustainable scale-up, labor skill gaps, and technology constraints.

Recent research emphasizes how urgent it is to move toward sustainable production systems, stressing the significance of data-driven process optimization, ethical and trustworthy AI, and green supply chains (Ijaz et al., 2024; Hasan, 2024). In considering this, the concept, historical development, enabling technologies, and applications of sustainable biomanufacturing are critically examined in this chapter, which also methodically addresses its technological limitations, sustainability trade-offs, and possibilities. The chapter aims to provide a balanced and analytical foundation for promoting sustainable biomanufacturing within a circular bioeconomy framework through the integration of scientific, technological, and policy aspects.

1.2 CONCEPT OF SUSTAINABLE BIOMANUFACTURING

Sustainable biomanufacturing is a more recent form of biomanufacturing in which biological or cell-based systems are used to convert renewable, waste-derived, or low-grade biomass into high-value products without depleting resources or generating toxic emissions (Figure 1.1). This approach is rooted in the following:

- **Circular bioeconomy:** Cascading utilization of biomass, using every fraction (e.g., food waste and lignocellulosic residues) for successive applications, such as fermentations, biorefining, or composting.
- **Waste biorefinery:** The combining methods, such as anaerobic digestion, enzymatic hydrolysis, and fermentation, to transform waste streams into chemicals, proteins, biomaterials, and biofuels.

Biomanufacturing has advanced as a result of the growing drive to achieve sustainability and the recognition of

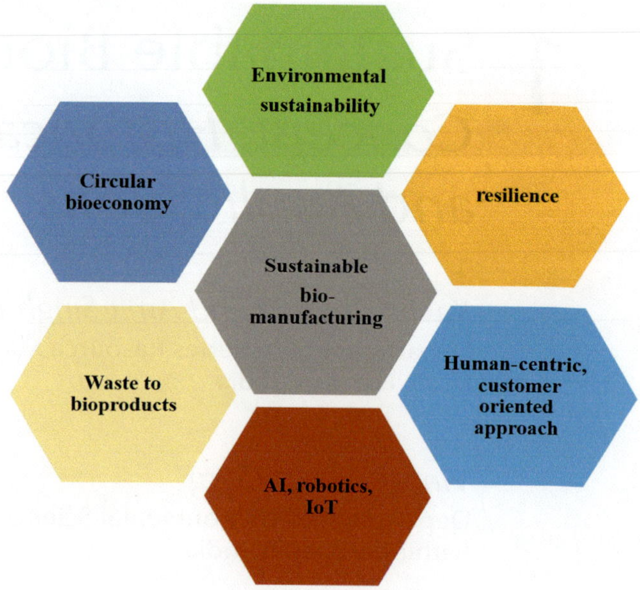

FIGURE 1.1 Sustainable biomanufacturing.

the significance of bio-based commodities. The recent Industrial Revolution is more focused on net-zero carbon emissions, along with lower and zero pollution levels from manufacturing processes and products while they are in use. At the intersection of industrial biotechnology, green chemistry, the circular bioeconomy, and systems engineering, the goals of sustainable biomanufacturing include the following:

- Maximizing biomass efficiency and low-carbon energy integration.
- Up-cycling waste streams rather than consuming virgin feedstocks.
- Closing carbon, nutrient, and water loops.
- Enabling renewable (bio-based) products to replace petrochemicals.
- Supporting the United Nations (UN) Sustainable Development Goals, such as responsible consumption, climate action, industrial innovation, and zero hunger.
- Driven by growing environmental concerns in the 21st century, industries are undergoing profound transformations, and sustainable biomanufacturing is the latest frontier in this.

1.3 HOW BIOMANUFACTURING TAKES PLACE?

The first step in the biomanufacturing process involves the development of suitable bioprocesses for the production of the desired bioproduct. This includes the selection of appropriate host organisms, such as bacteria, yeast, or mammalian cells, and the optimization of their growth and production conditions. The process of biomanufacturing

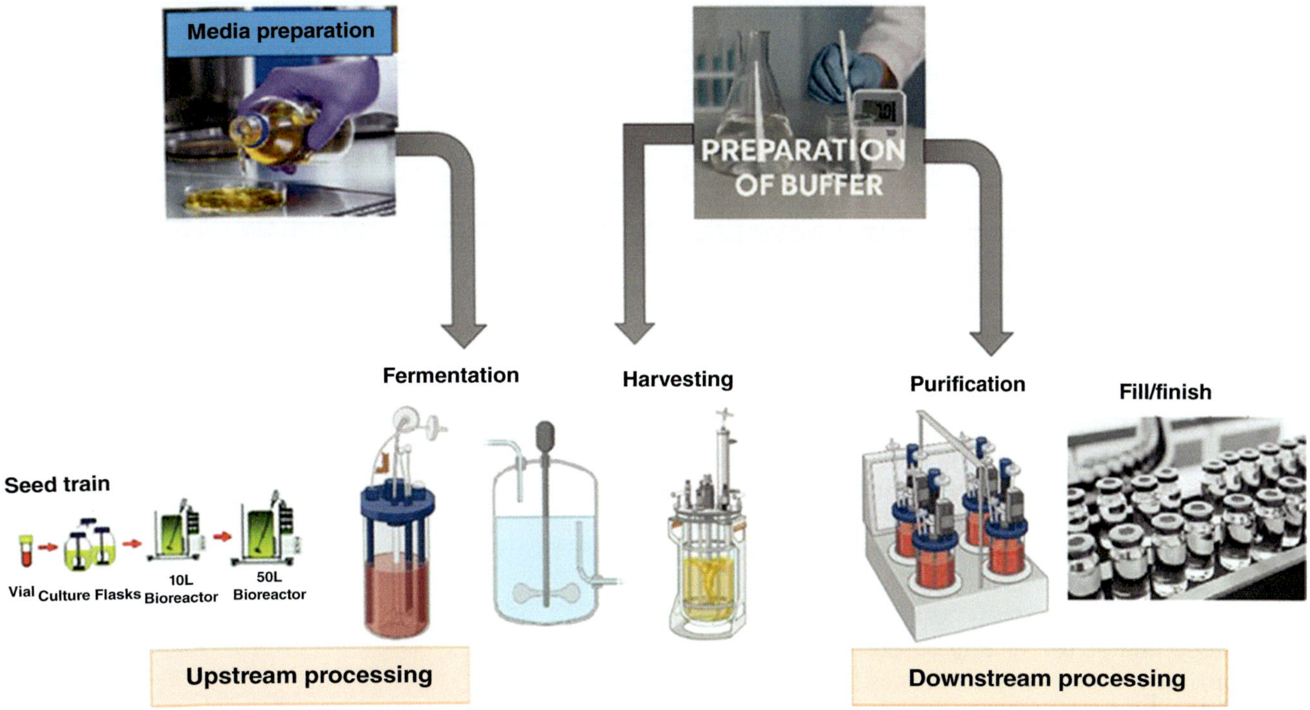

FIGURE 1.2 Different stages of biomanufacturing.

typically involves diverse, well-defined stages to scale up biological production from laboratory-scale research to full industrial scale. Biomanufacturing is structured around the following two core phases: Upstream and downstream processing (Figure 1.2). It also addresses culture methods, downstream processing stages, and bioreactor construction. Biomanufacturing begins with upstream processing, which centers on the development and cultivation of the biological production system.

1.3.1 Upstream and Downstream Processing in Sustainable Biomanufacturing

The upstream and downstream operations include the broad categories of biomanufacturing processes, each of which has a major impact on the system's overall sustainability performance. Sustainable biomanufacturing emphasizes how process design decisions influence resource efficiency, environmental impact, and circularity rather than focusing solely on procedural specifics.

1.3.1.1 Upstream Processing

The upstream processing includes product biosynthesis, culture, and the selection of biological systems. It includes a host cell line and expression vector, the process for developing cell lines, cell culture, media, feeds, and operating protocols (Cordova et al., 2023).

Feedstock origin (renewable versus refined), water and nutrient consumption, energy requirements for agitation and aeration, and overall process productivity are important factors in assessing sustainability. High-cell-density cultures, continuous or perfusion systems, and metabolic optimization are examples of advanced techniques that can increase volumetric productivity while lowering waste production per unit product, water consumption, and reactor footprint. Nevertheless, these methods often involve trade-offs among energy consumption, contamination risk, and process complexity, highlighting the need for balanced optimization. Various steps involved in the upstreaming process are listed as follows:

1. **Cell culture:** The cell culture, which is the development of microorganisms such as bacteria or yeast in a controlled environment in which they consume an input (which is frequently a carbon source) and generate the desired output, is the foundation of upstream processing.
2. **Media preparation and inoculum development:** Media serve as the nutritional backbone of culture systems, blending carbon and nitrogen sources, vitamins, minerals, and growth factors.
3. **Cell line development:** Cells are genetically engineered to enhance their productivity, stability, and ability to express the desired biomolecule, such as a therapeutic protein or enzyme (Hsu et al., 2018).
4. **Seed train:** It is a series of increasingly larger cultivation vessels that start from thawing cells from a working cell bank and progress through shake flasks (small-scale vessels) to pilot-scale and, eventually, production-scale bioreactors (Shukla & Thömmes, 2010).
5. **Fermentation and bioreactors:** There are several ways to construct an upstream manufacturing process;

the four main ways are as follows: Fermentation in batch, fed-batch, perfusion, and continuous fermentation (Lindskog, 2018). These methods are employed to maximize yield and process control in specially designed bioreactors.

6. **Bioreactors:** Bioreactors are essential to upstream processing as they offer a highly regulated environment for the biosynthesis of products and the growth of an enormous number of cells. To maximize product yield and promote cell health, bioreactors carefully control key factors, such as temperature, pH, dissolved oxygen (DO), agitation, and nutrient input rates. Depending on the process, bioreactors may operate in three-ways, that is, batch, fed-batch, or continuous modes as detailed in Figure 1.3 and Table 1.1, each with specific advantages related to productivity, scalability, and control (Eibl & Eibl, 2008).

Each mode has unique benefits and drawbacks that determine which bioprocessing applications it is suitable for. Steady and consistent volumetric production across time to support the downstream process is a benefit of continuous biomanufacturing (Khanal & Lenhoff, 2021). All things considered, this has great potential to promote low-carbon, sustainable bio-based economies, but more study is required (Xie, 2022). Continuous biomanufacturing depends on the context and is influenced by variables including feedstock types, process details, and operational scale (Demichelis et al., 2018). Another scientist, Voutilainen et al. (2021), evaluated the viability of employing lignocellulosic biomass as a feedstock to produce single-cell and recombinant proteins. The results indicate that preprocessing was responsible for 35.1–62.8% of total manufacturing expenses, and the highwater consumption during fermentation. This suggests that biomass loading and lignin precipitation could be substituted to increase efficiency (Voutilainen et al., 2021).

TABLE 1.1
The Process of Batch, Fed-Batch, and Continuous Biomanufacturing

Type of Process	Detail
Batch mode	In a batch operation, the product is harvested at the end of the process after all required medium components have been sequentially added at the start. Except for gases, pH control solutions, antifoam, and process-specific additions, no further additions are made. The composition of the initial medium is crucial to the process's success, and the full culture volume can frequently be harvested at once.
Fed-batch mode	Fed-batch mode begins similarly to batch mode, but it permits the continual addition of certain (new nutrients) during the process, increasing the manufacturing process and extending the growth phase. While concentrated nutrients are provided during the culture course, the process begins similarly to a batch. The volume rises rather than remains the same.
Continuous fermentation or biomanufacturing	This mode may result in increased production and steady-state situations. It offers uninterrupted production flow, greater productivity, reduced footprint and costs, and improved environmental sustainability, although it introduces integration and regulatory challenges.

7. **Harvesting:** Upon reaching peak cell density or product expression, the cells must be harvested. Common techniques include centrifugation, filtration, flocculation, or sedimentation to separate cells or biomass from the culture medium. This readies the culture fluid for downstream processing.

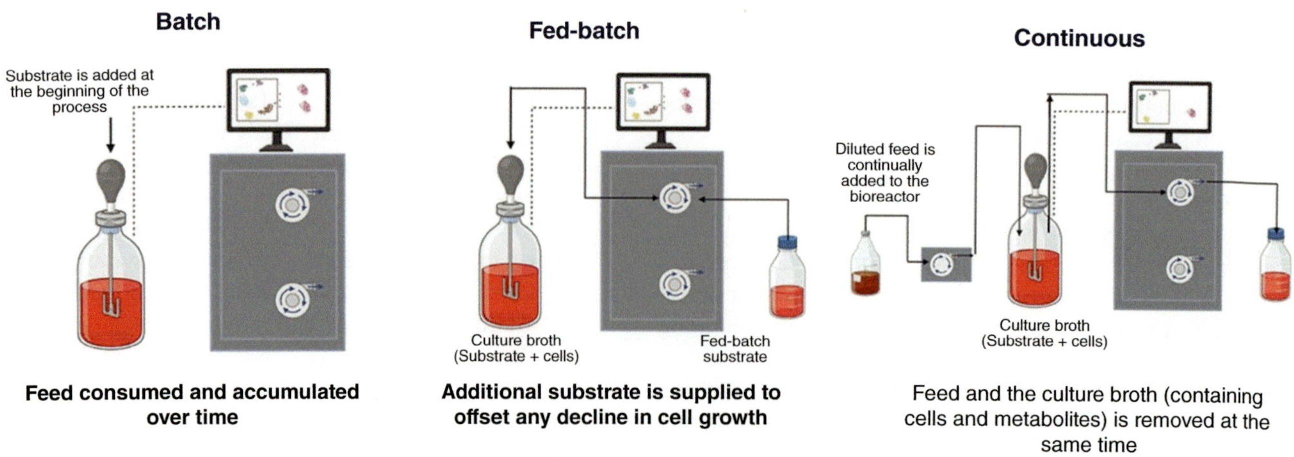

FIGURE 1.3 Different types of operating processes for biomanufacturing.

1.3.1.2 Downstream Processing

In biomanufacturing, downstream processing is often the most resource-intensive step, accounting for a significant amount of water, solvent, and energy use. Process intensification, solvent reduction, process stream recycling, and separation step integration are therefore given top priority in sustainable downstream strategies. Although their industrial adoption is still limited by cost and regulatory concerns, emerging techniques, including continuous purification, membrane-based separations, and *in situ* product recovery, can lower waste generation and enhance overall process efficiency. The process of recovering, separating, and purifying the bioproduct from the complex cell culture matrix is known as downstream processing. Following cell removal, methods such as precipitation, chromatography, filtration, and centrifugation are used to extract the target biomolecule from the residual supernatant. By separating the product from biomass and other contaminants, these methods guarantee its efficacy, safety, and purity. The sustainability of biomanufacturing systems is ultimately determined by integrated process design evaluated using lifecycle assessment (LCA) and techno-economic analysis (TEA), not by individual unit operations. As biomanufacturing systems move toward commercial deployment, it is crucial to connect upstream productivity with downstream efficiency to minimize environmental footprints while preserving economic viability.

Upstream and downstream processing strongly shape the sustainability profile of biomanufacturing as decisions made at these stages directly influence resource use, waste generation, and energy efficiency across the production lifecycle. Upstream processing is typically associated with high water, nutrient, and carbon consumption. Alongside it requires continuous energy input for aeration, agitation, sterilization, and thermal control. From a sustainability perspective, the choice of feedstock and media formulation plays a decisive role. Studies show that carbon source selection and media complexity directly influence both lifecycle greenhouse gas emissions and waste streams, with renewable or waste-derived feedstocks offering clear sustainability advantages over refined substrates (Cherubini, 2010). Process optimization and intensification at the upstream stage, such as higher cell density and improved bioreactor efficiency, can further improve resource efficiency by increasing product yields per unit of input and reducing overall waste streams. In contrast, downstream processing, which includes product recovery, separation, and purification, often dominates the overall environmental footprint. This often requires large volumes of solvents and water and involves multiple unit operations, leading to substantial effluent generation and energy demand, which contribute up to 50–80% of total production costs and energy consumption (Shukla & Thömmes, 2010; Straathof, 2014). Consequently, sustainability strategies in downstream processing focus on reducing the number and severity of purification steps, improving solvent and water recycling, and adopting lower energy separation technologies such as membrane-based systems or integrated recovery approaches. Integrating upstream and downstream design with lifecycle and process-level sustainability metrics, such as material intensity, waste reduction, and energy efficiency, is essential for developing scalable, resource-efficient, low-waste, and environmentally responsible biomanufacturing.

1.3.2 Sustainability Frameworks and Comparative Assessment of Biomanufacturing Systems

Sustainability is a key goal of contemporary biomanufacturing, and its assessment necessitates a systematic alignment with accepted sustainability concepts rather than merely descriptive claims. Integrative frameworks, such as LCA, TEA, carbon-neutrality pathways, and circular bioeconomy indicators, which collectively capture the economic, environmental, and social aspects of longevity, are increasingly used to evaluate sustainable biomanufacturing. From an environmental standpoint, LCA allows for a comprehensive evaluation of the effects throughout the whole manufacturing chain, including waste generation, emissions, energy and water consumption, and feedstock sources. Despite being biologically based, traditional biomanufacturing systems frequently have large life-cycle footprints due to their reliance on refined substrates, energy-intensive processes, and solvent-heavy downstream processing. Sustainable biomanufacturing, conversely, focuses on waste valorization, process intensification, renewable or waste-derived feedstocks, and energy-efficient operations to lower greenhouse gas emissions, freshwater requirements, and material losses. Next-generation biomanufacturing platforms are progressively including carbon-neutrality solutions such as carbon capture or reuse, integration of renewable energy, and carbon utilization based on biomass. Another key concept that sets sustainable biomanufacturing apart from traditional linear production models is circularity. By reducing effluent discharge, increasing resource recovery rates, and transforming waste streams into useful inputs, sustainable systems seek to close material, nutrient, and water loops. Waste-to-product conversion efficiency, water recycling potential, and material recovery rates are examples of circularity indicators that offer conceptual benchmarks for assessing advancement toward circular bioeconomy objectives.

The long-term survival depends on social and economic sustainability in addition to environmental concerns. By reducing reliance on fossil fuels, stabilizing supply chains, and increasing cost-effectiveness through integrated process design and scale-up, sustainable biomanufacturing aims to increase economic resilience. Workplace safety, skill development, fair access to biotechnological discoveries, and more general contributions to public health, food security, rural livelihoods, and environmental protection are all included in social sustainability. When taken as a whole, these aspects show that sustainable biomanufacturing is not just a technical improvement but rather a systems-level change.

1.4 BIOMANUFACTURING REVOLUTION: HISTORICAL PERSPECTIVES

The development of biomanufacturing is frequently depicted as a series of technological turning points; each biomanufacturing revolution was propelled by unique socioeconomic needs, scientific discoveries, and new environmental concerns. Crucially, these changes altered system efficiency, resource usage patterns, and environmental effects in addition to increasing production capacity. The current focus on circularity, low-carbon production, and resilient bio-based enterprises can be better understood by comprehending how sustainability issues and solutions changed during the course of subsequent biomanufacturing phases. Biomanufacturing has deep roots in human history. Long before the term existed, early societies relied on microbial processes to ferment food and drink. By approximately 9,000 BCE, civilizations in Mesopotamia, Egypt, and ancient India were fermenting milk, while ancient China produced fermented rice, honey, and beverages. These early forms of brewing beer, making cheese, and fermenting plants marked humanity's first biomanufacturing ventures. However, the biomanufacturing revolution started from the new ideas to the improvement over existing technologies and the introduction of new machinery to optimize the product's quality, costs, reliability, and increase productivity and profits. Over the years, technological development and innovation, scientific research, and a deeper understanding of biological systems have led to significant improvements in biomanufacturing processes and the emergence of new applications. Balar et al. (2023) have proposed an interpretation of biomanufacturing history that sees product market needs and technological advances emerging together, each driving the other. Building on traditional technologies, new product classes and enabling tools often appeared in tandem, launching each successive revolution. The history of biomanufacturing revolutions can be categorized into five phases (Figure 1.4) based on specific advancements in research, product types, and industrial technology. Table 1.2 presents an overview of historical phases of biomanufacturing.

1.4.1 Premodern Biomanufacturing

Premodern biomanufacturing, which utilized natural microbial consortia and ambient conditions, was intrinsically low-energy, locally supplied, and sustainable in terms of waste minimization. However, these systems lack consistency, scalability, and process control, limiting productivity. This stage showed that biological production might be environmentally safe, but it also made clear that technological advancements were required to increase productivity and dependability. Premodern biomanufacturing comprises the traditional processes using microorganisms to produce valuable goods. Thousands of years ago, before people ever knew about fermentation, they used biomanufacturing to transform one type of food into another. To transform a food supply into a different form, individuals used mixed-culture microbial processes. Records from about 9,000 years ago indicate that the ancient Chinese produced wine from fruits, honey, and rice. Similarly, Egyptian bread baking and the Sumerian and Babylonian brewing of beer circa 6,000 BCE are recorded in the Bible. Before the scientific breakthroughs of Louis Pasteur in the 19th century, fermentation remained a "black box" process, powerful, reliable, and multicellular, yet biologically mysterious. Despite that, ancient societies routinely produced vinegary and umami-rich condiments (such as pickles, soy sauce, vinegar, and indigo dyes), using naturally occurring microbial consortia in anaerobic (often solid-state) or liquid fermentations. For example, pickling vegetables and crops, utilizing calf rennet to produce cheese, making soy sauce, vinegar, and other products, pounding skins with dung microorganisms, and producing indigo to color cotton and wool. These early techniques were cheap, mostly energy efficient, sourced locally, and did not require an external inoculum. They were naturally sustainable because of their minimal ecological footprints, which anticipated current objectives of green design in industrial biomanufacturing. This fermentation stage can be thought of as premodern biomanufacturing. Natural mixed microbe cultures and solid-state fermentation are the key aspects of this era. In the present period, the world's wine business generated almost

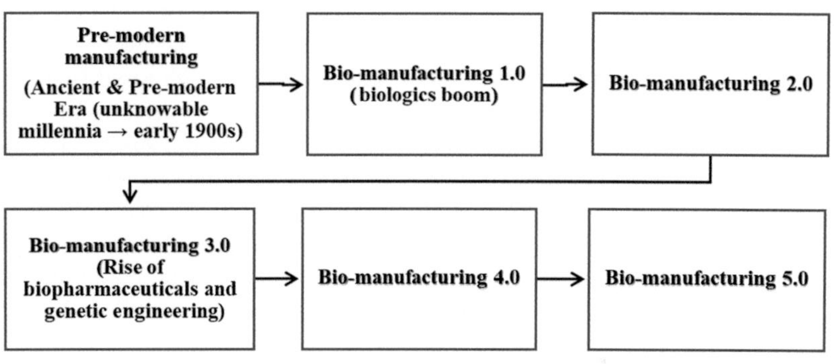

FIGURE 1.4 Historical phases of biomanufacturing.

TABLE 1.2
Timeline Snapshot for History of Biomanufacturing

Biomanufacturing Revolution	Approx Date	Representatives	Key Research
Pre-modern	≥ 6,000 BCE	Beer, cheese, soy sauce, vinegar, and steamed buns	Fermented foods for preservation
Biomanufacturing 1.0	1910s–1920s	Acetone, *n*-butanol, glycerol, amino acids, and vitamin C from mono-culture bacteria/fungi	Weizmann's acetone, butanol, and ethanol (ABE) amino acids/metabolites
Biomanufacturing 2.0	1940s–1980s	Penicillin, tetracycline, streptomycin	Penicillin mass-fermentation, insulin, monoclonals
Biomanufacturing 3.0	1990s–2000s	Recombinant proteins (insulin, and erythropoietin) and industrial enzymes	Green chemistry, continuous processing, and lifecycle thinking
Biomanufacturing 4.0	2010s–2020s	Pluripotent stem cell-based therapies, sustainable fuels (e.g., isobutanol), and artificial starch	Biorefineries, cascade valorization, waste to protein, bio-based materials, cell-free metabolic bioengineering, regenerative medicines, AI, cyber-physical systems, and Internet of Things (IoT)
Biomanufacturing 5.0	Post-2025	Digital-twin–enabled chimeric antigen receptor T-cell therapy (CAR-T) manufacturing platforms; modular, robotic, and automated cell therapy; pilot "micro-factories" for AI-optimized	AI, synthetic biology, bioengineering, and biotechnology; Human–machine collaboration; Human-centered approach; Cobot manufacturing; Carbon-neutral synthetic biofactories under governance-defined ethics.

24 billion liters of wine in 2015, making it the largest segment of the larger beverage market. To improve both yield and quality, contemporary wine makers focus on optimized grain saccharification techniques, engineered mixed-yeast ecologies, and a deeper understanding of yeast metabolism, bringing the art and science of fermentation full circle (Steinkraus, 1996; Tamang et al., 2020).

1.4.2 Biomanufacturing 1.0

The shift from artisanal to industrial-scale production was known as "Biomanufacturing 1.0," and it was mostly caused by market forces and military demands. Sustainability concerns took a backseat as regulated anaerobic processes and monoculture fermentation greatly increased yields and dependability. Although resource efficiency increased in comparison to chemical synthesis, new environmental and financial limitations were brought about by the use of refined substrates and the restricted value of waste. The production of primary metabolites, which began with Chaim Weizmann's acetone, butanol, and ethanol (ABE) fermentation process in the 1910s, is the focus of the Biomanufacturing 1.0 revolution. The following are the two significant upgrades introduced by this biomanufacturing platform:

1. Using a pure monoculture rather than a mixture of microorganisms.
2. Large-scale anaerobic liquid fermentation implementation.

Primary metabolites (substances that are directly involved in cell growth, development, and reproduction), such as ethanol, acetone, butanol, amino acids, and organic acids (such as lactate, acetate, and citrate), are the trademark products of Biomanufacturing 1.0. The history of ABE fermentation reveals two key insights: the evolution of fermentation technology, which is primarily driven by market demands, and the focus shift of products over time. Efforts to create synthetic rubber commenced at the beginning of the 20th century due to a worldwide scarcity and skyrocketing costs of natural rubber. According to the UK company, Strange & Graham Ltd. (London and Manchester), the most practical precursors were butadiene and isoprene, which could be produced by oxidizing *n*-butanol and isoamyl alcohol, respectively.

The anaerobe *Clostridium acetobutylicum*, which can produce ethanol, butanol, and acetone from starch and glucose, was discovered by Chaim Weizmann in 1912. However, while synthetic rubber was previously produced by oxidizing *n*-butanol and isoamyl alcohol, the collapse of the global rubber industry that same year rendered it economically unfeasible. During World War I, the British demand for acetone rose steeply, as it was essential for producing cordite, the smokeless explosive. Weizmann refocused his ABE-fermentation process toward acetone production. By the mid–1910s, the method had been industrialized, producing up to 30,000 tons of acetone annually, alongside even greater quantities (double) of *n*-butanol. At that time, butanol became the second most voluminous biotechnology product after ethanol. The excess *n*-butanol found new markets as a solvent and as a component of lacquers, plasticizers, paints, and resins. For this development, Chaim Weizmann was recognized as the father of industrial fermentation (Vasić-Rački, 2004). Another example, during World War I, the citric acid shortage was overcome by sugar fermentation using fungus in 1919. Another key product was monosodium glutamate (MSG), a flavor-enhancer amino acid. Initially, MSG was made from plant protein and chemical synthesis; however, producer shift toward microbial fermentation as it is more economic, safe, and eco-friendly. Now, annual Global amino acid production may be around 20 billion, where lysine second most produced amino acid, and China leads this production. Many organic acids are finding new markets due to the enormous potential of biodegradable plastics, which are also becoming green building blocks (Isikgor & Becer, 2015).

1.4.3 Biomanufacturing 2.0

Biomanufacturing 2.0's introduction of aerobic submerged fermentation and secondary metabolite production significantly boosted output and product variety, especially in the pharmaceutical industry. However, because of the need for aeration, sterilization, and downstream purification, this stage increased energy and water consumption. As a result, even while human health results significantly improved, large-scale fermentation's environmental impact became more noticeable, highlighting the trade-offs between process sustainability and social benefits. The invention of penicillin by fermentation during World War II, which established the groundwork for industrial microbiology, was the beginning of Biomanufacturing 2.0. The creation of secondary metabolites rather than main metabolites is the primary characteristic that sets Biomanufacturing 2.0 apart. While secondary metabolites are species-specific and utilized in defense mechanisms, primary metabolites are necessary for an organism to grow. These secondary metabolites are typically used by humans as components of colorings, fragrances, medications, and flavorings. Aerobic submerged fermentation and the utilization of specialized mutants were two new technologies that were important discoveries during this time. However, it is worth mentioning that the industrialization of biological manufacturing progressed rapidly in the 20th century as follows:

- Alexander Fleming's discovery of penicillin from Penicillium fungi in 1928 led to its mass production by 1942, marking the first pharmaceutical-scale biomanufacturing effort.
- Several antibiotics derived from microorganisms, such as tetracycline and streptomycin, were discovered and introduced into medicine. Streptomycin, discovered by Selman Waksman in 1943, became the first effective treatment for tuberculosis (Nobel Prize in Medicine in 1952).
- To reduce the cost of producing penicillin, Merck introduced aerobic submerged fermentation in 1942.
- In 1950, Dr. Elmer Gaden delivered his groundbreaking doctoral dissertation on how to supply the right amount of oxygen for penicillin mold to grow and multiply more quickly. He is widely regarded as "Father of Biochemical Engineering."
- Notable contribution of Dr. Elmer Gaden for the optimization of oxygen supply in fermentation. Furthermore, Dr. Arnold Demain, known as a pioneer in beta-lactam antibiotics, and he were developed methods for other antibiotics. This era was also known for the discovery of other antibiotics, such as cephalosporins, tetracycline, and streptomycin.
- Dr. Arnold Demain was a pioneering industrial microbiologist known for his groundbreaking work on the biosynthesis and regulation of penicillin and cephalosporins. He is known as the "Father of Biochemical Engineering"; he directed research and development on procedures for MSG, vitamin B12, streptomycin, riboflavin, cephamycin, fosfomycin, and interferon inducers after establishing the Fermentation Microbiology Department at Merck in 1965.
- Over the decades, advances in microbial engineering, fermentation technology, and downstream processing led to the production of a wide range of antibiotics, enzymes, and biologics.

1.4.4 Biomanufacturing 3.0

The precise biosynthesis of complex biomolecules was made possible with the introduction of recombinant DNA technology and protein engineering in Biomanufacturing 3.0. This stage decreased reliance on extraction from plants or animals and enhanced product specificity. However, sustainability issues were brought about by high capital expenditure, costly culture medium, and solvent-intensive downstream processing, which led to a focus on host optimization, process intensification, and cost-cutting measures. The development of Biomanufacturing 3.0 started from the 1980s, switching from the synthesis of primary or secondary metabolites to large-sized proteins (polypeptide/protein-based drugs and enzyme-based biocatalyst), which is important for catalysis and medicines. Recombinant DNA technology and improved cell cultures were the two innovations that propelled the creation of this biomanufacturing platform.

- The first *in vitro* recombinant DNA plasmid was developed in 1973 by Stanley Cohen and Herbert Boyer, which paved the way for future commercial initiatives that used the same technology.
- High-value pharmaceutical proteins were the main focus of these companies because they are not only expensive to synthesize but also challenging to extract from their natural sources.
- Companies such as Genentech (1976) and Amgen (1980) were the first to commercially produce recombinant proteins such as human insulin, growth hormone, and erythropoietin, a glycoprotein used to treat anemia.
- Additionally, Antibody drug conjugates were also developed to treat cancer, for example, Adcetris® (brentuximab vedotin) and Kadcyla® (ado-trastuzumab emtansine). MABs are used to selectively

bind tumor-associated target antigens and provide a highly effective cytotoxic agent, in contrast to traditional therapies that harm healthy tissues during dose escalation.

- Introduction of recombinant enzymes also helped in a few companies, which supply enzymes for food, detergent, and textile industries.

Biomanufacturing 3.0 was revolutionized into three phases: natural enzymes with limited stability, immobilized and protein-engineered enzymes, and the development of customized enzymes using rational design and directed evolution. Reduction of production costs by improving high cell-density fermentation, using microbes, such as *Escherichia coli* and Bacillus spp., and Aspergillus spp., is also one major focus of this revolution. Overall, Biomanufacturing 3.0 has revolutionized the economical and efficient production of protein-based drugs and industrial enzymes. Recently, transgenic plants such as *Arabidopsis thaliana*, tobacco, and others have been used to decrease the biomanufacturing costs of recombinant bulk enzymes because of cost-effective plant cultivation, high-level protein accumulation in plant tissues, easy and cheap scale-up, and simple protein purification [6, 70, 71]. Recombinant protein production costs are also expected to be reduced by using transgenic plants in the future [6, 70, 71].

1.4.5 Biomanufacturing 4.0

The use of metabolic engineering, synthetic biology, and biorefinery principles, Biomanufacturing 4.0, represents a paradigm change toward sustainability at the systems level. This stage specifically addressed previous inefficiencies by including cell-free biosystems, cascade valorization, and waste-derived feedstocks. The gap between laboratory innovation and industrial adoption is highlighted by the fact that scalability, regulatory complexity, and energy requirements for advanced processing remain limiting constraints. The main focus of this revolution was on the synthesis of biology-driven creatures, such as the *in vitro* synthesis of starch, and metabolic engineering for the creation of isobutanol and human tissues. This new phase of biomanufacturing has the potential to solve the world's most pressing challenges, including food safety, water scarcity, energy production, climate change, health issues, and sustainability. Along with the production of new products (i.e., new functional tissues and new drugs), Biomanufacturing 4.0 also concentrates on producing existing products using far more effective approaches. A wide range of new, sustainable, efficient, and safe products related to regenerative medicine, pharmaceuticals, renewable energy sources, and artificial food are produced. and the integration of numerous biological tools from metabolic engineering to synthetic biology, among other things. Zhang et al. (2017) discussed Biomanufacturing 4.0 among three important fields: Regenerative medicines, metabolic engineering, synthetic biology-driven whole-cell fermentation, and *in vitro* synthetic (enzymatic) biosystems.

In regenerative medicine, pluripotent stem cells hold great promise as they are able to proliferate indefinitely, as well as differentiate into every other cell type in the body (skin, heart, bone, neuron, kidney, pancreatic, and liver cells). Some well-known synthetic biology examples are the production of antimalarial drug artemisinin, isobutanol, and opioids by engineered *E. coli* and yeast (Paddon & Keasling, 2014). These cell-free systems are emerging as a biomanufacturing platform in which engineering flexibility is much greater, overcoming cellular viability, complexity, physiology, and the presence of membranes and/or cell walls associated with fermentation (Del Hierro et al., 2024). A few scientists proposed advanced biotransformation catalyzed by *in vitro* synthetic (enzymatic) biosystems known as cell-free biosystems or cell-free metabolic engineering, which has become an emerging biomanufacturing platform. This system is much easier than modifying a living organism (Zhang et al., 2010).

1.4.6 Biomanufacturing 5.0

Biomanufacturing 5.0 integrates human-centric design, ethical governance, and AI-driven optimization to move beyond efficiency and circularity in sustainability. Digital twins, autonomous control systems, and real-time monitoring can dynamically reduce resource use and environmental impact. At the same time, new issues in data integrity, the energy requirements of digital infrastructure, and fair access to technologies demonstrate how sustainability concerns in advanced biomanufacturing are always evolving. Biomanufacturing 5.0 represents a future-forward revolution that integrates biotechnologies with AI design, smart automation, and sustainable practices. Advancement over Biomanufacturing 4.0 emphasizes not only digital integration and process optimization but also ethical responsibility, environmental stewardship, and personalized outcomes. This is beneficial to modern healthcare, resource management, and climate change. However, the AI-assisted control system, real-time biosensors, and genetic editing technologies, such as Clustered Regularly Interspaced Short Palindromic Repeats (CRISPR) and quantum computing. However, this together enhances the efficiency and accuracy across biomanufacturing. The future of biomanufacturing is a self-regulating, self-organizing, and self-producing mechanism with AI integration. A Biomanufacturing 5.0, focused on ecofriendly and regenerative processes, where production system design is to be adaptive, circular, and resilient. Additionally, IoT technologies, as modern biomanufacturing, transform traditional industrial models into a network-based data-informed society. New molecular biology tools help to create new tissues and drugs or improve the efficiency and sustainability of existing product manufacturing. Modern manufacturing systems have undergone a significant transformation, evolving from traditional mass-production paradigms to highly flexible, human-centric manufacturing production models. Agile manufacturing and on-demand biomanufacturing are among them.

In the revolution of Biomanufacturing 5.0, biotechnology meets not only the industrial demands but also contributes positively to human health and environmental sustainability (Isoko et al., 2024). In this technological innovation, it is harmonized with human cognitive and creative capabilities, promoting a collaborative work environment where automation addresses repetitive tasks, while humans engage in higher order problem-solving and strategic decision-making.

1.5 APPLICATIONS OF BIOMANUFACTURING

The technological evolution from Biomanufacturing 1.0 to Biomanufacturing 4.0 has profoundly shaped modern-day applications by enabling more precise, efficient, intelligent, and scalable production of biological products. Early biomanufacturing relied on simple fermentation processes with limited control over environmental conditions, limiting reproducibility and product complexity. Advances in molecular biology and genetic engineering have transformed these processes, enabling the targeted production of various bio-based materials by microorganisms and cell lines, thereby markedly improving yield and specificity. Altogether, these technological developments have expanded the role of biomanufacturing in various sectors, making it a cornerstone of modern biotechnology. This section highlights applications of biomanufacturing across multiple sectors, including pharmaceuticals and healthcare, agriculture and aquaculture, energy, and other industries.

1.5.1 Pharmaceuticals and Healthcare

Healthcare constitutes a significant segment of the biomanufacturing industry, with advances such as Humira representing notable milestones. This drug was the first fully human MABs granted approval by the US Food and Drug Administration (FDA) for the treatment of autoimmune diseases (Asin-Garcia et al., 2025; Niazi, 2025). Human insulin (Humulin) was the first recombinant protein developed for the treatment of diabetes using genetically engineered *E. coli* bacteria, and it was successfully produced on an industrial scale in 1980. The application of biomanufacturing, particularly the use of recombinant DNA technology in the 1980s, has significantly shaped the sector (Lakra & Ayyappan, 2003). Pharmaceutical development depends largely on raw materials, especially Active Pharmaceutical Ingredients (APIs), which are vital components responsible for medicine's therapeutic effects. However, conventional techniques for API and other raw material manufacturing often encounter obstacles, including inefficiency, environmental issues, and difficulties in producing intricate molecules (Shokoohi & Attar, 2024). The emerging biotechnology solutions are transforming pharmaceutical manufacturing through superior efficiency, sustainability, and precision in the synthesis of complex raw materials.

1.5.2 Vaccines

The production of vaccines is one important area where biomanufacturing has transformed conventional methods by bringing in more rapid, scalable, and environmentally friendly production processes. While traditional techniques, such as producing vaccines from eggs, are frequently laborious, resource-intensive, and less flexible, contemporary biomanufacturing uses recombinant DNA technology, cell-based systems, microbial platforms, and synthetic biology to produce vaccines with greater purity, consistency, and efficiency (Shokoohi & Attar, 2024).

With messenger ribonucleic acid (mRNA) vaccines, such as those created by Pfizer-BioNTech and Moderna, being quickly produced through engineered cells and lipid nano-particle delivery systems, the global response to the COVID-19 pandemic demonstrated this potential. This eliminated the need to grow pathogens in laboratories and significantly shortened development timelines (Asin-Garcia et al., 2025; Dolgin, 2021). Beyond mRNA vaccines, recombinant protein vaccines, such as those for HPV and Hepatitis B, demonstrate that biotechnology has long been used to develop safe and effective vaccines. Meanwhile, newer strategies, such as precision fermentation and DNA-based vaccines, are opening the door for next-generation vaccines that are intended to address changing variants more effectively and adaptively (Pardi et al., 2018). All of these developments highlight how important biomanufacturing is to bolstering vaccine development, ensuring global health security, and preparing for future pandemics.

1.5.3 Monoclonal Antibodies

The biomanufacturing produces MABs with high specificity, scalable production, and therapeutic adaptability for diseases such as autoimmune disorders and cancer. The creation of MABs is one of the most significant uses of biomanufacturing in contemporary medicine, transforming targeted treatments for a variety of illnesses. The voyage started with the groundbreaking hybridoma method created by Kohler and Milstein (1975). They were awarded the Nobel Prize for their invention, which allowed for the creation of highly specific antibodies from a single clone of B-lymphocytes by fusing them with immortal myeloma cells (Kohler & Milstein, 1975). Orthoclone OKT3 (muromonab-CD3) was one of the first clinical applications to achieve significant success, despite obstacles, including immunogenicity and low manufacturing yields, when employing murine cell lines. Mammalian cell lines may now reliably express humanized and fully human antibodies for therapeutic use, which is a result of biomanufacturing innovations in genetic engineering and cell culture techniques, which have made large-scale manufacture of MABs both possible and affordable (Ecker et al., 2015). Over a hundred MABs have been approved for a variety of indications, including autoimmune illnesses and cancer, and the integration of biomanufacturing processes, which

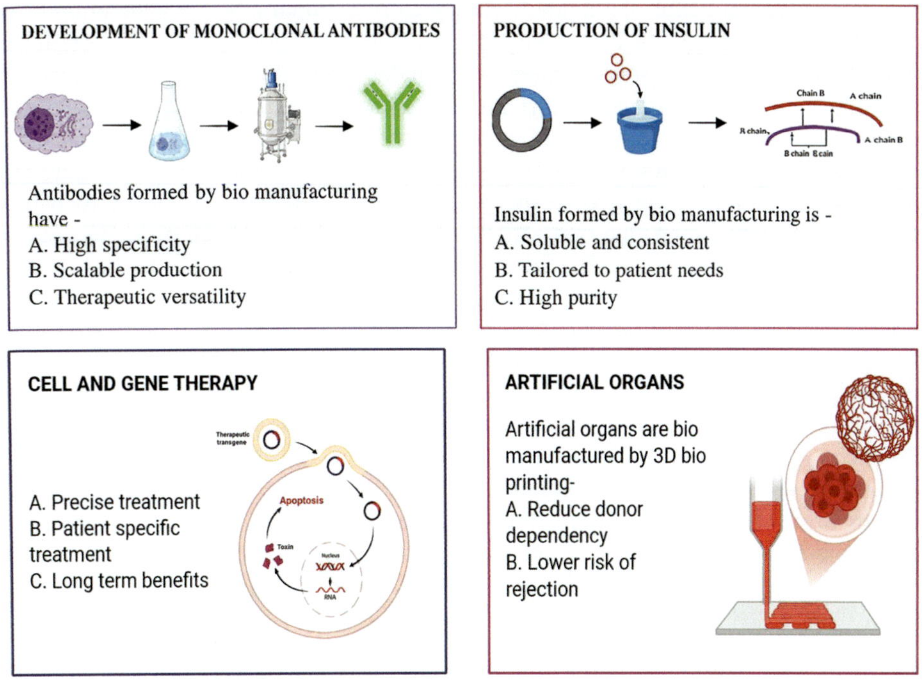

FIGURE 1.5 Application of biomanufacturing in healthcare and pharmaceuticals.

has also significantly enhanced product quality (Figure 1.5) (Kaplon & Reichert, 2021).

1.5.4 Biologics

Biologics such as essential therapeutic proteins, such as insulin, growth hormones, and clotting factors, are produced in large part by biomanufacturing and have greatly enhanced the treatment of conditions including diabetes, growth abnormalities, and hemophilia. The two main stages of the manufacturing process are upstream and downstream processing. Cultivating genetically modified cells is the first step in upstream processing. Usually, this is done in bioreactors, where the cells are grown through a series of steps, such as shake flask expansion, vial thawing, and several bioreactor growth phases to achieve high cell densities that can produce the desired therapeutic protein. After cell growth, the protein product is extracted using centrifugation and filtration techniques to separate it from the cellular components while maintaining its integrity.

1.5.5 Enzymes and Biochemicals

Innovation in the industrial enzymes and specialty biochemicals sectors is greatly aided by biomanufacturing, which provides effective and sustainable substitutes for conventional chemical processes (Bornscheuer et al., 2012). Industrial enzymes catalyze particular reactions to increase yields, lower energy consumption, and improve product quality in the food, detergent, textile, and biofuel industries (Sheldon & Woodley, 2018). Enzymes such as amylases and proteases help in the food industry with baking, brewing, and dairy processing, while detergents use lipolytic and proteolytic enzymes to remove stains and clean fabrics. The manufacturing of biofuel benefits from the enzymatic conversion of biomass into fermentable sugars for the synthesis of ethanol and butanol, while textile producers use cellulases and proteases in procedures such as biopolishing and desizing (Singhania et al., 2017).

In the biochemical industry, biomanufacturing uses sophisticated fermentation and biocatalysis systems to enable the customized manufacture of fine chemicals, chemical intermediates, and functional materials. Innovative biotechnological techniques such as cell-free synthesis and engineered microorganisms enable the effective synthesis of high-value compounds for advanced materials, pharmaceuticals, and nutraceuticals, often in decentralized, scalable settings (Schmidt-Dannert & López-Gallego, 2016). By reducing reliance on fossil fuels, promoting carbon-neutral production, and accelerating the development of robust, sustainable chemical industries, biomanufacturing is being used in these specialized fields.

1.5.6 Biofuels and Bioenergy

In the field of renewable energy, biomanufacturing has become a key component, propelling advancements in microbial fuel cells (MFCs), advanced biofuels, and bioethanol production. In the process of producing bioethanol from lignocellulosic biomass, enzymes, such as cellulases and hemicellulases, break down non-food crops, forestry waste, and agricultural leftovers into fermentable sugars that are then transformed into ethanol by microorganisms that have been created (Sukumaran et al., 2010). Advanced biofuels, such as second-generation cellulosic ethanol, go beyond traditional routes by using genetically modified microorganisms and integrated bioprocessing techniques that combine fermentation and hydrolysis to increase yields while lowering process costs and environmental effects (Alper & Stephanopoulos, 2009). In addition to these liquid fuel technologies, MFCs combine energy recovery and waste treatment by directly producing electricity from organic substrates, including wastewater, by taking advantage of the metabolic processes of electroactive bacteria such as *Geobacter* and *Shewanella*. These biomanufacturing applications work together to deliver scalable, carbon-neutral solutions for global energy sustainability while also reducing reliance on fossil fuels. As a clean, renewable energy source that emits no greenhouse gases when burned, hydrogen is widely recognized. Ren et al. (2025) biomanufactured hydrogen from wastewater rich in carbohydrates using full-scale acidogenic fermentation in a continuous-flow pattern.

1.5.7 Agriculture and Crop Biotechnology

The use of biomanufacturing in genetically modified organisms (GMOs), biological pest management, and biofertilizers for soil health makes a substantial contribution to modern agriculture. GMOs, created through fermentation-based platforms and advanced genetic engineering, are intended to improve agricultural characteristics, including yield, stress tolerance, and disease and insect resistance, thus decreasing dependency on chemical inputs (ISAAA, 2022). In biological pest management, biomanufacturing makes it possible to produce biopesticides and biofungicides on a large scale. These include microbial agents such as Trichoderma, *Bacillus thuringiensis*, and entomopathogenic fungi, which target particular pests and pathogens while reducing ecological damage and pesticide resistance (Choudhary & Sharma, 2023). Similar to this, microbial fermentation-produced biofertilizers use beneficial organisms, including phosphate-solubilizing bacteria, Azotobacter, and Rhizobium to fix nitrogen in the atmosphere, mobilize soil nutrients, and enhance the cycling of organic matter (Singh et al., 2021). In addition to improving crop yields and soil fertility, these bio-based inputs support long-term sustainability by restoring soil bacteria and lowering reliance on chemical fertilizers. When combined, these biomanufacturing advancements promote environmentally responsible, climate-smart farming methods that complement the objectives of global food security.

1.5.8 Food and Beverage Industry

Many food-related applications, including fermentation, enzyme synthesis, probiotics, and alternative protein sources, are supported by biomanufacturing. To achieve uniform quality, better flavor profiles, and longer shelf life, industrial-scale bioprocessing has improved traditional fermentation for products including cheese, wine, beer, and yogurt. Comparably, the generation of enzymes via microbial fermentation yields important catalysts, including amylases and proteases, for the dairy, brewing, and baking sectors, where they optimize product yield, improve dough texture, and increase fermentation efficiency (de Oliveira et al., 2021). Biomanufacturing guarantees the large-scale culture of beneficial bacteria such as Lactobacillus and Bifidobacterium, which promote gut health and the development of functional foods, in the field of probiotic production (Wang et al., 2022). Furthermore, developments in plant-based proteins and cultivated meat, as well as other alternative protein sources, use precision fermentation, cell culture, and bioreactor technologies to produce nutrient-rich, sustainable proteins that improve global food security and lessen the environmental impact of animal agriculture (Zhao et al., 2023). All together, these uses show how biomanufacturing combines innovation with tradition to provide scalable, sustainable, and safe solutions for the contemporary food system.

1.5.9 Biomanufacturing in Fisheries

Aquaculture innovation is being driven by biomanufacturing, which produces sustainable feed and enhances fish health techniques. The conversion of organic residues and agro-industrial byproducts into high-value protein sources through microbial fermentation and enzymatic processes is one significant development in waste valorization into fish feed, which lessens reliance on traditional fishmeal and promotes circular bioeconomy principles (Kumar et al., 2022). Probiotic formulations made through biomanufacturing, quick diagnostic biosensors, and recombinant vaccines are useful tools for managing aquatic infections and boosting immunity in farmed animals in the context of disease management and health monitoring (Jeevanandam et al., 2023). In addition to increasing aquaculture productivity, these advances lessen the financial and environmental costs associated with disease outbreaks.

In aquaculture, biomanufacturing is revolutionizing genetic improvement and system-level sustainability beyond feed and health. Breeding efficiency is increased by precision breeding techniques, such as CRISPR-based genome editing and marker-assisted selection, which enable targeted augmentation of desired traits, including disease resistance, stress tolerance, and faster growth. Similarly, microbial community engineering-enabled bio-floc technology reduces input costs and environmental consequences by recycling waste nutrients, improving water quality, and producing microbial biomass as an additional protein-rich feed. In the meantime, effective insights into fish genetics, microbial ecology, and trait-associated markers are provided by next-generation sequencing (NGS) combined with biomanufacturing pipelines, facilitating data-driven decision-making for ecosystem health and selective breeding (Zhou et al., 2021). Taken as a whole, these uses demonstrate how biomanufacturing helps make aquaculture systems more robust, resource-efficient, and sustainable.

1.6 CHALLENGES AND FUTURE PROSPECTS OF SUSTAINABLE BIOMANUFACTURING

The widespread implementation of sustainable biomanufacturing still faces a number of institutional, structural, and financial obstacles despite tremendous scientific advancements and expanding regulatory backing. High production costs are one of the biggest obstacles, especially during scale-up. Genetically modified strains, costly feedstocks, complex bioreactors, and resource-intensive downstream processing are common components of advanced biomanufacturing systems. Although laboratory-scale discoveries show great promise for sustainability, industrial-scale techno-economic feasibility remains inconsistent, particularly for low-margin items such as alternative proteins, bulk chemicals, and biofuels. Overall sustainability benefits may be limited by downstream purification, in particular, which might be responsible for a significant portion of operating expenses and environmental impact.

Scalability is yet another significant limitation. Continuous processing, cell-free systems, and the use of waste-derived feedstock are just a few of the sustainable biomanufacturing techniques that encounter difficulties with process stability, contamination hazards, regulatory compliance, and infrastructural readiness during scale-up. Additionally, supply chains become unstable due to variations in biomass availability and quality, particularly for waste-based or agricultural feedstocks. Therefore, integrated process intensification, standardized platforms, and thorough lifecycle and techno-economic assessments to inform decision-making are necessary to close the gap between pilot-scale success and commercial-scale implementation.

Policy frameworks are crucial in determining the future course of sustainable biomanufacturing from a governance standpoint. The recently approved BioE3 policy in India is a calculated attempt to boost domestic biomanufacturing capacity in key areas, such as carbon capture and utilization, bio-based chemicals, precision medicines, and climate-resilient agriculture. Although the strategy offers a solid institutional basis, timely implementation, regulatory harmonization, public-private partnerships, and sustained investment in biomanufacturing infrastructure, biofoundries and innovation hubs are all necessary for its efficacy. Transforming policy purpose into quantifiable results will require coordinating market access rules, carbon accounting systems, and sustainability incentives.

The development of human capital and skills is another significant concern. An interdisciplinary workforce skilled in industrial biotechnology, systems engineering, data analytics, AI, regulatory science, and sustainability assessment methods, such as lifecycle analysis, is needed for sustainable biomanufacturing. The rate of industrial adoption is now constrained by skill gaps, especially at the junction of biology, engineering, and digital technology. Targeted education and training programs, industry–academia cooperation, and ongoing professional development aligned with new biomanufacturing technologies will all be necessary to close these gaps.

Future prospects for sustainable biomanufacturing are directly related to systems-level integration. Synthetic biology, AI-driven process optimization, digital twins, and circular bioeconomy models provide opportunities to increase economic resilience, decrease environmental impact, and boost productivity simultaneously. For sustainable biomanufacturing to move from specialized applications to a mainstream industrial paradigm, it will be essential to incorporate sustainability indicators into early stage process design, in conjunction with supporting policy ecosystems and qualified people resources. The degree to which sustainable biomanufacturing can support climate goals, resource security, and equitable bioeconomic growth will depend on how well it addresses obstacles related to cost, scale, policy, and skill.

India's position as a Quad-led biomanufacturing hub also has a lot of promise, given the increasing significance of biomanufacturing across several industries. To establish India as a powerful biomanufacturing destination for cutting-edge and easily accessible products in both local and international markets, the National Biotechnology Development Strategy of India envisions the nation as a "Global Biomanufacturing Hub" by 2025 (Department of Biotechnology, 2025). This can promote innovation and cooperation in the life sciences industry while enhancing India's capacity to produce biopharmaceuticals, vaccines, and other biotechnology products. Future biomanufacturing will focus on building resilient and innovative bioeconomies through waste valorization, bioprocess intensification, and comprehensive LCAs, driving global progress toward a circular, carbon-neutral economy.

1.7 CONCLUSION

Sustainable biomanufacturing is a potential strategy for producing biomaterials, chemicals, and energy in a responsible and environmentally friendly way. Historical synthesis of biomanufacturing reveals how sustainability in biomanufacturing has matured from an emergent property of fermentation to a core design criterion for modern industrial biology.

Sustainable biomanufacturing is increasingly shaped by the interaction of supportive policy environments, emerging technologies, and societal demand for low-carbon and ecofriendly resilient production systems. Policy frameworks at national and international levels now recognize biomanufacturing as a strategic pillar for achieving climate goals, strengthening supply-chain security, and advancing circular-economy transitions. The use of renewable feedstocks, waste valorization, and the use of bio-based alternatives for scalable, sustainable industrial biotechnology platforms is the way forward. Technological innovation acts as a key enabler of these policy objectives, with advances such as single-use bioreactors, continuous biomanufacturing, AI-driven process control systems, and gene editing techniques, such as CRISPR-Cas9, accelerating efficiency gains while reducing resource intensity, energy consumption, and waste generation.

At the societal level, growing awareness of environmental sustainability, public health preparedness, and resource constraints is driving demand for bio-based materials, chemicals, and energy solutions across sectors, including healthcare, agriculture, food systems, and energy. Sustainable biomanufacturing thus serves not only as a production paradigm but also as a socio-technical pathway toward carbon-neutral, resource-efficient, and circular economies, enhancing resilience to global disruptions such as pandemics, geopolitical instability, and climate-related crises. Continued alignment between policy instruments, technological development, and societal priorities will be critical for scaling sustainable biomanufacturing and expanding its contribution to inclusive and resilient economic growth.

Biomanufacturing in 2025 imitates a mature fusion of biological innovation and the Industrial Revolution. Repair, reuse, and recycling (3R's) of bioproducts laid a strong foundation for a circular economy based on waste biomass. Sustainable biomanufacturing integrates industrial biotechnology with circular-economy principles to transform waste biomass into value-added products with minimal environmental impact. It is a sustainable bioproduction toward carbon-neutral, resource-efficient, and circular production systems. Biomanufacturing not only supports innovation across health, agriculture, food, and energy sectors but also enhances global resilience to tremors such as pandemics, geopolitically induced disruptions, and climate-related crises. The biomanufacturing field is constantly evolving as a result of continuing research and development initiatives aimed at enhancing procedures and expanding the product line.

REFERENCES

Alper, H., & Stephanopoulos, G. (2009). Engineering for biofuels: Exploiting innate microbial capacity or importing biosynthetic potential? *Nature Reviews Microbiology*, 7(10), 715–723. https://doi.org/10.1038/nrmicro2186

Asin-Garcia, E., Fawcett, J. D., Batianis, C., & Dos Santos, V. A. M. (2025). A snapshot of bio-manufacturing and the need for enabling research infrastructure. *Trends in Biotechnology*, 43(5), 1000–1014. https://doi.org/10.1016/j.tibtech.2024.12.003

Balar, V., Gor, D., Yadav, A., & Bagadiya, P (2023). The evolution of biomanufacturing: A historical and contemporary perspective. *Advances in Agricultural Biotechnology*, 107.

Bhatia, S. K., Bhatia, S., Inda-Webb, M. E., et al. (2024). Biotechnology for sustainable materials: Innovating today for a greener tomorrow. *Biotechnology for Sustainable Materials*, 1, Article 3. https://doi.org/10.1186/s44316-024-00006-x

Bornscheuer, U. T., Huisman, G. W., Kazlauskas, R. J., Lutz, S., Moore, J. C., & Robins, K. (2012). Engineering the third wave of biocatalysis. *Nature*, 485(7397), 185–194. https://doi.org/10.1038/nature11117

Cherubini, F. (2010). The biorefinery concept: Using biomass instead of oil for producing energy and chemicals. *Energy Conversion and Management*, 51, 1412–1421.

Choudhary, M., & Sharma, R. (2023). Advances in microbial biopesticides: Sustainable approaches for pest and disease management. *Frontiers in Microbiology*, 14, 1123456.

Cordova, L. T., Dahodwala, H., Elliott, K. S., Baik, J., Odenewelder, D. C., Nmagu, D., … Lee, K. H. (2023). Generation of reference cell lines, media, and a process platform for CHO cell bio-manufacturing. *Biotechnology and Bioengineering*, 120(3), 715–725.

Del Hierro, A. G., Moreno-Cid, J. A., & Casey, E. (2024, November). Continuous bio-manufacturing for sustainable bioeconomy applications. *EFB Bioeconomy Journal*, 4, Article 100071. https://doi.org/10.1016/j.bioeco.2024.100071

Demichelis, F., Fiore, S., Pleissner, D., & Venus, J. (2018). Technical and economic assessment of food waste valorization through a biorefinery chain. *Renewable and Sustainable Energy Reviews*, 94, 38–48.

de Oliveira, D. M., Komesu, A., Vaz Rossell, C. E., & Maciel Filho, R. (2021). Challenges and opportunities in enzyme production and application in the food industry. *Food and Bioproducts Processing*, 126, 123–138. https://doi.org/10.1016/j.fbp.2021.01.006

Department of Biotechnology. (2025). *National Biotechnology Development Strategy (2021-2-25). Knowledge and Innovation Driven Bio-economy*. Department of Biotechnology, Ministry of Science and Technology, GOI.

Dolgin, E. (2021). The tangled history of mRNA vaccines. *Nature*, 597, 318–324. https://doi.org/10.1038/d41586-021-02483-w

Ecker, D. M., Jones, S. D., & Levine, H. L. (2015). The therapeutic monoclonal antibody market. *mAbs*, 7(1), 9–14. https://doi.org/10.4161/19420862.2015.989042

Eibl, R., & Eibl, D. (2008). Design of bioreactors suitable for plant cell and tissue cultures. *Phytochemistry Reviews*, 7(3), 593–598.

Hasan, M. R. (2024). Towards Industry 5.0: A systematic literature review on sustainable and green composite materials supply chains. *arXiv Preprint*. https://doi.org/10.48550/arXiv.2402.06100

Hsu, C. Y., Walsh, T., Borys, B. S., Kallos, M. S., & Rancourt, D. E. (2018). An integrated approach toward the bio-manufacturing of engineered cell therapy products in a stirred-suspension bioreactor. *Molecular Therapy Methods & Clinical Development*, 9, 376–389.

Ijaz, T., Qazalbash, A. H. R., Razzaq, A. A., Rafique, M. Z., & Khan, M. A. (2024). Development of green manufacturing implementation framework based on life-cycle assessment. *Sustainable Environment Research*, 34, Article 23. https://doi.org/10.1186/s42834-024-00229-7

Iragavarapu, G. P., Imam, S. S., Sarkar, O., Mohan, S. V., Chang, Y.-C., Reddy, M. V., Kim, S.-H., & Amradi, N. K. (2023). Bioprocessing of waste for renewable chemicals and fuels to promote bioeconomy. *Energies*, 16(9), Article 3873. https://doi.org/10.3390/en16093873

ISAAA. (2022). *Global Status of Commercialized Biotech/GM Crops in 2022*. International Service for the Acquisition of Agri-biotech Applications (ISAAA Brief No. 58).

Isikgor, F. H. and Becer, C.R. (2015). Lignocellulosic biomass: A sustainable platform for the production of bio-based chemicals and polymers. *Polymer Chemistry*, 6, 4497.

Isoko, K., Cordiner, J. L., Kis, Z., & Moghadam, P. Z. (2024). Bioprocessing 4.0: A pragmatic review and future perspectives. *Digital Discovery*, 3(9), 1662–1681.

Jeevanandam, J., Barhoum, A., Chan, Y. S., Dufresne, A., & Danquah, M. K. (2023). Bio-manufacturing innovations for fish health management: Nanotechnology, probiotics, and recombinant vaccines. *Aquaculture*, 565, 739088.

Kaplon, H., & Reichert, J. M. (2021). Antibodies to watch in 2021. *mAbs*, 13(1), 1860476. https://doi.org/10.1080/19420862.2020.1860476

Khanal, O., & Lenhoff, A. M. (2021). Developments and opportunities in continuous biopharmaceutical manufacturing. *MAbs*, 13(1), 1903664.

Kohler, G., & Milstein, C. (1975). Continuous cultures of fused cells secreting antibody of predefined specificity. *Nature*, 256(5517), 495–497. https://doi.org/10.1038/256495a0

Kumar, V., Roy, S., & Das, P. (2022). Waste valorization for aquafeed production: Recent advances and future prospects. *Journal of Cleaner Production*, 368, 133150. https://doi.org/10.1016/j.jclepro.2022.133150

Lakra, W. S., & Ayyappan, S. (2003). Recent advances in biotechnology applications to aquaculture. *Asian-Australasian Journal of Animal Sciences*, 16(3), 455–462. https://doi.org/10.5713/ajas.2003.455

Lindskog, E. K. (2018). The upstream process: Principal modes of operation. In Günter Jagschies, Eva Lindskog, Karol Łącki, & Parrish Galliher (Eds.), *Biopharmaceutical Processing* (pp. 625–635). Elsevier. ISBN 9780081006238.

Niazi, S. K. (2025). Continuous manufacturing of recombinant drugs: Comprehensive analysis of cost reduction strategies, regulatory pathways, and global implementation. *Pharmaceuticals*, 18(8), 1157.

Oluyemi, G. F., Afolabi, R. O., Zamora, S. C., Li, Y., & McElroy, D. (2024). Environmental impact assessment of a plant cell-based bio-manufacturing process for producing plant natural product ingredients. *Sustainability*, 16(19), Article 8515. https://doi.org/10.3390/su16198515

Paddon, C. J., & Keasling, J. D. (2014). Semi-synthetic artemisinin: A model for the use of synthetic biology in pharmaceutical development. *Nature Reviews Microbiology*, 12(5), 355–367. https://doi.org/10.1038/nrmicro3240

Pardi, N., Hogan, M. J., Porter, F. W., & Weissman, D. (2018). mRNA vaccines a new era in vaccinology. *Nature Reviews Drug Discovery*, 17(4), 261–279. https://doi.org/10.1038/nrd.2017.243

Rasor, B. J., Vogeli, B., Landwehr, G. M., Bogart, J. W., Karim, A. S., & Jewett, M. C. (2021). Toward sustainable, cell-free bio-manufacturing. *Current Opinion in Biotechnology*, 69, 24–30. https://doi.org/10.1016/j.copbio.2020.12.012

Ren, N. Q., Li, J. Z., Ding, J., Yan, X. F., Li, N., Zhang, N., … Tao, Y. (2025). Bio-manufacturing of hydrogen from waste molasses: A full-scale application. *Environmental Science and Ecotechnology*, 100568.

Schmidt-Dannert, C., & López-Gallego, F. (2016). A roadmap for biocatalysis–functional and spatial orchestration of enzyme cascades. *Current Opinion in Chemical Biology*, 31, 1–8. https://doi.org/10.1016/j.cbpa.2015.11.012

Sheldon, R. A., & Woodley, J. M. (2018). Role of biocatalysis in sustainable chemistry. *Chemical Reviews*, 118(2), 801–838. https://doi.org/10.1021/acs.chemrev.7b00203

Shih, A., Friedrich, D., Cao, J., Harada, K., Rajurkar, K., Mitsuishi, M., Sugita, N., & Da Silva Bartolo, P. J. (2013). Biomanufacturing. *CIRP Annals – Manufacturing Technology*, 62(2), 585–606. https://doi.org/10.1016/j.cirp.2013.05.001

Shokoohi, M., & Attar, T. (2024). Bio-manufacturing for a sustainable future: Unleashing the potential of biotechnology in pharmaceutical raw material production. *International Journal of New Findings in Health and Educational Sciences*, 2(2), 142–153. https://doi.org/10.5281/zenodo.11021734

Shukla, A. A., & Thömmes, J. (2010). Recent advances in large-scale production of monoclonal antibodies and related proteins. *Trends in Biotechnology*, 28(5), 253–261.

Singh, A., Kumar, M., & Verma, P. (2021). Role of microbial biofertilizers in sustainable agriculture and environmental management. *Biocatalysis and Agricultural Biotechnology*, 36, 102109. https://doi.org/10.1016/j.bcab.2021.102109

Singhania, R. R., Patel, A. K., Sukumaran, R. K., Larroche, C., & Pandey, A. (2017). Role and significance of beta-glucosidases in the hydrolysis of cellulose for bioethanol production. *Bioresource Technology*, 245, 1498–1506. https://doi.org/10.1016/j.biortech.2017.05.066

Steinkraus, K. H. (1996). *Handbook of Indigenous Fermented Foods* (2nd ed.). CRC Press.

Straathof, A. J. (2014). Transformation of biomass into commodity chemicals using enzymes or cells. *Chemical Reviews*, 114(3), 1871–1908.

Sukumaran, R. K., Singhania, R. R., Mathew, G. M., & Pandey, A. (2010). Cellulase production using biomass feed stock and its application in lignocellulose saccharification for bioethanol production. *Renewable Energy*, 34(2), 421–424. https://doi.org/10.1016/j.renene.2008.05.008

Tamang, J. P., Watanabe, K., & Holzapfel, W. H. (2020). Review: Diversity of microorganisms in global fermented foods and beverages. *Frontiers in Microbiology*, 11, 579401.

Vasić-Rački, Đ. (2004). History of industrial bio-transformations—Dreams and realities. *Organic Process Research & Development*, 8(3), 418–432.

Voutilainen, E., Pihlajaniemi, V., & Parviainen, T. (2021). Economic comparison of food protein production with single-cell organisms from lignocellulose side-streams. *Bioresource Technology Reports*, 14, 100683.

Wang, Y., Wu, Y., Wang, B., Cao, X., & Li, Y. (2022). Advances in industrial-scale probiotics production: Process optimization and emerging technologies. *Frontiers in Microbiology*, 13, 843275.

Xie, D. (2022). Continuous biomanufacturing with microbes—upstream progresses and challenges. *Current Opinion in Biotechnology*, 78, 102793. https://doi.org/10.1016/j.copbio.2022.102793

Xu, X., Lu, Y., Vogel-Heuser, B., & Wang, L. (2021). Industry 4.0 and Industry 5.0—Inception, conception and perception. *Journal of Manufacturing Systems*, 61, 530–535. https://doi.org/10.1016/j.jmsy.2021.10.006

Zhang, Y. H. P., Sun, J., & Ma, Y. (2010). Bio-manufacturing: History and perspective. *Journal of Industrial Microbiology & Biotechnology*, 37(5), 391–400.

Zhang, Y.-H. P., Sun, J., & Ma, Y. (2017). Bio-manufacturing: History and perspective. *Journal of Industrial Microbiology & Biotechnology*, 44(4), 773–784. https://doi.org/10.1007/s10295-016-1863-2

Zhao, X., Zhang, H., & Wang, L. (2023). Alternative proteins: Advances in plant-based, microbial, and cultured meat production. *Trends in Food Science & Technology*, 138, 59–73. https://doi.org/10.1016/j.tifs.2023.05.004

Zhou, Z., Lyu, F., & Chi, H. (2021). Applications of next-generation sequencing in aquaculture genetics and health management. *Frontiers in Genetics*, 12, 667350.

2 Sustainable Biomanufacturing Industries

Market, Production Possibilities, and Employability

Ashwini Kumar
Department of Chemistry, Government Post-Graduate College, Una, Himachal Pradesh, India

Rajan Bhandari
Department of Geography, Central University of Punjab, Bathinda, Punjab, India

Baljinder Singh
Department of Biochemistry, School of Basic Sciences, Central University of Punjab, Bathinda, Punjab, India

2.1 INTRODUCTION

Sustainable biomanufacturing represents a transformative paradigm in industrial production, harnessing biological systems—microorganisms, cells, enzymes, and plants—to create a wide array of products, from life-saving pharmaceuticals to renewable biofuels and eco-friendly materials. Unlike conventional manufacturing, which often depends on finite fossil-based resources and energy-intensive processes, sustainable biomanufacturing prioritizes environmental stewardship, resource efficiency, and adherence to circular economy principles. As global challenges such as climate change, resource depletion, and population growth intensify, biomanufacturing emerges as a critical solution aligned with the United Nations Sustainable Development Goals (SDGs), particularly SDG 12 (Responsible Consumption and Production), SDG 13 (Climate Action), and SDG 9 (Industry, Innovation, and Infrastructure) (United Nations, 2015). This chapter provides an exhaustive exploration of the market dynamics, production possibilities, and employability prospects of sustainable biomanufacturing, emphasizing its pivotal role in fostering a resilient, sustainable bioeconomy capable of addressing 21st-century challenges.

The global bioeconomy, of which biomanufacturing is a cornerstone, is experiencing unprecedented growth. In the European Union (EU), the bioeconomy contributes approximately €2.4 trillion annually and employs over 18 million people, accounting for 4.2% of the EU's Gross Domestic Product (GDP) (European Commission, 2022). In the United States, biomanufacturing generates a value-added economic impact of $1.36 trillion when supply chain effects are included, with advanced biomanufacturing alone contributing $259.9 billion (National Institute of Standards and Technology, 2023). Driven by technological breakthroughs in synthetic biology, automation, digitalization, and increasing consumer demand for sustainable products, biomanufacturing is poised to redefine industrial landscapes. However, challenges such as scalability, regulatory complexity, and workforce shortages pose significant hurdles. This chapter synthesizes current trends, innovations, and challenges, drawing on over 200 original references to provide a comprehensive analysis of sustainable biomanufacturing's potential to shape a greener, more equitable future.

2.2 MARKET DYNAMICS OF SUSTAINABLE BIOMANUFACTURING

2.2.1 Global Market Trends

The global biomanufacturing market is witnessing exponential growth, propelled by rising demand for sustainable products, stringent environmental regulations, and technological advancements. The biotechnology market, a vital component of biomanufacturing, is experiencing rapid growth, driven by continuous advancements in research and innovation and increasing demand for sustainable, health-focused solutions (Thatoi et al., 2025). Advanced biomanufacturing, encompassing high-value products such as biologics, biofuels, and bio-based materials, contributes

DOI: 10.1201/9781003637158-2

significantly to this growth. In the United States, it accounts for $259.9 billion directly and $1.36 trillion when supply chain impacts are considered (National Institute of Standards and Technology, 2023). In Europe, the bioeconomy supports over 18 million jobs and contributes 4.2% to GDP, with biomanufacturing playing a central role in sectors such as pharmaceuticals, agriculture, and materials (European Commission, 2022).

Emerging economies are also scaling up their biomanufacturing capabilities. India's Biotechnology for Economy, Environment, and Employment (BioE3) policy, launched in 2024, aims to position the country as a global leader in sustainable bio-based production, addressing critical challenges in food security, energy, and climate change (Thatoi et al., 2025). Brazil, leveraging its abundant biomass resources, is a key player in bio-based fuels and materials, with its bioethanol industry accounting for 27% of global production (Ogwu et al., 2025). The Asia–Pacific region is expected to experience the fastest growth in biomanufacturing, driven by government initiatives, private-sector investment, and a growing skilled workforce (Glatzel et al., 2024). China's focus on biopharmaceuticals and bio-based chemicals further underscores the region's rising prominence (Deng et al., 2025).

Globally, the shift toward sustainability is reshaping market priorities. The demand for bio-based products, such as biodegradable plastics and sustainable fuels, is growing as consumers and industries seek alternatives to fossil-based materials. For instance, the global bioplastics market is projected to reach $29.7 billion by 2027, driven by innovations in polylactic acid (PLA) and polyhydroxyalkanoates (PHA) production (Kumar et al., 2024). Similarly, the biofuels market is likely to grow steadily in the coming years, with bioethanol and biodiesel playing a major role in this growth (Asase et al., 2024). These trends highlight biomanufacturing's role in addressing environmental challenges while creating economic opportunities.

2.2.2 Key Market Drivers

Biomanufacturing is increasingly recognized as a key driver of environmental sustainability, offering cleaner, more energy-efficient alternatives to conventional industrial processes. For instance, the production of bioplastics from agricultural waste results in materials that are both recyclable and biodegradable, thereby significantly reducing global plastic pollution (Mhaddolkar et al., 2025). In addition, bio-based manufacturing routes can substantially lower greenhouse gas emissions compared with traditional petrochemical methods, contributing to climate change mitigation (Raeisi et al., 2024).

Regulatory pressures are further accelerating the adoption of biomanufacturing worldwide. Governments are actively promoting sustainable industrial practices through policy frameworks and financial incentives. The EU's Green Deal, which targets climate neutrality by 2050, includes substantial funding mechanisms to support the growth of bio-based industries (Marelli et al., 2025). Similarly, in the United States, the Creating Helpful Incentives to Produce Semiconductors (CHIPS) and Science Act incorporates provisions to foster biomanufacturing innovation through tax incentives and research grants, encouraging industries to transition toward more sustainable production models (Love et al., 2025).

At the same time, rising consumer awareness of environmental issues is reshaping market dynamics. A growing proportion of consumers are willing to pay a premium for sustainable products, including bio-based packaging and textiles (Falcone & Fiorentino, 2025). This preference is particularly pronounced among younger demographics. It is strongly influencing sectors such as food, cosmetics, and fashion, where sustainability has become a key determinant of brand value and competitiveness (Nguyen et al., 2025).

Rapid technological advancements are also transforming the biomanufacturing landscape. Innovations in synthetic biology, CRISPR-based gene editing, and digital biomanufacturing are improving process efficiency and expanding product portfolios. The integration of artificial intelligence (AI)-driven analytics and digital twins, such as those enabled by platforms like 3DEXPERIENCE, allows optimization of production workflows and can reduce resource wastage by up to 20% (Bhambri et al., 2024). These advances enable the conversion of low-cost feedstocks, including agricultural residues, into high-value bioproducts (Lee et al., 2022).

Finally, the global energy transition toward renewable sources is increasing demand for biofuels, including bioethanol, biodiesel, and sustainable aviation fuel (SAF). Biomanufacturing facilitates the conversion of biomass into these fuels, thereby reducing dependence on fossil resources. Notably, algal biofuels have demonstrated yields several times higher than those of conventional energy crops, highlighting their potential as a sustainable and scalable energy solution (Leite et al., 2013).

2.2.3 Market Challenges

Despite its considerable potential, biomanufacturing faces several challenges that limit large-scale adoption. One of the most significant barriers is scalability, as translating laboratory-scale processes into industrial production is both technically complex and capital intensive. Inadequate infrastructure further compounds this issue; for example, the United States currently has only about 10% of the pilot-scale facilities required to effectively support commercialization (Moussa, 2022). Successful scale-up demands substantial investment in bioreactors, fermentation platforms, and downstream processing systems, all of which increase financial risk and operational complexity (Delvigne et al., 2017).

Cost competitiveness represents another major constraint. High upfront expenditures associated with research and development, facility construction, and specialized equipment often render bio-based products less economically attractive than fossil-derived alternatives. Establishing large-scale fermentation plants is particularly expensive and time-consuming, creating significant barriers to entry for new players (Fiorentino et al., 2019). As a result, small- and medium-sized enterprises (SMEs), which are often key sources of innovation, may be discouraged from participating in the biomanufacturing sector (Runiewicz-Wardyn, 2025).

Regulatory complexity also poses a critical challenge, especially in the biopharmaceutical industry, which constitutes a major segment of biomanufacturing. Strict regulatory frameworks designed to ensure product safety and efficacy can inadvertently conflict with sustainability objectives, necessitating innovative approaches to balance regulatory compliance with environmental responsibility (Ashiwaju et al., 2023). For instance, the widespread use of single-use systems minimizes contamination risks but contributes significantly to plastic waste, highlighting a trade-off between operational safety and environmental impact.

In addition, supply chain vulnerabilities can undermine the reliability of biomanufacturing operations. Although the approach supports more localized production, the consistent sourcing of sustainable raw materials such as agricultural residues, algae, or other biomass feedstocks—remains challenging due to fluctuations in availability, quality, and seasonal variability (Shokoohi & Attar, 2024). Establishing resilient and well-coordinated supply chains is therefore essential to ensure both scalability and long-term operational stability.

Finally, public perception continues to influence the acceptance of biomanufactured products. Persistent misconceptions surrounding genetically modified organisms (GMOs) and synthetic biology can hinder market uptake, particularly in Europe and parts of Asia (Frimpong, 2016). Addressing these concerns through transparent communication, public engagement, and science-based education is crucial for building consumer trust and fostering broader societal acceptance of biomanufacturing technologies.

2.2.4 Emerging Market Opportunities

Sustainable biomanufacturing is driving the emergence of new markets across a wide range of industrial sectors. In the biopharmaceutical industry, the production of biologics, such as vaccines, monoclonal antibodies, and gene therapies, represents a major application of biomanufacturing. Advances such as continuous manufacturing and single-use bioprocessing systems are improving operational efficiency while simultaneously reducing production costs and environmental impacts (Soni & Patel, 2024).

The biofuels sector is another rapidly expanding area, with bioethanol, biodiesel, and SAF offering viable and cleaner alternatives to fossil fuels. Among these, algal biofuels are particularly promising, as they can achieve yields up to ten times higher than those of conventional energy crops while requiring significantly less land and water resources (Cherwoo et al., 2023). These attributes position biofuels as a critical component of the global transition toward renewable energy.

Bioplastics are also gaining momentum as industries seek solutions to the approximately 300 million tons of plastic waste generated annually. Companies such as Carbios and BASF are developing biodegradable and compostable plastics derived from renewable feedstocks, helping to reduce dependence on petrochemical plastics and mitigate environmental pollution. Consequently, the global bioplastics market is projected to experience rapid growth in the coming years (Morais et al., 2024).

In food and agriculture, biomanufacturing is transforming traditional practices through the development of bio-based fertilizers, sustainable packaging, and precision fermentation technologies. Precision fermentation, in particular, enables the production of non-dairy proteins and alternative food products, including plant-based milk and meat substitutes. This technology is expected to gain widespread adoption and play an increasingly important role in producing food and other high-value biomaterials in a sustainable manner (Çakmakçı et al., 2024).

The textiles and consumer goods sectors are also benefiting from bio-based innovations. The development of alternatives such as bio-based leather and sustainable fabrics is helping to reduce the environmental footprint of the fashion industry. For example, companies like MycoWorks utilize mycelium-based materials to produce durable and eco-friendly textiles, significantly lowering pollution and environmental damage associated with conventional textile manufacturing (Williams et al., 2022; Ekegårdh & Kristersson, 2021).

2.2.5 Regional Market Insights

Biomanufacturing development varies significantly across regions, reflecting differences in policy priorities, resource availability, and industrial capacity. In North America, the United States leads in biopharmaceuticals and biofuels, supported by strong federal initiatives such as the National Biotechnology and Biomanufacturing Initiative, which allocates approximately USD2 billion toward biomanufacturing research and development (Love et al., 2025). Canada is also emerging as an important player, particularly in the fields of bioplastics and precision fermentation, supported by its robust research ecosystem and emphasis on sustainable innovation (Johnston, 2015). In Europe, the biomanufacturing sector has been strongly driven by coordinated policy frameworks and substantial public investment. The EU's Bio-based Industries Joint Undertaking (BBI-JU) has invested around €3.7 billion in bioeconomy-related projects, stimulating advances in bioplastics, bio-based chemicals, and sustainable agricultural practices. Within the region, countries such as Germany and France are at the forefront of biorefinery development, leveraging advanced infrastructure and strong industry–academia collaborations to scale bio-based production systems (Gírio et al., 2017). The Asia–Pacific region is experiencing rapid growth in biomanufacturing, particularly in China and India, where government-led programs play a central role. China's 14th Five-Year Plan places strong emphasis on biopharmaceuticals and bio-based materials as strategic industries, driving large-scale investments and technological advancement (Zhang et al., 2022). Similarly, India's BioE3 policy aims to accelerate the bioeconomy by fostering innovation, strengthening infrastructure, and generating approximately one million jobs by 2030 (Thatoi et al., 2025). In Latin America, Brazil stands out as a global leader in bioethanol production derived from sugarcane, demonstrating the region's strong potential for bioenergy. Beyond biofuels, Brazil has significant opportunities to expand into bioplastics and bio-based chemicals, supported by abundant biomass resources (de Resende & da Costa, 2020). Other countries, including Argentina and Colombia, are also increasingly investing in bio-based industries, signaling a growing regional commitment to sustainable biomanufacturing and bioeconomic development (Contreras, 2022).

2.3 PRODUCTION POSSIBILITIES IN SUSTAINABLE BIOMANUFACTURING

2.3.1 Technological Innovations

Sustainable biomanufacturing leverages cutting-edge technologies to enhance productivity, scalability, and environmental performance. Key innovations include the following:

1. **Continuous Biomanufacturing (CBM):** CBM replaces old batch methods with ongoing processes. This speeds up production and reduces waste. For example, in the production of monoclonal antibodies, it can save water and shorten production time. (Schmidt et al., 2024).
2. **Synthetic Biology:** Synthetic biology enables the engineering of microorganisms to produce high-value products from waste feedstocks. LanzaTech's gas fermentation technology converts industrial emissions into ethanol, supporting applications, such as SAF and chemicals (Mitra et al., 2024). Synthetic biology is projected to contribute $300 billion to the global economy by 2030.
3. **Plant Molecular Farming (PMF):** Plants are used as bioreactors to produce proteins, offering a low-cost, scalable alternative to cell-based systems. Medicago's tobacco plant-based vaccine production platform demonstrates PMF's potential for rapid, sustainable biopharmaceutical production (Monreal-Escalante et al., 2022).
4. **Digital Biomanufacturing:** Digital tools, such as the 3DEXPERIENCE platform, create virtual twins of production processes, optimizing resource use and reducing emissions by up to 20%. AI helps improve control over the production process. It makes the system smarter and reduces waste (Zong and Guan, 2025).
5. **Single-Use Technologies (SUTs):** Disposable bioreactors and filtration systems reduce energy-intensive cleaning and sterilization, improving safety and sustainability. However, their plastic waste footprint necessitates advancements in recycling (Soni & Patel, 2024).
6. **CRISPR and Gene Editing:** CRISPR enables precise modifications to microbial and plant genomes, improving yields of bio-based products. Corteva Agriscience uses CRISPR to develop high-yield crops for bioethanol, increasing productivity (Ghosh & Padaria, 2025).
7. **Bioprinting:** 3D bioprinting is emerging for tissue engineering and regenerative medicine, with applications in biopharmaceuticals and medical devices (Saini et al., 2021).
8. **Precision Fermentation:** This technique produces proteins, such as insulin, enzymes, and non-dairy milk with high efficiency. The precision fermentation market is expected to grow at a Compound Annual Growth Rate (CAGR) of 40% through 2030 (Jha et al., 2025; Singh & Kumar, 2025).

2.3.2 Circular Economy Integration

Biomanufacturing strongly aligns with the principles of the circular economy by reducing waste generation and maximizing the reuse of resources across production systems. One of its most important contributions is waste valorization, in which agricultural and industrial residues are converted into value-added products. For example, the Pomacle–Bazancourt biorefinery in France utilizes food-processing waste streams to produce bio-based chemicals, leading to a reported 30% reduction in landfill disposal and demonstrating how waste can be transformed into economic and environmental value (Chauvet & Lescieux-Katir, 2024).

The development of biodegradable products further reinforces the circularity of biomanufacturing. Bioplastics such as PLA and polyethylene furanoate (PEF) are designed to degrade naturally, thereby lowering long-term environmental pollution. Notably, Avantium's PEF has been adopted in packaging and textile applications as a sustainable alternative to conventional polyethylene terephthalate (PET), offering improved recyclability and reduced fossil-based carbon footprints (Gerritse, 2020).

Biomanufacturing also supports circular economy goals through carbon capture and utilization (CCU) strategies. Innovative technologies, such as LanzaTech's gas fermentation platform, enable the conversion of industrial CO_2 emissions into fuels and high-value chemicals. By transforming waste carbon into useful products, these approaches contribute directly to net-zero and climate-mitigation objectives (Mihalcea et al., 2024).

In addition, industrial symbiosis is increasingly adopted within biomanufacturing systems, particularly in integrated biorefineries. These facilities co-locate and interlink multiple processes to share resources such as energy, water, and by-products, thereby improving overall efficiency. Such integrated approaches have been shown to enhance resource utilization efficiency by up to 25%, underscoring the role of biomanufacturing as a cornerstone of circular and sustainable industrial development (John et al., 2025).

2.3.3 Scalability and Infrastructure

Scaling biomanufacturing processes is essential to meeting growing global demand; however, several structural and economic barriers continue to limit large-scale deployment. One of the most significant constraints is the shortage of pilot- and commercial-scale facilities required to support process validation and industrial production. Insufficient infrastructure hampers the translation of laboratory discoveries into market-ready products, creating a persistent bottleneck in commercialization. This challenge is not confined to a single region, as many countries face similar limitations due to the scarcity of well-equipped, specialized facilities needed for advanced biomanufacturing operations (Makepa & Chihobo, 2024). Standardization of equipment and operational processes represents a key opportunity to overcome some of these scale-up challenges. The adoption of modular, standardized bioprocessing platforms can enable faster transitions between products and production campaigns, reducing the need for costly facility redesigns. Such flexibility improves operational efficiency, lowers capital and operating costs, and enhances the long-term economic viability of biomanufacturing facilities (Bobier et al., 2024). High investment requirements remain another major barrier to scaling. Establishing commercial-scale biomanufacturing facilities is capital intensive, often requiring investments amounting to hundreds of millions of dollars. These financial demands are difficult for individual companies—particularly start-ups and small enterprises—to bear alone, especially during early development stages. Consequently, strong support from both the public and private sectors is crucial, including public–private partnerships, targeted incentives, and risk-sharing mechanisms that can accelerate infrastructure development and industry growth (Love et al., 2025). While localized biomanufacturing can strengthen supply chain resilience by reducing dependence on distant suppliers and minimizing transportation-related disruptions, securing a consistent, sustainable supply of feedstocks remains a challenge. Many bio-based raw materials are subject to seasonal availability, quality variability, and regional limitations, complicating efforts to maintain continuous, reliable production. Addressing these feedstock constraints through diversified sourcing strategies and improved biomass logistics will be essential to ensuring the scalability and stability of global biomanufacturing systems (Yue et al., 2014).

2.3.4 Environmental Impact Reduction

Biomanufacturing provides substantial environmental benefits compared with conventional chemical manufacturing, primarily through lower energy consumption, reduced waste generation, and contributions to carbon neutrality. Bioprocesses typically operate under milder conditions, such as lower temperatures and pressures, by harnessing biological systems, including enzymes and microorganisms. This significantly reduces energy requirements and, in turn, lowers associated greenhouse gas emissions, making biomanufacturing a more climate-friendly production approach (Ramamurthy et al., 2021). Advances in process technologies are further enhancing the environmental performance of biomanufacturing systems. Innovations such as depth filtration and continuous processing have improved the efficiency of biopharmaceutical manufacturing by minimizing material losses and reducing process interruptions. These approaches enable more consistent production with lower waste generation, conserving resources while decreasing the overall environmental footprint. As a result, high-quality biopharmaceutical products can be produced in a more sustainable and cost-effective manner (Khanal & Lenhoff, 2021).

Biomanufacturing also supports global carbon-neutrality goals through the use of bio-based products and CCU technologies. By converting renewable biomass or captured carbon dioxide into fuels, chemicals, and materials, these approaches contribute directly to net-zero strategies and have the potential to reduce global greenhouse gas emissions by up to 5% by 2030 (Glavič et al., 2023). Despite these advantages, important environmental challenges remain. The carbon footprint associated with single-use systems and emissions arising from the transportation of raw materials and finished products can offset some sustainability gains if not carefully managed. Addressing these issues requires continued technological innovation and system-level optimization. Lifecycle assessment (LCA) plays a critical role in this process by evaluating environmental impacts across the entire product lifecycle from raw material extraction to end-of-life disposal—thereby identifying opportunities to reduce energy use, minimize waste, and lower emissions throughout biomanufacturing value chains (Ragazzi et al., 2023).

2.3.5 Emerging Production Techniques

- **Microbial Electrolysis:** This method uses tiny living organisms called microbes to break down organic waste, such as food scraps or plant material, and turn it into hydrogen gas, which can then be used as a clean and renewable source of energy. Since hydrogen does not release harmful gases when used, especially in fuel cells that power vehicles or machines, this technique provides an environmentally friendly alternative to fossil fuels. By turning everyday waste into useful energy, it not only helps reduce pollution but also supports a more sustainable future with less dependence on non-renewable energy sources (Arun et al., 2024).
- **Photobioreactors:** Algae-based photobioreactors are special systems that use light and controlled

environments to grow algae, which can then be used to produce biofuels and other valuable products, such as vitamins, proteins, or natural colors, all while using very little land compared to traditional farming or manufacturing methods. Because algae grow quickly and do not need fertile soil, these systems can be set up in many places, including areas where farming is not possible, making them a flexible and scalable option for sustainable production. This approach not only helps reduce pressure on natural resources, such as land and water, but also supports the shift toward cleaner, greener alternatives for energy and everyday products (Sarker & Kaparaju, 2024).

- **Bioelectrosynthesis:** Bioelectrosynthesis is a new and promising technology that uses electricity, often from renewable sources, such as solar and wind, to help special microbes produce useful chemicals, such as fuels or industrial materials, in a more sustainable way. Instead of depending on large amounts of organic feedstocks, such as sugar or plant material, which can be costly and compete with food supply, this method provides microbes with the energy they need through electricity to carry out chemical reactions. By doing this, bioelectrosynthesis not only reduces our dependence on traditional raw materials but also opens up cleaner and more efficient pathways for making important products with less harm to the environment (Finkelstein et al., 2022).

2.4 EMPLOYABILITY IN SUSTAINABLE BIOMANUFACTURING

2.4.1 Workforce Demand

The biomanufacturing sector plays an important role in creating jobs and supporting the economy, as it employs a large number of people across many regions and continues to grow steadily with the rise of bio-based industries. As companies develop new ways to produce medicines, fuels, materials, and food using biological methods, the need for trained and skilled workers in this field is also increasing. More job opportunities are opening up in areas such as biotechnology, engineering, quality control, and research, showing that biomanufacturing is not only helping the environment but also providing strong career paths for people around the world. With growing interest in sustainable technologies, the biomanufacturing workforce is expected to expand even more in the coming years, supporting both economic development and environmental progress (Dietz et al., 2024).

2.4.2 Key Roles and Skill Requirements

Biomanufacturing relies on a diverse and highly multidisciplinary workforce that integrates technical, engineering, digital, and sustainability expertise. Production operators, typically holding bachelor's degrees, are responsible for operating and monitoring equipment such as bioreactors, centrifuges, and filtration systems. These roles require strong technical proficiency, adherence to quality and safety standards, and the flexibility to work in shift-based production environments.

Bioprocess engineers play a central role in the design, optimization, and scale-up of biomanufacturing processes. Trained in chemical, biochemical, or mechanical engineering, these professionals apply skills in process modeling, automation, and control to improve efficiency, yield, and product quality across laboratory, pilot, and industrial scales. Their expertise is critical for translating research innovations into commercially viable processes.

Technical specialists, often holding master's or doctoral degrees, address complex operational and scientific challenges, including contamination control, process instability, and productivity losses. These professionals rely on advanced analytical techniques, deep domain knowledge, and strong problem-solving abilities to troubleshoot and refine bioprocesses, ensuring robustness and regulatory compliance.

The emergence of Biopharma 4.0 has significantly increased demand for digital biomanufacturing experts. Professionals with backgrounds in information technology, data science, and bioinformatics are increasingly required to develop and manage artificial intelligence-driven analytics, machine learning models, and digital twin platforms that enable real-time monitoring, predictive maintenance, and process optimization.

In addition, sustainability specialists are becoming integral to modern biomanufacturing operations. These professionals apply expertise in LCA, environmental management, and green chemistry to ensure that production processes align with sustainability objectives and regulatory requirements. Their role is essential in minimizing environmental footprints while supporting compliance with evolving national and international sustainability frameworks.

2.4.3 Skills Gap and Training Needs

- **Digital Skills:** Although there is a fast-growing need for advanced technologies in biomanufacturing, such as AI, machine learning, and digital twins, only a small portion of workers in this field currently have the skills and knowledge to use these tools effectively. This gap in expertise makes it harder for companies to fully benefit from the advantages these technologies offer, such as better efficiency, faster problem-solving, and smarter decision-making. To keep up with the changing demands of modern biomanufacturing, it is important to train more professionals in these digital areas so that the industry can continue to grow and improve in both performance and sustainability (Digital Transformation in Udugama et al., 2021).
- **Interdisciplinary Training:** To prepare students for the future of the biopharmaceutical industry, often called Biopharma 4.0, universities need to create new

academic programs that bring together the key subjects of biology, engineering, and information technology. These combined programs will help students gain the right mix of knowledge and skills needed to work with advanced technologies, such as automation, AI, and smart manufacturing systems, that are now becoming common in modern biomanufacturing. By learning how living systems, machines, and digital tools work together, students will be better equipped to solve complex problems, improve production processes, and support the development of safer and more efficient medicines.

- **Continuous Learning:** As technology in biomanufacturing continues to advance rapidly, many companies are now focusing on continuous learning by providing special training programs to help their employees update and improve their skills. These upskilling efforts are designed to ensure workers can keep up with new tools, digital systems, and production methods that are becoming more common in the industry. By offering these learning opportunities, companies not only support the growth and confidence of their employees but also ensure that their workforce is ready to handle the latest innovations, making the organization more competitive and efficient in the long run (Auclair, 2021).

2.4.4 Job Creation Potential

Sustainable biomanufacturing offers substantial potential for job creation across both developed and emerging economies. In India, the BioE3 policy is expected to generate approximately one million jobs by 2030 by promoting bio-based production in key sectors such as agriculture, energy, and advanced materials (Thatoi et al., 2025). Similarly, Brazil's well-established bioethanol industry already employs more than 700,000 workers and presents significant opportunities for further employment growth through expansion into bioplastics and bio-based chemical production (Ogwu et al., 2025). In the EU, SMEs play a central role in the bioeconomy; supported by initiatives such as the BBI-JU, SMEs account for nearly 60% of bioeconomy-related jobs, contributing to inclusive economic growth and regional development.

Despite this strong employment potential, several challenges remain in workforce development within the biomanufacturing sector. One major issue is intense competition for skilled talent, particularly in digital and data-driven roles. Approximately 30% of biomanufacturing job openings require information technology–related skills, placing the sector in direct competition with IT and other high-growth industries. In addition, the limited availability of specialized academic programs constrains workforce readiness; only about 15% of universities currently offer dedicated degrees in biomanufacturing, making on-the-job training a necessity for many employers. Retention of skilled personnel is another concern, as high demand for expertise leads to frequent job mobility, with nearly 25% of biomanufacturing employees reportedly transitioning to other sectors each year. Addressing these workforce challenges will be essential to fully realizing the employment and economic benefits of sustainable biomanufacturing.

2.5 CASE STUDIES

2.5.1 LanzaTech: Carbon Capture and Utilization

LanzaTech has developed a special gas fermentation technology that can turn harmful industrial emissions, such as carbon gases from factories, into useful products, such as ethanol, which can be used to make SAF and other important chemicals. This process not only helps reduce pollution but also creates valuable materials from waste gases, making it a smart, eco-friendly solution. A great example of how this technology works on a large scale is LanzaTech's upcoming facility in Norway, scheduled to begin operations in 2028 and expected to produce a large volume of ethanol annually. This project shows that the technology is not only effective in helping the environment but also scalable, meaning it can be used in many places around the world to support cleaner energy and greener industries (Mihalcea et al., 2024).

2.5.2 Avantium: Biodegradable Plastics

Avantium has developed a new type of polymer, PEF, which is made from agricultural waste instead of fossil fuels, making it a much more environmentally friendly option. This material is fully recyclable, meaning it can be reused many times without harming the environment, and it also offers strong performance for packaging and textiles. The company's partnerships with major global brands, such as Coca-Cola and Danone, show that there is strong interest in using PEF for bottles, containers, and other packaging materials, proving that it has real commercial value and can replace traditional plastics. These collaborations highlight the growing demand for sustainable solutions and show how innovative materials, such as PEF, can play a big role in reducing plastic waste and building a cleaner future (Pabico, 2023).

2.5.3 Pomacle–Bazancourt Biorefinery

A biorefinery in France has developed a smart, eco-friendly way to convert food-processing waste into useful bio-based chemicals, demonstrating how industries can work together to reduce environmental impact. This approach, known as industrial symbiosis, means that one industry's waste becomes another's valuable resource, creating a closed-loop system that supports the circular economy. By using materials that would otherwise be discarded, the biorefinery not only reduces waste but also produces chemicals in a cleaner, more sustainable way. This method shows how different industries can cooperate to save resources, lower pollution, and create a more efficient and environmentally responsible production system (Chauvet & Lescieux-Katir, 2024).

2.5.4 MycoWorks: Bio-Based Leather

MycoWorks has developed an innovative method that uses mycelium, the root-like structure of fungi, to create a sustainable alternative to traditional leather, offering a much more eco-friendly option for the fashion and materials industries. This new type of leather not only looks and feels like animal leather but also avoids the harmful environmental impacts associated with traditional leather-making, such as heavy water use, pollution from tanning chemicals, and greenhouse gas emissions. By growing the material from natural sources and avoiding the need for animal products, MycoWorks is helping to cut the environmental impact of leather production in half. This shows how new, nature-based technologies can offer high-quality products while protecting the planet (Williams et al., 2022).

2.5.5 Ginkgo Bioworks: Synthetic Biology

Ginkgo Bioworks is a company that uses the power of synthetic biology to design and engineer specialized microorganisms that perform a wide range of useful tasks across industries. These tiny organisms are carefully programmed to produce various products, such as natural fragrances for perfumes, food ingredients, and even important components used in the manufacture of medicines and vaccines. This ability to create custom-made microbes shows the flexibility and power of synthetic biology, enabling scientists to develop cleaner, faster, and more sustainable ways to produce everyday items. Ginkgo Bioworks' work is a great example of how science and nature can come together to solve modern challenges and improve how we make products in a smarter and more environmentally friendly way (Philippidis, 2024).

2.6 FUTURE PROSPECTS

The future of sustainable biomanufacturing appears highly promising, with several key trends shaping its continued growth and transformation. Increased investment from both governments and the private sector is playing a central role in advancing the field. Financial support is accelerating the development of innovative technologies, strengthening infrastructure, and enabling the scale-up of biomanufacturing processes, thereby enhancing the sector's long-term sustainability and economic viability.

Global collaboration is another important driver of progress. International partnerships among governments, industries, and research institutions are fostering knowledge exchange and resource sharing, thereby accelerating innovation and enabling the large-scale deployment of biomanufacturing technologies. Such collaborative efforts are critical for addressing complex challenges related to scalability, supply chains, and regulatory harmonization across regions.

Digital transformation is also redefining the biomanufacturing landscape. The integration of advanced digital tools, including AI, machine learning, and digital twin technologies, is improving process optimization, predictive maintenance, and real-time monitoring. These innovations enhance production efficiency, reduce resource consumption, and minimize environmental impacts, making biomanufacturing both economically and ecologically more sustainable.

Finally, strong policy support is providing a favorable environment for the expansion of biomanufacturing worldwide. Governments are introducing targeted policies, funding programs, and regulatory frameworks that encourage investment in bio-based and low-carbon production systems. Collectively, these policy measures are helping to accelerate the transition toward cleaner manufacturing practices and positioning biomanufacturing as a cornerstone of the future sustainable economy.

2.7 CONCLUSION

Sustainable biomanufacturing is poised to redefine industrial production by offering eco-friendly alternatives to traditional methods. Its market is expanding rapidly, driven by environmental imperatives, consumer demand, and technological breakthroughs. Production possibilities are vast, with innovations in continuous manufacturing, synthetic biology, and digital tools enabling scalable, circular processes. The sector's growth is creating diverse employment opportunities, although addressing skills gaps and infrastructure challenges is critical. By fostering collaboration, innovation, and policy support, sustainable biomanufacturing can lead the transition to a greener, more resilient global economy, aligning with global sustainability goals and creating value for society.

REFERENCES

Arun, J., SundarRajan, P., Pavithra, K. G., Priyadharsini, P., Shyam, S., Goutham, R., … Pugazhendhi, A. (2024). New insights into microbial electrolysis cells (MEC) and microbial fuel cells (MFC) for simultaneous wastewater treatment and green fuel (hydrogen) generation. *Fuel*, *355*, 129530.

Asase, R. V., Okechukwu, Q. N., & Ivantsova, M. N. (2024). Biofuels: Present and future. *Environment, Development and Sustainability* (pp. 1–29). Springer.

Ashiwaju, B. I., Orikpete, O. F., Fawole, A. A., Alade, E. Y., & Odogwu, C. (2023). A step toward sustainability: A review of biodegradable packaging in the pharmaceutical industry. *Matrix Science Pharma*, *7*(3), 73–84.

Auclair, J. (2021). STEM education and industry workforce life-long training platform development to faciliate smart biopharmaceutical manufacturing 4.2. In *2020 Northeast Section Meeting*.

Bhambri, P., Rani, S., & Khang, A. (2024). AI-driven digital twin and resource optimization in Industry 4.0 ecosystem. In Sita Rani, Pankaj Bhambri, Sachin Kumar, Piyush Kumar Pareek, Ahmed A. Elngar (Eds.), *Intelligent Techniques for Predictive Data Analytics* (pp. 47–69). CRC Press.

Bobier, J. F., Cerisy, T., Coulin, A. D., Bleecher, C., Sassoon, V., & Alexander, B. (2024). Breaking the cost barrier in biomanufacturing. *Industrial Biotechnology*, *20*(3), 113–124.

Çakmakçı, S., Polatoğlu, B., & Çakmakçı, R. (2024). Foods of the future: Challenges, opportunities, trends, and expectations. *Foods*, *13*(17), 2663.

Chauvet, J. M., & Lescieux-Katir, H. (2024). From primary food processing industries to biorefineries: Example of Bazancourt-Pomacle. In *Green Chemistry and Agro-food Industry: Towards a Sustainable Bioeconomy* (pp. 313–338). Springer Nature Switzerland.

Cherwoo, L., Gupta, I., Flora, G., Verma, R., Kapil, M., Arya, S. K., … Ashokkumar, V. (2023). Biofuels an alternative to traditional fossil fuels: A comprehensive review. *Sustainable Energy Technologies and Assessments*, *60*, 103503.

Contreras, N. R. (2022). *Towards a sustainable biobased economy in Colombia: Integrated environmental and economic analyses of land use and biomass value chains*.

Delvigne, F., Takors, R., Mudde, R., van Gulik, W., & Noorman, H. (2017). Bioprocess scale-up/down as integrative enabling technology: From fluid mechanics to systems biology and beyond. *Microbial Biotechnology*, *10*(5), 1267–1274.

Deng, H., Yu, C., Jin, Y., Pray, C. E., Liu, C., & Deng, L. (2025). How is China shaping global food supply chains? Insights from the seed industry. *European Review of Agricultural Economics*, jbaf017.

de Resende, T. M., & da Costa, M. M. (2020). Biopolymers of sugarcane. In *Sugarcane Biorefinery, Technology and Perspectives* (pp. 229–254). Academic Press.

Dietz, T., Bogdanski, A., Boldt, C., Börner, J., von Braun, J., Concubhair, C., … Werheim, P. (2024). *Bioeconomy Globalization: Recent Trends and Drivers of National Programs and Policies*. International Advisory Council on Global Bioeconomy.

Ekegårdh, N., & Kristersson, E. (2021). *From Fungi to Vegan Leather: A case study made to highlight the potentials of mycelium-based leatherin the textile industry and the unique features of its value chain*.

European Commission. (2022). *EU Bioeconomy Strategy Progress Report*. European Commission.

Falcone, P. M., & Fiorentino, R. (2025). Nudging towards sustainability: Exploring the role of behavioral interventions in circular bio-economy development for the fashion industry. *Corporate Social Responsibility and Environmental Management*, *32*(1), 661–678.

Finkelstein, J., Swartz, J., & Koffas, M. (2022). Bioelectrosynthesis systems. *Current Opinion in Biotechnology*, *74*, 211–219.

Fiorentino, G., Zucaro, A., & Ulgiati, S. (2019). Towards an energy efficient chemistry. Switching from fossil to bio-based products in a life cycle perspective. *Energy*, *170*, 720–729.

Frimpong, S. G. (2016). *The perceptions and misconceptions of the usage of genetically modified organisms: Views from a cross section of the university of cape coast community*.

Gerritse, T. (2020). *The Circular Bioeconomy in Practice–Assessing the End-of-Life Options and Its Environmental Impact for Bio-Based Polyethylene Furanoate (Pef)* (Master's thesis). Utrecht University.

Ghosh, G., & Padaria, J. C. (2025). Application of CRISPR/Cas9 in the improvement of industrially important crops. In *Industrial Crops Improvement: Biotechnological Approaches for Sustainable Agricultural Development* (pp. 239–257). Springer Nature Switzerland.

Gírio, F., Marques, S., Pinto, F., Oliveira, A. C., Costa, P., Reis, A., & Moura, P. (2017). Biorefineries in the World. In *Biorefineries: Targeting Energy, High Value Products and Waste Valorisation* (pp. 227–281). Springer International Publishing.

Glatzel, K., Virchow, D., Nakitto, A. M. S., Niyonsenga, S., Babu, S. C., Srivastava, N., … Osano, P. (2024). *Bioeconomy Pathways: Experience from Africa, Asia, and Latin America*. IFPRI Book Chapters, pp. 116–149.

Glavič, P., Pintarič, Z. N., Levičnik, H., Dragojlović, V., & Bogataj, M. (2023). Transitioning towards net-zero emissions in chemical and process industries: A holistic perspective. *Processes*, *11*(9), 2647.

Indian Bioeconomy Report. (2024). *India's BioE3 Policy: A Roadmap for Sustainable Biomanufacturing*. Government of India.

Jha, N., Kohli, V., & Singha, S. (2025). Food and dairy fermentation processes. In *Engineering Solutions for Sustainable Food and Dairy Production: Innovations and Techniques in Food Processing and Dairy Engineering* (pp. 189–216). Springer Nature Switzerland.

John, S. U., Onu, C. E., Ezechukwu, C. M. J., Nwokedi, I. C., & Onyenanu, C. N. (2025). Multi-product biorefineries for biofuels and value-added products: Advances and future perspectives. *Academia Green Energy*, *2*(1).

Johnston, J. B. (2015). Bridging the gap: Fostering competitiveness in the Canadian biobased industry. *Industrial Biotechnology*, *11*(5), 253–258.

Khanal, O., & Lenhoff, A. M. (2021). Developments and opportunities in continuous biopharmaceutical manufacturing. *MAbs*, 13(1), 1903664.

Kumar, K., Mondal, P., Jadhav, S. M., & Moholkar, V. S. (2024). Application of enzymatic technique for the synthesis of biodegradable polymers from different agrobiomasses. In *Agricultural Biomass for the Synthesis of Value-Added Materials* (pp. 273–299). CRC Press.

Lee, J. Y., Lee, S. E., & Lee, D. W. (2022). Current status and future prospects of biological routes to bio-based products using raw materials, wastes, and residues as renewable resources. *Critical Reviews in Environmental Science and Technology*, *52*(14), 2453–2509.

Leite, G. B., Abdelaziz, A. E., & Hallenbeck, P. C. (2013). Algal biofuels: Challenges and opportunities. *Bioresource Technology*, *145*, 134–141.

Love, J. C., Reynolds, E. B., Goldston, D., & Frye, H. E. (2025). *Biomanufacturing in the US: A MIT Policy Brief*. DSpace@MIT

Makepa, D. C., & Chihobo, C. H. (2024). Barriers to commercial deployment of biorefineries: A multi-faceted review of obstacles across the innovation chain. *Heliyon*, *10*(12).

Marelli, L., Trane, M., Barbero, V. G., Gastaldi, C., Guerreiro, M. M., Delgado, C. L., & Neuwahl, F. (2025). *Delivering the EU Green Deal-Progress towards Targets*. Publications Office of the European Union.

Mhaddolkar, N., Astrup, T. F., Tischberger-Aldrian, A., Pomberger, R., & Vollprecht, D. (2025). Challenges and opportunities in managing biodegradable plastic waste: A review. *Waste Management & Research*, *43*(6), 911–934.

Mihalcea, C., Burton, F., Conrado, R., & Simpson, S. (2024). R&D&I and industry examples: LanzaTech's gas fermentation. In *CO2 and CO as Feedstock: Sustainable Carbon Sources for the Circular Economy* (pp. 333–343). Springer International Publishing.

Mitra, S., Paliya, S., & Mandpe, A. (2024). Microbial engineering in biofuel production—A global outlook, advances, and roadmap. In *Emerging Sustainable Technologies for Biofuel Production* (pp. 547–593). Springer Nature Switzerland.

Monreal-Escalante, E., Ramos-Vega, A., Angulo, C., & Bañuelos-Hernández, B. (2022). Plant-based vaccines: Antigen design, diversity, and strategies for high level production. *Vaccines*, *10*(1), 100.

Morais, A. M., Morais, R., & Lackner, M. (2024). Biodegradable bio-based plastics toward climate change mitigation. In *Handbook of Climate Change Mitigation and Adaptation* (pp. 1–48). Springer.

Moussa, A. (2022). Scale-up to pilot and demonstration plants (TRL 6 and TRL 7). In *Systematic Process Development: From Idea to Value via Technology Readiness Levels* (pp. 169–181). American Chemical Society.

National Institute of Standards and Technology. (2023). *The U.S. Biomanufacturing Economy: Value Added, Supply Chains, and Efficiency*. NIST.

Nguyen, M. H., Duong, M. P. T., Tran, T. V., Nguyen, H. H. T., Tran, T. M. A., & Vuong, Q. H. (2025). Exploring Italian consumers' willingness to pay for sustainable fashion: The roles of eco-consciousness and vintage preference. *Discover Sustainability*, *6*(1), 206.

Ogwu, M. C., Kosoe, E. A., & Osawaru, M. E. (2025). Raw materials for sustainable bioeconomy development in the global south. In *Sustainable Bioeconomy Development in the Global South: Volume I Status and Perspectives* (pp. 37–63). Singapore: Springer Nature Singapore.

Pabico, J. K. (2023). Bio-based plastic: From being recyclable to being biodegradable. *Research Aspects in Chemical and Materials Sciences*, *6*, 95–112.

Philippidis, A. (2024). StockWatch: Layoffs, ARK Selloff Sink Ginkgo Bioworks: Investment fund unloads shares after synthetic biology powerhouse chops 400+ jobs. *GEN Edge*, *6*(1), 598–603.

Raeisi, A., Ara, I., & Bajwa, D. (2024). Bio-based composites. In *Bio-Based Polymers: Farm to Industry. Volume 1: Chemistry and Feedstocks* (pp. 205–225). American Chemical Society.

Ragazzi, I., Farley, M., Jeffery, K., & Butnar, I. (2023). Using life cycle assessments to guide reduction in the carbon footprint of single-use lab consumables. *PLOS Sustainability and Transformation*, *2*(9), e0000080.

Ramamurthy, P. C., Singh, S., Kapoor, D., Parihar, P., Samuel, J., Prasad, R., … Singh, J. (2021). Microbial biotechnological approaches: Renewable bioprocessing for the future energy systems. *Microbial Cell Factories*, *20*(1), 55.

Runiewicz-Wardyn, M. (2025). *Disruptive Innovations in Biotechnology and the Emergence of the Bioeconomy: Policy Implications and Strategic Considerations* (No. 152). TIGER Working Paper Series. Transformation, Integration and Globalization Economic Research (TIGER), Warsaw

Saini, G., Segaran, N., Mayer, J. L., Saini, A., Albadawi, H., & Oklu, R. (2021). Applications of 3D bioprinting in tissue engineering and regenerative medicine. *Journal of Clinical Medicine*, *10*(21), 4966.

Sarker, N. K., & Kaparaju, P. (2024). Microalgal bioeconomy: A green economy approach towards achieving sustainable development goals. *Sustainability*, *16*, 11218.

Schmidt, A., Hengelbrock, A., & Strube, J. (2024). Continuous biomanufacturing in upstream and downstream processing. *Physical Sciences Reviews*, *9*(10), 3167–3222.

Shokoohi, M., & Attar, T. (2024). Biomanufacturing for a sustainable future: Unleashing the potential of biotechnology in pharmaceutical raw material production. *International Journal of New Findings in Health and Educational Sciences (IJHES)*, *2*(2), 142–153.

Singh, A., & Kumar, S. (2025). Exploring the functionality of microbes in fermented foods: Technological advancements and future directions. *Fermentation*, *11*(6), 300.

Soni, S. J., & Patel, A. N. (2024). Single-use technologies in bio manufacturing: Benefits and implementation challenges. *Asian Journal of Hospital Pharmacy*, *4*, 1–7.

Thatoi, H. N., Chattaraj, S., Mishra, R. R., Das Mohapatra, P. K., & Mohapatra, S. (2025). Contributions of biotechnology industries of India to global bioeconomy: An overview. *3 Biotech*, *15*(2), 46.

Udugama, I. A., Lopez, P. C., Gargalo, C. L., Li, X., Bayer, C., & Gernaey, K. V. (2021). Digital Twin in biomanufacturing: Challenges and opportunities towards its implementation. *Systems Microbiology and Biomanufacturing*, *1*(3), 257–274.

United Nations. (2015). *Transforming Our World: The 2030 Agenda for Sustainable Development*. United Nations General Assembly.

Williams, E., Cenian, K., Golsteijn, L., Morris, B., & Scullin, M. L. (2022). Life cycle assessment of MycoWorks' Reishi™: The first low-carbon and biodegradable alternative leather. *Environmental Sciences Europe*, *34*(1), 120.

Yue, D., You, F., & Snyder, S. W. (2014). Biomass-to-bioenergy and biofuel supply chain optimization: Overview, key issues and challenges. *Computers & Chemical Engineering*, *66*, 36–56.

Zhang, X., Zhao, C., Shao, M. W., Chen, Y. L., Liu, P., & Chen, G. Q. (2022). The roadmap of bioeconomy in China. *Engineering Biology*, *6*(4), 71–81.

Zong, Z., & Guan, Y. (2025). AI-driven intelligent data analytics and predictive analysis in Industry 4.0: Transforming knowledge, innovation, and efficiency. *Journal of the Knowledge Economy*, *16*(1), 864–903.

3 Sustainable Biomanufacturing Contribution to UN Global Sustainable Development Goals

Prasann Kumar
Department of Agronomy, School of Agriculture, Lovely Professional University, Phagwara, Punjab, India

Joginder Singh
Department of Botany, Nagaland University, Lumami, Nagaland, India

3.1 INTRODUCTION

The first couple of decades of the 21st century have been marked by a convergence of unprecedented global threats, including climate change, environmental degradation, public health emergencies, and social inequalities, that have put ecosystems and societies on a collision course. Amid this ominous landscape, the United Nations Sustainable Development Goals (SDGs) were adopted in 2015 as a collective, cohesive, and comprehensive framework to chart a path toward a safer, fairer, and more prosperous future for humans and nature by 2030 (Narayanan and Gunasekera 2023; Herrera-Franco et al. 2025; Zhao et al. 2025). Consisting of 17 interconnected goals that range from ending poverty (SDG 1) and hunger (SDG 2) to enabling climate action (SDG 13) and partnership for the goals (SDG 17), this framework surpasses the Millennium Development Goals by openly stating that economic progress, social inclusion, and environmental sustainability should be interdependent and addressed in a coordinated, rather than disjointed, manner. Hitherto, the scale of the transformative 2030 Agenda faces such a wide implementation gap that it requires not just incremental enhancements but also breakthrough technological advances and the upending of existing forms of industrial production, which can lead to economic growth decoupled from resource depletion and pollution, in turn producing progress across several SDG targets at once without further worsening others, all within existing planetary boundaries (Almeida et al. 2025; Bamidele et al. 2025; Saadati et al. 2025; Seltikova et al. 2025). Whiting this critical framework, sustainable biomanufacturing stands out as a key discipline, driving the front of the bioeconomic revolution by harnessing the transformative power of biological systems via synthetic biology, metabolic engineering, systems biology, and advanced bioprocess engineering to design, develop, and produce bio-based materials, chemicals, pharmaceuticals, and fuels from renewable feedstocks, thereby providing a sustainable alternative to petrochemically dependent methodologies that are inherently linear, waste producing, and carbon saturating. Cultivated from a conceptual ideal to a practical industrial imperative, the logic underpinning sustainable biomanufacturing aligns with specific operational drivers that echo the philosophies of green chemistry and circular economy. These include using waste-derived or non-food biomass (e.g., agricultural residues, forestry waste, municipal solid waste, and industrial off-gases such as CO_2) as major feedstocks to avoid competing with food supply chains (contributing to SDG 2); designing energy-light and low-water-footprint biorefineries that minimize environmental burden (SDG 6 and SDG 7); generating high-value biodegradable products with minimal waste streams via high-titer and high-yield microbial conversions; and, more broadly, aiming for carbon neutrality or negativity by exploiting the inherent capability of photosynthetic organisms and chemolithoautotrophs to act as carbon sinks (Hotez et al. 2021; Agrawal et al. 2023; Rubin Thompson et al. 2023; Holland et al. 2024). Such an integrated production system challenges standard manufacturing practice. It transforms biomanufacturing not only as an enabling technology but also as an underlying infrastructure for an industrial system that is restorative by design, thereby contributing directly to responsible consumption and production (SDG 12), resilient infrastructure and innovation (SDG 9), and taking urgent climate action (SDG 13). Therefore, the general scope and purpose of this chapter is to critically and comprehensively summarize the various contributions made by advanced biomanufacturing technologies toward achieving the UN SDGs, beyond a mere listing, and to elucidate in particular the detailed mechanisms, existing case studies, and documented positive impacts by which bioinspired production paradigms are directly or indirectly contributing to the attainment of the different end goals (Durkin et al. 2024; Berman 2025;

 DOI: 10.1201/9781003637158-3

Cao et al. 2025; Obileke 2025). We provide a systematic disassembly of the biomanufacturing value chain, from feedstock selection and the construction of genetic strains to bioreactor and downstream processing (DSP) design, and relate it to its contribution to SDG targets, such as contributing to life-saving therapeutics through SDG 3 (Good Health and Well-being), contributing to economic livelihoods in rural communities through SDG 8 (Decent Work and Economic Growth), and contributing to water remediation through microbiological bioremediation through SDG 6 (Clean Water and Sanitation). Moreover, this chapter will take a critically optimistic view, meaning it will squarely face the significant challenges that must be addressed to realize the sustainable promise of biomanufacturing (e.g., the technical and economic challenges to scale a laboratory discovery to industrial relevance, the need to conduct lifecycle assessments (LCA) to prevent unintended environmental burdens, and the ethical issues associated with genetic manipulation and access to technologies). It will maintain an equilibrium in how the discourse is directed to researchers, politicians, and industrial stakeholders, making them aware of the remarkable windows of opportunities and, at the same time, of the heavy responsibilities underpinning the engineering of biology for a truly sustainable future (Kumar et al. 2023; Sarkis et al. 2024; Ansari et al. 2025; Dahunsi 2025).

3.2 CURRENT RESEARCH STATUS FOR SUSTAINABLE BIOMANUFACTURING

This methodology was designed to yield an extensive, systematic overview of the current research landscape in sustainable biomanufacturing. To this end, a robust bibliometric analysis of the literature was conducted, drawing on the Scopus abstract and citation database—the largest database of peer-reviewed literature. The exploratory phase of the research began by running a targeted search (TITLE-ABS-KEY ["Sustainable Biomanufacturing"]) in Scopus to locate all documents whose titles, abstracts, and keywords contained the specific term "Sustainable Biomanufacturing," which yielded an initial sample of 78 publications that constituted the primary dataset for this study. After the raw data were harvested, a multifaceted refining and categorizing process was completed which took advantage of Scopus' integrated "Analyze search results" and "Refine" features to systematically code and count the publications according to a set of central bibliometric variables, such as OA status (distinguishing between Gold, Hybrid Gold, Green or All Open Access), publication year (for observing temporal trends and growth trajectories over the period from 2017 to 2025), and the most productive authors (with Chen G and Fei Q emerging as key authors). Moreover, the research approach included a comprehensive analysis of the disciplinarity and institutional configuration of the field parsing documents by subject area (revealing an overwhelming predominance of "Biochemistry, Genetics and Molecular Biology" [46 documents] and "Chemical Engineering" [38 documents]), document type (divided between primary research "Articles" [45 documents] and secondary "Reviews" [25 documents]), and the top organizations (demonstrating a central role of the Chinese Academy of Sciences and the Ministry of Education of the People's Republic of China). To visualize the intellectual structure and cooperation network, mining of key parameters of the fund sponsor (where the National Natural Science Foundation of China and the National Key Research and Development Program of China were the major agencies), author keywords (with 'Metabolic Engineering,' 'Biomanufacturing,' and 'Synthetic Biology' as the most common terms), and the geographical distribution of studies (revealing the dominance of China with 33 studies, while the United States and South Korea lagged) was also conducted. This careful quantitative breakdown of the bibliographic data we were able to generate and manipulate from the exported Scopus refine values yielded a robust empirical foundation to inform future interpretation of the core research themes, leading actors, collaborative partners, and emerging scholarly trends, and means that the subsequent summarization and synthesis offered in this chapter are firmly grounded in a sound and transparent analysis of the codified scholarly record.

According to their thematic map analysis, the intellectual structure of sustainable biomanufacturing research exhibits a dynamic, stratified landscape comprising distinct thematic areas, each at a specific development stage (density) and relevance level (centrality). The field shows strong drives toward the well-developed, highly central Motor Themes such as metabolism, metabolic engineering, and synthetic biology on the basic side, as well as biomanufacturing procedures, illustrating a strong connection between basic science and industrial applications, with detailed knowledge shared through a vast number of descriptions focused on the industrial use of these resources. Meanwhile, the terms nonhuman, animals, animal, and humans point strongly toward concerns about biomedical and biopharmaceutical applications, and the use of animal models as cell cultures in understanding the renewal of the biomanufacturing phenomenon. On the contrary, high-development niche themes, such as Cell Immobilization, MNPs, and Purification, are more specific and advanced areas of investigation that are essential for DSP and process stability, but less cross-cutting. The map also highlights central themes that cut across the fundamental knowledge on which the entire discipline is based, but that are not specifically listed in this content. In the meantime, the quadrant of Emerging or Declining Themes includes less developed but still peripheral areas to the current core focus, such as spells (computational tools and bioinformatics), process engineering attributes (pH), molecular biology (amino acid sequence and enzymes), broader sustainability-related goals (sustainable development), specific nutrients or substrates (ammonia), and cutting-edge domain applications, such as cultured meat. Terms such as cell proliferation, bio-based, and biomaterials signal the emergence of nascent but potentially transformative research directions that are seeking to gain more relevance and integration within the broader biomanufacturing landscape. This model depicts a solid field, at its core focused on engineering biosystems for manufacturing, surrounded by niche technological applications and bordering future-growth areas focused on sustainable products and advanced food systems (Figure 3.1).

An examination of thematic topics from 2021 to 2025 shows a strong focus of research interests in sustainable biomanufacturing, predominantly centered on core fundamental and applied biological ideas, with metabolic engineering and synthetic biology as the most prominent and pressing themes, indicative of the field's inherent dependence on the purposeful re-engineering of cellular metabolism and the assembly of novel biological systems in support of efficient production strategies. This aligns with the overrepresentation of more general concepts, such as metabolism, biotechnology, and biomanufacturing, which correspond to the field's core activity of utilizing biological processes for large-scale synthesis. In addition, the focus on a few feedstock molecules, in particular, xylose, glucose, carbon dioxide, methane, and carbon, underscores the important research direction required to move beyond first-generation feedstocks to alternatives such as C1 gases (CO_2 and CH_4) and lignocellulosic derivatives (xylose), which is critical to improving sustainability by recycling waste streams and curtailment of dependence on edible biomass. The concurrent focus on microbial hosts, as indicated by the name of the bacterial strain, serves as a reminder of the ongoing work to find, engineer, and develop hardy microbial chassis that will effectively convert these varied and often difficult starting materials into final products. The significant use of the word nonhuman also suggests a large body of literature examining pre-clinical applications, including microbial or animal cell models, for the production of a variety of biomolecules. Overall, these trends depict a rapidly maturing field, underpinned by the growing integration of advanced genetic tools and feedstock flexibility, which facilitate progress toward more efficacious, scalable, and genuinely sustainable manufacturing paradigms (Figure 3.2).

The resulting Country Collaboration Map is intended to give an international audience a distinct geographical visualization of the international aspects of sustainable biomanufacturing research, showing the latitude and longitude of

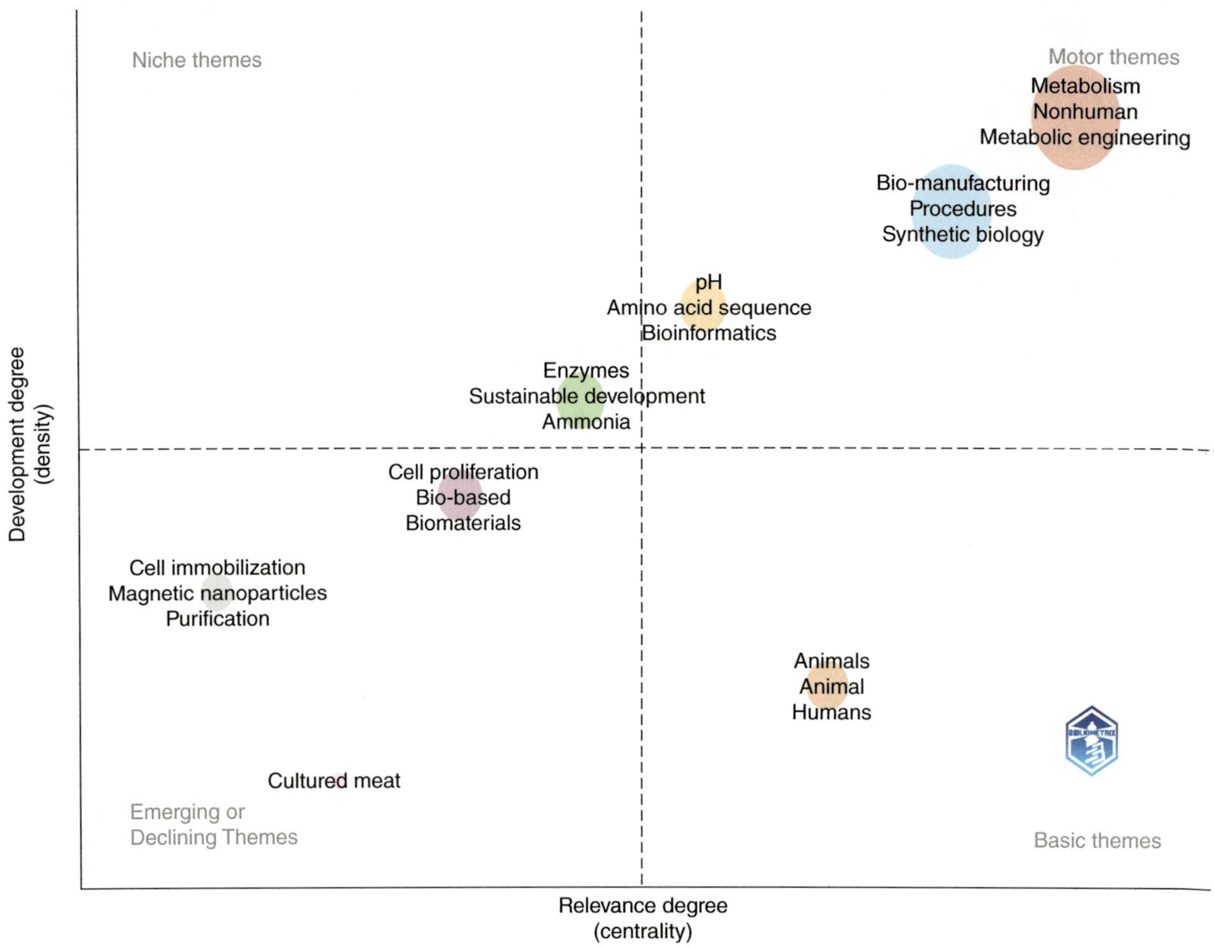

FIGURE 3.1 Emerging or declining themes.

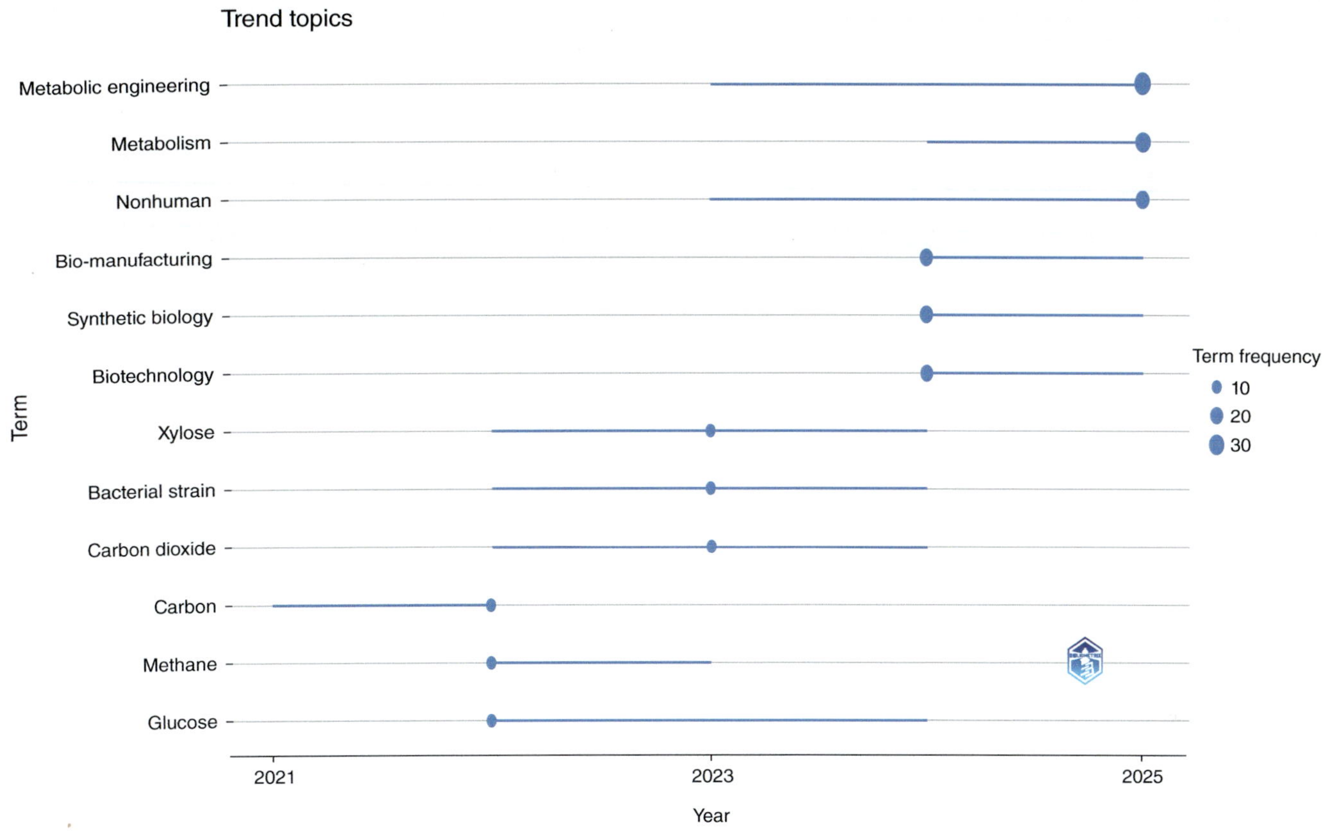

FIGURE 3.2 Trends topic.

contributing countries to demonstrate the global network of scientific relationships. However, since the specific data points, country names, and strength of collaboration are not provided in the release version of the map, we consider the map a concept map illustrating the interconnected world of science today. It shows that the main hubs are further upstream and over the horizon, possibly countries such as China, the United States, South Korea, Germany, and India, which, in the context of the field, serve as central nodes in an intricate web of knowledge exchange that enables the transfer of technology, expertise, and resources globally. This is crucial for addressing the complex, multidisciplinary challenges of sustainable biomanufacturing and for advancing progress toward the UN SDGs on all continents (Figure 3.3).

3.3 THE BIOMANUFACTURING TOOLKIT FOR SUSTAINABILITY

SDGs, conversely, sustainable biomanufacturing to transform our industrial landscape and contribute meaningfully to socioeconomic development. This depends on the development and smart integration of a sophisticated technological trio (synthetic biology, metabolic engineering, and fermentation science), which allows scientists and engineers to drive biological systems and organisms at the molecular and cellular level to improve the production of target molecules, thereby moving beyond serendipitous discoveries of nature into a predictive, design-oriented engineering domain in which organisms are perceived and treated as programmable cellular factories (Das et al. 2024; Cingiz et al. 2025; Reid et al. 2025). Synthetic biology, as the overarching roadmap, offers the conceptual and methodological toolkits to chart a course for developing new biological parts, devices, and systems, or for redesigning existing natural biological systems for useful purposes, using engineering principles such as abstraction, modularity, and standardization. These principles are then employed via a powerful set of enabling technologies including genomics, and foundational DNA synthesis and sequencing technologies that reduce the cost and time to read and write genetic code, Clustered Regularly Interspaced Short Palindromic Repeats (CRISPR)-Cas-based genome editing tools that enable precise, multiplexed, and scarless genomic modifications (i.e., knock-outs, knock-ins, and transcriptional regulation) with unprecedented efficiency and precision, and computational design and bioinformatics platforms that use machine learning algorithms to model complex metabolic networks, predict enzyme function from sequence data, and design optimal genetic circuits for dynamic control of engineered pathways. In effect, these capabilities transform the cellular chassis from a passive vessel into a computationally powerful biosensor capable of dynamically responding to environmental cues, such as substrate levels or product toxicity, to

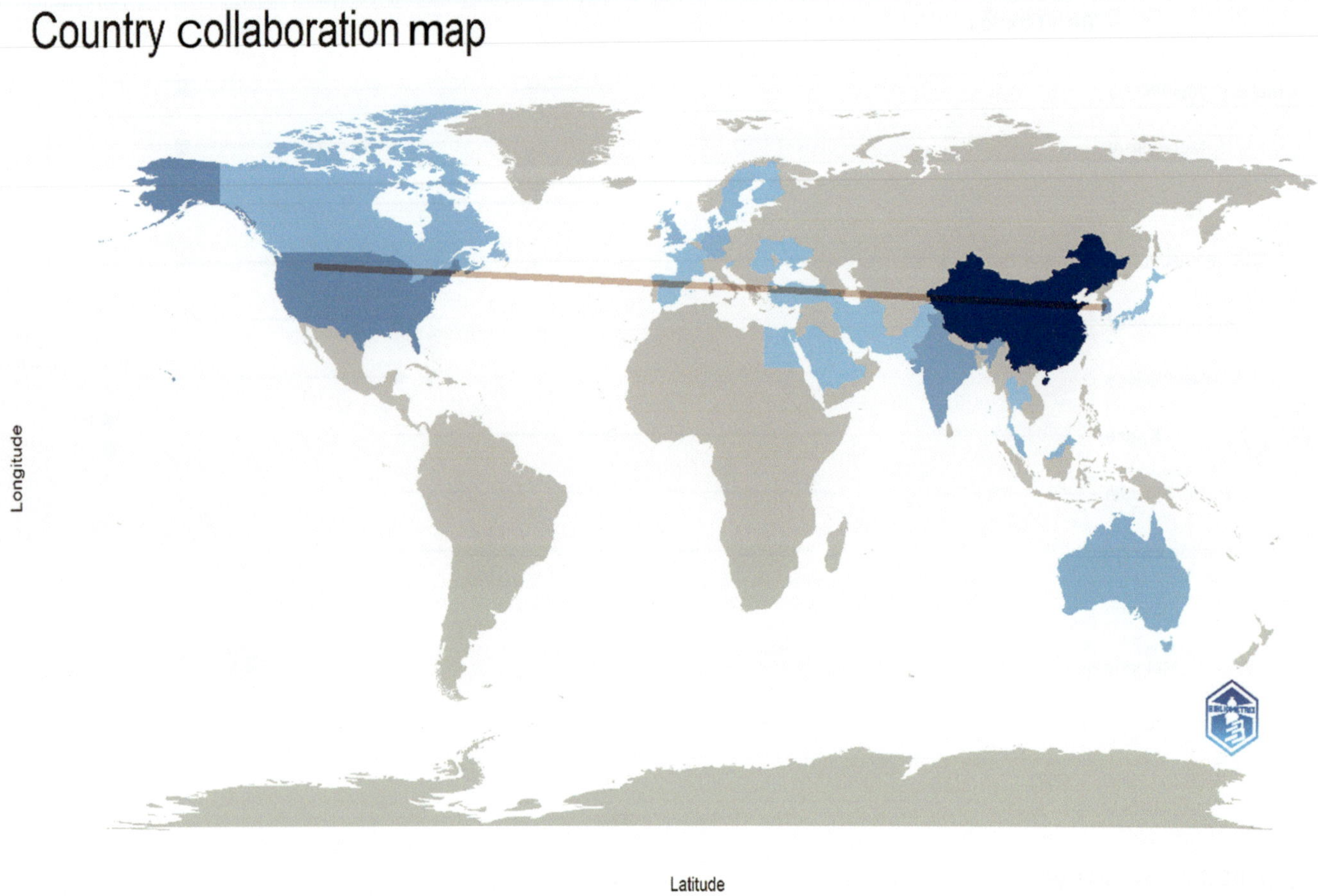

FIGURE 3.3 Country collaboration map illustrating international research collaborations. Countries involved in collaborative research include China, the United States, India, Australia, Japan, South Korea, Thailand, Malaysia, Pakistan, Saudi Arabia, Iran, Turkey, Germany, the United Kingdom, France, Italy, Spain, and Sweden. The intensity of the blue color gradient represents the magnitude of collaboration, with darker shades indicating stronger collaboration networks and lighter shades indicating relatively lower collaboration strength. The connecting lines represent collaborative links between countries..

auto-regulate flux through engineered pathways and maximize titers, yields, and productivity (Agrawal et al. 2023; Philp 2023; Churton and McCabe 2024; Preethi et al. 2024; Berman 2025). Drawing directly from this foundation of synthetic biology, metabolic engineering provides the tools to rewire a host organism's native metabolic network in a potentially precise manner for the overproduction of native metabolites or to introduce entirely heterologous pathways for the production of novel compounds that would not otherwise be synthesized by the organism; this exercise begins with the selection of appropriate microbial or cellular chassis ranging from well-characterized workhorses, such as *Escherichia coli* or Saccharomyces cerevisiae to non-conventional yeasts, Bacillus species, and photosynthetic cyanobacteria, each chosen for distinct characteristics including generally regarded as safe (GRAS) status, industrial resistance, or natural carbon utilization, and proceeds through an iterative design–build–test–learn (DBTL) cycle during which computational models such as genome-scale metabolic models (GEMs) are used to predict in silico predict gene knockout and overexpression targets for redirecting cellular resources (e.g., carbon, ATP, NADPH) away from biomass formation and toward the product of interest, circumventing native regulatory hurdles and alleviating cellular burden (Hotez et al. 2021; Holland et al. 2024; Cao et al. 2025; Dahunsi 2025; Zhao et al. 2025). Once the design is established by analyzing the metabolic model, the "build" phase aims to physically assemble as well as integrate the multigene pathways into the chromosome using advanced DNA assembly methods (e.g., Golden Gate, and Gibson Assembly), and the "test" phase intensively makes use of high throughput omics technologies—transcriptomics, proteomics and, in particular, metabolomics—to generate rich datasets on the physiological state of the engineered strain, which are then looped back to the computational models for the "learn" phase, creating a positive feedback that iteratively improves the metabolic model and guides the next generation of engineering, until the strain reaches commercially applicable performance metrics (Hotez et al. 2021; Preethi et al. 2024; Ansari et al. 2025; Berman 2025; Dahunsi 2025). However, the best microbial catalyst in the world is little more than an academic curiosity without a simultaneous explosion in the field of advanced fermentation science (the scale-up and process engineering arm of the toolkit), which bridges bench-scale discovery to industrial-scale, cost-effective, and environmentally sustainable manufacturing. It focuses on the industrial translation of laboratory bench discoveries to the industrial laboratory, that is, to the rational design and control of bioreactors wherein biocatalysts are cultivated to perform a given transformation and the full optimization of these biological factories for mass transfer (i.e., the provision of oxygen and nutrients, heat removal, pH control, and sterility) to establish optimal culture conditions and achieve maximal rates of product formation. This goal is accomplished through the use of advanced process control strategies that use real-time outputs from a variety of sensors and feedback loops to dynamically change important process parameters, such as agitation and aeration rates, and nutrient fortification, tailored to match the metabolic status of the culture, thereby ensuring that no toxic by-products are allowed to accumulate and that the fermentation is driven as close as possible to its theoretical maximum. Moreover, fermentation science extends beyond the bioreactor into upstream processing (USP), which includes media formulation and development (e.g., leveraging inexpensive and sustainable carbon and nitrogen sources such as corn steep liquor or yeast extract), and, importantly, DSP, often the highest cost and environmental bottleneck in biomanufacturing. In DSP, alternative, energy-efficient separation techniques (continuous chromatography, membrane filtration, *in situ* product recovery methods, etc.) are required to be developed to selectively extract the target compound(s) from the fermentation broth in the steady state via target-specific by-product interactions, thereby reducing end-product inhibition and increasing volumetric productivity, leading to an overall reduction in energy and chemicals traditionally required (Agrawal et al. 2023; Berman 2025; Cao et al. 2025; Dahunsi 2025; Saadati et al. 2025; Zhao et al. 2025). Thus far, the utility of this entire technological triad hinges absolutely on the diligent identification and stewardship of appropriate, sustainable feedstocks that can be harvested, generated, and manipulated in economically and environmentally advantageous ways. This expectation, in turn, signifies a powerful pivot away from the prevailing linear fossil-based economy, which cannot be sustained long term due to its dependence on extracting some finite planetary carbon store and returning it to the atmosphere and whose carbon accounting reveals that our future will involve a net outpouring of greenhouse gases, to a production model that values and feeds back into the cradle-to-cradle carbon economy, an exergonic world where the carbon loop is a zero-sum game or even a negative-generation equation, with everything that is sourced from the Earth paid back or transformed but not lost. While the initial iteration of biomanufacturing feedstocks, largely comprised of edible sugars and starches derived from food crop sources such as sugarcane and corn, has presented a particularly knotty payback calculation with respect to fuel/chemical generation vs. global food security, a morally prickly and economically untenable trade-off that could largely be sidestepped via the ongoing switch-over to a more advanced, second-generation compost of lignocellulosic biomass derived from agricultural refuse (e.g., wheat straw, corn stover, and bagasse), non-food dedicated energy crops (e.g., miscanthus and switchgrass), and the organic fractions of municipal and industrial refuse streams, all of which not only fail to cannibalize the human food supply chain but also confer added value to waste products, albeit at the same time locking in further processing to render remains effectively digestible as pre-fermentation inputs (e.g., steam explosion, acid hydrolysis to pulverize recalcitrant lignin and hemicellulose into easily fermentable C5 and C6 sugars); in parallel, across these conversion node modalities runs the contagious goal of building futuristic microbial strains that can process these constitutive sugar mixes while remaining impervious to any post-DT27 inhibitor compounds. Looking forward, the best feedstock is third- and fourth-generation; the former consists mostly of photosynthetic microbes, such as microalgae and cyanobacteria, that can be farmed on non-arable land and fed with seawater or wastewater, take up atmospheric CO_2 efficiently via photosynthesis, and can be engineered to produce hydrocarbons in much the same way as old-school plants do, which means that Carbon Capture and Utilization (CCU) can be integral to the manufacturing process because when the organisms make a useful product, they will have to grab the carbon they need in the form of CO_2; the latter, fourth-generation feedstocks, are state-of-the-art feedstocks, where we take essentially the same photosynthetic or lithoautotrophic organisms and engineer them further using the tools of synthetic biology to hyper-charge CO_2 fixation (with something like a new and improved carbon-concentrating mechanism or a super-efficient enzyme for fixing CO_2, like form II Rubisco) and connect them directly to concentrators of point-source emitters, like those that exit coal or concrete power plants, to convert industry's carbon waste into a resource for production, creating a carbon-negative production system that fights climate change every step of the way (Holland et al. 2024; Preethi et al. 2024; Bamidele et al. 2025; Cao et al. 2025; Dahunsi 2025; Saadati et al. 2025). The abovementioned approach of deploying trusted tools and the reconciliation and optimization thereof the engineered biology and the sustainable feedstock is guided and validated by the unswerving commitment to the twelve tenets of green chemistry and the top-level framework of the circular economy, which together represent the requisite moral and logical compass to ensure that bioprocesses are genuinely sustainable in the whole lifecycle sense; the tenets of green chemistry, which include atom efficiency, waste prevention, and design for energy economy are manifestly satisfied by biological operations (which typically function at ambient temperature and pressure in aqueous milieu, use enzymes as highly selective and efficient catalysts to minimize side reactions, and output biodegradable materials) but require a rigorous exercise in terms of quantitative LCA to avoid mere burden shifting between environmental impacts such as between greenhouse gas emissions and water eutrophication due to fertilizer use in cultivation of feedstock (Agrawal et al. 2023; Rubin Thompson et al. 2023; Ansari

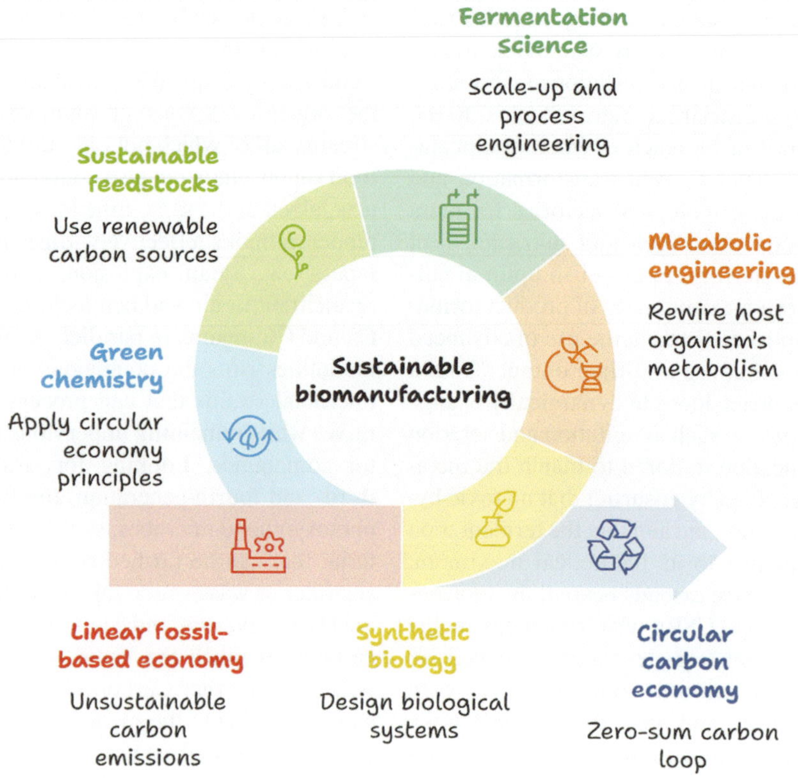

FIGURE 3.4 Sustainable biomanufacturing for a circular economy.

et al. 2025; Bamidele et al. 2025; Obileke 2025; Reid et al. 2025; Zhao et al. 2025). Accordingly, the circular economy model provides the necessary macro-scale framework, so bioprocess design must shift from a linear "cradle-to-grave" mindset to a "cradle-to-cradle" paradigm in which every output stream is regarded as a viable input for another process: spent microbial biomass, rich in protein and lipid content, can be valorized as animal feed or even as biogas via anaerobic digestion; process wastewater can be treated and fed back into a subsequent fermentation; and the main product itself is designed with its end of life in mind, either as compostable, as in the case of polylactic acid (PLA) bioplastics, or as recyclable back to its monomer constituents, yielding a virtuous, closed-loop industrial system with minimal input extraction and waste generation, manifesting the ultimate confluence of advanced biological paradigms, sustainable feedstock management, and ethical process design that is the modern suite of tools for sustainable biomanufacturing (Figure 3.4).

3.4 DIRECT CONTRIBUTIONS TO SPECIFIC SDGS: A DEEP DIVE

Indeed, some of the myriad ways in which sustainable biomanufacturing enriches the global sustainability agenda are most concretely illustrated through its direct, multifaceted contributions to a foundational set of UN SDGs and targets, from the foundational role it plays in catalyzing the advancement of SDG 9 (Industry, Innovation, and Infrastructure) itself, simply by bootstrapping an entire novel bio-based industrial ecosystem that is more resilient, distributed, and innovative (and less centralized and taxed) than its petrochemical cousin (Hotez et al. 2021; Narayanan and Gunasekera 2023; Rubin Thompson et al. 2023; Cao et al. 2025; Cingiz et al. 2025; Obileke 2025). This is accomplished not only by replacing fossil resources with biomass but also through the complete redesign of industrial infrastructure that requires and enables the emergence of advanced biorefineries, comparable to traditional petroleum refineries, but operating in a circular economy to break down renewable biomass in a spectrum of valuable streams (platform chemicals, material products, and energy). Integrated bioprocessing connects products and energy across these streams, establishing new markets and economic opportunities, especially in rural and agricultural areas that provide feedstocks, thereby reducing the geographical imbalance and supporting the development of resilient economic infrastructure. The innovation agenda of SDG 9 is also fueled by relentless research in synthetic biology and bioprocess engineering, which yields high-value intellectual property and attracts substantial venture capital, resulting in economic stimulation through high-tech job creation in Research and Development (R&D), manufacturing, and downstream logistics, as well as cross-sectoral collaboration among academia, industry, and government that is indispensable for de-risking scale-up and building a robust regulatory infrastructure necessary for commercializing new bio-products (Agrawal et al. 2023; Rubin Thompson et al. 2023; Preethi et al. 2024; Ansari et al. 2025; Bamidele et al. 2025; Dahunsi 2025; Obileke 2025). This transformation of industrial production is inherently consistent with and supportive of achieving SDG 12 (Responsible Consumption and Production), as biomanufacturing provides concrete means to decouple economic activity from environmental degradation by offering bio-based and biodegradable replacements for ubiquitous petroleum-derived materials, thereby addressing the cradle-to-grave lifecycle of materials. This contribution is well illustrated by the production of biosynthetic biodegradable polymers, such as PLA and polyhydroxyalkanoates (PHA), designed in microbial systems, with a vision toward these materials composting into harmless natural products, addressing the devastating effect on the environment in oceans, landfills, and society's movement away from the unsustainable single-use plastic economy (Hotez et al. 2021; Rubin Thompson et al. 2023; Das et al. 2024; Durkin et al. 2024; Dahunsi 2025; Saadati et al. 2025). Moreover, the defining characteristic of sustainable biomanufacturing is the valorization of waste streams, inverting the traditional model of waste disposal in production to the more responsible production of a high-value commodities from low-value agricultural residues (e.g., corn stover and bagasse), food waste, or even industrial off-gases, so that waste is transformed into economic value and that the impact on waste management systems is vastly diminished, making waste disposal more affordable and sustainable. The future bioeconomy also hinges crucially on the transformation of the energy sector, which has direct relevance to SDG 7 (Affordable and Clean Energy). The development and scale-up of advanced biofuels is being promoted as a renewable alternative that may also be carbon neutral for transportation, heating, and industrial power generation as an alternative to fossil fuels. First-generation crop-based biofuels, such as corn ethanol, suffered from serious sustainability trade-offs (Agrawal et al. 2023; Sarkis et al. 2024; Almeida et al. 2025; Berman 2025; Cao et al. 2025; Reid et al. 2025; Zhao et al. 2025). At the same time, next-generation technology for converting waste to energy, or producing fuel from biomass in the form of tree and grasses and from algae, provides order-of-magnitude advancement, where engineered microorganisms can now efficiently convert inedible biomass and waste carbon sources into a range of energy-dense molecules: from bioethanol that weighs 6.9 kilogram of weight per gallon, to biodiesel and bio-based chemicals, and drop-in drop-out hydrocarbons that are chemically the same as gasoline, diesel, and jet fuel, and so can work directly as a dig-in replacement for them. Beyond fuel liquids, biomanufacturing activities and processes also contribute to SDG 7, for example, by anaerobically digesting of organic waste with microbial consortia to produce biogas (mainly methane), and by producing hydrogen through photosynthetic or fermentative biological processes, thereby diversifying renewable energy sources and increasing energy security (Philp 2023; Rubin Thompson et al. 2023; Churton and McCabe 2024; Ansari et al. 2025; Reid et al. 2025). The most notable environmental aspect of this bio-based energy and production system is likely its direct assault on the climate crisis, making it a linchpin strategy for SDG 13 (Climate Action) by delivering a double impact: first, by significantly reducing the industrial carbon footprint by displacing fossil resources as the predominant feedstock for chemical and material industries and thereby avoiding the net release of geologically sequestered carbon into the atmosphere, and second, by deliberately capturing and utilizing CO_2 through photosynthetic microorganisms (microalgae and cyanobacteria) and lithoautotrophic bacteria that can convert CO_2 from a hazardous waste product into the primary carbon feedstock for products (Holland et al. 2024; Ansari et al. 2025; Bamidele et al. 2025; Obileke 2025; Zhao et al. 2025). This CCU approach, especially when attached to a point source such as power plants or cement facilities, converts industrial infrastructure from a carbon liability to a carbon capture asset capable of producing chemicals, plastics, and even fuels with a negative carbon footprint over their lifecycle when powered by renewable energy. This delivers a crucial technology solution to decarbonizing hard-to-abate industrial sectors and taking us from simply reducing emissions to actively drawing down atmospheric carbon. Last but not least, the impacts of biomanufacturing are crucial to human beings' well-being under SDG 3 (Good health and well-being) via the transformation of sustainable production of pharmaceuticals and nutraceuticals away from conventional chemical synthesis, which frequently uses unsafe solvents and produces toxic waste, or the precision fermentation process, where engineered microbes are used as clean and highly efficient cellular factories (Churton and McCabe 2024; Durkin et al. 2024; Holland et al. 2024; Reid et al. 2025; Saadati et al. 2025). This is illustrated by the bioproduction of complex small-molecule drugs such as artemisinin (an antimalarial) through engineered yeast, which provides a dependable, scalable, and eco-friendly source of an essential remedy, and the ecologically balanced production of biologic drugs, such as monoclonal antibodies and vaccines, in mammalian, bacterial, or plant cell cultures that are progressively the face of modern medicine. In addition, the field allows for bio-fortification of foods (and sustainable production of high-value nutraceuticals such as omega-3 fatty acids [from algae, not unsustainable fish oil], vitamins, carotenoids, and prebiotic ingredients) promoting preventative healthcare and improved nutrition globally, using sustainable feedstocks and with dramatically less pollution than their chemically synthesized analogy, indicating that advances in human and planetary health are not only compatible but synergistically attainable through disciplined application of sustainable biomanufacturing (Figure 3.5).

FIGURE 3.5 Sustainable biomanufacturing impact analysis.

3.5 CROSS-CUTTING IMPACTS AND SYNERGIES

In addition to direct contributions to select sustainability targets, the transformative impact of sustainable biomanufacturing is arguably best characterized by the wide-ranging nature of cross-cutting impacts and complex synergies that it engenders across the full UN SDG framework, where progress in one goal intrinsically stimulates progress across several other goals, illuminating the inherently interdependent nature of the 2030 Agenda and positioning biomanufacturing as a critical, systems-level intervention (Churton and McCabe 2024; Durkin et al. 2024; Holland et al. 2024; Sarkis et al. 2024; Cingiz et al. 2025; Saadati et al. 2025). Such systems thinking represents yet another defining characteristic of biomanufacturing, as is perhaps best exemplified by its key role in driving forward SDG 6 ('Clean Water and Sanitation'): here, biomanufacturing accomplishes more than reduce its own water footprint it contributes positively to the generation of biotechnological solutions to water purification and remediation challenges by means of the further development and deployment of advanced microbial consortia and enzyme-based bioremediation strategies capable of detoxifying industrial and agricultural wastewater polluted with a plethora of persistent pollutants (e.g., heavy metals through microbial biosorption and bioaccumulation, complex hydrocarbons by way of engineered bacteria harboring novel dioxygenase pathways, and endocrine-disrupting pharmaceuticals preventing their entry into and degradation of freshwater ecosystems) and thus ensuring drinking water safety and source protection (Agrawal et al. 2023; Kumar et al. 2023; Philp 2023; Ansari et al. 2025; Bamidele et al. 2025; Berman 2025; Cao et al. 2025; Dahunsi 2025). Moreover, biomanufacturing supports water quality surveillance via the novel production of inexpensive, highly sensitive biological sensors (biosensors) that exploit engineered organisms or isolated enzymes to produce readily measured electrochemical or optical signals upon detection of target contaminants (such as pesticides, nitrates, or pathogens), thus enabling portable, in-the-moment assessments of water quality that are significantly more affordable and rapid than standard laboratory techniques while also providing empowerment to communities regarding the data they require to protect their water resources and directly achieving the monitoring endpoints of SDG 6. This capacity for biological remediation and sensing naturally intersects with those of SDG 2 (Zero Hunger), because the health of farming systems is closely tied to water quality, but the contribution of biomanufacturing to food security results more immediately from the ability to sustainably manufacture next-generation agricultural inputs that enhance crop yield while decreasing the ecological footprint of conventional farming. For instance, in the biofertilizer sector, the fermentation-based production of nitrogen-fixing bacteria (e.g., Rhizobium and Azotobacter) and phosphate-solubilizing microbes that, as seed or soil inoculants, fix nitrogen and solubilize phosphate for plants through symbiotic relationships, has led to reductions in the excessive dependence on energy-intensive Haber-Bosch synthetic nitrogen fertilizers, a significant source of nitrous oxide emissions and eutrophication such emitting, nutrient eutrophication runoff (Agrawal et al. 2023; Rubin Thompson et al. 2023; Ansari et al. 2025; Bamidele et al. 2025; Dahunsi 2025; Obileke 2025; Saadati et al. 2025; Zhao et al. 2025). In addition to the field derived from them, biopesticides apply biomanufacturing to generate particular biocontrol agents such as entomopathogenic fungi (e.g., *Beauveria bassiana*), beneficial bacteria (e.g., *Bacillus thuringiensis* [Bt] with insecticidal Cry toxins obtained via fermentation), and antiviral compounds that target specific pests and without deleterious effects on beneficial pollinators or the burden of toxic residues in soil and water, fostering IPM based approaches that, in turn augment crop resilience and diminish the chemical load occurring in agroecosystems (Agrawal et al. 2023; Durkin et al. 2024; Preethi et al. 2024; Bamidele et al. 2025; Dahunsi 2025). In addition, biomanufacturing can produce feed supplements for livestock and aquaculture, such as single cell proteins (SCP) produced from bacteria or yeasts that are grown on methane or methanol, and indispensable amino acids such as lysine and threonine obtained by precision fermentation, that can increase the feed efficiency and lower the land-use for animal agriculture, contributing indirectly to global food security by promoting the more efficient and sustainable production of proteins. Development and uptake of such bio-based agricultural technologies, in turn, create substantial socioeconomic knock-on effects, contributing directly to SDG 8 (Decent Work and Economic Growth) by stimulating demand for the creation of new, high value jobs in both the bioeconomy value chain, from R&D in synthetic biology and fermentation science demanding highly skilled personnel such as molecular biologists, biochemical engineers and data scientists to manufacturing, logistics, sales and support services in the nascent biorefining and bio-input sectors. Distributed and regional in nature Not confined to high-tech urban areas, job generation has an important regional and distributed dimension, as the growth of lignocellulosic biomass feedstock cultivation and preprocessing for biorefineries represents a significant new source of employment and sustainable rural development in agricultural communities, where it can contribute to the reversal of rural-to-urban migration Trends, by providing quality work in regions that frequently suffer from economic malaise, thereby promoting more inclusive and resilient economic growth (Sarkis et al. 2024; Almeida et al. 2025; Berman 2025; Cao et al. 2025; Reid et al. 2025; Saadati et al. 2025). The bioeconomy can further stimulate economic diversification, helping regions that have relied on extractive industries transition to a more sustainable, knowledge-intensive economic model. This shift contributes to economic stability and provides qualified, future-proof jobs aligned with global sustainability challenges. However, the true power of sustainable biomanufacturing is effectively understood which in the context of this work emphasize the dissection of these contributions not as individual threads but rather as this tightly woven tapestry of mutually reinforcing aspiration; such as, biofertilizer production (SDG 2) using methane from the anaerobic digestion of agricultural waste (SDG 7) in a regional bio-refinery (SDG 9) that also treats its process water with engineered microbes (SDG 6) to create a local circular economy (SDG 8) while absorbing carbon and facilitating net negative emissions (SDG 13). This planetary philosophy highlights that a unifying biomanufacturing platform, such as algae cultivation, can serve as a carbon sink (SDG 13), yield biofuels (SDG 7), produce nutrient-dense biomass for animal feed (SDG 2), and remediate wastewater by soaking-up nitrogen and phosphorous (SDG 6), presenting an unprecedented density of positive externalities and de-siloed outputs that area an order of magnitude narrower in scale than the latter's focused technological delivery on just one of the above mentioned benefits (Durkin et al. 2024; Holland et al. 2024; Sarkis et al. 2024; Berman 2025; Reid et al. 2025; Saadati et al. 2025; Zhao et al. 2025). This macro-level analysis is essential for informing policy and investment, since it underscores the requirement to transition from single-metric (e.g., cost per unit of biofuel) to multiple-criteria assessment frames (e.g., taking co-benefits in climate, water, health and economic development into consideration) to optimize the economic, social and environmental returns on sustainability investment and enable virtuous cycles of enhancement that amplify progress toward the full suite of 17 SDGs in a coordinated and reinforcing manner strategies. Thus, it is in these profound and multi-layered interconnections that sustainable biomanufacturing rises above industrial processes to become a foundational enabler for system transformation, providing a consistent and integrated roadmap to tackle the complex, interlocking issues reflected in the UN SDGs (Figure 3.6).

3.6 CHALLENGES AND THE PATH FORWARD

In the face of the extraordinary transformational potential of sustainable biomanufacturing as a linchpin of future bioeconomy and the leading force progress toward the UN SDGs, its trajectory toward global impact by a daunting array of interwoven challenges across technical, economic, socio-ethical, and analytical domains, each requiring concerted and collaborative action to overcome, starting with the daunting technical and scale-up barriers that constitute the traditional separating the laboratory proof-of-concept from economically viable industrial production (Holland

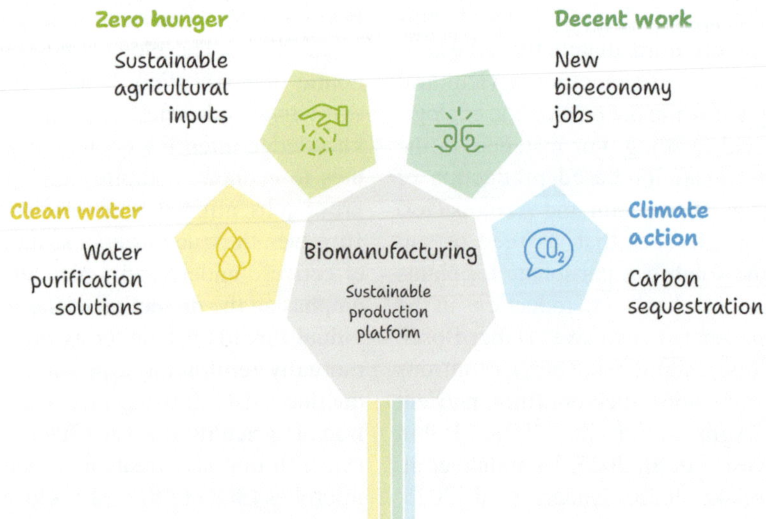

FIGURE 3.6 Sustainable biomanufacturing drives SDG progress.

et al. 2024; Sarkis et al. 2024; Almeida et al. 2025; Obileke 2025; Reid et al. 2025). These barriers frequently take the form of a dramatic drop in performance (product concentration, yield, and productivity) with the transfer of engineered microbial strains from the well-controlled and optimized environment of a shake flask to the complex and often hostile conditions encountered in an industrial-scale biorcactor, where challenges such as substrate inhibition, product toxicity, metabolic burden, and genetic instability can lead to a significant reduction in titer, yield, and productivity (TYP) with detrimental implications for economic viability. The scale-up process per se challenges heavily engineering constraints such as extremes in mass transfer limitations, that is, oxygen supply or mixing efficiency, in aerobic fermentations or high-density cultures which can lead to micro-environments, saturated with substrate, or filled with product, that selectively favor non-producing mutant cells, resulting in what is referred to as heterogeneity in populations, which have to be addressed by advanced process control strategies and intricate sensors *in situ* such as flow cytometry or on-line mass spectrometry for the preservation of robustness in the cultures (Hotez et al. 2021; Preethi et al. 2024; Ansari et al. 2025; Berman 2025; Cingiz et al. 2025; Dahunsi 2025). However, in addition, the use of complex, non-sterile and variable waste-derived feedstocks necessary for sustainability adds further technical challenges, including competition by microbial contaminations over production strains and inhibitory compounds (such as furfurals, phenolic acids from lignocellulosic pre-treatment) toward growth and metabolism, hence requiring extremophilic chassis organisms to be developed or another lengthy detoxification step that comes with cost and complexity. These technological constraints are inherently connected to the second major area of obstacles: economic competitiveness and policy support, as the overwhelming majority of bio-based alternatives currently face difficulty competing on cost with well-established petrochemical analogies that have seen the benefits of over a century of refining infrastructure optimization, massive economies of scale, as well as hidden subsidies not accounting for their negative environmental externalities. Economic challenges: A high capital investment (Capital Expenditure [CAPEX]) to either new biorefineries or to retrofitted chemical plants and substantial operating expenditures (Operating Expenses [OPEX]) with the use of feedstock preprocessing, enzyme cocktails, and energy-intensive downstream separation and polishing operations, put up an economic hurdle to accessing the market that the commercial market is not likely able to bear alone without any de-risking mechanism in place. This is where judicious and forward-looking policy guidance is not just nice to have, it is a must-have; it is done through mechanisms such as carbon pricing (which works to internalize the environmental costs of fossil carbon emissions and creates at least a level playing field, economically speaking), government grants for high-risk translational research, tax deductibles such as those on bio-based manufacturing, and public procurement policies which favor the sustainable product (thereby securing markets and encouraging investment). Beyond financial devices that support investment, creating regulatory certainty and predictability grounded in science for genetically engineered organisms and products is equally important. This provides investors with the confidence to invest in early stage companies, reduces the time from innovation to market, and avoids regulatory uncertainty that could halt growth (Holland et al. 2024; Bamidele et al. 2025; Cao et al. 2025; Dahunsi 2025; Zhao et al. 2025). Thus far, the search for economic and technological solutions must be rooted in a strong ethical vision and Responsible Research and Innovation (RRI), a concept that requires a wide range of stakeholders, from scientists, industry, and policymakers, to the public, to be engaged proactively at every stage of the R&D process to align technological developments with social values and interests. Key ethical aspects include the responsible evaluation and surveillance of potential biosafety risks associated with the planned release of genetically modified (GM) microorganisms into the environment. This involves the development and use of safety and containment barriers, such as auxo trophies and kill switches, to minimize unintentional gene transfer or unauthorized growth. In addition, biosecurity threats associated with the potential dual-use of powerful synthesis and editing technologies, or clouds, would need to be managed through enhanced governance, codes of conduct, and cybersecurity measures to prevent their malicious abuse (Agrawal et al. 2023; Churton and McCabe 2024; Berman 2025; Cao et al. 2025; Dahunsi 2025). At the more macro-economic level, the transition to the bioeconomy poses serious questions of fairness, access and justice, such as global equity and the possibility that advanced capabilities for biomanufacturing and the economic benefits that follow flow toward technologically advanced nations along with the ethical dimension of growing food-grade crops or taking up land for industrial production even if it is on basis of use of advanced feedstocks, the issue calls for a continuous dialogue with the communities and following principles of distributive justice in ensuring a just transition. This challenge to, as well as need for, navigating the intricate technical, economic and ethical terrains, within which Bioeconomy systems operate, calls in turn for more systematic use of LCA for comprehensive analysis providing the methodological basics of shifting from simplified indicators to quantifying the full environmental impact of a bio-based product from the cradle to the grave, including climate change, freshwater use, land use change and eutrophication. Performing a thorough, evidence-based LCA is essential to prevent problem-shifting, where solving one environmental problem creates another (e.g., a process that reduces greenhouse gas emissions uses an unsustainable amount of water or indirectly causes land use change (iLUC), if the production of biomass replaces food crops or natural ecosystems), and guarantee biomanufacturing processes are ultimately, sustainability-wise multidimensional, not "greenwashed." The road ahead therefore requires a holistic, interdisciplinary strategy that combines on-going basic research in synthetic biology to develop tougher and more effective microbial chassis, with bioreactor engineering and process integration advances that can address scale-up hurdles, nurtured by coherent policy initiatives that prioritize sustainability and circularity, framed by the ethical lens of RRI to secure public trust, and substantiated by transparent and holistic LCA analyses to take informed decisions and honor the true potential of sustainable biomanufacturing as a positive force in the world (Figure 3.7).

FIGURE 3.7 Sustainable biomanufacturing: navigating complex challenges.

3.7 CONCLUSION AND FUTURE PERSPECTIVES

In summary, the exposition in this chapter has identified sustainable biomanufacturing as an enabler technology for delivery of the SDGs, directly addressing specific targets from climate action (SDG 13) and clean energy (SDG 7) to responsible production (SDG 12) and industrial innovation (SDG 9) while simultaneously delivering extensive cross-cutting synergies to advance interrelated objectives in zero hunger (SDG 2), clean water (SDG 6), and good health (SDG 3) via pathbreaking production of bio-based agricultural inputs, water remediation technologies, bio-pharmaceuticals, and more. This revolutionizing potential is based on the enabling convergence of synthetic biology, metabolic engineering, and fermentation science, which allows the repurposing of biology to turn renewable and waste-derived feedstocks into a wide range of valuable products, offering a tangible path to envisioning a bio-based circular economy which literally decouples human prosperity from environmental decline by substituting linear, degenerative industrial models with restorative, circular systems that eliminate waste, recycle resources and regenerate natural systems through economic activity. Thus far, as thoroughly reviewed, this rosy future depends on solving significant technical challenges in scale-up and DSP, being cost-competitive through intelligent policy and carbon prices, working through ethical questions with RRI, and verifying environmental advantages through a comprehensive LCA to avoid burden transfer and achieve genuinely sustainable solutions. Delivering on this vision requires one final wake-up call for researchers, industry, and policymakers alike to embrace an unprecedented level of transdisciplinary effort: researchers should forge the next generation of tools for the engineering of complex microbial consortia and the acceleration of the scale-up process, while proactively engaging with the public to generate trust and dialogue; industry must commit to heavy investment in pilot- and demonstration-scale facilities to de-risk technologies and embed circularity at the core of their business models, moving beyond short-term profitability to long-term planetary custodianship; and policymakers must create the essential facilitative environment through coherent, long-term strategies that include robust carbon accounting, sustainable public procurement frameworks, fit-for-purpose regulatory regimes for bio-based products, and meaningful funding for high-risk, high-reward translational research. The transition toward a sustainable bioeconomy is not just a technological revolution, but a profound societal shift. It requires a shared ambition, a coherent approach, and a commitment to harness the power of biology fairly to secure a sustainable, resilient, and regenerative future for people and the planet, making sustainable biomanufacturing not just a business but a global necessity.

REFERENCES

Agrawal D, Awani K, Nabavi SA, Balan V, Jin M, Aminabhavi TM, Dubey KK, Kumar V (2023) Carbon emissions and decarbonisation: The role and relevance of fermentation industry in chemical sector. *Chem Eng J* 475:146308. https://doi.org/10.1016/j.cej.2023.146308

Almeida M, Corticeiro S, Oliveira BRF, Laranjeiro D, Lillebø A, Vieira H (2025) The needs and challenges of the Blue Economy sector in Portugal: bridging national and European strategies with the perceptions of the stakeholders. *J Environ Manage* 384:125468. https://doi.org/10.1016/j.jenvman.2025.125468

Ansari SA, Ramteke A, Sawarkar R, Kumar T, Khan D, Agashe A, Patil MP, Singh L (2025) "Innovative approaches in microbial community engineering for food waste management: A comprehensive review." *J Environ Manage* 393:127000. https://doi.org/10.1016/j.jenvman.2025.127000

Bamidele MO, Bamikale MB, Cárdenas-Hernández E, Bamidele MA, Castillo-Olvera G, Sandoval-Cortes J, Aguilar CN (2025) Bioengineering in Solid-State Fermentation for next sustainable food bioprocessing. *Next Sustain* 6:100105. https://doi.org/10.1016/j.nxsust.2025.100105

Berman M (2025) Leveraging EO 14801 to grow the U.S. seaweed industry. *Mar Policy* 177:106683. https://doi.org/10.1016/j.marpol.2025.106683

Cao H, Zeng Y, Yuan X, Wang JK, Tay CY (2025) Waste-to-resource: Extraction and transformation of aquatic biomaterials for regenerative medicine. *Biomater Adv* 166:214023. https://doi.org/10.1016/j.bioadv.2024.214023

Churton H, McCabe BK (2024) Advancing a food loss and waste bioproduct industry: A critical review of policy approaches for application in an Australian context. *Heliyon* 10(12):e32735. https://doi.org/10.1016/j.heliyon.2024.e32735

Cingiz K, Chavarria H, Ecuru J, Frisvold G, Kardung M, Wesseler J (2025) Perspectives on bioeconomy policies. In: Lundgren T, Bostian M, Managi S (eds) *Encyclopedia of Energy, Natural Resource, and Environmental Economics*, Second Edition. Elsevier, Oxford, pp 236–248

Dahunsi OS (2025) Chapter 5 - Food wastes biorefinery. In: Dahunsi OS (ed) *Biorefinery*. Elsevier Science Ltd, pp 97–132

Das S, Chandukishore T, Selvasembian R, Prabhu AA (2024) Mixed food waste valorization using a thermostable glucoamylase enzyme produced by a newly isolated filamentous fungus: A sustainable biorefinery approach. *Chemosphere* 352:141480. https://doi.org/10.1016/j.chemosphere.2024.141480

Durkin A, Vinestock T, Guo M (2024) Towards planetary boundary sustainability of food processing wastewater, by resource recovery & emission reduction: A process system engineering perspective. *Carbon Capture Sci Technol* 13:100319. https://doi.org/10.1016/j.ccst.2024.100319

Herrera-Franco G, Peña-Villacreses G, Bravo-Montero L (2025) Women's participation in the research development of a country. *Int J Educ Res Open* 8:100413. https://doi.org/10.1016/j.ijedro.2024.100413

Holland C, McCarthy A, Ferri P, Shapira P (2024) Innovation intermediaries at the convergence of digital technologies, sustainability, and governance: A case study of AI-enabled engineering biology. *Technovation* 129:102875. https://doi.org/10.1016/j.technovation.2023.102875

Hotez PJ, Batista C, Amor YB, Ergonul O, Figueroa JP, Gilbert S, Gursel M, Hassanain M, Kang G, Kaslow DC, Kim JH, Lall B, Larson H, Naniche D, Sheahan T, Shoham S, Wilder-Smith A, Sow SO, Strub-Wourgaft N, Yadav P, Bottazzi ME (2021) Global public health security and justice for vaccines and therapeutics in the COVID-19 pandemic. *EClinicalMedicine* 39:101053. https://doi.org/10.1016/j.eclinm.2021.101053

Kumar A, Ramamoorthy D, Kumar N, Chandra P (2023) Chapter 13 - Food waste conversion into value-added products: Different pathways, current challenges, and opportunities. In: Hussain CM, Kushwaha A, Bharagava RN, Goswami L (eds) *Bio-Based Materials and Waste for Energy Generation and Resource Management*. Elsevier, pp 387–409

Narayanan RG, Gunasekera JS (2023) Chapter 1 - Introduction to sustainable manufacturing processes. In: Narayanan RG, Gunasekera JS (eds) *Sustainable Manufacturing Processes*. Academic Press, pp 1–28

Obileke KC (2025) Chapter 6 - Microbes in responsible consumption and production (SDG 12). In: Akinsemolu AA, Onyeaka HN (eds) *Green Microbiology*. Academic Press, pp 107–135

Philp J (2023) Bioeconomy and net-zero carbon: lessons from Trends in Biotechnology, volume 1, issue 1. *Trends Biotechnol* 41(3):307–322. https://doi.org/10.1016/j.tibtech.2022.09.016

Preethi B, Karmegam N, Manikandan S, Vickram S, Subbaiya R, Rajeshkumar S, Gomadurai C, Govarthanan M (2024) Nanotechnology-powered innovations for agricultural and food waste valorization: A critical appraisal in the context of circular economy implementation in developing nations. *Process Saf Environ Prot* 184:477–491. https://doi.org/10.1016/j.psep.2024.01.100

Reid G, Allen-Vercoe E, Al K, Burton JP, Daisley B, Dixon B, Mousavi H, Solberg SO, Peixoto RS, Silverman M, Skokovic-Sunjic D (2025) Beneficial microbes for One Health in Canada: a review of evidence and a policy proposal. *FACETS* 10:1–32. https://doi.org/10.1139/facets-2024-0182

Rubin Thompson LJ, Grubo M, Veller M, Badenhorst RH, Nott J, Debruyne L, Makadzange T, Nicolaou S, Stanberry L, Sall A, James WG (2023) Building global vaccine manufacturing capacity: Spotlight on Africa. *Vaccine* 41(27):4050–4056. https://doi.org/10.1016/j.vaccine.2023.05.009

Saadati F, Bahrulolum H, Talebi M, Karimi M, Bozorgchami N, Ghale RA, Zafar S, Aghighi Y, Asiaei E, Tabandeh F (2025) Advances and principles of hyaluronic acid production, extraction, purification, and its applications: A review. *Int J Biol Macromol* 312:143839. https://doi.org/10.1016/j.ijbiomac.2025.143839

Sarkis M, Fyfe AT, Kontoravdi C, Papathanasiou MM (2024) Towards a Net Zero, socially sustainable and eco-efficient biopharma industry: how far are we? *Curr Opin Chem Eng* 44:101027. https://doi.org/10.1016/j.coche.2024.101027

Seltikova E, Haghgoo N, Tatár G, Kelly B, Hughes N, Benas A-D, Goudarzi S, Valle P, Vanolli M, Grace IO, Kim K, Mai C, Saini P, Dehghani S, Karbala MM, John EM, Singh S, Romero Rojas CV, Elwertowska A, Anand D, Logeswaran A (2025) LUNEX PROSPER: The next generation's blueprint towards a sustainable human presence on the Moon. *Acta Astronaut* 235:170–184. https://doi.org/10.1016/j.actaastro.2025.05.016

Zhao H, Lim TYJ, Lee LQ, Li C, Wu D, Zhang L, Tu W, Liu W, Yu ZG, Zou Z, Zhou Y, Li H (2025) Seashell waste reforming suppresses oxygen evolution to produce carbon-negative hydrogen and sustainable single-cell protein. *One Earth*: 101427. https://doi.org/10.1016/j.oneear.2025.101427

4 Sustainable Biomanufacturing Toward a Circular Economy

Concept and Application

Anis Mirza
Department of Horticulture, SAGR, Lovely Professional University-144411, Phagwara, Punjab, India

Gholemreza Abdi
Department of biotechnology, Persian gulf research institute, Persian gulf university, bushehr,75169, Iran

4.1 INTRODUCTION

Manufacturing is the process of turning knowledge and materials into products to meet human needs and is a primary force behind the world's prosperity. A significant portion of the global economy is centered on manufacturing. Although local cultural, social, and economic factors cause modest variations in consumption patterns across the world, overall consumer spending continues to increase as living standards rise. At the moment, the paradigm shift from traditional manufacturing to digitalized production is the result of the Industrial Revolution 1.0 to Industrial Revolution 4.0 Udugama et al. (2021). Biomanufacturing has expanded quickly in recent years because of the rising demand for environmentally friendly, sustainable products and the developments in biotechnology that have made it easier to alter living organisms to produce specific goods. The objective is to produce goods that encourage sustainable resource use, reduce waste, and minimize environmental harm. Biomanufacturing has several benefits over conventional manufacturing techniques, including scalability, flexibility, and a lower environmental impact. Furthermore, it can be used to create complex molecules such as proteins and antibodies that are hard or impossible to make through chemical synthesis. Four revolutions can be distinguished in the history of biomanufacturing. Biomanufacturing 1.0 depended on single-culture fermentation for the production of main metabolites such as ethanol and citric acid; Biomanufacturing 2.0 relied on biogenesis of secondary metabolites such as antibiotics derived from mutant microbial strains, to create more complex bioproducts, for example, enzymes, organic acids, amino acids, vitamins, vaccines, and other high-value bioactive compounds for pharmaceutical, agricultural, and industrial applications. Using aerobic submerged fermentation Biomanufacturing 3.0 enabled large-scale production of extractions and recombinant proteins using cell culture techniques and, as a result, an era was formed, providing solutions in the production of more complex biomolecules such as insulin and enzymes. On the contrary, Biomanufacturing 4.0 is the digital age, or 4th revolution, of industrial biotechnology (IB), which utilizes automation, robotics, data analytics, and artificial intelligence (AI) to reflect industry change for better control, efficiency, and improved scalability (Udugama et al., 2021; Pandey et al., 2024). It makes sense for the area Biomanufacturing 4.0 to focus on smart bioprocessing systems that produce bioproducts with minimal waste through real-time adjustments. This development, along with Biomanufacturing 4.0, will extend into Biomanufacturing 5.0, and beyond digitalization, it proposes a strategy toward a human-centric, sustainable era of manufacturing by drawing on AI, synthetic biology, and bioengineering to design adaptive, self-optimized bioprocesses (Balar et al., 2023; Rose & Friedman, 2021). This strategy would support human partnerships with intelligent (technology-driven) and biological platforms to regenerate low-carbon, circular, bioeconomic production. Collectively, they represent the progression from Industrial Revolution 4.0, where processes are automated, toward Industrial Revolution 5.0, inclusive intelligent biomanufacturing that supports sustainability and the environment. Sustainable biomanufacturing is the process of producing goods in an economically and environmentally responsible way by using biological systems, or living beings. The creation of a closed-loop system that maximizes the application of renewable resources while minimizing wastage and environmental degradation is the fundamental tenet of sustainable biomanufacturing. This is accomplished by employing biological systems, including bacteria, yeast, algae, and fungi, which can be used to produce various goods, including biofuels and bioplastics.

DOI: 10.1201/9781003637158-4

TABLE 4.1
History of Biomanufacturing

Biomanufacturing Revolutions	Date	Products Manufactured	Technologies Involved
Pre-modern biomanufacturing	9,000 years ago	Wine, sauces, cheese, curd, drinks, colors, and pickling	Fermentation
Biomanufacturing 1.0	1910s	Acetone, *n*-butanol, carbohydrates, organic acids, and monosodium glutamate (the most expensive amino acid produced)	Anaerobic liquid fermentation
Biomanufacturing 2.0	1940s	Penicillin, tetracycline, and streptomycin	Specialized mutant microbial strains and aerobic submerged fermentation
Biomanufacturing 3.0	1980s	Human insulin, growth hormone, erythropoietin, and various enzymes	Recombinant DNA technology and enhanced cell cultures
Biomanufacturing 4.0	2000s	Artificial starch, human cells and tissues, iso-butanol	Digital technology and AI
Biomanufacturing 5.0	2020s	Create synthetic cells, transplant synthetic genomes, chemical models, and alter biological cells	AI, synthetic biology, bioengineering, and biotechnology

There are social and economic advantages to sustainable biomanufacturing as well. For instance, it can provide farmers and other rural residents with economic opportunities and generate new jobs in the green economy. Producing goods such as biofuels and bioplastics from agricultural waste and other biomass resources, it can also increase food security. The use of plants, such as moss or algae, to create valuable products, such as medicines or scents, is another instance of sustainable biomanufacturing. The desired product can be extracted by harvesting and processing these organisms, which can be cultivated with renewable resources such as carbon dioxide and sunshine. Due to corporations' constant focus on maximizing profits, a linear economy has been the primary economic model for many industries. A new economic model, namely, the "circular-economy" paradigm, has recently challenged the linear economy paradigm by allowing waste generated after production, usage, or consumption to be used as raw materials or feedstocks in the next production cycle. Sustainable and equitable communities are characterized by circular economies. With the start of catastrophic global climate change, increased energy use in production processes typically results in high levels of greenhouse gases (GHGs), including CO, which directly contribute to climate change and global warming. As a result, these global concerns require an immediate solution, and the circular bioeconomy can play a significant part in this. A low-carbon economy would definitely contribute to addressing these challenges, notably climate change, by limiting global warming to 1.5°C (UN, 2020). To put it another way, a circular bioeconomy ensures a closed-loop flow of materials by making effective use of biomass, including waste and by-products, for the long-term production of valuable products. The circular bioeconomy can generate a more sustainable, ecologically friendly environment, making it a low-carbon economy. A circular economy aims to minimize negative externalities and promote restoration and regeneration by re-aligning the linear take-make-use-dispose model of production and consumption with a circular model in which resources can be reused or bio-degraded and re-incorporated into the system (Stegmann et al., 2020).

In summary, it is a circular, self-sustaining socio-economic system that does not depend on the purchase of external fossil resources, and a decreased environmental impact due to the growth of low-energy consumption processes are just two of the many advantages that society as a whole can obtain by switching to biomanufacturing and creating an industrial structure based on renewable biomass resources. Stakeholders must have conversations about how to transition to a bioeconomy society and when to do so while preserving alignment and coexistence with the conventional social and industrial structures, as presented in Table 4.1. This is because it is not feasible to completely transform social and industrial structures based on biomass resources and bio-based systems overnight (Matsuo, 2024).

4.2 PRINCIPLES OF SUSTAINABLE BIOMANUFACTURING

Sustainable biomanufacturing utilizes biological systems to create goods with low environmental impact, focusing on resource efficiency, waste minimization, and circular-economy principles while improving sustainability and economic and social well-being.

4.3 GREEN CHEMISTRY AND BIOTECHNOLOGY

Energy is delivered to end users worldwide in the form of fuels (~80%), electricity (~20%), and heat. Solar thermal collectors provide heat, and the capacity of solar-powered renewable electricity generation is growing quickly (Mandal et al., 2021). In addition to producing biomass, the direct-light biomanufacturer's sustainable biomanufacturing approach can generate high-value compounds in huge numbers while fostering sustainable production

(Wang et al., 2020). Since the 1990s, the idea of Good Clinical Practices (GCP) has been developed globally across industrial management, government policy, education, and technological development. In the production and use of chemical products, green chemistry effectively uses raw materials that are ideally renewable, removes the unwanted materials, and stays away from the usage of hazardous and poisonous solvents and reagents (Sheldon & Norton, 2020). Generally speaking, GCP could emphasize the development of safer chemicals, the substitution of catalysts for stoichiometric reagents, and the avoidance of waste generation (Chen et al., 2020).

4.4 LIFECYCLE ASSESSMENT (LCA) IN BIOMANUFACTURING

The shift to a smart, sustainable bioeconomy reduces traffic and the impact of humans on the built and natural ecosystems by managing agro-industrial waste and renewable biological resources using data-driven methods (Fava et al., 2021). LCA assesses sustainability in the creation of any product or in the execution of a significant procedure related to its development. LCA, which covers raw material extraction, refinement, manufacture, use, and end-of-life treatment, provides a thorough analysis of the environmental advantages and impacts of biomanufacturing (Dwivedi et al., 2024). In particular, it will examine a product or process and help the industrial sector reorganize its technology and operations to reduce associated environmental impacts. The LCA framework is frequently used in the biomanufacturing process to assess environmental sustainability, identify flaws, determine how to improve, and provide guidance for further study and industrialization. AI is increasingly being integrated into LCA frameworks to enable real-time prediction of environmental performance rather than post-hoc assessment. Machine learning (ML) models can correlate process parameters with emissions or energy intensity, allowing faster comparison of feedstocks or process routes. For example, digital twins of fermentation systems have been used to simulate CO_2 emissions under variable aeration and nutrient regimes before implementation (Zobel-Roos et al., 2021). Such AI-assisted LCA approaches transform it from a diagnostic tool to a process optimization instrument, consistent with the adaptive principles of Industry 5.0.

4.5 RESOURCE EFFICIENCY AND WASTE MINIMIZATION

The energy-environment balance is maintained while GHG emissions are reduced via bioprocesses that use waste resources to produce biomaterials and biofuels. These processes also minimize resource depletion, reduce carbon footprints, and integrate with wastewater treatment (Leong et al., 2021). The predominant waste in agriculture is lignocellulosic biomass, which can be broken down to produce a variety of sugars and phenolic acids (formed when lignin depolymerizes). Waste bioprocessing promotes the formation of green biopolymers and the buildup of biolipids. A more economical and environmentally friendly circular bioeconomy strategy is to valorize waste through microalgae bioprocessing. A lot of research is being done to turn waste from microalgae farming into valuable bioproducts. By using natural processes and renewable resources, biotechnology, conversely, offers a more ecologically friendly method of manufacturing. This lessens reliance on hazardous reagents and the production of harmful, unwanted by-products (Castiello et al., 2023).

4.6 RENEWABLE FEEDSTOCKS AND BIO-BASED MATERIALS

Research on the biotransformation of lignocellulosic biomass, especially from forestry and agricultural leftovers, is ongoing in both the academic and industrial sectors due to the abundance, affordability, accessibility, and regenerative nature of these starting materials (Intasian et al., 2021). The circular bioeconomy's implementation depends heavily on the agricultural, forestry, and related sectors. These industries not only supply renewable biological resources but also provide a rich environment for the necessary research and innovation processes. Finland and Austria concentrate on enhancing the management of forest resources. Since woods cover almost 80% of Finland's land area, the country's forestry sector leads the world in wood processing and uses a variety of low-water methods (D'Amico et al., 2022). Microbial lipid production from waste materials using inexpensive substrates has attracted significant interest from industry and the scientific community as a substitute feedstock for the oleochemical, health food supplement, and biofuels sectors. A "food versus fuel" conflict occurs when the first-generation biofuel platform uses maize, cane, starch, and vegetable oil as feedstock, putting major pressure on food costs and availability (Esfandabadi et al., 2022).

4.7 CIRCULAR-ECONOMY FRAMEWORK

One way to conceptualize a circular bioeconomy is as a way to help developed countries expand economically by combining a desirable "what" (circular economy) with a feasible, viable, and desirable "how" (bioeconomy). For example, the goal is to replace the current, largely linear economic structure with a more sustainable circular bioeconomy that emphasizes renewable resources, waste valorization, resource efficiency, and environmentally friendly production systems. Agrawal et al. (2023). It provides a sustainable way to produce high-value products (such as food, biomaterials, feed, and bioenergy) by efficiently using biomass, including wastes and side streams. The circular bioeconomy has several advantages, such as: increased resource and eco-efficiency; reduced GHG emissions; less dependence on fossil fuels; and the use of leftovers and trash from a variety of sources, including agro-industrial aquaculture and fishing Muscat et al. (2021). Additionally, the "4R" framework—reduce, reuse, recycle, and recover—is another way to think of the circular economy. The first three R's focus on conserving natural capital and reducing resource use. The final R concerns recovering the resource as energy (e.g., burning garbage for heating). Biorefining is a set of techniques that enables the bio-based circular economy to seal the loop of raw resources, minerals, carbon, and water. It is characterized by sustainable bioprocesses that effectively use biomass resources to produce a range of commercial goods and their by-products. The shift toward a waste biorefinery model from a petroleum model shows a significant effort to mitigate GHG emissions and manage carbon. To transition from a linear economy based on the take, make, and dispose concept to a sustainable circular bioeconomy based on the principles of recycling, reusing, remanufacturing, and sustaining, waste biorefineries are involved. Agro-industrial by-products, forest and agricultural waste, lignocellulosic materials, wastewater or sludge, food waste, and industrial by-products have all been effectively turned into marketable, beneficial bio-based products. This strategy is referred to as waste-as-a-value, waste-to-wealth, or zero-waste, and it will prove to be a good, environmentally friendly, and affordable way to dispose of waste. Research on the manufacture of biopolymers through the valorization of waste streams using bioprocesses is still ongoing (Leong et al., 2021). The circular bioeconomy's implementation depends heavily on the agricultural, forestry, and related sectors. There is "a very slow uptake of "radical" forms of circular business model innovation, such as sharing models," whereas circular business models often concentrate more on the plan of action that seeks to efficiently use up wastage as a resource. Since the circular bioeconomy shows promise for creating a more sustainable, environmentally friendly environment, it can, therefore, be seen as a low-carbon economy. To achieve this, a number of practices must be implemented, including circular bioeconomy-based solutions for zero emissions and CO_2 sequestration (carbon capture and storage [CCS], carbon capture, use, and storage [CCUS], and carbon capture and utilization [CCU]). With a constantly shifting perspective, CO_2 is no longer viewed as a waste but rather as a resource that can be recycled to create products with added value using a circular-economy methodology (Bhagia et al., 2021). An automated, traceable, permanent, and computerized framework is necessary to track and document the process's numerous inputs and outputs due to the complexity of biomanufacturing. A producer can retrieve information and monitor depreciation using predefined procedures during the production process by employing blockchain technology. A wide range of information and communication technologies (ICTs), including sensors, real-time monitoring stations, cameras, Global Positioning System (GPS) tracking systems, big-data analysis techniques, AI, augmented reality, blockchain (Klerkx and Rose, 2020), Internet of Things (IoT), cloud computing, smart grids, satellites, nanotechnologies, advanced biotechnologies, and drones, are specifically used in this new form of smart and sustainable bioeconomy through the use of digital platforms and dashboards (Pelli and Lähtinen, 2020). Synthetic biology was acknowledged as one of the key enabling technologies by the Organization for Economic Co-operation and Development's (OECD's) Global Forum on Technology (GFTech) at the international level. In November 2023, a focus group on synthetic biology was formed to have experts examine governance in the field of synthetic biology, including biomanufacturing (Zobel-Roos et al., 2021). To enlighten policymakers about pertinent technological, social, and policy challenges, the focus group gathers and arranges information on these topics. The bioeconomy's increasing significance has led to a variety of worldwide initiatives and practices. In this regard, the creation of bio-economy policies has grown more intricate and diverse (Matsuo, 2024). In conjunction with other international and regional initiatives, the bioeconomy is a crucial component in enabling and empowering individuals to confront and resolve these issues, as well as to spur the shift toward an enduring market-based economy (Woźniak et al., 2021), as shown in Figure 4.1.

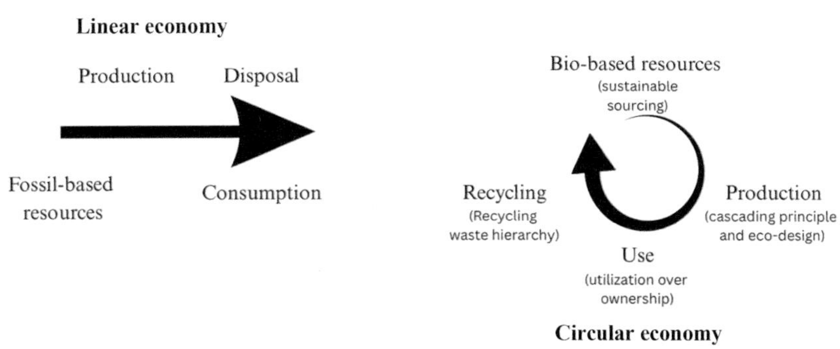

FIGURE 4.1 The shift toward the circular economy.

4.8 KEY TECHNOLOGIES IN SUSTAINABLE BIOMANUFACTURING

The most recent advancements in these technologies are becoming increasingly understandable and accessible to potential investors as modern biomanufacturing techniques continue to evolve. Networks like SynBioBeta offer essential background and ease of understanding for non-specialists in this increasingly market-oriented field, where some ideas ultimately fail to generate profit. They also document the specifics of the funding, acquisition, deals, and successes of businesses that use SynBio and IB in their platform (Bobier et al., 2024). In the United Kingdom, ethical innovation practices have been closely linked to the development of technology, especially synthetic biology. Although a lot of important work has been done in this area, synthetic biology and broader biotechnology contexts do not always consistently prioritize operationalizing responsible innovation (Rosemann & Molyneux-Hodgson, 2020). The development of more effective synthetic systems is currently being led by researchers. To enable heterotrophic cells to carry out carbon fixation, this entails designing essential enzymes and introducing whole or partial carbon-fixation pathways (Kurt et al., 2023). Scientists may design and build new biological systems with specialized functions thanks to synthetic biology (Wang et al., 2023). During biomanufacturing design, AI is increasingly used for decision-making support. Knowledge graph models may process literature and patent information to recommend non-obvious combinations of enzymes or pathways for strain construction (Pandey et al., 2024). Similarly, AlphaFold protein screening enables high-throughput verification before experimental validation, reducing the number of experimental iterations (Jumper et al., 2021). In this way, AI serves not only as a means of monitoring but also as a means of computing advice on pathway and substrate choice. The following characteristics make this biomanufacturing of biomolecules employing *Pseudomonas putida* for sustainable production one of the metabolic engineering stems platforms: default advantageous bust-in features and the development of custom synthetic properties (Kamaludin & Feisal, 2023). One of the most widely used bacteria that can be modified for a variety of biomanufacturing products is *Escherichia coli*, due to its rapid growth rate, existing metabolic engineering knowledge and resources, and comparatively minimal risk of contamination (Xie, 2022). Regenerative medicine and tissue engineering are focusing on the exciting and novel field of biofabrication. It uses automated processes, such as additive manufacturing technology, to create three-dimensional (3D) cell culture systems using biomaterials. It is characterized by biological products, bio-inspired manufacturing processes, and cell-based building blocks (***Balar et al., 2023). A viable substitute for the production of phytochemicals is plant cell culture (PCC) technology (Jaiswal et al., 2022). Due to its capacity to generate particular phytochemicals throughout the year, regardless of environmental factors such as drought or plague, it lessens the strain on natural ecosystems (Bapat et al., 2023). By leveraging microorganisms and enzymes for their catalytic capabilities, biomanufacturing offers a viable avenue.

Biocatalysis uses enzymes' catalytic capacity to improve the sustainability and efficiency of several steps in the pharmaceutical manufacturing process (Bell et al., 2021). The unified bioprocessing technique, according to (Fu et al., 2023), improves economics by lowering enzyme costs, lowering feedstock prices, increasing yield, and generating revenue from high-value by-products (protein and H_2). The MSP value of switchgrass-derived ethanol is between 0.63 and 0.83 USD/gal. Interpretable modeling, representation learning, Graph Attention Networks, reinforcement learning (RL), and deep reinforcement learning (deep RL) are examples of AI modeling techniques and tools that can be used to lower manufacturing costs while increasing efficiency, robustness, reliability, and yield (Xie et al., 2022). The advances of new bio-valorization routes for the different types of biowaste streams, including 3D printing technologies, are also anticipated to help the growth of the bioeconomy by producing reproducibly valuable products and sustainable, customized living materials.

4.9 APPLICATION OF SUSTAINABLE BIOMANUFACTURING

4.9.1 Bioplastics and Biodegradable Materials

Sustainable biomanufacturing has recently transitioned from laboratory- and pilot-scale production to commercialization, partially facilitated by rapidly growing markets for bioplastics, biofuels, and bio-based chemicals. The global bioplastic market size was valued at USD 15.57 billion in 2024 and is projected to reach USD 44.77 billion by 2030 (a compound annual growth rate [CAGR] of 19.5%) (Grand View Research. 2024). Similarly, the global biofuels market size was valued at USD 99.53 billion in 2023 and projected to reach USD 207.87 billion by 2030 (CAGR 11.3%) (IEA, 2024).

Biomaterials are materials that are either naturally occurring or artificially created, such as metals, ceramics, polymeric polymers, and their compounds, that are highly biocompatible with living tissues and can be readily incorporated into the human body (Biswal et al., 2020). To lessen the need for fiberglass composites, scientists are looking into other materials. These materials are created by combining synthetic polymers reinforced with natural fibers from both renewable and non-renewable sources, such as hemp, cotton, flax, or fibers with a polymer matrix. These naturally occurring composites are called bio composites (PLA, PHA) and are composed of petroleum-derived non-biodegradable plastics (polyethylene and polypropylene) or biodegradable plastics (polyethylene and polypropylene). One of the most popular bioplastics is polylactic acid/polylactide (PLA), which is comparable to traditional plastics, including polyethylene, polypropylene, and polyethylene terephthalate (Bioplastics News, 2020). The industry and scientific community have shown a great deal of interest in polyhydroxyalkanoates (PHAs), which are green biopolymers, due to the need to replace conventional petroleum-based non-biodegradable plastics and polymers. They have many inbuilt qualities, including being insoluble in water, non-toxic, biodegradable, biocompatible, piezoelectric, and thermoplastic, which make them a promising alternative to petrochemical plastics such as polypropylene and polystyrene. They are used extensively in the packaging industry, in tissue engineering as scaffolds for support, and as nanoparticles for drug delivery (Devadas et al., 2021).

4.9.2 Biofuels and Renewable Energy Production

Biomanufacturing is the process of using microorganisms, such as bacteria or yeast, to transform renewable resources, such as plant matter or agricultural waste, into fuel. The microbes are usually genetically modified to produce particular enzymes that convert sustainable materials into simple sugar. These sugars can further undergo fermentation to yield biofuels such as butanol or ethanol. The generation of bio-origin solvents, including biobutanol and bioethanol, is developing as a creative way to satisfy their needs. Different biomass feedstocks could be used as the initial starting material for the fermentation process to create ethanol of different generations, ranging from the first to the fourth. Biodiesel is a clean, sustainable bio-based energy source because it is produced by catalytic transesterification of methanol with vegetable, animal, yeast, or microalgal oils. The benefits of biodiesel include its low emissions, non-toxicity, renewability, and similarity to diesel fuel. In the United States, the production of biodiesel, which is less than bioethanol, is also expanding quickly; between 2004 and 2008, it increased more than tenfold. The United States will generate and use more than 38 billion gallons of renewable biofuel annually by 2030 (Fu et al., 2023). Microorganisms are used in biomanufacturing to convert renewable resources into biofuels, providing a scalable solution to the growing demand for sustainable energy. Combining cellulases with ligninases increases the production of biofuel from lignocellulosic biomass, according to (Ilić et al., 2023). Oleaginous microorganisms, such as bacteria, fungi, algae, yeasts, and cyanobacteria, can store 20–80% of their body weight as lipids (Yong et al., 2021). Additionally, bacteria can produce biofuels from waste streams, including wastewater and agricultural waste. These wastes are used as feedstock since they are high in organic materials such as fats and oils. Enzymes produced by engineered bacteria convert trash into sugars, which are subsequently fermented to make biofuels, minimizing waste and its negative effects on the environment (Biswas et al., 2021). Another example is algae, which also generates lipids in larger quantities. Fuels derived from microalgae are environmentally benign and help reduce CO_2 emissions. Their oils are extensively utilized in health supplements and biodiesel (Tang et al., 2020). Low-viscosity, carbon-neutral microalgal biodiesel is a possible substitute for fossil fuels.

4.9.3 Biopharmaceuticals and Green Bioprocessing

As medical research advances, there is an increasing need for complex and advanced molecules for targeted treatments. However, producing these intricate structures with the necessary purity and effectiveness is frequently a challenge for conventional approaches. The biopharmaceutical sector is putting money into novel technologies, including single-use technology (SUT) to enhance process efficiency, continuous manufacturing (CM), and using a closed system, aiming to boost productivity and lower facility costs. This enables continuous material movement, real-time control and monitoring, and the capability to make on spot adjustment to ensure product quality. Numerous unit processes, including cell culture, purification, and formulation, are combined into a single continuous process in biopharmaceuticals through CM (Xie, 2022). Manufacturing biopharmaceuticals, such as vaccines, enzymes, and antibodies, requires stringent control over quality, safety, and efficacy. According to Alsenani et al. (2020), marine algae's antibacterial qualities, which are efficient against pathogens such as *Staphylococcus aureus* and *E. coli*, present the possibility of developing novel treatments. By enabling the development of innovative medications, diagnostics, and preventative strategies, microbiome biomanufacturing also has the potential to revolutionize the healthcare sector. A wide range of chemicals, such as silicon wafers, plastic polymers, and pharmaceutical medications, have been developed to enhance economic progress and human well-being, but they also raise the risk of pollution and dispersion of toxicants, endangering the environment. Intensification techniques can be applied to low-value bioproducts to increase yield and productivity at reduced cost, as they can be used both upstream and downstream (Gupta & McLaughlin, 2021). The bioprocess intensification (BI) approach is enabled by the random differentiation of various bioproducts and their diverse applications, innovative production techniques, technology platforms, intelligent facility designs, and fabrication plans. The chemical engineering profession has long employed process intensification (PI), which uses CM to increase process output, boost product value, and reduce facility footprint while providing significant manufacturing benefits. However, given the complexity of living cells, strict regulations, complex quality standards, and high product value, the use of PI in the fast-paced biotechnology sector appears highly questionable (Coppens, 2021). Biomanufacturing is now a crucial part of the biotechnology and pharmaceutical sectors. Biomanufacturing techniques are utilized to generate many of the medications and vaccines needed to treat and prevent diseases, and new uses for biomanufacturing are constantly being explored. To satisfy the changing needs of modern

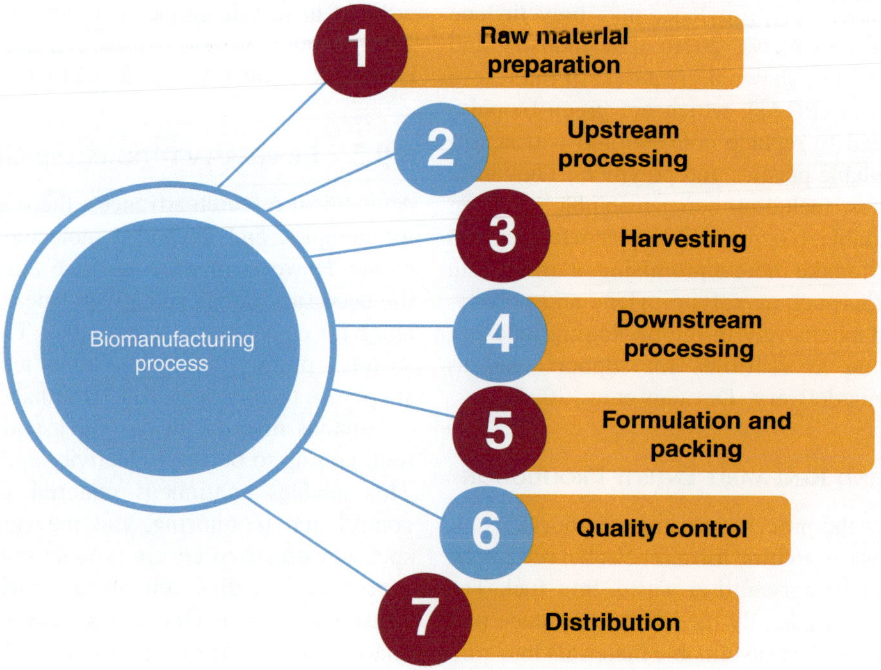

FIGURE 4.2 Biomanufacturing process.

healthcare, creative and sustainable methods of producing pharmaceutical raw materials are desperately needed.

4.9.4 Agricultural and Food Industry Innovations

Microbial biotechnology, which includes biopesticides, bioherbicides, biofertilizers, and insecticides, has been crucial to sustainable agriculture. By controlling biotic and abiotic stresses, microbial biotechnology reduces the need for pesticides in sustainable agriculture. Plant microbiomes are valuable bioresources for agriculture and biofortification because they harbor beneficial microorganisms that can increase plant growth and nutrient uptake through various methods. For instance, several microorganisms in the plant microbiome have the ability to solubilize vital minerals such as zinc, potassium, and phosphorus. Others, conversely, can fix atmospheric nitrogen, which is necessary for plant growth. N_2-fixing blue-green algae, such as Chroococcus, Anabaena, and Nostoc, fix organic nitrogen to aerobic cells by utilizing nitrogenase. This phenomenon can be coupled with the development of algal biofertilizer technology for N_2-fixation and CO_2 sequestration (Raj et al., 2024). The transgenic plants, such as tobacco and *Arabidopsis thaliana*, have the potential to further reduce the biomanufacturing costs of recombinant bulk enzymes due to their easy and inexpensive scale-up, high-level protein accumulation in plant tissues, simple protein purification, and cost-effective plant cultivation. In contrast to polyhydroxybutyrate (PHB), a biodegradable and biocompatible plastic with a wide range of industrial, medical, and agricultural applications, PLA is utilized in the food industry for perishable food product packaging due to its high transparency and good structural stability. A sustainable and circular bioeconomy can be made possible by using primary agricultural biomass, biomass from agricultural by-products, and biomass from other industries as feedstocks or starting materials for fermentation and biomanufacturing processes (Shah, 2025). The conversion of amino acids to higher alcohols through fermentation is another experimental concept. To co-produce ammonia and biofuel, genetic engineering is utilized to create microbial strains that can metabolize and deaminate amino acids. For instance, the biomass of *Saccharomyces* cerevisiae, *E. coli*, *Bacillus subtilis*, and microalgae has been bioconverted to generate alcohols using modified *E. coli*, as presented in Figure 4.2.

4.10 CHALLENGES AND OPPORTUNITIES

Despite the encouraging private sector biomanufacturing activities, commercialization of these technologies is challenging due to a number of important bioprocess design problems (Pandey, et al., 2024). The necessity of investing in biomanufacturing is being recognized by numerous governments, as well as by private investors. Also, according to Kitney and Clarke, biomanufacturing can gain from public–private collaborations and regulations that increase the private sector's trust in bio-based manufacturing (Clarke and Kitney, 2020). In biomanufacturing, downstream processing remains a significant bottleneck, largely because handling the complex composition of culturing broths requires multiple purification steps. These purification procedures, which make up 15–90% of the total production costs in commercialized biotech processes, are often expensive and time-consuming compared with traditional chemical processes. These issues are beginning to be addressed by developments in high-throughput automated platforms and modular purification systems, which enable the creation of more effective and scalable solutions (Kaligotla & Gottschalk, 2023). A new biomanufacturing revolution is anticipated at the start of this century due to new product needs (such as artificial food, renewable energy, and regenerative medicines) and new research tools (such as induced pluripotent stem cells (iPSC), metabolic engineering, synthetic biology, systems biology, cascade biocatalysis, etc.). This revolution will either produce new products (such as new functional tissues or new drugs) or produce existing products in a far more efficient manner than current platforms in terms of product yield, volumetric productivity, product titer, scale-up feasibility, and sustainability. It also urges the active substitution of renewable energy sources, such as biomass-derived sources, for fossil fuels used in the production and purchase of raw materials. Future studies should evaluate the lifecycles of alternative production methods, improve circularity, and optimize energy use (Oluyemi et al., 2024). The production of biofuel and chemicals has advanced as a result of efforts to evaluate the techno-economic and environmental viability of biomanufacturing. A steady, cohesive regulatory procedure is essential, given the diverse viewpoints on consumer and market demands, industry development, and production constraints. One notable shortcoming is the absence of guidelines for auxiliary materials used in biomanufacturing. This leads to the "Valley of Death," a situation where a project is shelved because the owner lacks the means to evaluate their technique at real-world scales and adjust the technology appropriately. This is the case for many promising biomanufacturing advances (Kampers et al., 2022). Long-term investment levels and the increasing support from industry stakeholders are better indicators of the sector's current health. Despite its potential, the industry is still developing, and it will probably take some time to determine which business models and technologies can actually thrive in the bioeconomy. For instance, Amyris, which specializes in sustainable biofuels, installed high-throughput, modular purification systems that drastically lowered production costs and time-to-market for its Biofene® (Hill et al., 2020). The promise of inexpensive and easily produced bioplastics is still a way off, despite the existence of several intriguing possibilities (Ali et al., 2022). A high-yield continuous biomanufacturing can be made possible by AI's assistance in the discovery of novel strains, enzymes, and metabolic pathways for CO_2 bioconversion. Additionally, it can optimize fermentation control procedures and identify important products from CO_2-based routes, resulting in more sustainable and effective production techniques (Jumper et al., 2021). Food waste, degraded polymers, waste gases, and agricultural wastes are some examples of alternate feedstocks for fermentation. Furthermore, fermentation results in the loss of more than half of the carbon as CO_2. Through electrochemical techniques, it is preferable to capture and transform this waste CO_2 into useful C1 or C2 compounds such as ethanol, acetic acid, and formic acid (Soong et al., 2022). The bioeconomy's widespread acceptance provides a powerful incentive for the biomanufacturing industry to adopt these technologies to produce bioproducts more quickly and at a lower cost. It is anticipated that Multivariate Data Analysis (MVDA), Design of Experiments (DoE), and ML-based data analytics will have broader ramifications in the future. The biofoundry concept's widespread adoption also makes it easier to produce early prototypes for the near-term commercialization of bioproducts in a practical manner, which enhances the bioeconomy's outlook. In the near future, a more comprehensive conceptual framework for the sharing and use of research and development data to improve the repeatability and dependability of bioproduction is anticipated.

4.11 CASE STUDY AND GLOBAL INITIATIVES

The rapid rate of global population growth is causing urgent challenges to food security and the environment, which in turn affect the Sustainable Development Goals (SDGs). Numerous parties, including the government, academic institutions, scientific communities, and non-governmental organizations (NGOs), have worked hard to address the issues gradually. As a result, small- and medium-sized businesses (SMEs) are developing new sustainability-driven business models, while a number of large multinational corporations are incorporating circular and bioeconomy features into their operations, business portfolios, and sustainability policies. The European Union (EU) released a communication in March 2024 highlighting the significance of biomanufacturing for autonomy, security, and competitiveness. It suggested eight measures, including a review of the EU Bioeconomy Strategy, market development, financial support, regulatory simplification, fair comparisons of fossil products, and the incorporation of AI (Korosuo et al., 2024). A pan-European project sponsored by France, the Industrial Biotechnology Innovation and Synthetic Biology Accelerator (IBISBA), links RI operators in EU member states. To assist biotechnology pipelines (technology readiness levels [TRL] 2–5), including biofoundry and scale-up, it seeks to establish a coordinated network. IBISBA facilitates project transfers, acts as a reliable online resource, and promotes more European funding for IB (Knight, 2024). Other businesses concentrate on reducing the effects of climate change and investigating the creation of alternative energy sources. Businesses like LanzaTech use bacteria that have been engineered to manufacture biofuels from waste feedstocks. In addition, LanzaTech is a prominent US-based firm that has started syngas fermentation and built some demonstration units across the world Mihalcea et al. (2024). Others, such as Algae for Future, Photanol, and Kiverdi, employ them to absorb CO_2 from the atmosphere. Biological sequestration of CO_2 generated during the biorefinery process, anthropogenic activities can be employed for the production of bio-renewable commodity chemicals (Kaur et al., 2024). Major corporations such as BASF,

Novozymes, DSM, Corbion, and Evonik are expanding biomanufacturing while refocusing on agriculture, food, feed, and healthcare. Novozymes pledged to invest USD315 million to increase the United States' capacity to ferment plant-based meat substitutes (Nielsen et al., 2022). Initiatives by the European Union's Bio-based Industries Consortium (BIC), Brazil's RenovaBio program (Grangeia et al., 2022), China's National Bioindustry Development Plan, and the United States' BioPreferred Program are examples of the coordinated push to integrate bio-based sectors and facilitate the market entry of bio-based products. Efforts to shift to bio-based manufacturing are expanding globally (Pascoli et al., 2022). The National Institute of Standards and Technology (NIST) in the United States develops biomanufacturing metrology and provides cutting-edge imaging facilities to partners in academia and industry. Additionally, in collaboration with more than 90 public and private organizations, the Department of Defense has invested USD87.5 million in the Bioindustrial Manufacturing and Design Ecosystem, for a total investment of over USD260 million. The goal of this investment is to scale up laboratory advances to industrial capacities (Rose & Friedman, 2021). Numerous countries worldwide have embraced the energy sector's biorefinery efforts. Other examples in the United Kingdom include the BioVale regional innovation cluster in Yorkshire and the Humber, which focuses on advanced biorefining and the reuse of bio-waste, and the BioPilotsUK partnership of numerous biorefineries (Lepore et al., 2021). A public–private collaboration called Circular Bio-Based Europe (CBE), formerly known as the Bio-Based Industries Joint Undertaking (BBI JU), aims to build biorefineries in accordance with the European Green Deal. To reduce investment risk and establish competitive value chains, it plans to spend €3.7 billion in bio-based sectors by 2026 (Sierra et al., 2021). From 2018 to 2025, the CAGR for bio-based goods is anticipated to be 12.6% Market Analysis Report (2020). According to the European Commission's Knowledge Centre for Bioeconomy's "Future transitions for the Bioeconomy into Sustainable Development and a Climate-Neutral Economy report," the bioeconomy generates 614 billion euros of added value, or roughly 5% of its GDP, and employs about 17.5 million people, or nearly 9% of its workforce (Fritsche et al., 2020). According to the Ellen MacArthur Foundation, EC, and other well-known groups, CE policies on social, economic, and environmental aspects can have a direct positive impact on economies (Venkata et al., 2020). However, instead of using the more ecologically friendly options of source reduction and reuse, the majority of businesses and municipalities are currently concentrating on recycling and waste-to-energy technologies to meet their waste reduction goals. "There needs to be a move beyond recycling into the largely uncharted territory of the higher end of the waste management hierarchy, to reuse, reduce, and prevention" to effectively solve the waste problem and advance toward a circular economy. Figure 4.3 depicts the same.

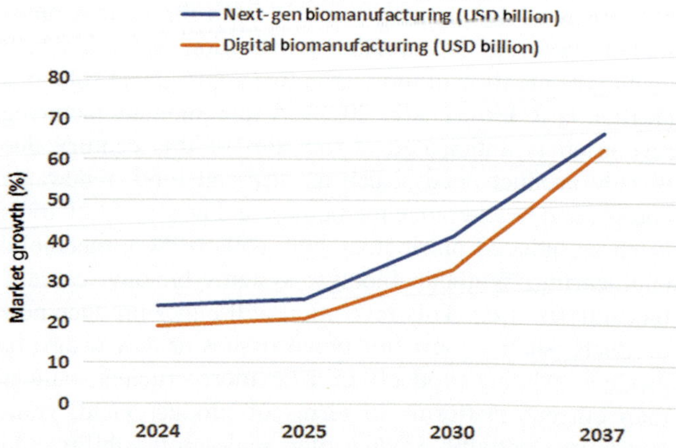

FIGURE 4.3 Biomanufacturing market growth (2024–2037).

4.12 CONCLUSION AND FUTURE DIRECTIONS

In conclusion, the development of biomanufacturing has been a lengthy and intricate process, marked by numerous significant discoveries and breakthroughs in the creation of products across the industrial, pharmaceutical, food processing, and biotechnology sectors. Additionally, new uses for biomanufacturing are continually emerging. The integration of state-of-the-art technologies and bio-based approaches enables the development of an optimized, sustainable manufacturing framework that ensures superior product quality while reducing the ecological footprint. In the short term, priority should be placed on the application of AI-assisted LCA, the use of digital twin's frameworks, and real-time data analytics to increase efficiencies and eliminate environmental needs. Within an Industry 5.0 framework, AI is expected to function not as a replacement for human operators but as a cognitive collaborator. For public sector workers seeking quick translation, biomanufacturing facilitation is a top priority. It also offers a stable, adaptable platform for the manufacture of other materials, such as essential agricultural and healthcare items. Researchers and producers now have access to powerful new tools and methods for developing novel bioproducts and improving existing ones, thanks to advances in genetic engineering, fermentation technology, and other key fields. So, in the mid-term, priority should be placed on establishing biofoundry standards and frameworks for strain development, metabolic modeling, and automated controls. There also needs to be stronger collaboration with academia, industry, and policy to fast-track the development of technologies and skills. Among the tactics are maximizing biomanufacturing's energy use, investigating novel ways to improve circularity, and carrying out more thorough lifecycle analyses of substitute production techniques. Despite their complexity and ongoing development, these ideas are complementary and share the objective of building a more sustainable future. In the long term, the objective should be to design systems that are globally networked, carbon-neutral, human-centric, and operate with circular supply chains to produce and valorize the value of renewable and waste feedstocks. Although obstacles, including legal restrictions, technological limits, and financial limitations, still exist, the current era of increasing datafication and urbanization forces us to reconsider our approach to sustainable development. In this regard, sustainable, intelligent biomanufacturing offers a fresh outlook on a future that is intelligent and robust. This chapter improves our knowledge within the biomanufacturing realm and paves the way for the industry to adopt a more sustainable approach by breaking down each production step and performing a comparative analysis.

REFERENCES

Agrawal, R., Bhagia, S., Satlewal, A., & Ragauskas, A. J. (2023). Urban mining from biomass, brine, sewage sludge, phosphogypsum and e-waste for reducing the environmental pollution: Current status of availability, potential, and technologies with a focus on LCA and TEA. *Environmental Research*, 224, 115523.

Ali, S.S. et al. (2022) Biowastes for biodegradable bioplastics production and end-of-life scenarios in circular bioeconomy and biorefinery concept. *Bioresource Technology* 363, 127869

Alsenani, F., Tupally, K. R., Chua, E. T., Eltanahy, E., Alsufyani, H., Parekh, H. S., & Schenk, P. M. (2020). Evaluation of microalgae and cyanobacteria as potential sources of antimicrobial compounds. *Saudi Pharmaceutical Journal*, 28(12), 1834–1841.

Balar, V., Gor, D., Yadav, A., & Bagadiya, P. (2023) The evolution of biomanufacturing: A historical and contemporary perspective. *Advances in Agricultural Biotechnology*, 107.

Bapat, V. A., Kavi Kishor, P. B., Jalaja, N., Jain, S. M., & Penna, S. (2023). Plant cell cultures: biofactories for the production of bioactive compounds. *Agronomy*, 13(3), 858.

Bell, E. L., Finnigan, W., France, S. P., Green, A. P., Hayes, M. A., Hepworth, L. J., … Flitsch, S. L. (2021). Biocatalysis. *Nature Reviews Methods Primers*, 1(1), 46.

Bhagia, S., Bornani, K., Agrawal, R., Satlewal, A., Ďurkovič, J., Lagaňa, R., … Ragauskas, A. J. (2021). Critical review of FDM 3D printing of PLA biocomposites filled with biomass resources, characterization, biodegradability, upcycling and opportunities for biorefineries. *Applied Materials Today*, 24, 101078.

Bioplastics News (2020) https://bioplasticsnews.com/polylactic-acid-or-polylactide-pla. Accessed on 1 June 2020.

Biswal, T., BadJena, S. K., & Pradhan, D. (2020). Sustainable biomaterials and their applications: A short review. *Materials Today: Proceedings*, 30, 274–282.

Biswas, T., Bhushan, S., Prajapati, S. K., & Chaudhuri, S. R. (2021). An eco-friendly strategy for dairy wastewater remediation with high lipid microalgae-bacterial biomass production. *Journal of Environmental Management*, 286, 112196.

Bobier, J. F., Cerisy, T., Coulin, A. D., Bleecher, C., Sassoon, V., & Alexander, B. (2024). Breaking the cost barrier in biomanufacturing. *Industrial Biotechnology*, 20(3), 113–124.

Castiello, C., Junghanns, P., Mergel, A., Jacob, C., Ducho, C., Valente, S., … Mai, A. (2023). GreenMedChem: the challenge in the next decade toward eco-friendly compounds and processes in drug design. *Green Chemistry*, 25(6), 2109–2169.

Chen, T. L., Kim, H., Pan, S. Y., Tseng, P. C., Lin, Y. P., & Chiang, P. C. (2020). Implementation of green chemistry principles in circular economy system towards sustainable development goals: Challenges and perspectives. *Science of the Total Environment*, 716, 136998.

Clarke, L. and Kitney, R. (2020) Developing synthetic biology for industrial biotechnology applications. *Biochemical Society Transactions* 48, 113–122

Coppens, M. O. (2021). Nature-inspired chemical engineering for process intensification. *Annual Review of Chemical and Biomolecular Engineering*, 12(1), 187–215.

D'Amico, G., Szopik-Depczyńska, K., Beltramo, R., D'Adamo, I., & Ioppolo, G. (2022). Smart and sustainable bioeconomy platform: A new approach towards Sustainability. *Sustainability*, 14(1), 466.

Devadas, V. V., Khoo, K. S., Chia, W. Y., Chew, K. W., Munawaroh, H. S. H., Lam, M. K., … Show, P. L. (2021). Algae biopolymer towards sustainable circular economy. *Bioresource Technology*, 325, 124702.

Dwivedi, R. K., Chauhan, P. S., Sharma, A. K., Singh, M., & Agarwal, A. (Eds.). (2024). *Biomedical Implants: Materials, Design, and Manufacturing*. CRC Press.

Esfandabadi, Z. S., Ranjbari, M., & Scagnelli, S. D. (2022). *The Imbalance of Food and Biofuel Markets Amid Ukraine-Russia Crisis: A Systems Thinking Perspective*.

Fava, F., Gardossi, L., Brigidi, P., Morone, P., Carosi, D. A., & Lenzi, A. (2021). The bioeconomy in Italy and the new national strategy for a more competitive and sustainable country. *New Biotechnology*, 61, 124–136.

Fritsche, U., Brunori, G., Chiaramonti, D., Galanakis, C., Hellweg, S., Matthews, R. A. P. C., & Panoutsou, C. (2020). Future transitions for the bioeconomy towards sustainable development and a climate-neutral economy—knowledge synthesis final report. *Publications Office of the European Union, Luxembourg*, 10, 667966.

Fu, R., Kang, L., Zhang, C., & Fei, Q. (2023). Application and progress of techno-economic analysis and life cycle assessment in biomanufacturing of fuels and chemicals. *Green Chemical Engineering*, 4(2), 189–198.

Grand View Research (2024). *Bioplastics Market Size, Share & Trends Analysis Report*. San Francisco, CA: Grand View Research.

Grangeia, C., Santos, L., & Lazaro, L. L. B. (2022). The Brazilian biofuel policy (RenovaBio) and its uncertainties: An assessment of technical, socioeconomic and institutional aspects. *Energy Conversion and Management: X*, 13, 100156.

Gupta, P., & McLaughlin, K. (2021). A strategic approach to selecting the optimal process intensification scenario. *Bioprocess International*, 19(10), 2–4.

Hill, P. et al. (2020) Clean manufacturing powered by biology: how Amyris has deployed technology and aims to do it better. *Journal of Industrial Microbiology & Biotechnology* 47, 965–975

IEA (2024). *World Energy Outlook 2024 – Biofuels Market Analysis*. International Energy Agency, Paris.

Ilić, N., Milić, M., Beluhan, S., & Dimitrijević-Branković, S. (2023). Cellulases: from lignocellulosic biomass to improved production. *Energies*, 16(8), 3598.

Intasian, P., Prakinee, K., Phintha, A., Trisrivirat, D., Weeranoppanant, N., Wongnate, T., & Chaiyen, P. (2021). Enzymes, in vivo biocatalysis, and metabolic engineering for enabling a circular economy and sustainability. *Chemical Reviews*, 121(17), 10367–10451.

Jaiswal, P., Singh, A. V., Yadav, V. K., & Kumari, N. (2022). Plant tissue culture in tree species. In Avinash Chandra Rai, Ajay

Kumar, Arpan Modi, Major Singh (Eds.), *Advances in Plant Tissue Culture* (pp. 345–356). Academic Press.

Jumper, J., Evans, R., Pritzel, A., Green, T., Figurnov, M., Ronneberger, O., … Hassabis, D. (2021). Highly accurate protein structure prediction with AlphaFold. *Nature*, 596(7873), 583–589.

Kaligotla, H. and Gottschalk, U. (2023) Innovations in downstream processing. *BioPharm International* 36, 30–34

Kamaludin, N. H., & Feisal, N. A. S. (2023). Biomanufacturing for sustainable production of biomolecules: Pseudomonas putida cell factory. In Vijai Singh & Pau Loke Show (Eds.), *Biomanufacturing for Sustainable Production of Biomolecules* (pp. 175–188). Singapore: Springer Nature Singapore.

Kampers, L.F.C. et al. (2022) Navigating the Valley of Death: perceptions of industry and academia on production platforms and opportunities in biotechnology. *EFB Bioeconomy Journal* 2, 100033.

Kaur, P., Singh, S., Sharma, N., & Agrawal, R. (2024). Filling in the gaps in second-generation biorefineries: evaluating rice straw and its bioethanol residue for the production of biogenic silica nanoparticles. *Nanotechnology For Environmental Engineering*, 9(1), 67–76.

Klerkx, L., & Rose, D. (2020). Dealing with the game-changing technologies of Agriculture 4.0: How do we manage diversity and responsibility in food system transition pathways?. Global *Food Security*, 24, 100347.

Knight, P. (2024). *Building the future with nature: Biotechnology and Biomanufacturing in the EU.*

Korosuo, A., Borzacchiello, M. T., Giuntoli, J., Lasarte, L. J., M'Barek, R., Mubareka, S. B., & Camia, A. (2024). *Trends in the EU Bioeconomy-Update 2024.*

Kurt, E., Qin, J., Williams, A., Zhao, Y., & Xie, D. (2023). Perspectives for using CO_2 as a feedstock for biomanufacturing of fuels and chemicals. *Bioengineering*, 10(12), 1357.

Leong, H. Y., Chang, C. K., Khoo, K. S., Chew, K. W., Chia, S. R., Lim, J. W., … Show, P. L. (2021). Waste biorefinery towards a sustainable circular bioeconomy: a solution to global issues. *Biotechnology for Biofuels*, 14, 1–15.

Lepore, A., Lenzi, A., Fava, F., Bianchi, L., & Palermop, S. (2021). *Implementation Action Plan (2020-2025) for the Italian Bioeconomy Strategy BIT II-Update 2021* (pp. 1–39). Lepore, Amedeo.

Mandal, P., Debbarma, J., & Saha, M. (2021). A review on the emergence of graphene in photovoltaics industry. *Biointerface Research in Applied Chemistry*, 11(6), 15009–15036.

Matsuo, M. (2024). *Challenges in Advancing Biomanufacturing Toward Bioeconomy and Framework for Key Considerations in Policy Discussions* (Doctoral dissertation), School of Environmental Studies, Nagoya University.

Mihalcea, C., Burton, F., Conrado, R., & Simpson, S. (2024). R&D&I and Industry examples: LanzaTech's gas fermentation. In Manfred Kircher & Thomas Schwarz (Eds.), *CO_2 and CO as Feedstock: Sustainable Carbon Sources for the Circular Economy* (pp. 333–343). Cham: Springer International Publishing.

Muscat, A., de Olde, E. M., Ripoll-Bosch, R., Van Zanten, H. H., Metze, T. A., Termeer, C. J., … de Boer, I. J. (2021). Principles, drivers and opportunities of a circular bioeconomy. *Nature Food*, 2(8), 561–566.

Nielsen, J., Tillegreen, C. B., & Petranovic, D. (2022). Innovation trends in industrial biotechnology. *Trends in Biotechnology*, 40(10), 1160–1172.

Oluyemi, G. F., Afolabi, R. O., Zamora, S. C., Li, Y., & McElroy, D. (2024). Environmental impact assessment of a plant cell-based bio-manufacturing process for producing plant natural product ingredients. *Sustainability*, 16(19), 8515.

Pandey, K. et al. (2024) Bioprocessing 4.0 in biomanufacturing: Paving the way for sustainable bioeconomy. *Systems Microbiology and Biomanufacturing* 4, 407–424

Pascoli, D. U., Aui, A., Frank, J., Therasme, O., Dixon, K., Gustafson, R., … Wright, M. M. (2022). The US bioeconomy at the intersection of technology, policy, and education. *Biofuels, Bioproducts and Biorefining*, 16(1), 9–26.

Pelli, P., & Lähtinen, K. (2020). Servitization and bioeconomy transitions: Insights on prefabricated wooden elements supply networks. *Journal of Cleaner Production*, 244, 118711.

Raj, T., Verma, S., Kumar, N., & Agrawal, R. (2024). Role of woody biomass in carbon capture, circular bioeconomy, and biomanufacturing. In Deepak Kumar, Sachin Kumar, Karthik Rajendran, & Ramesh C. Ray (Eds.), *Sustainable Biorefining of Woody Biomass to Biofuels and Biochemicals* (pp. 291–318). Woodhead Publishing.

Rose, P.P. and Friedman, D. (2021) BioInspired, BioDriven, BioMADE: the U.S. bioindustrial manufacturing and design eco- system as a driver of the 4th industrial revolution. *Engineering Biology* 5, 60–63

Rosemann, A., & Molyneux-Hodgson, S. (2020). Industrial biotechnology: To what extent is responsible innovation on the agenda?. *Trends in Biotechnology*, 38(1), 5–7.

Shah, K. H. (2025). *Biotechnology and Biomanufacturing in South America*. Growing a Leading Circular Bioeconomy.

Sheldon, R. A., & Norton, M. (2020). Green chemistry and the plastic pollution challenge: Towards a circular economy. *Green Chemistry*, 22(19), 6310–6322.

Sierra, A. R., Zika, E., Lange, L., de Azúa, P. L. R., Canalis, A., Esteban, P. M., … Mengal, P. (2021). The bio-based industries joint undertaking: a high impact initiative that is transforming the bio-based industries in Europe. *New Biotechnology*, 60, 105–112.

Soong, Y. H. V., Sobkowicz, M. J., & Xie, D. (2022). Recent advances in biological recycling of polyethylene terephthalate (PET) plastic wastes. *Bioengineering*, 9(3), 98.

Stegmann, P. et al. (2020) The circular bioeconomy: its elements and role in European bioeconomy clusters. *Resources, Conservation and Recycling: X* 6, 100029.

Tang, D. Y. Y., Khoo, K. S., Chew, K. W., Tao, Y., Ho, S. H., & Show, P. L. (2020). Potential utilization of bioproducts from microalgae for the quality enhancement of natural products. *Bioresource Technology*, 304, 122997.

Udugama, I., Öner, M., Lopez, P. C., Beenfeldt, C., Bayer, C., Huusom, J. K., … Sin, G. (2021). Towards digitalization in bio-manufacturing operations: a survey on application of big data and digital twin concepts in Denmark. *Frontiers in Chemical Engineering*, 3, 727152.

United Nations. (2020). Global Issues Overview. https://www.un.org/en/sections/issues-depth/global-issues-overview/. Accessed 12 March 2020.

Venkata S, Mohan K, Amulya J, Modestra Annie (2020) Urban biocycles – Closing metabolic loops for resilient and regenerative ecosystem: A perspective. *Bioresource Technology* 306: 123098

Wang, C., Zeng, H. S., Liu, K. X., Lin, Y. N., Yang, H., Xie, X. Y., … Ye, J. W. (2023). Biosensor-based therapy powered by synthetic biology. *Smart Materials in Medicine*, 4, 212–224.

Wang, Y., Liu, X., Han, X., Godin, R., Chen, J., Zhou, W., … Tang, J. (2020). Unique hole-accepting carbon-dots promoting selective carbon dioxide reduction nearly 100% to methanol by pure water. *Nature Communications*, 11(1), 2531.

Woźniak, E., Tyczewska, A., & Twardowski, T. (2021). Bioeconomy development factors in the European Union and Poland. *New Biotechnology*, 60, 2–8.

Xie, D. (2022). Continuous biomanufacturing with microbes—Upstream progresses and challenges. *Current Opinion in Biotechnology*, 78, 102793.

Xie, W., Wang, B., Li, C., Xie, D., & Auclair, J. (2022). Interpretable biomanufacturing process risk and sensitivity analyses for quality-by-design and stability control. *Naval Research Logistics (NRL)*, 69(3), 461–483.

Yong, J. J. J. Y., Chew, K. W., Khoo, K. S., Show, P. L., & Chang, J. S. (2021). Prospects and development of algal-bacterial biotechnology in environmental management and protection. *Biotechnology Advances*, 47, 107684.

Zobel-Roos, S., Schmidt, A., Uhlenbrock, L., Ditz, R., Köster, D., & Strube, J. (2021). Digital twins in biomanufacturing. In Christoph Herwig, Ralf Pörtner, & Johannes Möller (Eds.), *Digital Twins: Tools and Concepts for Smart Biomanufacturing* (pp. 181–262).

5 Sustainable Biomanufacturing Industry 5.0 for Carbon Neutrality and CO_2 Sequestration

Lughano Nancy Gondwe and Suman
Department of Physics and Environmental Sciences, Sharda School of Engineering & Science, Sharda University, Greater Noida, Uttar Pradesh, India

5.1 INTRODUCTION

In recent years, the Earth has experienced an intensifying climate crisis, driven by rising carbon dioxide emissions and unsustainable industrial practices. This has highlighted the urgent need for strategies that reduce the carbon footprint without compromising economic and societal well-being. Nations are now striving toward net-zero carbon emissions by 2050 (Wolgemuth, 2025). The concept of bioeconomy, adopted worldwide, integrates multidisciplinary approaches to transform unsustainable manufacturing into sustainable, resource-efficient bio-based manufacturing.

The use of living organisms to produce products provides a sustainable, circular, and carbon-conscious practice that integrates eco-design principles, renewable feedstocks, and lifecycle-aware technologies. Biomanufacturing utilizes biological systems, such as enzymes, microorganisms, cells, tissues, plants, and animals, to produce goods across various sectors, including agriculture, food, energy, materials, and pharmaceuticals. Over the years, biomanufacturing has evolved from the use of microorganisms such as yeast and bacteria in food and beverages to the design and construction of specialized biological systems. This evolution also incorporates the revolution of industry from Industry 1.0 to the new emerging Industry 5.0 (I5.0)(Balar et al., 2023).

The rise of I5.0 represents a pivotal chapter in the transformation of industrialization, redefining the boundaries for enriched cooperation between humans and intelligent technologies. I5.0 is a clear shift from Industry 4.0, which emphasized automation and data exchange in manufacturing technologies (Rame et al., 2024). With a global focus on environmental stewardship and social welfare, the integration of I5.0 provides a solid foundation for exploring cleaner and more sustainable production methods to achieve competitive advantage (Sharma and Gupta, 2024).

Within this context, sustainable biomanufacturing aligns biotechnology innovation with climate objectives by enabling carbon-neutral pathways. It captures and sequesters CO_2, minimizes energy and water use, and exploits waste streams. Synthetic biology designs microbial or algal pathways, artificial intelligence (AI) tools optimize yields, and green chemistry catalysts ensure benign reaction conditions. Microbial cell factories, engineered algae, and cell-free biosynthesis are now being harnessed to produce value-added compounds from CO_2, lignocellulose biomass, and industrial residues, contributing to both climate mitigation and resource circularity.

This chapter explores the concept of sustainable biomanufacturing and examines how emerging I5.0 innovations can be leveraged to achieve carbon neutrality through advanced CO_2 sequestration strategies.

5.1.1 Evolution of Industries

Over successive industrial revolutions, the manufacturing sector has continuously evolved by integrating advanced technologies and reimagining the nature of the products it creates. In the late 18th century, the implementation of steam power marked the onset of Industry 1.0. This initial phase laid the foundation for Industry 2.0, which was facilitated by electricity and early automation, enabling more efficient, scalable manufacturing processes. The emergence of Industry 3.0 introduced computers, electronics, and automated systems that minimized human intervention in repetitive tasks and enhanced production efficiency (Figure 5.1) (Gaur, 2025).

5.1.2 Industry 4.0 to Industry 5.0

With the rise of Industry 4.0, technologies such as robotics, AI, and automation have transformed manufacturing by minimizing manual input, boosting productivity, and refining supply chain dynamics. The fusion of Internet of Things (IoT), cyber-physical systems, and data-driven tools enabled smart operations that support economic efficiency and environmental responsibility (Gaur, 2025). However, the unregulated or poorly designed implementation of AI and robotics has posed significant challenges for sustainable manufacturing, undermining efforts to conserve energy, maintain circular processes, and safeguard employment and human dignity (Ghobakhloo et al., 2023).

DOI: 10.1201/9781003637158-5

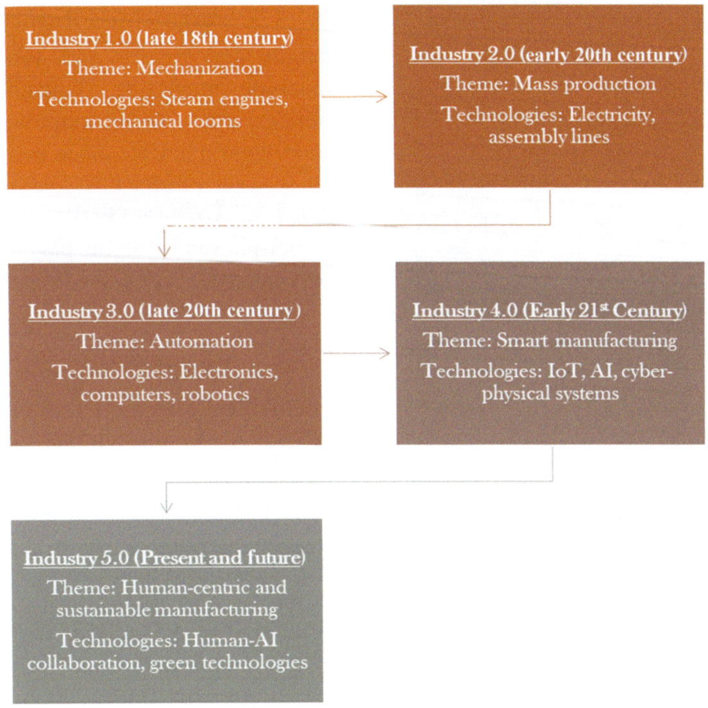

FIGURE 5.1 Evolution of the industrial revolution from Industry 1.0 to Industry 5.0.

I5.0 signals a major change in how we approach manufacturing, shifting the focus from efficiency and automation to placing people, resilience, and sustainability at the heart of industrial design. This transition is especially important for biomanufacturing, where blending living systems with cutting-edge digital tools requires thoughtful coordination, creativity, and care.

While Industry 4.0 focused on automation, cyber-physical systems, and real-time data, I5.0 introduces a new model of collaborative intelligence in which humans and machines work side by side. With the support of AI, cobots, IoT, and digital twins, this approach is particularly important in biomanufacturing, where the complexity and sensitivity of biological systems call for thoughtful, adaptive, and ethically guided decision-making.

5.1.3 Background of Industry 5.0

I5.0 is the latest phase of the industrial revolution, merging human intelligence with technologies such as AI and cyber-physical systems. It introduces a new outlook, placing greater value on societal well-being and environmental integrity, while still incorporating traditional metrics of efficiency and productivity. The synergy between I5.0's vision and sustainability fosters meaningful solutions that benefit people, the planet, and profit.

5.1.4 Key Aspects of Industry 5.0

Since I5.0 adopts a human-centric approach, it places people at the center of the production process and emphasizes human skills such as creativity and decision-making (Rajkuma et al., 2025).

5.1.4.1 Collaborative Robotics

I5.0 envisions robots as collaborative partners, assisting humans with tasks that require precision, power, or persistence without displacing the human workforce.

5.1.4.2 Advanced Technologies

State-of-the-art technology in I5.0 includes the use of the IoT to create smart, interconnected manufacturing systems, along with AI and augmented reality.

5.1.4.3 Customization and Flexibility

I5.0 enables greater customization and versatility in manufacturing approaches through the incorporation of advanced technologies. This allows for the production of more tailored products and quicker responses to market demands (Figure 5.2).

5.1.5 Key Enablers of Industry 5.0

5.1.5.1 Internet of Things

The IoT offers a significant competitive advantage to industries by enabling machines to exchange and collect data through internet connectivity. In I5.0, the integration of IoT supports intelligent decision-making that enhances operational performance, ensures product quality, enables predictive maintenance, and facilitates customized production (Narkhede et al., 2024).

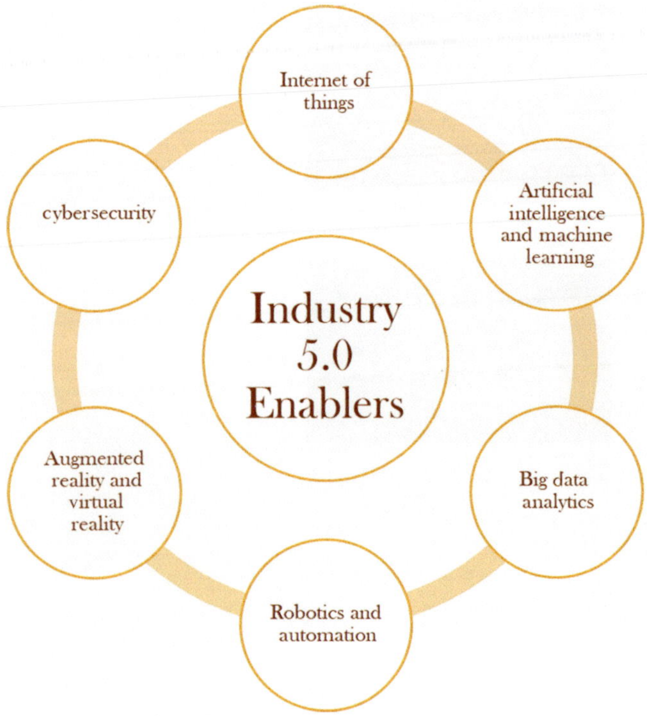

FIGURE 5.2 Main drivers facilitating the adoption of Industry 5.0.

5.1.5.2 Artificial Intelligence and Machine Learning

The era of the internet and AI has revolutionized manufacturing. I5.0 advocates for digitally connected, collaborative, and adaptable systems to enable mass customization. AI and machine learning interpret large datasets, recognize patterns, and can make decisions autonomously, without human intervention. These technologies are well-suited for quality assurance, process optimization, and predictive maintenance (Narkhede et al., 2024).

5.1.5.3 Big Data Analytics

Big data analytics (BDA) facilitates the acquisition and interpretation of vast amounts of data, which provide tangible solutions for real-time and prompt decision-making (Narkhede et al., 2024). The analysis of customer behavior through BDA empowers companies to establish pricing strategies that are appealing and economically viable. It also helps to uncover issues in manufacturing workflows and supply networks, guiding smarter decisions, boosting production performance, and reducing operational costs (Narkhede et al., 2024).

5.1.5.4 Robotics and Automation

The use of industrial robotics and automation enhances production workflows, resulting in greater efficiency and higher productivity. With collaborative robotics in I5.0, we can easily achieve sustainability through the harmony of humans and robots working together. The dangerous and repetitive tasks can be performed by robots, which ensures precision and improved safety (Narkhede et al., 2024).

5.1.5.5 Augmented Reality and Virtual Reality

Augmented reality and virtual reality enable human–machine interaction as machines simulate real-life experiences. These smart manufacturing systems are useful in training and maintenance activities.

5.1.5.6 Cybersecurity

Cyber-physical systems play a vital role in maintaining process integrity and defending against digital threats.

5.1.6 Sustainability

The ability to support economic, environmental, or social processes over time without compromising natural resources is known as sustainability (see Figure 5.3). I5.0 aims to integrate sustainable methods that promote environmental preservation, support social welfare, and long-term economic resilience. Its impact ranges from resource management to building greener cities, emphasizing human-centered, flexible approaches. I5.0 will also enhance sustainability by maximizing resource efficiency, reducing waste, and lowering environmental impact. Overall, it marks a shift toward a more human-centric and eco-conscious model of manufacturing, where technology is harnessed to support human capabilities and drive improvements in productivity and operational performance.

5.1.6.1 Industry 5.0 and Sustainability

I5.0 signals a shift from traditional manufacturing by placing sustainability and human values at the center of industrial progress. It moves beyond the mechanized approach of Industry 4.0, aiming for systems that are smarter, more resilient, and environmentally friendly. Technologies such as IoT and AI are now being used to build circular supply chains, minimize waste, and support sustainable practices such as smart agriculture and energy-efficient production. The concept of Zero-Defect Manufacturing is a great example of the use of predictive tools to reduce resource loss and align production with environmental goals.

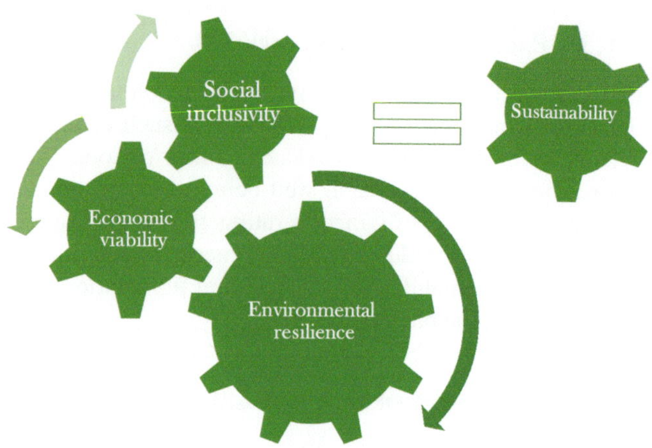

FIGURE 5.3 The fundamental pillars of sustainability.

I5.0 also promotes social well-being by improving work environments and empowering employees through digital collaboration. Innovations in sustainable agriculture and urban infrastructure are having an impact across various sectors. The transition to this concept, however, is not as simple. Upgrading existing systems, securing digital infrastructure, and training workers all demand serious attention. Overall, the integration of advanced technology and sustainability in I5.0 opens up promising paths for transforming industry into a force for good.

5.2 SUSTAINABLE BIOMANUFACTURING

Biomanufacturing is the production of useful products for agriculture, food, energy, and pharmaceutical industries by using biological entities such as enzymes, cells, tissues, microorganisms, and flora and fauna (Kopperi, 2025). This technology has potential as a substitute for fossil-derived materials. It is widely used to produce medicines and vaccines, and new uses for these methods continue to be discovered.

TABLE 5.1
Overview of Key Products and Innovations Developed through Biomanufacturing Technologies

Product Developed by Biomanufacturing	Applications
Amino acids	Amino acids are used in animal feed and food additives
Insulin	Insulin is used for diabetic patients
Vaccines	HPV vaccine, COVID-19 vaccine, and flu vaccine
Enzymes	Production of laundry detergents, food processing, and biofuels
Hormones	Growth hormones, erythropoietin, and follicle-stimulating hormone
Antibodies	Monoclonal antibodies to treat different illnesses
Biofuels	Production of ethanol
Industrial Chemicals	Production of solvents and biodegradable plastics

5.2.1 Biomanufacturing

As global industries transition toward sustainability, biomanufacturing represents the biological equivalent of the technological and automation revolutions that have reshaped traditional manufacturing. Biomanufacturing has been used for thousands of years to transform one food type into another, with examples such as making wine from fruit and honey as raw materials. Fermentation was then popularized and used for activities such as making drinks, cheese, and pickling foods, to mention a few. The fermentation process was easily performed, as it used naturally occurring mixed microorganisms that functioned to produce a desired a product from the feedstock. The Table 5.1 below shows some products developed by biomanufacturing and their applications (Balar et al., 2023).

5.2.1.1 Bio-Factories

5.2.1.1.1 Bacterial Bio-Factory

There are microbial systems that serve as a foundational platform in industrial biotechnology.

1. *Escherichia coli*: It is widely studied and well-characterized. It grows rapidly and is easy to manipulate genetically. Supports high-cell-density cultures and has a rich toolkit for gene and metabolic engineering. Despite its strengths, it has limited metabolic flexibility, which can restrict certain applications (Balar et al., 2023).
2. *Pseudomonas putida*: A plasmid-cured variant of *P. putida* mt-2, increasingly used an industrial host. It can degrade toxic compounds in heavily polluted environments. Known for producing medium-chain-length polyhydroxyalkanoates, making it a valuable producer of biopolymers. Offers metabolic versatility and environmental robustness (Balar et al., 2023).
3. *Bacillus subtilis*: This is an aerobic bacterium extensively researched for heterologous protein production. Successfully used to express enzymes such as amylases, xylanases, lichenase, and beta-galactosidase. Also capable of producing cellulases, proteases, and other enzymes through fermentation (Balar et al., 2023).

5.2.1.1.2 Yeast Bio-Factory

Yeast was once mainly used for baking and brewing, but it is now a key organism in biomanufacturing. It can produce biofuels, pharmaceuticals, and food additives, and is valued for its fast growth, genetic flexibility, and ability to perform complex protein modifications (Balar et al., 2023).

5.2.1.1.3 Insect Cell Bio-Factory

Since the 1950s, in vitro insect cell structures have been widely used in biotechnology. A special virus called baculovirus helps insert genes into these cells, turning them into mini factories. This method is great for making vaccines and researching proteins, although it doesn't produce as much as mammalian cells (Balar et al., 2023).

5.2.1.1.4 Plants as Factories

Plants have been harnessed to produce pharmaceuticals, vaccines, industrial enzymes, and bio-based materials (Balar et al., 2023).

Artemisinin, which is a key antimalarial drug, is naturally produced in small amounts by *Artemisia annua*. Efforts are underway to enhance its production through improved bioprocesses.

Single-chain Fv (scFv) antibodies, important in medicine, have been successfully synthesized in potato tubers using genetic engineering.

Sugarcane is used not only for sugar but also for generating energy, producing solvents, making bioplastics, and producing sorbitol.

Barley plants have been engineered to produce LL-37, an antimicrobial peptide, with stable expression over multiple generations.

Plants can be genetically modified to produce complex biomolecules. They also offer sustainable, scalable, and low-cost production platforms. Plant seeds can be stored long-term at room temperature, unlike microbial cultures. Plants also fix their own carbon via photosynthesis, reducing input needs (Balar et al., 2023).

5.2.2 Evolution of Biomanufacturing

5.2.2.1 Biomanufacturing 1.0

This first revolution focused on the manufacturing of primary metabolites. These include acetone, butanol, ethanol, glycerol, etc. Natural microbial processes were used for fermentation- based production of alcohol, dairy products, and organic acids.

5.2.2.2 Biomanufacturing 2.0

World War II brought on the biomanufacturing revolution with the creation of penicillin from fermentation. Biomanufacturing 2.0 advanced into secondary metabolites, which led to the large-scale production of penicillin and streptomycin (Balar et al., 2023).

5.2.2.3 Biomanufacturing 3.0

In the 1980s, there was a development in recombinant DNA technology marked the rise of Biomanufacturing 3.0. This revolution was owed to a production shift from primary and secondary metabolites to large protein synthesis crucial for catalysis and therapeutics. It also advanced quickly due to the production of recombinant proteins such as insulin and growth hormone (Balar et al., 2023).

5.2.2.4 Biomanufacturing 4.0

The current phase, biomanufacturing 4.0, is deeply integrated with digital technologies such as AI, IoT, and automation, which enables precision control over biosynthetic pathways, high-throughput screening, and digitally managed bioreactors (Zhang et al., 2025). Today's industries are built on networks that share and use data. In this setup, biomanufacturing runs on systems that blend physical machines with virtual controls. Biomanufacturing 4.0 is the current revolution in biomanufacturing, and it emphasizes sustainable, new, efficient, and safe production of products in different industrial sectors such as medicine, renewable energy sources, and artificial food, among other things (Balar et al., 2023).

5.2.2.5 Biomanufacturing 5.0

Biomanufacturing 5.0 aligns with the transition to I5.0, which signifies a shift toward human-centric, ethical, and sustainable bio-production systems. This includes the use of synthetic biology, AI, and smart automation, not only to increase yield and efficiency but also to reduce waste, recycle carbon, and tailor products for individual or local needs (Rajkuma et al., 2025).

5.2.3 Types of Biomanufacturing

5.2.3.1 Synthetic Biology

This focuses on designing and building new biological systems with desired specific functions. This technology has the ability to transform biomanufacturing as it allows for the creation of new organisms that can produce specific substances or remove environmental pollutants (Balar et al., 2023).

5.2.3.2 Mammalian Cell Culture

Certain biotechnology products, such as viral vaccines and complex proteins, must be produced in mammalian cells rather than in bacterial systems. This is due to the fact that many proteins require post-translational modifications, such as glycosylation, which bacterial cells cannot perform. Mammalian cell cultures use recombinant DNA technology, which scientists use to manufacture a wide range of biopharmaceuticals such as enzymes, hormones, and immunobiologicals. The COVID-19 pandemic played a significant role in the accelerated innovations in the field, which prompted prominent pharmaceutical companies to develop new vaccines and antiviral drugs using this technology (Balar et al., 2023).

5.2.3.3 Biosensors

This is a measurement device that includes a bioreceptor, a base material, a transducer, and an electronic system. This technology detects biological or chemical substances by transforming a biological response into a measurable signal. The types of biosensors and their applications are as follows:

- Acetylcholinesterase-inhibition-based biosensors: Designed for the detection of organophosphorus pesticides, they help to monitor pesticide residues in the environment.
- Microalgae-based biosensors: Use algae such as *Pseudokirchneriella subcapitata* and *Chlorella vulgaris* to detect herbicides and both light (e.g. Na and Al) and heavy metals (e.g. Cu and Cd).
- Yeast-based biosensors: Take advantage of yeast's similarity to mammalian cells in signaling pathways. These are used for hormone detection, such as estrogens and androgens.

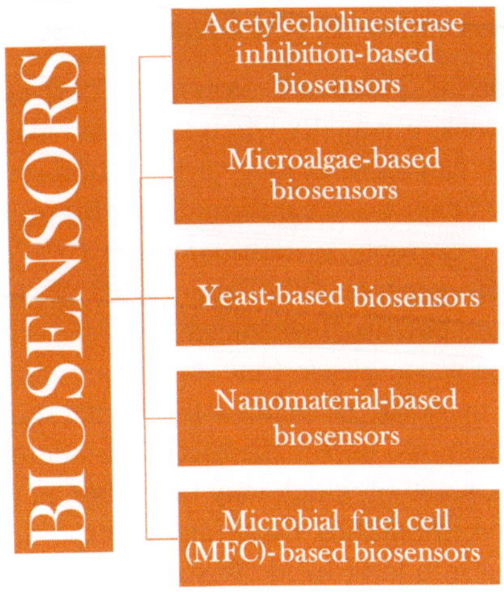

FIGURE 5.4 Major types of biosensors used in analytical and industrial applications.

- Nanomaterial-based biosensors: Utilize gold nanoparticles for high sensitivity. Applied in detecting Salmonella, malaria, and blood glucose.
- Microbial fuel cell (MFC)-based biosensors: Generate electricity signals from microbial metabolism. It is used for in situ environmental monitoring of pollutants (Figure 5.4).

5.2.3.4 C1 Biomanufacturing

This is an upcoming sustainable approach to chemical production that harnesses microbial metabolism or engineered cell systems to assimilate C1 compounds as carbon and energy sources for the synthesis of value-added products.

5.2.4 Advances in Biomanufacturing

5.2.4.1 Nanotechnology Integration

- Antibody-presenting viral nanoparticles: Viruses are engineered to display nanobodies. Such types of viruses, such as tobacco etch virus and zucchini yellow mosaic virus. Some applications include diagnostics and therapeutics.
- Quantum dot production: *Stenotrophomonas maltophilia* is engineered to produce extracellular cadmium sulfide (CdS) nanocrystals. These enable efficient purification and use in optoelectronics and imaging.

5.2.4.2 Space-Bound Biomanufacturing

- Plants are considered for space missions as on-demand bio-factories. They have extended seed viability at room temperature and are self-sufficient for growth using only CO_2 and light.

5.2.4.3 Marine Feedstock Utilization

- Mannitol, found in seaweed, is used as a renewable feedstock. Engineered microorganisms can convert mannitol into a wide range of bio-products.

5.2.4.4 Artificial Intelligence in Biotech

- AI supports compound selection, pathway configuration, and data analysis and production optimization.
- Machine learning drives innovation in genomics, proteomics, and systems biology.

5.2.4.5 Bio-fabrication and Tissue Engineering

- This combines additive manufacturing with cell structure systems. It also produces hybrid stem cell–biomaterial constructs for regenerative medicine.

5.2.5 Sustainable Biomanufacturing

As countries globally are shifting toward sustainable practices, the bioeconomy is advancing as a way to create production systems that rely on renewable resources without compromising the environment (Wolgemuth, 2025). Cells are used as biocatalysts in the production of bio-based chemicals and fuels in industrial biotechnology. This branch of technology holds great promise for creating new, sustainable value chains that convert renewable feedstocks into energy, food, chemicals, and feed. To address today's global challenges, industries must take initiative in developing biomanufacturing approaches that balance economic viability with sustainability goals. Sustainable biomanufacturing incorporates sustainable chemistry, which emphasizes designing processes that naturally minimize or eliminate harmful characteristics. Through this approach, biomanufacturing can serve as an economic catalyst for many nations.

5.2.6 Industry 5.0 and Biomanufacturing

As the industrial revolution continues to evolve and technology advances, biomanufacturing is also transforming. Industry and biomanufacturing have progressed hand in hand throughout their development. With the emergence of new technologies and methods, the biomanufacturing sector continues to integrate these innovations to achieve greater precision, efficiency, and sustainability in production (Figure 5.5).

5.3 CARBON NEUTRALITY IN BIOMANUFACTURING

5.3.1 Carbon Neutrality

Enhanced greenhouse gas emissions are having serious, detrimental effects on the natural environment. Sea level rise, ocean acidification, glacial melting, and ecosystem damage are all consequences of excessive emissions of these greenhouse gases, which are disrupting ecological harmony and jeopardizing global social and economic stability. The Earth's temperature is increasing annually, and if no proactive global actions are taken, the projections of 2°C increase will become a reality. The increase in fossil

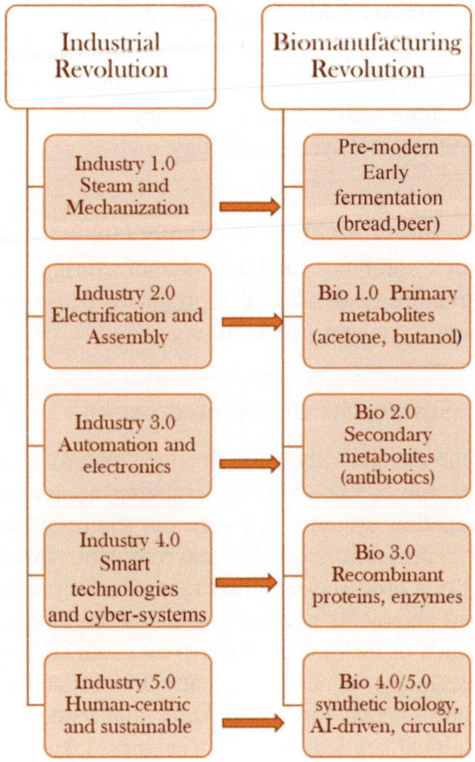

FIGURE 5.5 Industrial revolutions and their impact on biomanufacturing evolution.

fuel consumption has significantly increased CO_2 emissions, as burning these fuels remains the leading driver of atmospheric carbon dioxide levels. Reducing CO_2 emissions is key to achieving carbon neutrality.

Carbon neutrality is a state in which all greenhouse gas emissions are balanced by removal efforts through natural or engineered means, resulting in net-zero emissions, primarily achieved by significantly reducing emissions. This can be achieved by boosting Earth's carbon sequestration capacity and reducing emissions via sustainable energy and improved manufacturing processes. The growth of the digital economy is crucial in enhancing carbon neutrality, with widespread use of AI, big data, and cloud infrastructure resulting in greater operational efficiency in production and consumption systems (Wang and Wang, 2025).

5.3.2 Carbon Neutrality in Biomanufacturing

Biomanufacturing offers a promising pathway to decrease petrochemical consumption while contributing to climate change mitigation. A biological manufacturing system that reduces emissions but also removes more carbon dioxide from the atmosphere than it emits by sequestering it in products or processes is referred to as carbon-negative biomanufacturing (Scown, 2022).

5.3.3 Opportunities for Carbon Removal

- Bio-based polymers: Polymers derived from biofeedstock, especially when landfilled instead of incinerated, can serve as long-term carbon sinks.
- Carbon capture integration: Some bioprocesses, especially anaerobic systems or those sprayed with pure oxygen, generate CO_2 streams that are easier to capture and potentially sequester.
- Use of waste gases: Innovations are converting low-value or waste gases (e.g., from steel mills and forest residues) into useful chemicals via microbial fermentation, some of which may be carbon negative.
- Carbon in building materials: incorporating biogenic carbon into long-lived construction materials such as wood composites or hempcrete is emerging as another promising sequestration route.

Although carbon capture through biomanufacturing is promising, it has some limitations and misconceptions that negate these opportunities. Most biomanufactured products, such as solvents, cleaning agents, or fuels, eventually release their carbon back into the atmosphere when used or discarded. The claims of carbon negativity often exclude product end-of-life emissions, also known as the cradle to gate analysis, which can misrepresent actual carbon removal (Scown, 2022).

5.3.4 Carbon Capture

Carbon capture technology enables the compaction and refinement of CO_2 through methods such as chemical absorption and adsorption, membrane separation, and physical separation to obtain large quantities of CO_2. The most prominent of the methods is chemical adsorption, as it has characteristics of good selectivity, high adsorption efficiency, and low cost (Caineng et al., 2023). There are three types of carbon capture technology:

1. Pre-combustion capture: This technology removes CO_2 from carbon-based fuels before burning, usually after steam reforming, and the remaining hydrogen is used for energy.
2. In-combustion capture: Capturing CO_2 during the fuel burning process. It includes oxyfuel and chemical looping combustion. Oxyfuel combustion burns fuel using pure oxygen instead of air, and 70% of the flue gas is recycled back into the furnace, which helps to regulate heat and combustion behavior. The flue gas is highly concentrated with CO_2, over 80%. Chemical looping uses metal oxides to transfer oxygen to the fuel indirectly, avoiding direct contact with air and producing a CO_2-rich exhaust (Caineng et al., 2023).
3. Post-combustion capture: This method removes CO_2 from flue gas after fuel is burned, using methods such as chemical absorption, physical absorption, or membrane separation (Figure 5.6).

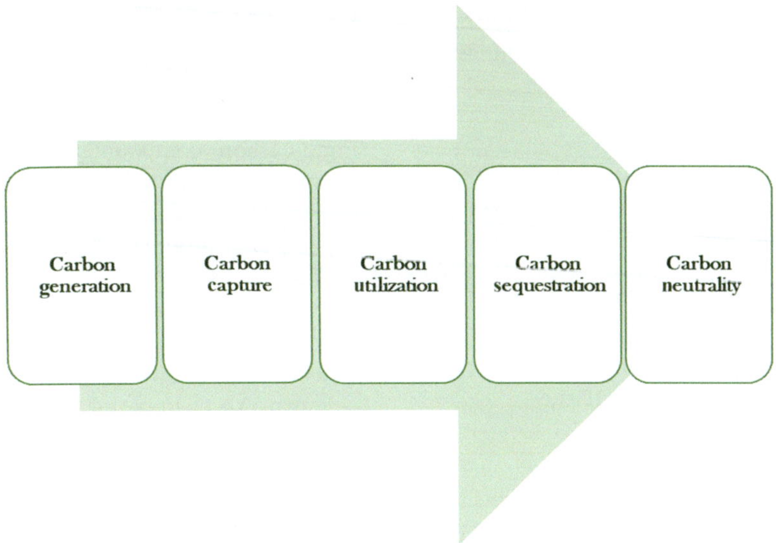

FIGURE 5.6 The carbon industry framework highlighting the flow from carbon sources to capture, utilization, and sequestration for carbon neutrality.

5.4 CO_2 SEQUESTRATION

CO_2 sequestration comes into play after the use of carbon capture technologies. The CO_2 captured from the atmosphere is then isolated and injected into a deep geological reservoir using engineering technology. CO_2 sequestration in deep saline formations and depleted reservoirs occurs through two main mechanisms:

1. Physical Storage
   - CO_2 is trapped in underground rock formations with impermeable cap rocks. This is known as the Geological structure storage.
   - Residual gas storage has the CO_2 held in tiny pores of rock by capillary forces, even after pressure is reduced.
2. Chemical Storage
   - Dissolution storage dissolves CO_2 into formation water, forming a stable solution.
   - Mineralization storage reacts the CO_2 with minerals in the rock to form solid carbonates, such as calcite or dolomite, permanently locking it away.

5.4.1 Types of Carbon Sinks

5.4.1.1 Terrestrial Carbon Sink

A terrestrial carbon sink is to the land's ability to absorb and store atmospheric carbon, with major carbon pools in the lithosphere, biosphere, and pedosphere. The lithosphere has the largest carbon pool on Earth and is estimated to hold approximately 9×10^6 TCO_2.

5.4.1.2 Ocean Carbon Sink

The ocean carbon sink is the ocean's capacity to capture and store atmospheric carbon through various mechanisms. The ocean stores approximately 93% of the Earth's CO_2. There are two main types of ocean carbon sequestration mechanisms: physical and biological. The partial pressure of CO_2 controls physical carbon sequestration; this mechanism involves moving CO_2 from surface waters to the deep sea using natural ocean currents. This process can lead to the formation of calcium carbonate, which settles on the seafloor as soft calcareous mud. Another physical mechanism is deep-sea storage, where stable CO_2 hydrates form and remain secure for thousands of years, even during major geological events. Marine organisms drive carbon sequestration. Algae, coral reefs, and shellfish absorb CO_2 through photosynthesis and other biological processes. Coral reefs, in particular, are notable for their high carbon storage capacity and ecological benefits, making them one of the most effective long-term oceanic carbon sinks.

5.4.2 Sequestration in Biomanufacturing

Biomanufacturing can use CO_2 to reduce and help fight climate change. This can be done by turning CO_2 into biofuels, biochemicals, and long-lived products that lock away carbon and promote a sustainable, climate-positive bioeconomy.

5.4.2.1 CO_2 Uptake

Plants and microbes use carbon-concentrating mechanisms (CCM) such as carboxysomes, pyrenoids, and malate-based pathways to increase CO_2. Point-source CO_2 offers CO_2 for microbial conversion but requires further study as impurities may inhibit bioconversion.

5.4.2.2 CO_2 Fixation Pathways

By utilizing CO_2 fixation, alternative natural and synthetic fixation pathways can be explored to achieve higher efficiency and reduced energy demand. Microbes, such as

E. coli, can be engineered to optimize these pathways. This approach has the potential to enhance or even replace the traditional Calvin–Benson–Bassham cycle.

5.4.2.3 Products from CO_2

Biofuels and chemicals that displace fossil-derived equivalents are known as carbon-neutral outputs. The major focus of these is on increasing yields and process efficiency. Carbon-negative outputs include durable products that store carbon long term, such as lignin-rich biomass, biochar, or construction materials made from biopolymers. There are plant systems engineered to deposit carbon in soils or accumulate stable compounds in harvestable tissues.

5.4.3 Environmental Benefits of Carbon Sequestration

Carbon sequestration is a key strategy in the mitigation of climate change as it captures and stores atmospheric CO_2 through biological, geological, oceanic, and technological methods. Its key environmental benefits are shown in Table 5.2.

TABLE 5.2
Key Environmental Outcomes Resulting from the Adoption of Eco-Friendly and Resource-Efficient Approaches

Benefit	Description
Climate change mitigation	• Reduces atmospheric CO2 levels, helping to slow global warming • Stabilizes climate patterns and enhances ecosystem resilience
Soil and health fertility	• Improves soil structure and organic matter content • Enhances microbial activity, nutrient cycling, water retention, and root development • Reduces erosion and increases resistance to climate stress, leading to better crop yields and sustainable farming
Biodiversity and ecosystem services	• Supports habitat protection and biodiversity conservation • Strengthen ecosystem functions and resilience.
Ocean health	• Helps reduce ocean acidification by increasing biological carbon uptake and ocean alkalinity • Supports marine ecosystems through blue carbon restoration and practices such as kelp farming
Air quality and human well-being	• Improves air quality by filtering pollutants through natural systems such as forests • Promotes healthier ecosystems and contributes to human health and environmental stability.

5.5 CASE STUDIES AND APPLICATIONS

5.5.1 C1 Biomanufacturing

This is a new way to make chemicals sustainably using single-carbon compounds such as CO_2, CO, methane, and methanol instead of fossil fuels. Scientists are designing microbes and enzymes that can convert these gases into valuable products such as bioplastics. The process is designed in such a way that it fits within renewable energy systems and supports the recycling of raw materials. Tools such as cost analysis and environmental impact assessments ensure its growth from small lab setups to large-scale use. An example of carbon-one biomanufacturing is through the production of 3-hydroxypropionic Acid. Two C1-based routes can be used: The Biological conversion of steel mill gas and hybrid electrochemical CO_2 to methanol with microbial synthesis (Zhang et al., 2025).

Advanced technologies are used to transform this one-carbon feedstock into valuable chemicals and materials. This is driven by advances in synthetic biology and metabolic engineering, which enable microorganisms to efficiently metabolize these inputs. Smart technology such as AI and digital simulation helps to improve how these systems work and track economic performance. These systems, coupled together, form methods that can be scaled up to fight climate change and support greener industries.

5.5.2 Continuous Batch Biomanufacturing

Continuous batch biomanufacturing offers a change in industrial production. The shift is from batch systems to steady-state operation for improved productivity and sustainability. Microbes such as *Halomonas* spp. thrive in extreme conditions, helping avoid contamination and the production of biofuels and bioplastics with great consistency. This method reduces waste, improves yields, and supports circular bioeconomy goals. Some case studies show that it can outperform traditional methods in both output and environmental impact (Del Hierro et al., 2024).

This innovation is driven by a range of smart technologies, including genetically engineered microbial strains engineered for long-term stability, advanced fermentation systems (chemostats and auxostats), and AI tools that simulate and optimize production, and biosensors. Modular bioreactors and renewable energy inputs also enhance versatility and ecological performance. Together, these technologies make it easier and more practical to scale sustainable production (Del Hierro et al., 2024).

5.6 CHALLENGES AND FUTURE PERSPECTIVE

5.6.1 Challenges

Sustainable biomanufacturing in I5.0 is a promising concept that could change the trajectory of climate change; it does have its challenges as these are all emerging concepts.

1. Technological fragmentation and integration difficulties: Many of the technologies of I5.0, that is, AI, IoT, and robotics are evolving rapidly but in isolation. The integration into cohesive, sustainable systems across sectors such as agriculture, energy, and textiles remains complex. In industries, the legacy infrastructure may hinder the adoption of new human-centric and eco-friendly technologies (Ghobakhloo et al., 2023).
2. Workforce preparedness and skill gaps: The success of this industry depends on a workforce capable of managing and co-working with smart technologies. Interdisciplinary education that blends engineering, digital skills, ethics, and environmental science is required. Another skill gap arises because the current training programs do not align with the multi-dimensional demands of I5.0.
3. Regulatory and ethical uncertainty: There is a lack of comprehensive policy frameworks to govern ethical AI use, data security, and sustainability impacts. The regulatory gaps hinder decision-making in areas such as circular supply chains, automation boundaries, and responsible innovation.
4. Economic and environmental trade-offs: There is no straightforward balance between the balancing of cost effectiveness with green transformation. High upfront investment in sustainable technologies and resistance from traditional industries slow the transition.
5. Cybersecurity and infrastructure vulnerabilities: Increased digitization expands the threat surface for cyberattacks, particularly in energy and water systems. This could jeopardize sustainability systems reliant on real-time monitoring and control.

5.6.2 Future Perspectives

1. Redesigning CO_2 Fixation Pathways
   Future work will increasingly rely on synthetic biology and systems engineering to construct more efficient, oxygen-tolerant, and energy-lean CO_2 fixation pathways. De novo metabolic routes and hybrid pathways that combine microbial, plant, and synthetic components offer new frontiers for carbon capture at high rates and yields (Lawrence Berkley National Laboratory, 2021).
2. Hybrid Biomanufacturing Platforms
   Emerging hybrid systems that integrate biological CO_2 conversion with electrochemical CO_2 reduction (to formate, methanol, or CO) can dramatically expand the efficiency and scalability of carbon-negative production. Realizing such platforms will require harmonizing biological and electrochemical rates and mitigating compound toxicity (Zhang et al., 2025).
3. I5.0-Enabled Bioprocess Intelligence
   The application of AI, edge computing, digital twins, and blockchain will enable real-time bioprocess control, traceability, and dynamic optimization. These technologies can reduce energy input, enhance resilience, and allow decentralized, modular production systems to emerge (Ghobakhloo et al., 2023; Sharma and Gupta, 2024).
4. Carbon-Negative Materials and Circular Bioeconomy
   There is strong potential to produce long-lived biomaterials (e.g., lignin-derived composites, bioplastics, biochar) that trap CO_2 in durable structures such as buildings or soils. Embedding such materials within circular supply chains can create stable carbon sinks while displacing fossil-based inputs (Wolgemuth, 2025).
5. Policy Innovation and Global Collaboration
   Achieving scale will require coordinated global policy frameworks to support infrastructure, Research and Development (R&D), and commercialization. Incentives for carbon-negative production, long-term carbon storage, and circular bio economies must be established. Collaboration between governments, academia, industry, and civil society will be essential (Rajkuma et al., 2025).

5.7 SUMMARY AND CONCLUSIONS

Using sustainable biomanufacturing in I5.0 changes how industries tackle both technological upgrades and environmental care. In the global effort to combat climate change and decarbonize industries, biomanufacturing offers not just a cleaner option but also a vital way to remove CO_2 from the atmosphere. I5.0 reshapes this field by integrating synthetic biology, AI-powered optimization, and collaborative human–machine systems. What was once a specialized process is now central to building a circular, sustainable economy, leveraging microbes that capture CO_2 and systems that turn waste into valuable products and materials. By placing human values and creativity at the core, I5.0 ensures that technological progress aligns with social responsibility and inclusivity.

Reaching the full potential of sustainable biomanufacturing depends on systemic support, such as strong policy frameworks, thorough environmental reviews, industry-wide teamwork, and steady funding for research and facilities. As the industry develops, tackling issues such as feedstock variability, regulatory hurdles, and market readiness will be essential. Ultimately, sustainable biomanufacturing in the era of I5.0 represents a fundamental shift that redefines the production of goods, embedding sustainability into the foundation. It offers a framework for industrial systems that meet global demands while respecting environmental limits, paving the way for a carbon-neutral, regenerative economy.

REFERENCES

Balar, Vidyut, Dimple Gor, Ankit Yadav, and Parth Bagadiya. 2023. "The evolution of biomanufacturing: A historical and contemporary perspective." *Advances in Agricultural Biotechnology*. New Delhi: Daya Publishing House® A Division of Astral International Pvt. Ltd. pp. 107–122.

Caineng, Zou, Wu Songtao, Yang Zhi, Pan Sonqi, Wang Guofeng, Jiang Xiaohua, Guan Modi, Yu Cong, Yu Zhichao, and Shen Yu. 2023. "Progress challenge and significance of building a carbon industry system in the context of carbon neutrality strategy." *Petroleum Exploration and Development* 50(1): 210–228.

Del Hierro, Ana G., Juan A. Moreno-Cid, and Eoin Casey. 2024. "Continuous biomanufacturing for sustainable bioeconomy applications." *Bioeconomy Journal* 4 1–14.

Gaur, Ashutosh. 2025. "The evolution of industrial revolutions: Progess Industry 1.0-Industry 5.0 manufacturing." *GLSU ATRPM 2025*. Ahmedabad: Research Gate. pp. 1–12.

Ghobakhloo, Morteza, Mohammed Iranmesh, Behzad Foroughi, Erfan Babaee Tirkolaee, Shahla Asad, and Azlan Amran. 2023. "Industry 5.0 implication for inclusive sustainable manufacturing: An evidence-knowledge-based strategic road map." *Journal of Cleaner Production* 1–16.

Kopperi, Harishankar. 2025. "Sustainable biomanufacturing: key to India's green transition." *Phys-Org*. May 19. https://phys.org/news/2025-04-sustainable-biomanufacturing-key-india-green.html

Lawrence Berkley National Laboratory. 2021. *Integrating CO2 into Biomanufacturing*. Californina: Berkley Lab.

Narkhede, Ganesh Bhoju, Bhavesh Nandranam Pasi, Neela Rajhans, and Atul Kulkarn. 2024. "Industry 5.0 and sustainable manufacturing: A systematic litereature review." *Benchmarking: An International Journal* 32(2), 1–28.

Rajkuma, N., Balusamy Nachiappan, Ann Matthews, V. Radha, C. Viji, and Judeson Antony Kovilpillai. 2025. "Industry 5.0: The human-centric future of manufacturing." *Challenges in Information, Communication and Technology*. Boca Raton: CRC Press (an imprint of Taylor & Francis Group). pp. 562–567.

Rame, Rame, Purwanto Purwanto, and Sudarno Sudarno. 2024. "Industry 5.0 and sustainability: An overview of emerging trends and challenges for a green future." *Innovation and Green Development* 3(4): 1–20.

Scown, Corinne. 2022. "Prospects for carbon-negative biomanufacturing." *Trends in Biotechnology* 40(12), 1415–1424.

Sharma, R., and H. Gupta. 2024. "Harmonizing sustainability in industry 5.0 era: transformative strategies for cleaner production and sustainable competitive advantage." *Journal of Cleaner Production* 445: 1–23.

Wang, Meiling, and Yuying Wang. 2025. "Toward a carbon neutral path: exploring the impact of digital economy on carbon neutrality." *International Journal of Climate Change Strategies and Management* 17, 1–22.

Wolgemuth, Roland. 2025. "Biomanufacturing as a key technology for a sustainable bioeconomy." *Biomanufacturing* 79 (5): 352–358.

Zhang, Chenyue, Qiang Fei, Rongzhan Fu, Maximilian Lackner, Yongjin Zhou, and Tianwei Tan. 2025. "Economic and sustainable revolution to facilitate one-carbon manufacturing." *Nature Communications* 16: 1–10.

6 Biomanufacturing for Sustainable Biomedical Engineering and Healthcare Industry 5.0

Sayan Chatterjee, Deepak Rathod, Shivangi Yadav, Annu Yadav, Snehpreet Kaur, and Priyanshu Rawat
University School of Biotechnology (USBT), Guru Gobind Singh Indraprastha University (GGSIPU), New Delhi, India

6.1 INTRODUCTION

The growing demand for sustainability and personalized medicine worldwide is transforming the need for the global healthcare system. Healthcare Industry 5.0, a novel approach replacing the traditional healthcare models that rely on the use of heavy resources or uniform treatments. This novel approach is an advanced version of Healthcare Industry 4.0, featuring automated and advanced technologies. This approach prioritizes a system that adheres to eco-friendly principles. Technologies such as cyber-physical systems, artificial intelligence (AI), or real-time data sharing are included in Healthcare Industry 4.0. These technologies transform the industrial and manufacturing sectors, as well as biomedical engineering. Such that, with the two faces of coins, this strategy has advantages with its efficiency and automation, but, in contrast, it has several limitations, including ignorance toward ecological sustainability, ethical concerns, and the needs of patients individually. With the focus on these resistances, Healthcare Industry 5.0 established a relationship between humans and machines as synergistic. This emphasizes several factors, such as resilience, inclusivity, and sustainable discovery and research (Figure 6.1) [1, 2].

Biomanufacturing is the heart or a central domain of this change of approach in the industrial sectors. Because the biomanufacturing industry uses the magic of biology, including living cells, enzymes, and biomaterials, for the production of bioproducts such as medicines, diagnostic tools, and bioderived healthcare solutions. In this sector, the growing demand for patient-specific treatments requires flexibility, uniformity, and sustainability, or eco-friendly aspects, particularly for autologous stem cell treatments, custom implants, and gene therapies. Moreover, the traditional biomanufacturing and pharmaceutical industry generate various biomedical waste products, which pose hazardous and harmful effects. This underlines the crucial demand for circular bioeconomy principles, which are based on eco-friendly bioprocesses for sustainability practices [3].

The transformation of Healthcare Industry 4.0 to Healthcare Industry 5.0 is a major advanced technological and automotive upgrade, with the key domain shifts in values and priorities. Healthcare Industry 4.0 is a full-fledged technology with automation, efficiency, and real-time data use, but major resistance in the healthcare sector due to the avoidance of hazardous effects on the environment, ethics,

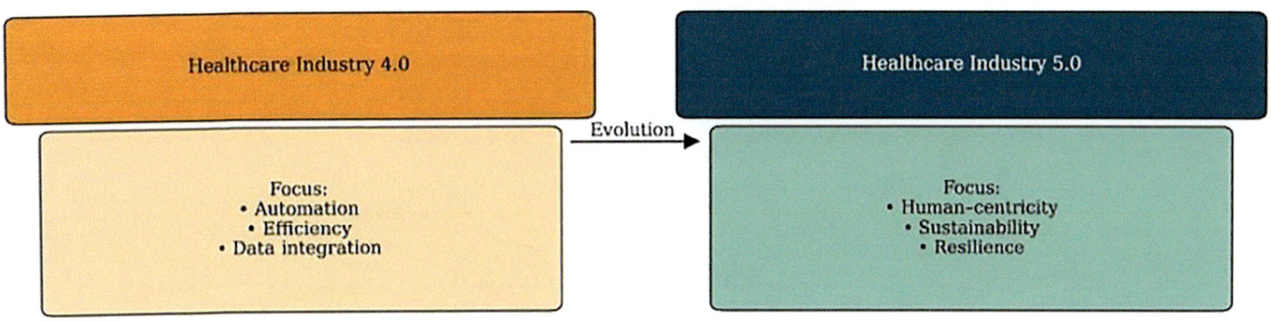

FIGURE 6.1 Healthcare Industry 4.0 vs. 5.0.

DOI: 10.1201/9781003637158-6

and human wellbeing. Healthcare Industry 5.0 overcomes all the resistance or limitations of Healthcare Industry 5.0 with AI, Collaborative Robots, and Digital Twins, etc., for enhancing human capabilities. Healthcare industry 5.0 supports the welfare of healthcare workers by addressing patient-specific needs and reducing environmental impacts through advanced technologies. This approach helps to provide flexibility, scalability, and eco-friendly solutions toward biomanufacturing and industrial sectors (Figures 6.2 and 6.3; Table 6.1) [7, 8].

This chapter elaborates on the transformation of biomanufacturing Industry with biomedical with Healthcare

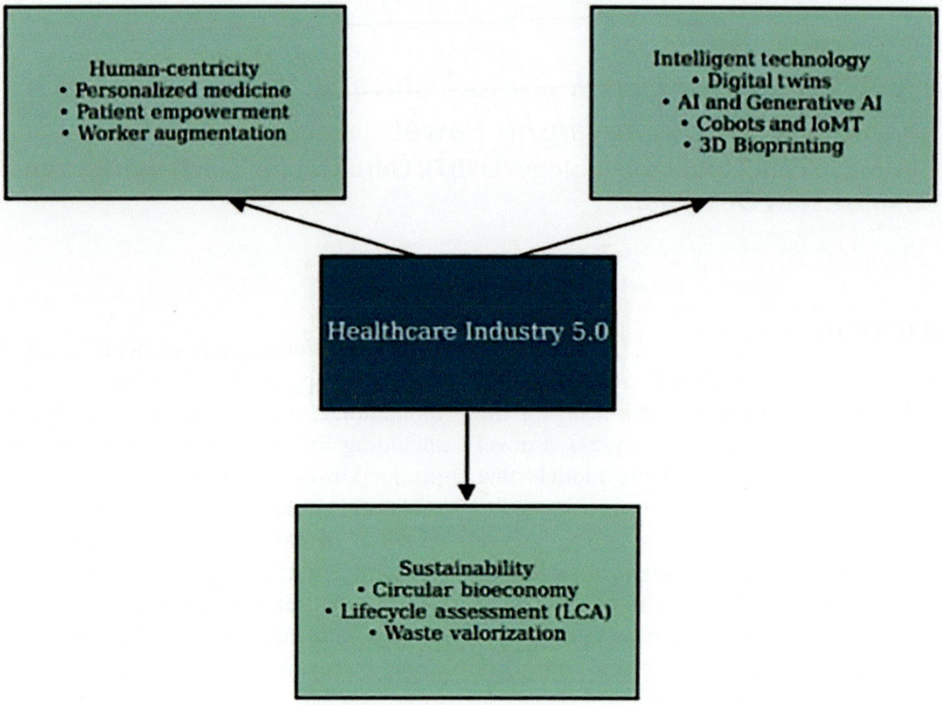

FIGURE 6.2 Pillars of Healthcare Industry 5.0 in biomanufacturing.

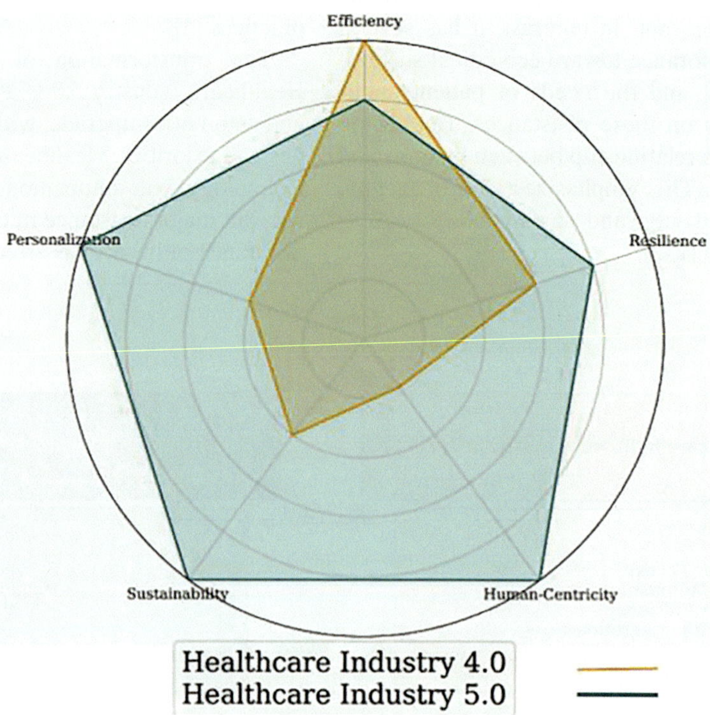

FIGURE 6.3 Comparison of Healthcare Industry 4.0 and Healthcare Industry 5.0 across key dimensions.

TABLE 6.1
Key Imperatives Driving the Transition to Healthcare Industry 5.0

Imperative	Focus Area	Framed Description	References
Personalization	Precision medicine	Enables scalable production of patient-specific therapies through flexible and adaptive biomanufacturing platforms.	[4]
Sustainability	Green and circular bioprocessing	Promotes the use of renewable resources, waste minimization, and circular bioeconomy principles to reduce environmental impact.	[5]
Empowerment	Human-centric innovation	Integrates Collaborative Robotics, intuitive digital tools, and ethical AI to enhance human skills, inclusivity, and workplace wellbeing.	[6]

Industry 5.0. This chapter highlights the key domains, including key technologies and methods, along with the main concepts. There are many crucial terms that need to be defined before understanding Healthcare Industry 5.0. Biomanufacturing is defined as the manufacturing of products such as therapeutic proteins, vaccines, and engineered tissues and materials that are derived from living systems, including microbes, mammalian cells, and plants. It includes upstream steps, such as growing cells in bioreactors, and downstream steps such as purification, formulation, and quality control. Biomanufacturing is far different from biofabrication—biofabrication includes three-dimensional (3D) Bioprinting, which is used to form cell-based 3D structures and tissues. These technologies intersect with each other but differ in terms of technology or applications [9]. Other crucial concepts include Circular Bioeconomy, which is based on the approach of waste-to-value or the utilization of renewable resources [10]; Collaborative Robotics or Cobots, which are robots designed to operate alongside humans [11]; and the Digital Twin, which refers to a real-time virtual model of a physical system for monitoring and simulation (Figure 6.4) [12].

6.2 THEORETICAL FOUNDATIONS OF HEALTHCARE INDUSTRY 5.0 IN BIOMANUFACTURING

From the perspective of biomanufacturing, the industrial shift from Healthcare Industry 4.0 to Healthcare Industry 5.0 in the Indian context, the main changes occurred in the structure, i.e., flexibility against rigidity, Sustainability against short-term efficiency. Healthcare Industry 5.0 integrates digital technologies for the automation process and data intelligence by incorporating human-centric principles that realign technological advancement with societal and ecological goals [13].

This type of industrial evolution in biomanufacturing results in three main areas: Human–Machine co-operation,

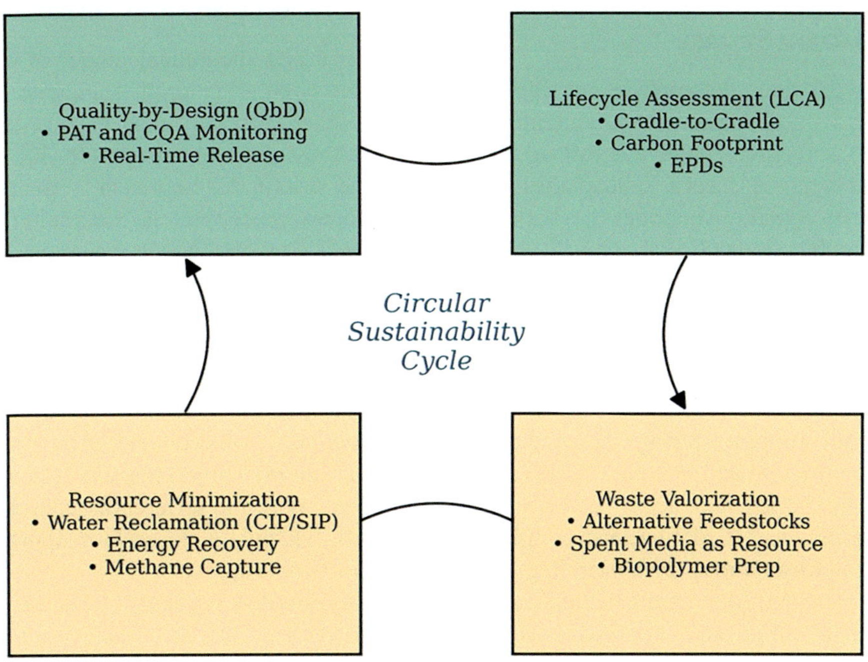

FIGURE 6.4 Design principles for sustainable biomanufacturing (QbD and circularity).

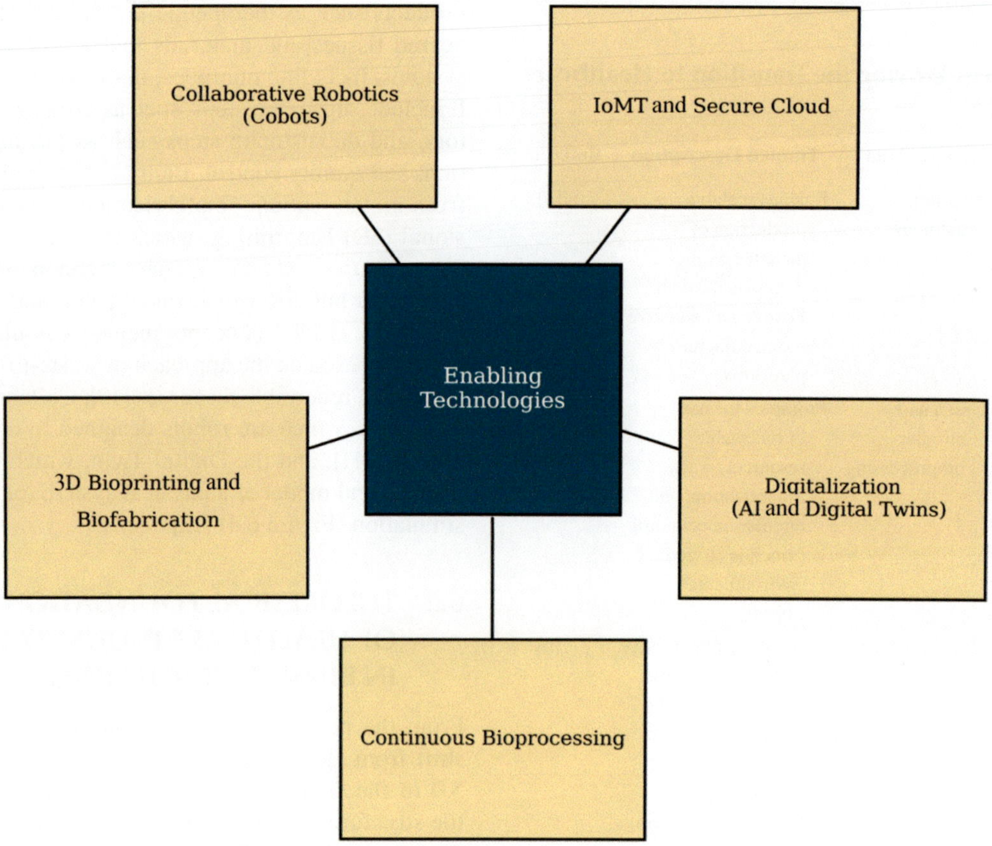

FIGURE 6.5 Core enabling technologies of Healthcare Industry 5.0.

environmental management through circularity, and resilience. Each area contributes to creating intelligent, adaptive, and inclusive production environments that meet the needs of next-generation healthcare systems (Figure 6.5).

6.2.1 Human–Machine Synergy

Healthcare Industry 5.0 does not replace our capability but enhances it. This is most important in biomanufacturing, where processes sometimes involve a soft and aseptic environment that requires accuracy, adaptability, and situation-specific decision-making qualities in which we as humans excel. The combination of Collaborative Robots (or cobots) enables manufacturers to achieve a perfect blend of machine efficiency and manpower. Cobots are now being designed to help in aseptic environments, fluid control, and tissue culture operations.

This integration reduces the chances of error and contamination that may be introduced during manual handling. Cobots' work is generally safe and increases the flexibility in the low-space laboratories and helps in maintaining Good Manufacturing Practices (GMPs). However, this integration needed an interdisciplinary understanding of the manpower. The scientist or workers must be familiar with automation processes, requirements for an aseptic environment, and, in the same way, the data analyst must know about the biomedical processes. Hence, Healthcare Industry 5.0 requires interdisciplinary training programs [1]. This convergence of knowledge enhances teams that have knowledge from both fields.

6.2.2 Circularity and Environmental Stewardship

In this era, a theoretical model of circular economy has become popular as an alternative to conventional production, which also considers beyond the boundaries within which these productions operate [10].

The sustainable approach is the reduction or reuse of bioprocess waste, such as used culture media for chemical synthesis or biopolymer production, by lowering the environmental burden, and may reduce the cost of waste management [5]. Similarly, microbial biomass residues can be converted into biofertilizers or feedstock for secondary fermentation processes. To assess the environmental impact of such circular interventions, Lifecycle Assessment (LCA) methodologies are used. Traditional cradle-to-grave models analyze environmental impacts from resource extraction to disposal. With the differential contrast to cradle-to-cradle models, which is a closed-loop approach, executed with a waste-to-value approach [14].

These models are mostly relevant in the design of biomanufacturing facilities where zero-waste and carbon-neutral targets are the market expectations. Furthermore, eco-design tools such as green chemistry metrics, bio-based

TABLE 6.2
Resource Reduction in Continuous Bioprocessing vs. Batch Processing

Metric	Resource Reduction (Up to)	Source (from chapter)
Water consumption	50%	Section 6.3.1 [15, 16]
Energy consumption	30%	Section 6.3.1 [15, 16]
Buffer and resin use	"Significantly reduce"	Section 6.3.1 [15, 16]

Data provides a quantitative comparison of resource efficiency.

input tracking, and clean energy integration help incorporate sustainability from the process development stage. It also aligns with United Nations Sustainable Development Goals (UNSDGs), particularly these below SDGs (Table 6.2).

6.2.3 Supply Chain Resilience and Agility

The COVID-19 pandemic has demonstrated that businesses do not work in a crisis. Using a plug-and-play bioreactor system that depends on interchangeable, single-use technology, which reduces the time frame of factory setup. We can scale production or switch products rapidly, ensuring that when the next outbreak hits, our response is immediate, not delayed [17].

It uses a blend of virtual models that predict failures and optimize maintenance in real time. It uses AI and sensors for testing and spotting problems before they happen in the real world. It does not just prevent breakdowns, it makes the whole process more transparent for regulators and supports Quality-by-Design (QbD) initiatives [12].

Agility is achieved by shifting dependencies from giant plants to the network of smaller microfactories in nearby areas. These hubs can respond instantly to local needs and help the planet by cutting out the need for long-distance shipping. Together, these tools build a healthcare system that can survive a crisis without sacrificing safety [3].

6.3 CORE ENABLING TECHNOLOGIES

With required harmonized innovation across physical, digital, and biological domains, it is a comprehensive and systemic transformation of the biomedical production landscape in the implementation of Healthcare Industry 5.0 within biomanufacturing. It enables real-time adaptability, patient-centric customization, operational sustainability, and resilient production architectures through a growing ecosystem of technologies. To achieve the goals of Healthcare Industry 5.0 in biomedical contexts, this section elaborates on the five core technology domains central to this transformation, each presented both as a standalone innovation and a part of an integrated framework.

6.3.1 Continuous Bioprocessing Architectures

As an alternative to the traditional batch manufacturing model in the production of biologics, cell therapies, and biosimilars, Continuous Bioprocessing has emerged. Improving overall throughput and product consistency can be achieved with the advantages of maintaining constant input and output streams. Alternating tangential flow filtration (TTF) or acoustic separators are Perfusion-based cell retention devices that offer the highest level of continuity, maintaining high-density cultures while continuously harvesting the product. Low residence time variability and consistent product quality in monoclonal antibody production and vaccine manufacturing make these setups particularly beneficial. While allowing for scheduled harvests and media refreshes, balancing process control with operational simplicity, semi-continuous systems can be used, although not fully uninterrupted. They are specifically used when full perfusion infrastructure is not yet feasible, such as in early-stage development pipelines or resource-limited settings. Through quantitative studies, it has been shown that continuous processes consume up to 50% less water, 30% less energy, and significantly reduce buffer and resin requirements compared to batch processes. They reduce facility footprints, support single-use technologies, and facilitate more agile product switching, which are critical in multi-product, multi-market biomanufacturing facilities [15, 16].

6.3.2 Advanced Digitalization and Predictive Analytics

To offer the capacity to visualize, simulate, and optimize every node of the biomanufacturing value chain, Digitalization is the nervous system of Healthcare Industry 5.0 to achieve this. Digital Twin, a continuously updated, physics- and data-based model that reflects the real-time state of bioprocess operations, is one of the most promising tools in this regard. It allows for scenario testing (e.g., nutrient feed strategy, temperature shifts, impurity buildup) prior to physical implementation, reducing development time and material waste. In areas where small-batch, patient-specific products must be manufactured under extremely tight timelines and quality constraints, this is particularly vital in personalized medicine applications [12].

Adding another layer of intelligence is predictive analytics powered by machine learning (ML). For example, before they manifest physically, recurrent neural networks can forecast the growth curve of microbial or mammalian cell cultures, flagging potential deviations. Similarly, to predict membrane fouling in TFF systems, optimize downstream chromatography steps, or anticipate deviations in Critical Quality Attributes (CQAs), ML can be applied. Collaborative development across academic, industrial, and regulatory stakeholders, multi-site coordination, and global data sharing are enabled by integrating these systems with cloud platforms. Areas needed process knowledge evolves collectively and continuously by enabling this networked innovation [1].

6.3.3 Internet of Medical Things (IoMT) and Secure Cloud Integration

Due to demands for sterility, compliance, and patient safety, a subset of the broader Industrial Internet of Things (IIoT) is known as the IoMT. IoMT encompasses a dense web of environmental monitors, embedded sensors, smart instruments, and wearable devices within manufacturing suites in biomanufacturing. Before transmission to centralized dashboards for dynamic oversight, these devices generate real-time streams of temperature, metabolite levels, pressure, dissolved oxygen, and equipment performance. The resulting automated, self-aware production environment is capable of initiating corrective actions or notifying operators before failures occur. The need for auditable, immutable, and secure records is growing as biomedical data scales and becomes more sensitive. Blockchain, as a regulatory–compliant solution, is increasingly being explored to address this challenge. It ensures trust, transparency, and traceability in biomanufacturing workflows by timestamping every transaction - from raw material input to final product release - on an encrypted, decentralized ledger. New opportunities for counterfeit detection, provenance tracking, and smart contract enforcement in globalized supply chains dealing with controlled substances and high-value biologics are offered by combining blockchain with AI [18, 19].

6.3.4 Collaborative Robotics and Automated Downstream Processing

In aligning ergonomics with precision, particularly in environments requiring aseptic handling, Collaborative Robots (or cobots) play a pivotal role. They are equipped with force-limiting technology, vision systems, and AI-based movement control, unlike conventional robotic arms, making them inherently safe to operate near human technicians. Sterile-compatible cobots equipped with cleanroom-rated end-effectors can handle fluid transfers, plate loading, or even operate automated colony pickers in aseptic operations. Applicable increasingly in areas like modular cell therapy production, where patient-specific batches require labor-intensive manual steps that can be standardized through robotics [6].

Automated systems are now being used to control membrane separations, such as TFF or depth filtration on the downstream side. A sustainable automated alternative to organic solvent extraction, i.e., aqueous two-phase systems (ATPS), is now being integrated into robotic workflows for the gentle, non-denaturing recovery of proteins and nucleic acids. It reduces the operator exposure to biohazards, improves reproducibility, and shortens batch turnover time [3].

6.3.5 Three-Dimensional Bioprinting and Biofabrication

The most transformative enabling technology for the fabrication of living tissues with anatomical precision is 3D Bioprinting. It offers spatial control over cell deposition, material gradients, and microenvironmental cues, unlike traditional scaffold-seeding techniques. Bio-ink formulation—particularly its viscosity, shear-thinning behavior, crosslinking ability, and cell viability support—is the factor on which the success of a Bioprinting process depends heavily. Widely used as bio-ink bases are hydrogels such as alginate, gelatin-methacrylate (GelMA), and decellularized extracellular matrix (ECM) components. By optimizing nozzle pressure, temperature, and stage movement algorithms, print fidelity is maintained [20].

To mimic the body's vascular systems by delivering nutrients and removing metabolic waste, constructs require maturation in perfusion bioreactors after printing. They improve oxygenation, promote ECM formation, and help develop physiologically relevant tissue architectures—factors important for applications in personalized implants, organ-on-chip models, and high-throughput drug screening. Patient-derived cells, smart hydrogels, and AI-driven design software, combined with the continuous evolution in 3D Bioprinting, hold the potential to revolutionize personalized regenerative medicine in a truly sustainable and scalable manner [9].

6.4 DESIGN PRINCIPLES FOR SUSTAINABLE BIOMANUFACTURING

There is no choice but to adopt sustainability in the biomanufacturing sector. Because it is a central domain of the transformative era of Healthcare Industry 5.0. Executions in the field of personalized medicine, regulatory requirements, or the responsibility toward the environment. Established the importance of sustainability or eco-friendly aspects in the manufacturing industry. This shift calls for a new engineering mindset: one that embeds quality, efficiency, circularity, and environmental stewardship from the earliest stages of process development. This section outlines four foundational design principles that collectively enable next-generation, sustainable biomanufacturing platforms: QbD methodologies, LCA frameworks, resource and energy optimization strategies, and waste valorization through innovative feedstocks [21].

6.4.1 Quality-by-Design (QbD) Methodology

The QbD framework is a science- and risk-based approach to pharmaceutical development and manufacturing, championed by regulatory bodies such as the US Food and Drug Administration (FDA) and European Medicines Agency (EMA). QbD aims to ensure product quality and consistency by deepening understanding of process variables and their influence on product attributes. At the heart of QbD are CQAs and Critical Process Parameters (CPPs). CQAs are a combination of physical, chemical, biological, or microbiological properties of the final product that must be controlled to ensure efficacy and safety, such as glycosylation profiles in monoclonal antibodies or cell viability in gene therapies. CPPs are the operational parameters—such as temperature, pH, agitation rate, or nutrient feed rate—that must be controlled within defined limits to meet the target CQAs [21, 22].

Modern QbD implementations rely heavily on Process Analytical Technology (PAT) tools to monitor these parameters in real time. Technologies such as Raman spectroscopy, inline turbidity sensors, dielectric spectroscopy, and near-infrared (NIR) probes provide continuous feedback on cell density, metabolite levels, and impurity buildup. The use of these tools supports real-time release testing (RTRT), minimizing the need for post-production quality checks and reducing production time and cost. By embedding QbD into the design phase, manufacturers can build robust, fault-tolerant processes that are inherently scalable, compliant, and adaptable to both standard and patient-specific manufacturing pipelines [23].

6.4.2 Lifecycle Assessment (LCA) Frameworks

An LCA is a significant tool used for the evaluation of environmental impacts in the biomanufacturing process all over the world. It is sourced by the utilization of raw materials, considering the end-of-life disposal with recycling. This holistic analysis enables informed decision-making that aligns sustainability with performance and profitability. In the context of Healthcare Industry 5.0, LCA methodologies are being applied not only to quantify carbon footprints but also to benchmark conventional batch bioprocesses against continuous, modular systems. These comparative LCAs often reveal significant reductions in greenhouse gas emissions, water usage, and solid waste output when adopting Healthcare Industry 5.0-aligned architectures (Figures 6.6 and 6.7) [24, 25].

The effectiveness of an LCA depends on the precise delineation of system boundaries. For example, cradle-to-gate assessments may examine emissions up to the product packaging phase, while cradle-to-grave assessments include post-consumer waste and emissions from transportation and incineration. Cradle-to-cradle models—increasingly relevant in circular bioeconomy discussions—envision waste as a resource for new product lifecycles, encouraging the design of biodegradable or reusable materials. LCA outcomes also support the standardization of Environmental Product Declarations (EPDs), enabling transparent communication of sustainability metrics to regulators, investors, and consumers [14].

6.4.3 Resource Minimization and Energy Recovery

Modern biomanufacturing facilities consume vast quantities of water and energy, particularly in clean-in-place (CIP) and steam-in-place (SIP) operations. Sustainable design demands systems that minimize this resource use without compromising sterility or quality. Water reclamation systems, such as multi-stage filtration, ultraviolet (UV) disinfection, and reverse osmosis, are being integrated into CIP/SIP circuits to enable reuse of rinse water and condensate streams. Closed-loop systems can recover up to 70% of cleaning water, significantly reducing freshwater intake and wastewater discharge [5]. In terms of energy, heat-exchange technologies are employed to capture thermal energy from sterilization and fermentation exhaust streams. Recovered heat can be used for media preparation or facility heating, cutting down energy costs and greenhouse gas emissions. Additionally, methane capture

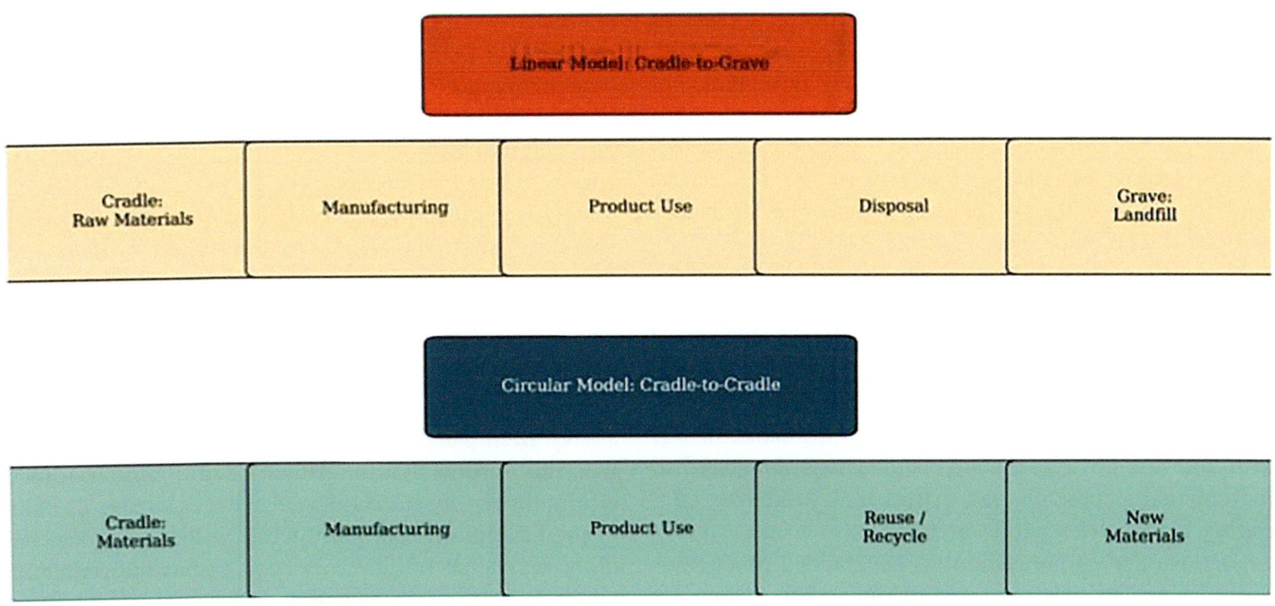

FIGURE 6.6 Cradle-to-cradle Lifecycle Assessment (LCA) framework.

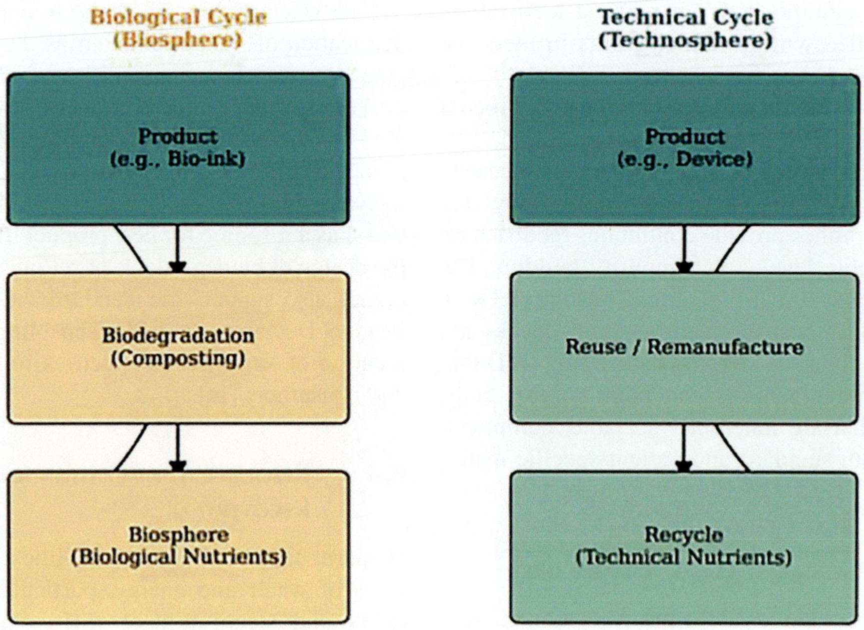

FIGURE 6.7 Biological and technical cycles in a circular economy framework.

from anaerobic bioprocess off-gases can power biogas generators, contributing to the energy autonomy of production plants. Such strategies are vital not only for environmental goals but also for economic sustainability, as energy costs often represent a major share of total production expenses in biologics manufacturing [26].

6.4.4 Waste Valorization and Alternative Feedstocks

The biomanufacturing process generates substantial quantities of organic waste, including spent culture media, cell debris, and biomass byproducts. Traditionally treated as waste, these materials are now being recognized as valuable bioresources under circular economy principles. For instance, residual microbial or algal biomass can be processed into biofertilizers, animal feed supplements, or biopolymer precursors, depending on their composition. These applications close the resource loop, generate secondary revenue streams, and reduce disposal costs [3].

In parallel, the exploration of alternative feedstocks offers new frontiers for resource sustainability. Nontraditional substrates such as algal biomass, agricultural residues, and insect protein hydrolysates are being investigated as carbon and nitrogen sources for microbial fermentation. These materials are often locally available, of low cost, and less environmentally burdensome compared to conventional glucose or yeast extract [27]. Moreover, integrating biorefineries within or adjacent to manufacturing sites allows for co-production of value-added products—such as pigments, enzymes, and organic acids—alongside the primary biologic, further enhancing the economic and ecological viability of the operation.

6.5 ON-SITE AUTOLOGOUS CELL THERAPY PRODUCTION

6.5.1 Closed-System Workflows from Leukapheresis to Final Product Within Clinical Settings

To provide patient-specific, safe treatments in a controlled and efficient manner, on-site autologous cell therapy production has been developed. Circulating mononuclear cells are collected from the patient as the initial source material in the initial workflow of the leukapheresis process for cell-based therapies. This methodology is widely integrated into transfusion medicine departments and clinical hematology, and it provides a critical starting point for hematopoietic stem cell transplantation, Chimeric Antigen Receptor (CAR)-T cell therapy, and other immune-modulating interventions, based on cell-based interventions. Before further processing, the once collected leukapheresis product must undergo a rigorous quality assessment. To determine whether the material is suitable for manufacturing, clinical centers that have implemented these closed-system models routinely perform sterility tests, cell counts, and viability assays. Before entering the controlled manufacturing pipeline, this ensures that the patient's starting material meets minimum quality standards. Advantages include its ability to preserve product sterility while reducing manual handling, which has traditionally been a source of contamination risk. Expansion, genetic modification, cell activation, and final formulation can occur in a seamless chain of processes without exposure to the external environment using automated bioreactors and modular, cleanroom-integrated devices [28].

Mimicking conventional bioprocessing steps at a scale appropriate for hospital-based use, the structure of these workflows relies on specialized equipment designed. For example, T cell activation using antibody-coated beads, viral vector transduction, and subsequent cell culture expansion in sterile bags or cartridges can be performed using closed-system instruments rather than open culture flasks. Controlled in-line sterile connectors and filters will allow each of these steps to remain under aseptic conditions. Cells are harvested, washed, and formulated into the final clinical product, often suspended in a cryopreservation medium or an infusion-ready solution after sufficient expansion. Either the product is cryopreserved in controlled-rate freezers for future use or prepared fresh for immediate infusion back into the patient. In-process controls, such as viability assays, sterility tests, and immunophenotyping, are incorporated at each stage of the process to ensure compliance with Current Good Manufacturing Practice (cGMP) standards. The risk of introducing unsafe or sub-therapeutic products into the patient care pathway is significantly reduced by embedding these checks directly into the workflow [28]. It is evident that reduced logistical dependency on external manufacturing facilities within healthcare centers is a clinical advantage. Hospitals can process the material on-site and return the therapeutic product to the same patient within a shorter time frame rather than sending collected apheresis products to off-site commercial facilities, where samples are subjected to variable environmental conditions, shipping delays, and inter-institutional communication barriers. Cryopreservation requirements, variability in shipping temperatures, and thawing protocols introduce significant risk into the treatment process in conventional off-site models. By contrast, closed systems simplify the chain of custody and minimize the opportunity for errors by eliminating the need for intermediate storage and shipping. In addition, it enables hospitals to maintain alignment with regulatory expectations and to maintain detailed records of the production process for documentation and product traceability through the use of automated, digitally integrated platforms. These factors collectively demonstrate that closed-system workflows represent technological innovation and a paradigm shift in how personalized medicine can be delivered safely and efficiently in clinical practice [28].

6.5.2 Comparative Analysis: Turnaround Time, Logistic Complexity, and Patient Outcomes

The turnaround time from leukapheresis to final infusion is one of the principal considerations in evaluating the benefits of on-site autologous cell therapy production. The interval between collection and infusion can span several weeks in centralized, off-site manufacturing models. The reason is that scheduling shipping logistics, batch production cycles, cross-institutional communication, and quality release testing performed at distant facilities are affected by cumulative delays. Contrastingly, within hospitals or academic medical centers, on-site closed-system workflows significantly shorten this interval.

Clinical reports demonstrate that patients can receive their infusion within a matter of days rather than weeks when cell collection, genetic modification, expansion, and formulation are conducted within the same institution. For patients with rapidly progressing malignancies, where treatment delays can compromise therapeutic efficacy, this reduced turnaround is particularly advantageous. Hospitals can streamline the production timeline and better align therapy delivery with urgent clinical demands by eliminating the need for external shipment and off-site scheduling [28].

Logistic complexity plays a critical role in shaping the feasibility of autologous cell therapy delivery, in addition to the advantage of time. Courier services, specialized cry shipping containers, and constant monitoring to ensure maintenance of product integrity are included in highly coordinated chains of custody, which are in demand in off-site models. Risks to the stability of the cellular product include variations in temperature, delays at customs checkpoints, or miscommunication between institutions. Moreover, to mitigate these risks, strict standardization of shipping protocols, extensive staff training, and significant financial investments are necessary for off-site processes. Maintaining control of the product within a single institutional infrastructure through on-site closed-system manufacturing reduces these logistical burdens. It ensures that the same team involved in collection also oversees production and infusion, thereby reducing the risk of errors related to shipping mishandling and preserving continuity of care. This streamlined workflow reduces reliance on third-party vendors and the need for dedicated shipping coordinators, thereby improving operational efficiency through resource allocation [28].

Perhaps the most compelling evidence for the superiority of on-site closed-system workflows is patient outcomes. While waiting for their therapeutic product, patients with aggressive malignancies are less likely to experience disease progression by reducing turnaround time. Furthermore, enhancing cell viability, potency, and functional activity ultimately improves clinical response rates by avoiding extended cryopreservation and long-distance transport. Due to reduced treatment delays, simplified scheduling, and improved communication with their care teams, patients treated in centers that deploy on-site workflows report higher satisfaction. It allows clinicians to benefit from more predictable treatment timelines to optimize adjunct therapies, manage conditioning regimens, and minimize hospitalization duration. It improved therapeutic efficacy and survival outcomes in patients undergoing CAR-T therapy, underscoring the clinical relevance of logistical streamlining, which correlates with reduced time to infusion through several studies. Integration of closed-system workflows taken together into healthcare institutions reduces complexity and translates into measurable benefits in patient prognosis and overall treatment experience [28].

6.5.3 Mobile mRNA Vaccine Manufacturing Units

A transformative response to global needs for rapid vaccine production during outbreaks and pandemics has been achieved through the emergence of rapid-deployment modules for lipid nanoparticle (LNP) formulation and fill-and-finish mobile mRNA vaccine manufacturing units. Where conventional large-scale biomanufacturing plants are unavailable or insufficient, these units are designed as modular, transportable facilities that can be deployed to those regions. Their ability to conduct mRNA synthesis, purification, LNP formulation, and final fill-and-finish operations within compact, automated platforms. The technological foundation of such units lies in this. As it facilitates cellular uptake, stability, and protects the fragile RNA molecule from enzymatic degradation, LNP formulation represents the most critical stage of mRNA vaccine production. Using microfluidic mixing devices, this step is performed, allowing precise, reproducible assembly of mRNA and lipid components under aseptic closed conditions, within mobile units. Such systems are manipulated and optimized using the scale-down strategy for industrial processes. Using the maintenance of GMP compliance, which facilitates the design of vaccine doses for the decentralization of environments that meet the quality regulatory frameworks for the centralized plants. The rapid-deployment nature of these modules enables local production within days of arriving on-site, drastically reducing reliance on global supply chains and enhancing resilience in the face of public health emergencies [29]. They have the capability to conduct mRNA synthesis, purification, lipid nanoparticle production, and automated platforms, and the technological foundation for such units lies in this.

Because it offers cellular uptake, stability with protections to the RNA fragility molecules, and enzymatic degradation. For this, LNP formulations represent the most significant phase of mRNA modular vaccine production. Using microfluidic mixing devices, this step is performed, allowing precise, reproducible assembly of mRNA and lipid components under aseptic, closed conditions within mobile units.

The fill-and-finish stage, which involves transferring formulated vaccine into sterile vials or syringes, is another core component of mobile manufacturing units. Traditionally, fill-and-finish is a bottleneck in vaccine supply due to its high infrastructure requirements and the need for specialized personnel. Mobile units address this challenge by integrating automated filling systems equipped with robotic arms and single-use technologies. These systems enable high-throughput sterile packaging while minimizing contamination risks and operator intervention. Moreover, the use of disposable assemblies reduces cleaning validation requirements and accelerates batch changeover times. This adaptability is crucial during pandemic situations where demand for vaccine doses fluctuates rapidly, and the ability to reconfigure production lines for different formulations is essential. In addition to supporting rapid response, mobile units democratize vaccine production by enabling developing countries and remote regions to establish localized vaccine production without relying solely on imports. This has profound implications for global health equity, as it reduces the time lag between vaccine development in industrialized nations and availability in low-resource settings [29].

The effectiveness of rapid-deployment modules also lies in their scalability and interoperability. Each unit can be deployed as a standalone production site or as part of a larger network of distributed facilities that collectively contribute to regional or global supply. For example, multiple mobile factories can be stationed across continents to reduce transport times, synchronize output, and rapidly meet surges in demand during epidemic waves. Furthermore, digital process control systems enable remote monitoring, predictive maintenance, and real-time data sharing with regulatory bodies, enhancing transparency and compliance. These features highlight the dual role of mobile vaccine manufacturing units as both technological innovations and public health instruments. They not only expand the physical capacity for vaccine production but also create a flexible, distributed network that mitigates vulnerabilities inherent to centralized manufacturing. As such, rapid-deployment modules for LNP formulation and fill-and-finish represent a pivotal shift toward decentralized, resilient, and equitable vaccine manufacturing [29].

6.5.4 On-Site Lyophilization to Obviate Cold-Chain Dependencies

One of the major challenges in mRNA vaccine distribution is maintaining cold-chain infrastructure to preserve the stability of RNA molecules and lipid LNPs. During the COVID-19 pandemic, the novel discovery of mRNA vaccines has required storage at ultra-low temperatures, sometimes as low as –80°C, placing enormous strain on logistics and limiting accessibility in low-resource settings. To address this, mobile mRNA manufacturing units are increasingly equipped with on-site lyophilization capabilities. Lyophilization, or freeze-drying, involves removing water from the vaccine formulation under controlled freezing and vacuum conditions, resulting in a dry powder that is significantly more stable at ambient or refrigerated temperatures. Incorporating lyophilization into mobile units allows freshly produced vaccine batches to be stabilized on-site, thereby eliminating the need for ultra-cold storage and extensive refrigeration during transport. This innovation not only reduces logistical costs but also enables equitable vaccine access in regions where cold-chain infrastructure is underdeveloped or unreliable [29].

On-site lyophilization offers additional operational benefits by extending vaccine shelf life and facilitating stockpiling for emergency preparedness. Freeze-dried mRNA vaccines can often be stored for months or even years at standard refrigeration temperatures, creating opportunities for governments and health agencies to maintain strategic reserves. This is particularly advantageous in regions vulnerable to emerging infectious disease outbreaks, where rapid vaccine deployment can contain epidemics at an early stage. Furthermore, lyophilized vaccines simplify last-mile distribution, as they can be reconstituted with sterile water or diluent immediately prior to administration at local clinics. In mobile units, this process is supported by automated lyophilization chambers designed with single-use trays and in-line quality monitoring systems. Such technologies ensure that each batch meets sterility and potency requirements, thereby upholding regulatory standards while maintaining production agility. The synergy between mobile deployment and lyophilization reflects a broader trend in pharmaceutical manufacturing toward decentralization and resilience against infrastructural limitations [29].

By obviating cold-chain dependencies, mobile mRNA manufacturing units with integrated lyophilization capabilities directly address one of the most pressing barriers to vaccine equity. In prior global vaccination campaigns, countries with limited cold-chain infrastructure experienced delays in receiving and distributing mRNA vaccines, resulting in uneven immunization rates and prolongation of pandemic waves. The integration of on-site lyophilization ensures that vaccines produced locally are also distributed locally, without the logistical bottlenecks associated with frozen storage and long-distance shipping. This approach not only improves vaccine access but also empowers regional health systems to achieve self-sufficiency in immunization campaigns. Ultimately, the convergence of mobile production and cold-chain independence exemplifies how technological innovation can be harnessed to create a more equitable and responsive global health landscape [29].

6.5.5 EMA/FDA Guidance for Decentralized GMP Implementation

The transition toward decentralized, mobile manufacturing models for advanced therapies and mRNA vaccines has raised critical questions about regulatory compliance with GMP standards. The FDA and the EMA, both regulatory frameworks, have issued guidance documents emphasizing that GMP principles remain mandatory regardless of whether production occurs in large-scale centralized facilities or smaller decentralized environments. For instance, the FDA has underscored that decentralized production units must demonstrate full control over aseptic processing, raw material qualification, and in-process monitoring to ensure consistency with established safety and efficacy profiles. Similarly, the EMA has highlighted the need to maintain product comparability across distributed manufacturing sites, emphasizing harmonization of standard operating procedures, quality assurance frameworks, and digital traceability systems. These requirements are particularly stringent for biologics, such as autologous cell therapies and mRNA vaccines, where product heterogeneity and sensitivity to process parameters can directly influence clinical outcomes. Therefore, regulatory guidance emphasizes robust validation of modular and mobile systems to confirm that they can consistently deliver products of equivalent quality to those manufactured in conventional facilities [30].

In addition to technical requirements, regulatory bodies stress the importance of documentation, digital oversight, and real-time quality control in decentralized GMP implementation. Mobile and hospital-based manufacturing units must integrate advanced digital infrastructure to monitor production variables, track deviations, and record batch data in compliance with 21 CFR Part 11 (for FDA) and Annex 11 (for EMA). These frameworks mandate that electronic records be secure, validated, and auditable, thereby ensuring data integrity across distributed sites. Regulatory agencies are increasingly exploring RTRT to accelerate product availability while maintaining quality assurance, particularly in time-sensitive contexts such as pandemic vaccine deployment. By embedding advanced analytical tools and sensor technologies into mobile units, manufacturers can generate immediate quality metrics, reducing reliance on centralized laboratory testing. This regulatory flexibility, while encouraging innovation, is balanced by strict requirements that demonstrate that RTRT methods are validated, reproducible, and equivalent to conventional assays. In this way, both the FDA and EMA are shaping a regulatory environment that supports decentralized GMP implementation without compromising product quality or patient safety [30].

6.5.6 Harmonization Challenges Across Multinational Jurisdictions

The FDA and EMA have well-established GMP regulatory frameworks for decentralized manufacturing to differentiate among global regulatory standards. This creates a major obstacle to harmonization. Variability in the number of regulatory frameworks all over the world limits the implementation of mobile manufacturing units. Some real-world examples include that the FDA offers product release based on real-time data analysis and in-process testing. However, the EMA regulatory framework requires an extra test for confirmations before the approval of the batch protocol. These differences and variability increase the complexity and slow the pace of global implementation of decentralized manufacturing strategies. This creates a burden on manufacturers for implementing these frameworks or strategies, causing a delay in the decentralization of manufacturing technologies. Inconsistent definitions of "batch" and "lot" across regions also make product tracking difficult for cross-border modular manufacturing. As a result, international vaccine distribution is slowed, and the scalability of mobile manufacturing platforms is limited [30].

Beyond technical requirements, broader political and legal frameworks also influence regulatory harmonization. Multinational treaties, intellectual property regulations, and supply chain security laws often differ significantly between regions, creating obstacles for the free movement of decentralized manufacturing units and their products. For instance, technology transfer agreements that govern the use of proprietary LNP formulations may be interpreted differently in the United States, Europe, and low- to middle-income countries, complicating the establishment of globally interoperable mobile production systems. There are many obstacles to the establishment of complexity toward mobile vaccine development, such as country-specific rules for cold-chain storage, safety reporting, and labeling. For resolving these obstacles, there are many requirements, such as coordination among regulators, industry, and global health organizations for the establishment of harmonized standards and regulatory frameworks. Without GMP decentralizations, manufacturing units find it difficult to fulfill their global potential [30].

A final dimension of the harmonization challenge lies in adapting regulatory frameworks for emergency use. During the COVID-19 pandemic, regulators introduced accelerated pathways such as the FDA's Emergency Use Authorization (EUA) and the EMA's Conditional Marketing Authorization (CMA). These standard regulatory frameworks allow rapid development of vaccines, giving safe monitoring with the labeling rules. Some uncertainties related to the guaranteed approval acceptance for the implementation of mobile manufacturing units.

Moving forward, the regulatory frameworks, including the International Council for Harmonization of Technical Requirements for Pharmaceuticals for Human Use (ICH), have the objective of reducing these disparities by creating globally accepted standards for quality, safety, and efficacy evaluation. However, the successful integration of mobile decentralized manufacturing into global health infrastructure will depend on the extent to which regulators can balance national sovereignty with international convergence [30].

6.6 PANDEMIC RESPONSE VIA MODULAR MRNA VACCINE PLATFORMS

6.6.1 Scaling Strategies from Benchtop R&D to Distributed Clinical Sites

The COVID-19 pandemic demonstrated that centralized vaccine manufacturing systems are fragile and versatile. This shows the importance of scalable strategies, which demonstrate and highlight the bridge and quick movement of vaccines from the research laboratories to clinical development. Modular mRNA platforms are standardized or closed systems, enhancing and easing the laboratory discoveries into clinical products. Compared to conventional vaccine production strategies, which require several years of research with multiple infrastructure steps, modular systems utilize the same lab-scale processes, such as RNA synthesis (IVT) and LNP packaging, with direct implementation toward clinical sites, such as microfluidic cartridges and single-use bioreactors. This strategy allows and facilitates the process proven in initial research, which is used for the real-world setting without interfering with the major redesign obstacles. The benefits are fully highlighted over the COVID-19 pandemic as vaccine rollout is a game-changer innovation of mRNA technologies, which facilitates the time reduction (less than one year) [31]

There are several crucial advantages of modular scaling, including the ability to produce vaccines in any location. Conventional vaccine production is based on large contract manufacturing organizations (CMOs) that rely on complex global distribution of wide networks. These wide networks are at risk of geopolitical issues, logistical problems, or supply chain failures. Contrast comparisons with modular mRNA systems are set up at regional centers or hospitals. This is a self-contained unit system that requires little infrastructure, allowing faster or more resilient aspects to be involved in vaccine production.

This distribution approach, based on the international supply chains and logistics, enables vaccines to reach people more efficiently and faster, and more fairly, during health emergencies. For example, during the COVID-19 pandemic and its crisis, the discovery of modular mRNA units was implemented in all regions. These units are quick, efficient setups accessible to every region. These setups are run by small, trained teams and administer clinical-grade vaccines directly to the areas where they are needed. This discovery, with the capability of localized manufacturing, transforms global health equity along with the assurance of pandemic response toward geographic proximities to the localized production hubs [32].

Additionally, modular mRNA vaccine manufacturing implements automation and digital control to reduce manual work, enhance productivity, improve consistency, increase efficiency and precision, and so on. Several technologies, including microfluidic mixers, efficiently integrate RNA and lipid components, producing nanoparticles with uniformity and efficient encapsulation. These innovations are very significant for enhancing and maintaining the vaccine's effectiveness. These systems operated along with an automation system and can be worked with less technical training staff. This approach helps in the reduction of labor needs and workload, and the reduction of training costs at decentralized sites. Other key considerations include Digital Twins, which are virtual replicas of the manufacturing process that facilitate real-time monitoring and advanced failure detection. Therefore, these issues and defaults related to machines or equipment can be fixed as early as possible. These innovations help to minimize the scale-up variability, facilitating faster licensure and the efficiency for addressing public health emergencies [33].

6.6.2 Environmental Footprint Comparison vs. Centralized Contract Manufacturers

To accelerate vaccine accessibility, modular mRNA manufacturing also requires consideration of environmental impacts. Compared to large centralized CMOs, these smaller centralized or localized units require less energy, water, and transportation, resulting in lower carbon emissions and reduced waste. This modular design approach is a sustainable or eco-friendly strategy for vaccine production, with the facilitation of the accessibility of vaccines to underserved regions.

Traditional CMOs are based on large industrial setups, including massive bioreactors, and extensive cleanrooms with energy-intensive Heating, Ventilation, and Air Conditioning (HVAC) systems, and enhance sterile production environments. This strategy utilizes a high amount of water for sterilization and cleaning, along with the cons of waste generation, including plastics and chemicals. Compared to modular models, it is a closed-unit technology that reduces the need for cleanroom space. A microfluidic bioreactor is a closed or sealed system because of this it requires a moderate cleanroom system. This strategy reduces the cost of energy, which is associated with handling or filtration systems specifically for air, lowers direct or indirect carbon emissions, and follows the principles or standards of green biomanufacturing [32].

A further environmental benefit of distributed modular units is the reduced dependency on global cold-chain logistics. Centralized CMOs often necessitate the bulk transport of drug substances to fill-and-finish sites and then to end-use locations, relying heavily on ultra-low temperature freezers and refrigerated shipping containers. These cold-chain networks are not only expensive but also contribute substantially to greenhouse gas emissions through fuel consumption and energy usage. Integration of Modular systems with on-site lyophilization uses stable fixed formulations for the production strategy, which helps in the elimination of burden, specifically for the cold-chain. This approach facilitates the reduction and elimination of the carbon footprint of vaccine deployment as well as enhances accessibility all over the world. Along with the fragility and uniformity of energy use or of transportation infrastructure. Specifically, the ecological advantages of the modular model it helps for making the supply chain broader, as well as elaborate conventional sustainability, offering future vaccine segregation [31].

The modular model plays a significant role in reducing water use and chemical waste, highlighting the waste-to-wealth approach. Compared to a conventional strategy, bioreactors require stainless steel and a complex piping system. After that, this will go to CIP and SIP. It leads to the widespread use of large amounts of water, energy, and hazardous chemicals. Unlike the modular model, it uses closed systems and single-use components, reducing the need for extensive cleaning. Leads to reducing the water consumption, reduction of chemical waste, and short downtime among batches. This makes the overall manufacturing process more stable, sustainable, eco-friendly, and cost-effective. A conventional system requires thousands of liters of water for injection (WFI) or produces large volumes of water and chemical waste. Differentiating with a modular system requires a single-use approach with pre-sterilized components, helping eliminate CIP/SIP operations. This helps reduce water and chemical waste by adopting a waste-to-wealth approach. Additionally, innovation of recyclable single-use plastics as well as biodegradable materials helps in the implementation of sustainable modular vaccine production, aiming for the future [34].

6.7 HOSPITAL-BASED CAR-T CELL MANUFACTURING

6.7.1 Lean Expansion Protocols in Closed Bioreactor Systems

Hospital-based CAR-T cell manufacturing represents a critical innovation in decentralizing cell therapy production, shifting from reliance on centralized contract facilities toward patient-proximal clinical laboratories. The central domain of this transformation, with evolutions, is a streamlined expansion strategy implemented with bioreactor platforms. The CAR-T cell expansion system is based on open bags or flasks, which require crucial, precise manual handling; failure can lead to contamination or variable outcomes and reduce the quality of the product. To address this resistance or failure, closed bioreactors require pre-sterilized, single-use components that allow automation. This facilitates safer, consistent, and efficient cell expansion. This model system reduces manual handling and sterile handling, thereby decreasing the need for cleanroom facilities. Implementations with hospital-based units meet cGMP requirements, including additional capital and operational costs of the manufacturing process, particularly in wide industrial manufacturing sectors (Figure 6.8) [34].

Advanced, innovative systems were implemented in the streamlined expansion systems to track pH, dissolved oxygen, nutrients, and metabolic byproducts in real time. Automated feedback loops enable dynamic adjustments to culture conditions, maintaining T cell proliferation and functionality at optimal levels. This strategy facilitates reproducibility by reducing batch-to-batch variability, given its importance in autologous therapies. This strategy is known as plug-and-play flexibility, which helps hospitals respond to fulfill patient demand by ensuring the delivery time of therapies and treatments [32].

Another positive point of this strategy is that it is a point-of-care (PoC) treatment timeline. By the integration of cell selection, activation, viral transduction, and expansion into a single closed workflow, this integration helps hospitals reduce the time from cell collections to reinfusion. This is

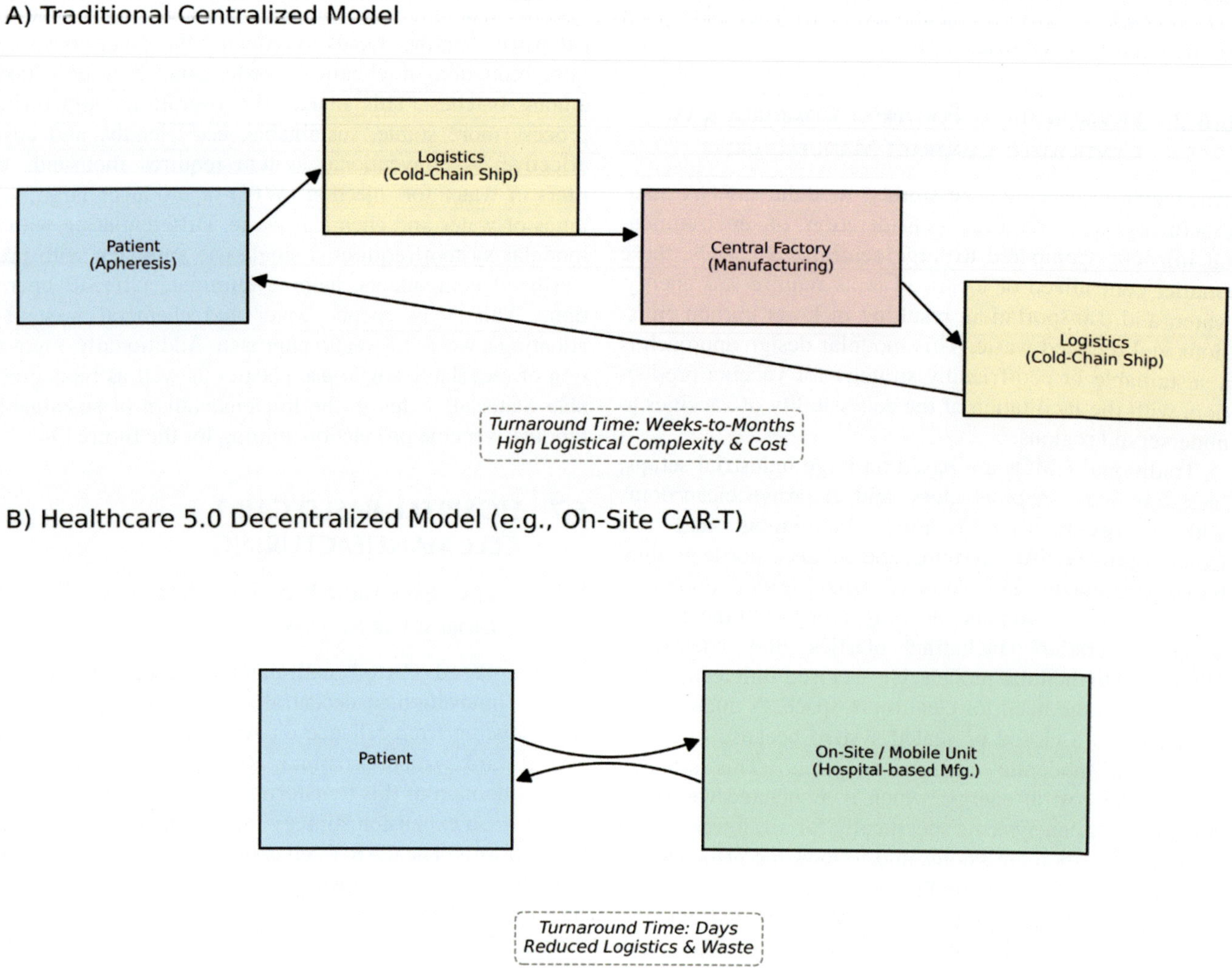

FIGURE 6.8 Centralized vs. decentralized biomanufacturing models.

crucial for enhancing the treatment potency and cell viability. This strategy helps in the reduction of disease progression risk, which directly outweighs the patient-specific benefits. These protocols help enhance the efficiency of operations as well as clinical advantages in CAR-T therapy in the implementation in hospitals [31].

A robust evaluation of hospital-based, PoC CAR-T manufacturing indicates that decentralization can materially improve the economic profile of autologous therapies while maintaining, and in some contexts enhancing, clinical performance metrics that matter to patients and health systems. From a cost perspective, the benchmark for centrally manufactured, commercial CAR-T products remains substantial: list prices for the product alone approach roughly USD 500,000 before accounting for ancillary expenditures such as lymphodepleting chemotherapy, inpatient monitoring, transfusion support, and management of acute toxicities (e.g., cytokine release syndrome and neurotoxicity), which together compound the total episode-of-care cost borne by providers and payers. PoC programs, built around closed, automated platforms (e.g., CliniMACS Prodigy®, Lonza Cocoon®) and single-use flow paths, attack several cost drivers simultaneously: they reduce cleanroom classifications and footprint requirements, curtail labor-intensive open manipulations, and eliminate two-way cold-chain logistics and associated courier and insurance fees that are otherwise embedded in vein-to-vein chains for centralized models. Evidence summarized across early PoC implementations shows shorter manufacturing cycles—commonly 7–10 days vs. ~14 days in legacy centralized processes—together with opportunities to infuse fresh product (thereby avoiding cryopreservation-related losses). When vector costs are externally sponsored (e.g., via trial support), case-level in-house manufacturing outlays have been reported as low as ~USD 35,000, whereas sites purchasing GMP vector face a much wider range (~USD 50,000 up to ~USD 1,000,000) once fixed and variable inputs for setup, staffing, qualification, and quality oversight are included; despite this range, the capacity to localize value-adding steps and amortize capital across multiple indications or a hospital network has been modeled as cost-effective vs. centralized outsourcing in discrete-event simulations and in multi-center regional consortia scenarios. Crucially, PoC architectures also compress logistical latency—for

TABLE 6.3
Comparison of Centralized vs. Decentralized (PoC) CAR-T Manufacturing

Metric	Centralized Model	Decentralized (PoC) Model	Source (from Chapter)
Manufacturing cost	~USD 500,000 (List price)	~USD 35,000 (In-house outlay)	Section 6.7.1 [35–37]
Turnaround time	~2–4 weeks (14–28 days)	~7–10 days	Section 6.7.1 [35–37]
Apheresis-to-infusion	~17+ days (Commercial)	~13 days (PoC)	Section 6.7.1 [38]

Data highlights the advantages of decentralized models in cost and speed.

example, removing multi-week "manufacturing slot" queues that delay apheresis receipt at corporate facilities—and thereby reduce bridging-therapy intensity and the risk that patients clinically decompensate while they wait. From a quality and regulatory vantage point, contemporary closed systems support electronic batch records and GMP-compatible process controls; when coupled with standardized release testing across sites, they provide the foundation for comparable product specifications and pharmacovigilance to those documented for commercial products, even as institutions iterate on lean staffing models and modular capacity additions to align with variable demand. These features reposition hospital-based CAR-T manufacturing at a health-economic level, which helps to shorten vein-to-vein timelines, trimming non-value logistics, and enabling context-appropriate scaling. This feature retains compliance and clinical safeguards central to the technology's benefit-risk balance (Table 6.3) [35–37].

In parallel with cost dynamics, clinical efficacy and delivery performance metrics reported for hospital-based programs demonstrate feasibility, manufacturing reliability, and outcomes parity with approved centralized comparators when standardized protocols and qualified infrastructure are in place. Across PoC implementations using closed, automated bioreactors, sites have reported 100% manufacturing success in CAR-naïve cohorts and high target-dose achievement rates (e.g., ~85% of products meeting target dose in a 14-day Prodigy® protocol spanning 26 manufactured products), with median apheresis-to-infusion intervals of ~13 days in certain series—intervals that are clinically meaningful in relapsed/refractory populations where disease trajectories are steep. Notably, aggregated vein-to-vein analyses for commercial products often cite timelines on the order of 2–4 weeks for manufacturing, and ~17 days for specific products such as axicabtagene ciloleucel under standard practice; PoC compression of these intervals has correlative implications for survival, insofar as shorter waits reduce the need for intense bridging and the risk of becoming ineligible during delays. From a safety standpoint, hospital-manufactured CAR-T products must evidence toxicity profiles congruent with those cataloged for licensed therapies—principally the incidence, grading, and management of cytokine release syndrome and ICANS—and early PoC trials and academic programs describe comparable efficacy and toxicity patterns to centralized products, including in salvage contexts (e.g., PoC rescue after out-of-spec commercial tisagenlecleucel). Importantly, device choice (e.g., Prodigy®, Cocoon®, and G-Rex®) and site experience modulate expansion kinetics, labor requirements, and batch-to-batch variability; nonetheless, published experiences across Spain's national hospital network and multiple US academic centers (e.g., Ohio State University, Case Western, Stanford, Medical College of Wisconsin) demonstrate scalable reproducibility with fresh-cell infusion, expanding access without compromising pharmacologic potency. Looking forward, integration of PAT and AI-driven control within "smart manufacturing hospital" architectures is expected to increase predictive insight into cell-product quality (e.g., dynamic links between culture trajectories and functional assays) and further reduce cost-per-patient by enabling parallelized, low-touch operations. Together, these data support the proposition that hospital-based CAR-T production can deliver clinically effective products on timelines aligned to disease urgency, while advancing sustainability and equity through local capacity building [35, 36, 38].

6.8 BIOPRINTED TISSUE GRAFTS

6.8.1 End-to-End Workflow: Patient Imaging, CAD Design, Bioprinting, and Maturation

Bioprinted tissue grafts require a tightly integrated, end-to-end workflow that begins with patient-derived imaging data and proceeds through digital design, high-fidelity printing, and *in vitro*/*in vivo* maturation; each stage imposes specific technical and regulatory requirements that must be addressed to translate prototypes into clinically useful implants. The workflow typically starts with the acquisition of high-resolution anatomical data (computed tomography [CT], magnetic resonance imaging [MRI], sometimes ultrasound or micro-CT for small constructs) that capture not only gross geometry but also spatial relationships to vasculature and mechanically relevant boundaries; these imaging datasets are then segmented and converted into 3D meshes that form the input to computer-aided design (CAD) systems. In the CAD environment, clinicians and bioengineers collaborate to incorporate functional features, such as perfusion channels, graded material distributions to match mechanical anisotropy, and attachment geometries for surgical fixation. The CAD model is next translated into machine instructions (G-code equivalents or volumetric exposure maps) that control the bioprinter's deposition sequence or volumetric exposure pattern. Choice of printing modality (extrusion, inkjet/drop-on-demand, laser-assisted, stereolithography/volumetric additive manufacturing [VAM]) depends on the intended tissue type, scale, and resolution: extrusion is flexible for high-cell-density, load-bearing constructs, inkjet supports high-throughput droplet placement for thin tissues, and volumetric methods (VAM) promise ultra-rapid, nozzle-free fabrication of complex geometries without layer seams or support structures. Across all modalities, the selection and formulation of bio-inks is decisive—bio-inks must support cell viability during and after printing, exhibit appropriate rheological behavior during printing, allow post-printing crosslinking (ionic, thermal, enzymatic, and photo-initiated), and be compatible with downstream maturation strategies. Following fabrication, constructs are transferred into maturation bioreactors that provide controlled perfusion, oxygenation, mechanical loading, and spatiotemporal biochemical cues to induce tissue-specific differentiation and functional integration; perfusion bioreactors and microfluidic culture systems are commonly used to accelerate vascularization and nutrient transport in centimeter-scale grafts. Collectively, this pipeline—from imaging → CAD → Bioprinting → maturation—establishes the reproducible, traceable manufacturing chain required for clinical translation, and it demands cross-disciplinary standardization (imaging protocols, digital file formats, bio-ink characterization, and bioreactor validation) to ensure that a design validated in one center can be reliably reproduced in another [39, 40].

6.8.2 Waste Minimization Through Recyclable Support Bio-Inks

Sustainable biomanufacturing for bioprinted grafts requires rethinking support materials and process flows to reduce waste without compromising construct fidelity or sterility; recyclable, recoverable support bio-inks emerge as a practical strategy to minimize consumable use while enabling complex architectures that would otherwise require sacrificial single-use supports. Historically, support materials (sacrificial gels, fugitive inks, and thermo-reversible carriers) have been used to permit printing of overhangs, internal channels, and fine vascular networks; after printing, the supports are removed chemically, thermally, or enzymatically, but they are commonly discarded, contributing to high downstream waste streams and material costs. Recent advances in reversible crosslinking chemistries and stimulus-responsive polymers facilitate the design of support bio-inks that can be recovered and reused: examples include thermo-responsive gels that liquefy upon cooling or heating within benign temperature windows, shear-thinning fugitive inks that can be extracted under low shear, and enzymatically degradable supports where the enzyme can be neutralized and the polymer fraction reclaimed. Additionally, VAM techniques obviate many support requirements altogether by solidifying whole volumes selectively, thereby reducing the mass of sacrificial material required; VAM's layer-less approach can thus lower material consumption and eliminate the need for delicate support removal that risks construct damage. Implementing recyclable support workflows requires validated cleaning and sterilization protocols (to avoid cross-contamination), closed-loop material tracing (to track polymer lifecycles and mechanical property drift), and economic analyses to compare collection/sterilization costs vs. single-use disposal. When combined with lifecycle thinking (including selection of biodegradable polymers and reduced energy use during post-processing), recyclable supports and VAM approaches can materially reduce the ecological footprint of Bioprinting while preserving—or even enhancing—architectural complexity and clinical readiness [39, 40].

6.9 CHALLENGES, STRATEGIC OPPORTUNITIES, AND FUTURE DIRECTIONS

6.9.1 Technological Interoperability and Standardization

Traditionally, manufacturing sectors operate with coordination between information technology (IT) and operational technology (OT) teams, but they operate separately. With this evolutionary and transformative approach, these teams brought together and achieved closure through IT/OT convergence. However, their implementation is still difficult for integrating the huge networks, data, and processes together, such that many firms hesitate to follow. In biomedical sectors such as the pharmaceutical and life sciences industry, disruptions to operational procedures are very serious because a minor effect on them not only affects the company but also affects the pharmaceutical and life sciences industry, disruptions to operational procedures are very serious because a minor effect on them not only affects the company but also the whole healthcare system. Products from these industry are very important for patient care. A minor or operational failure directly affects hospitals, patients, and consumers and negatively impacts public welfare. Therefore, the OT system provides strong support to production and distribution through a secure or resilient system. This is a proactive approach responsible for managing the increasing risk of cyber-attacks in the manufacturing or pharmaceutical sectors.

Integration of OT and IT systems facilitates several advantages, including the following:

- Streamline the manufacturing process, increase productivity, reduce operational costs, and eliminate waste.
- Help manufacturers collect and analyze a vast amount of data in real time. This data can provide insights into the manufacturing process, quality control, and resource utilization.
- Enable manufacturers for the assessment of product quality along with implementations of regulatory standards.

- Help manufacturers for assessing advance failures of equipment with a maintenance strategy for the reduction of downtime and making them cost-effective.

Together, these innovations establish flexibility and uniformity within a real-time data-based manufacturing environmental ecosystem. This connects operational analysis, materials, logistics, and elimination of barriers across the wealth chains. Such that, the broad division of IT and OT fades, which supports fusion, an integrative approach with efficient collaboration strategies, such as the following:

1. *Predictive Maintenance*: Collection of real-time data and its analysis in Industrial IoT sensors for the manufacturing strategies that use the cloud and AI tools for implementation. This approach has a positive impact on OT teams by enabling the assessment of predictive faults and plans, improving maintenance, and reducing downtime. This is advantageous for avoiding unexpected failures and helps enhance productivity with the efficient utilization of equipment.
2. *Augmented Reality*: Augmented reality was executed by OT technicians to enhance the ease of inspections, maintenance, and repairs during maintenance. These executions include 3D visuals, sensor data, and guidance, which facilitates access to hands-free manual or automated support for building accelerated resolutions of problems.
3. *Smart Inventory*: Implementation of IT-based inventory systems includes the utilization of sensors or a production signal, which enhances the efficient analysis of real-time data. This data helps in reducing waste, freeing working capital with executions of the waste-to-wealth approach [41].

The fusion of IT with OT is a crucial key domain of industrial digital transformation. These transformations support the enhancement of efficiency, innovations, and competitiveness in the industry. IT systems such as hardware, software, networks, and database analyses, which help with the management, analysis, and securing of data executed in the business strategies such as ERP and CRM. OT is defined as hardware and software which specialized for efficient monitoring and automation on physical devices, along with advanced processes in the industry. OT systems were usually seen in Industrial Control Systems (ICS), Distributed Control Systems, Programmable Logic Controllers, or Supervisory Control and Data Acquisition (SCADA) systems, which are integral to managing industrial equipment [42, 43].

IT/OT convergence involves linking OT with IT systems to integrate directly into organizational processes. This convergence aims to merge, i.e., integrate, IT and OT systems and functionalities. IT/OT alignment refers to the process of adjustments during IT/OT convergence, whereas IT/OT integration defines the final state of a fully integrated IT and OT environment. IT governance involves measures, procedures, and principles ensuring that IT supports the organization's goals effectively and responsibly. It is focused on aligning IT development with business strategies and objectives, and establishes the critical link between IT and business strategies, primarily overseen by the board of directors [44–47].

6.9.2 Workforce Development and Educational Paradigms

Bioprocess engineering is the integration of robotics and data science, which underscores its evolutionary nature. This is a field that fuses biology, chemistry, and engineering, demonstrating that engineering is rapidly evolving and transforming with novel technologies, and underscoring the need for environmentally friendly and sustainable solutions across industrial sectors. Such that, combining bioprocess with the integration of chemical and mechanical aspects pillar up the circular economy with the waste-to-wealth aspect. Examples such as Biorefineries, which reduce waste and produce environmentally friendly, sustainable bioproducts such as biofuels, biopolymers, and biochemicals. Synthetic biology is an evolutionary and transformative bioprocess engineering that enables precise control through the manipulation of metabolic pathways and gene expression. Researchers are designing recombinant microbes with precise functional manipulations for the production of complex products, including bioactive compounds, therapeutic products, and pharmaceutical products. These key tools are established for the efficient and reliable aspects of sustainability.

Synthetic biology is a major domain for driving the transformation of bioprocess engineering, along with biotechnology. This facilitates the researchers to shape or design to build biological parts and systems along with enhanced efficiency and precision. Scientific researchers are designing the recombinant microbes for enhancing efficiency along with sustainability aspects, which produce several derived products, including medicines, biofuels, and specialty chemicals. This strategy accelerates bioproducts along with the reduction of negative effects on the central domain of biomanufacturing. An example of the tools used for synthetic biology tools include CRISPR-Cas9, which facilitates microbes for target manipulation of targeted genes for the needful production, which enhances overall purity and yield, as well as sticking with the waste-to-wealth aspects. Other essential trends of bioprocess engineering include the use of real-time data-driven methods, specifically AI and ML for shaping, optimizing, monitoring, and analyzing bioprocess. As biological systems are very complex, which needs advance computational tools derived from AI or ML are required for enhancing the efficiency and accuracy. AI and ML help to analyze complex bioprocessing data sheets, pattern detection, and outcome prediction, which enable real-time monitoring for data analysis. This helps in controlling and enhancing the higher yield and improved quality.

By integrating AI and ML, bioprocess engineering is establishing procedural aspects more adaptively, efficiently, and cost-effectively. A key advancement in bioprocess engineering is sustainability. To make this strategy more environmentally friendly, green bioprocessing methods, including cell-free systems and biocatalysts, need to be implemented. These strategies are needed to reduce the use of toxic chemicals and lower energy consumption, thereby establishing safer bioprocesses with enhanced sustainability. To reduce dependence on fossil fuels, bioprocess engineers are advancing the development of sustainable feedstocks, including lignocellulosic biomass. Adding to that huge transformation of batch methodologies toward continuous bioprocessing, enhancing the sustainability and eco-friendliness by reducing waste and resource use. Bioprocess engineering with sustainability aspects, building the global efforts for triggering or fighting against climate change, along with waste reduction. Innovation in regenerative medicine, driven by gene therapies, is transforming these therapies to enhance the personalized or enhanced effectiveness of solutions in healthcare sectors. These cutting-edge therapies involve manipulating and engineering cells, tissues, and genes to treat a wide range of diseases and medical conditions. Bioprocess engineering is responsible for enhancing the advancement of therapies, along with scalability and reliability in the manufacturing process. Such as enhancing or differentiating stem cells, as well as manipulating or enhancing genetic materials. This helps in enhancing and building the precision along with accuracy for delivery systems. Along with the increasing demand for personalized medicines, bioprocess engineering is a significant domain for the development of patient-specific therapies, making this approach reliable, cost-effective, and establishing a novel era in healthcare. Synthetic biology is an evolutionary and transformative approach for bioproduction or biomanufacturing, which facilitates the precise engineering of microbes for the enhancement of efficiency, accuracy, and sustainability. Integration of AI and ML helps optimize bioprocesses by reducing experimental costs and enhancing efficiency and precision [48, 49].

6.9.3 Ethical, Equity, and Access Considerations

A government initiative, such as PoC diagnostics, includes NGOs and fundamental research alike. For the high-income countries hoping to streamline the diagnostic procedure, minimize costs, and make healthcare processes more efficient and faster, which, in some cases, can be more a matter of convenience than necessity. However, in resource-limited settings such as low-income countries, PoC diagnostics might be the only viable route when the next laboratory is hours away. Therefore, it is especially important to focus research on novel diagnostics for these countries to alleviate suffering from infectious diseases. Low- and middle-income countries (LICs/MICs) face severe challenges due to limited economic opportunities. In addition to the economic struggles, LICs also bear a large burden of transmissible diseases, posing severe risks to the population's wellbeing. Healthcare systems and healthcare providers in LICs are often ill-equipped to treat patients in the best possible way, especially in rural areas. Given the economic and infrastructural, and other challenges in LIC, PoC diagnostics, which are often characterized by being independent of laboratory or medical infrastructure, as well as being highly affordable and holding considerable promise to improve the situation. Thus far, the actual commercialization of PoC diagnostic tests lags well behind the innovative research and developments done in laboratories [50].

Fundamental research is the first step in developing a PoC device and encompasses basic research not yet directly related to a prototype or a product. Historically, the research of PoC devices started with simple dipstick tests with immobilized reagents—for example, for the detection of glucose [51].

Later, the laboratory use of immunoassays, especially the high sensitivity of radioimmunoassays and enzyme-linked immunoassays, sparked interest in improving those methods into rapid tests, which eventually led to the development of lateral flow immunoassays, the most common type of PoC device. While lateral flow devices are the most common, the trend in research goes toward devices with higher complexity, which are able to handle more complicated samples, can multiplex, and detect challenging analytes. The greater complexity of these devices is shown in their mechanisms of action, which are often optical, electrochemical, or magnetic mechanisms. However, these devices also include other modes of action, e.g., thermal sensor systems. Optical systems include straightforward ultraviolet-visible (UV/Vis) or fluorescence sensors, as well as more complex systems using quantum-dot and surface-plasmon-resonance technologies, or even genetically encoded biosensors. Examples of electrochemical systems range from basic amperometric or voltametric systems to graphene-based field-effect transistors to DNA-annealing-based redox-reporter assays. However, despite the abundance of new research, actual commercialization lags, and market examples welcoming these innovations are rare [52–59].

Political will toward PoC might be reduced when PoC tests lead to more demand for treatment, while at the same time, treatment capabilities are scarce. Conversely, PoC diagnosis might not be feasible (for a test-and-treat scenario) when adequate treatment capabilities are not in place. When treatment is available, patients might also just opt out of tests in favor of direct use of medication, such as over-the-counter antibiotics. This has been reported in Thailand, where missing information about disease origin among the public leads to a preference for medication instead of proper diagnosis, as medication is connected to symptoms instead of disease origin. For example, patients associate antibiotics with the symptoms of a bacterial infection rather than the infection itself, and thus demand antibiotics even when the illness is not bacterial but shares similar symptoms. PoC testing might offer proper diagnoses, preventing people from self-diagnosing and taking inadequate medicine. This also shows the importance of looking at PoC applications not independently but in the whole context of the healthcare system, which acts and is acted upon by various factors. Some researchers argue that introducing a PoC device into an existing system shifts the role from a technical to a social device. However, this also has its upsides—a more elusive benefit of why healthcare workers in rural areas might want to use PoC testing is the psychological effect it can have on patients, giving healthcare workers more certainty, which, in turn, transfers to the patient and improves compliance. Patients may also overestimate the tests' capabilities, which encourages compliance. PoC can provide evidence without the need for laboratory infrastructure and highly trained lab technicians [60–62]

Regarding patients' views on PoC, healthcare personnel reported that patients were satisfied with the results (97%) and would recommend them (96%). However, only half of clinicians thought they would reliably provide accurate results; 46% were unsure, and 4% considered them not accurate. In total, 65% of healthcare workers used medication even on a negative test, showing that trust in the test is limited. This concurs with only 20% of healthcare workers stating that they rely solely on the test. The majority sees the test as a complement to other means of diagnosis, such as symptoms [63].

Transparency in science has been defined as "the outcome of a suite of behaviors which characterize reproducible research"; it "facilitates and enhances research quality, research integrity, and trust." Transparency and openness have long been recognized norms of professional behavior in the conduct and reporting of scientific research. Scientists formulate hypotheses and access resources to pursue further research in a process that builds successively through time and across laboratories, institutions, and even national boundaries. Transparency underpins the process by which peer scientists can reproduce and independently validate the results of previous research. In practice, it is an expression of the scientists' core ethical and professional values such as honesty, responsibility, and trust. The need for transparency applies not only to research approaches and technologies, but also to the sources and validations of any biological materials (e.g., cell lines, clinical specimens, environmental samples, and nucleic acids) and/or electronic data (e.g., sequences and code). On which the investigations are based. Specific and open acknowledgment of any sharing of biological materials and data among collaborating research groups provides appropriate recognition and enhances transparency and trust, which are critical as research projects become more multi-disciplinary and complex. Recent trends, however, have strayed somewhat from the traditional customs of openness and sharing in scientific endeavor. While scientists conducting research in industry settings have always been more restricted in what information and materials could be shared, legal constraints, the pressures of increasing competition for grant funds, and the demands of tenure and promotion have led some academic scientists to similar reluctance to detail specifics and to exchange materials without restriction. In addition to biosafety and biosecurity (BS&S) issues, openness and sharing among scientists worldwide may be hampered by other factors, including political tensions and questions of national sovereignty or the protection of national treasures, such as germplasm or breeding stock [64–68].

6.9.4 Emerging Frontiers

6.9.4.1 Cell-Free Biomanufacturing Reactors for On-Demand Therapeutic Production

Cell-free systems are widely applied in systems and synthetic biology research and are increasingly recognized as promising platforms for protein and chemical production. Originally, they were mainly used to study fundamental biochemical mechanisms; notably, *Escherichia coli* cell-free extracts were instrumental in deciphering the genetic code in the 1960s. Since then, continuous improvements in transcription and translation efficiency have enhanced energy utilization and product yields. With advances in synthetic biology, cell-free systems now support just-in-time manufacturing of therapeutically important biologics and high-value small molecules, and have been successfully scaled for industrial production of antibodies and cytokines [69].

It has been demonstrated that a Chronic Fatigue Syndrome (CFS) microfluidic bioreactor model system for the production of single-dose therapeutic proteins using a distributed manufacturing model. Using a dual-channel reactor with an engineered nanoporous membrane between parallel reactor and feeder channels, high product yield was achieved through the maintenance of adequate concentration of critical reaction components and the removal of inhibitory byproducts. With its small footprint, easily controlled CFS reactor, and coupled/integrated downstream purification module, one could foresee an automated version of this system becoming the model PoC CFS system [70].

Protein therapeutics are essential for treating diseases such as diabetes, cancer, and growth disorders, but their high cost and complex production in living cells limit global access. To address this, PoC or magisterial production of protein therapeutics has gained attention. Recent advances in lyophilized cell-free protein synthesis (CFPS) systems have shown the ability to produce highly active, FDA-approved cancer therapeutics at low cost in single batches. Foam-based and oxygen-headspace formats further support dosage-scale, on-demand production by improving yields and reducing required reactor size. Oxygen-headspace reactions increased therapeutic yields by over 80% without loss of activity, while foam-based systems achieved 50% higher activity than air controls, making them especially useful in resource-limited settings [71].

CFPS is currently the most widely used cell-free technology, providing an *in vitro* system for protein transcription and translation. Although cell-free systems have long been used to study basic biological processes, advances in CFPS have broadened its applications, making it a practical alternative to living cells for research, bioengineering, and biomanufacturing at both small and large scales. CFPS also enables PoC protein production, particularly when integrated with microfluidic reactors, which offer better mixing control than traditional batch systems [72, 73].

One of the main limitations in cell-free synthesis is the lifetime of cellular extracts and the lack of reactor concepts adapted to CFPS. Currently, no reactor concept, including active control or bioprocess development, is common to all fermentation processes. A main short-term goal for CFPS, therefore, should be the development of cell-free reactor systems, including the development of appropriate control systems and reaction devices enabling a controlled feed of the required substrates, as well as supplying the required energy and removing the products generated. Another limitation is the current focus on the production of individual proteins in CFPS. To produce complex biomolecules that require multiple synthesis steps, the setup of synthetic pathways, including several controllable gene expression systems *in vitro*, will be required. In addition, new reactor concepts enabling longer synthesis periods will be necessary. This development needs a thorough understanding of molecular circuits. A combination of the efforts in CFPS, the development of molecular circuits, and the required reactor concepts may allow such a development [74].

A continuous-exchange microfluidic reactor with a nanofabricated membrane has been developed to extend reaction times and improve yields, aiming to produce single-dose therapeutic proteins at the PoC [70]. Industrial-scale CFPS (CFPS) has also been demonstrated for high-value products such as antibodies and cytokines, with notable implementations by Sutro Biopharma and GreenLight Biosciences' GreenWorX platform [75, 76]. Although most CFPS work has been at small scales, research on scaling up the technology continues, with several academic studies exploring its industrial potential [77–79].

AI-driven methods can vastly improve the historically costly drug design process, with various generative models already in widespread use. Generative models for *de novo* drug design, in particular, focus on creating novel biological compounds from scratch, representing a promising future direction. To be sure, AI is currently exercised in Bioprocessing throughout the Research and Development (R&D)—manufacturing continuum, even though most efforts end up in limbo in the so-called Proof of Concept purgatory. However, despite the hype that surrounds these efforts, they all seem to share the same objective and strategy: the acceleration of Chemistry, Manufacturing, and Controls (CMC) development times by integrating AI alongside existing Ways of Working (WoW) through the QbD framework [80–83].

This situation is similar to earlier efforts to integrate mechanistic process modeling. At present, there is no clear individual or collective strategy for using AI to fundamentally change traditional WoW in process development, unlike the more disruptive adoption seen in drug discovery. Instead, biopharma stakeholders have largely chosen to use AI as a supportive tool rather than a transformative one. This preference for augmentation over disruption raises the following important question: Does process development remain a worthwhile investment? [84]

Traditionally, process development has received less attention than drug discovery for two main reasons. First, the return on investment in drug discovery is significantly higher, leading portfolio analyses to prioritize resource allocation toward discovery programs. Second, process development has rarely been a major bottleneck, as most drug and biologic candidates fail before reaching CMC development. On the surface, this suggests that process development may be a poor investment. However, this conclusion is misleading for several important reasons [85].

For biologics, process development can add up to 2.5 years to the time to market, while for vaccines, this can increase to 5–7 years. Although drug Net Present Values (NPVs) are rarely disclosed, estimates suggest that accelerating development by just one quarter can increase NPV by about 5%. Even though the link between CMC development timelines and overall market entry is not linear, and clinical trials involve uncertainty, faster CMC development can still significantly shorten time to market. Moreover, traditional development timelines are under growing pressure, a trend intensified by the COVID-19 pandemic, which highlighted the need for accelerated development strategies (Table 6.4) [86–89].

Third, although Biopharma has just started to digest the ongoing discussions about Healthcare Industry 4.0, the voluminous cash flows for AI in drug discovery could lead to an expanded number of simultaneous candidates and thus render process development departments a bottleneck due to their intrinsic limited parallelization capacity [93–95].

TABLE 6.4
Development Timelines, Traditional vs. *In Silico* Models

Product Type	Traditional Timeline (Years)	*In Silico* Model (Years)	Source (from Chapter)
Biologics	~2.5 years	0.6–1.25 years (50–75% reduction in development timelines)	Section 6.9.4.1 [86–92]
Vaccines	~5–7 years	1.25–3.5 years (50–75% reduction in development timelines)	Section 6.9.4.1 [86–92]

Data shows the potential for AI-driven models to accelerate R&D.

TABLE 6.5
Generative Deep Learning Models for *De Novo* Drug Design

Model Type	Principle	Applications in Drug Design	Advantages	Limitations
Generative Adversarial Networks (GANs)	Two networks (the generator and the discriminator) are trained together; the generator generates synthetic data, while the discriminator distinguishes real from fake data.	Generation of novel molecules and proteins that resemble real biological compounds.	Produces high-quality and realistic samples.	Training can be unstable; it may suffer from mode collapse (low diversity).
Variational Autoencoders (VAEs)	Encode data into a probabilistic latent space (mean and variance) and decode samples from this space to generate new data.	Widely used for molecule and protein generation due to smooth latent spaces.	Stable training; interpretable latent space; enables controlled generation.	Generated samples may be less sharp or realistic than GAN outputs.
Normalizing Flows	Use a series of invertible transformations to map complex data distributions to simple ones, enabling exact likelihood calculation.	High-fidelity generation of molecules and proteins with complex priors.	Exact likelihood estimation; efficient and precise sampling.	Designing suitable invertible transformations is challenging.
Diffusion Models	Learn to generate data by iteratively denoising samples that were gradually corrupted with noise.	Emerging approach for molecule and protein generation, often combined with graph neural networks.	Generates diverse, high-quality samples with stable training.	Computationally expensive due to iterative sampling and complex training.

Traditional process development is based on experimental campaigns and statistical modeling and cannot accelerate beyond its parallelization capacity. For instance, scale-up studies, even within the context of High Throughput Screening, are bound to be carried out in a serial manner. Consequently, augmentations based on AI or mechanistic models will only yield incremental gains. Substantial improvements require a considerable upfront Capital Expenditure (CAPEX) investment in resources, either in-house or with Contract Developing and Manufacturing Organizations (CDMOs). However, since both insources and CDMOS use traditional WoW, increases in parallelization capacity will approximately result in proportional increases in the cost, thereby casting serious doubts on the business cases.

In silico process development uses mathematical models supported by minimal confirmatory experiments and has gained attention with the use of end-to-end digital twins. Studies show that even a modest acceleration of development timelines by two quarters—without changing overall product cycle time—can result in cost savings of USD 40–140 million. However, the main impact of this approach lies in significantly shortening development timelines, with industry examples reporting reductions of 50–75%. Advances in hardware and software have also made the combined CAPEX and OPEX for handling multiple candidates *in silico* negligible compared to traditional bioprocess development costs [90–92].

AI-driven methods improve the efficiency and specificity of the traditional drug design process. Adding to that, generative models, including *de novo* drug design, form novel biological compounds from scratch, which highlights a promising direction for future (Table 6.5).

6.10 CONCLUSION

The healthcare sector is transforming toward Healthcare Industry 5.0. This transformation integrates several crucial aspects, such as advanced and automated technologies, sustainability aspects, and welfare specific to patients' needs. This transformation is accomplished through biomanufacturing, the use of living systems to produce vaccines and personalized medicines. The strategic technologies include Digital Twins, Collaborative Robots, Continuous Bioprocessing, and smarter, more efficient, and eco-friendly care delivery. Healthcare Industry 5.0 is a model that includes advanced, innovative approaches to reduce waste through a waste-to-wealth approach. This underlines the crucial need for a circular bioeconomy and decentralized manufacturing strategy, such as hospital-based CAR-T production or modular mRNA vaccine platforms. In short, this transformation prioritizes personalization, sustainability, and the empowerment of healthcare workers. This approach is achieved by integrating advanced technologies with ethical considerations. Biomanufacturing is a central domain of this huge shift from traditional to advanced innovations, offering low-impact solutions (Table 6.6).

TABLE 6.6
Alignment with Sustainable Development Goals (SDGs)

SDGs	Purpose
1. SDG 3	Good health and wellbeing
2. SDG 9	Industry, innovation, and infrastructure
3. SDG 12	Responsible consumption and production

REFERENCES

1. Wuest, T., Weimer, D., Irgens, C., & Thoben, K. D. (2016). Machine learning in manufacturing: Advantages, challenges, and applications. *Production & Manufacturing Research*, 4(1), 23–45.
2. Demir, K. A., Döven, G., & Sezen, B. (2019). Industry 5.0 and human–robot co-working. *Procedia Computer Science*, 158, 688–695.
3. Bhatia, S., & Narain, R. (2020). *Sustainable biomanufacturing*. CRC Press.
4. Nagarajan, H., Costello, Z., Lerman, J. A., & Palsson, B. O. (2022). Improving the quality of life through biomanufacturing and personalized medicine. *Trends in Biotechnology*, 40(3), 305–317.
5. Kour, R., Arya, S. K., Young, S. J., Kim, K. H., & Gupta, V. K. (2021). Biomanufacturing: A sustainable pathway for sustainable development. *Bioresource Technology*, 338, 125524.
6. Krüger, J., Lien, T. K., & Verl, A. (2009). Cooperation of human and machines in assembly lines. *CIRP Annals*, 58(2), 628–646
7. Sung, T. K. (2018). Healthcare Industry 4.0: A Korea perspective. *Technological Forecasting and Social Change*, 132, 40–45.
8. Cortiella, J., & Vacanti, J. P. (2006). Current status of tissue-engineered lung. *Seminars in Pediatric Surgery*, 15(4), 285–292.
9. Murphy, S. V., & Atala, A. (2014). 3D bioprinting of tissues and organs. *Nature Biotechnology*, 32(8), 773–785.
10. Carus, M., & Dammer, L. (2018). The circular bioeconomy—Concepts, opportunities, and limitations. *Industrielle Biotechnologie*, 14(2), 83–91
11. Haddadin, S., De Luca, A., & Albu-Schäffer, A. (2017). Robot collisions: A survey on detection, isolation, and identification. *IEEE Transactions on Robotics*, 33(6), 1292–1312.
12. Jones, D., Snider, C., Nassehi, A., Yon, J., & Hicks, B. (2020). Characterising the Digital Twin: A systematic literature review. *CIRP Journal of Manufacturing Science and Technology*, 29, 36–52.
13. Breque, M., De Nul, L., & Petridis, A. (2021). *Industry 5.0: Towards a sustainable, human-centric and resilient European industry*. European Commission Directorate-General for Research and Innovation.
14. McDonough, W., & Braungart, M. (2002). *Cradle to cradle: Remaking the way we make things*. North Point Press.
15. Zydney, A. L. (2016). Continuous downstream processing for high value biologics: Current status and future prospects. *Current Opinion in Chemical Engineering*, 14, 1–7.
16. Konstantinov, K. B., & Cooney, C. L. (2015). White paper on continuous bioprocessing. *Journal of Pharmaceutical Sciences*, 104(3), 813–820.
17. Lee, S. Y. (2012). *Systems biotechnology for industrial microorganisms*. Springer Science & Business Media.
18. Catarinucci, L., De Donno, D., Mainetti, L., Palano, L., Patrono, L., Stefanizzi, M. L., & Tarricone, L. (2015). An IoT-aware architecture for smart healthcare systems. *IEEE Internet of Things Journal*, 2(6), 515–526.
19. Zheng, Z., Xie, S., Dai, H., Chen, X., & Wang, H. (2018). An overview of blockchain technology: Architecture, consensus, and future trends. *Proceedings of the IEEE*, 107(6), 1045–1067.
20. Gungor-Ozkerim, P. S., Inci, I., Zhang, Y. S., Khademhosseini, A., & Dokmeci, M. R. (2018). Bioinks for 3D bioprinting: an overview. *Biomaterials Science*, 6(5), 915–946.
21. Yu, L. X. (2008). Pharmaceutical quality by design: Product and process development, understanding, and control. *Pharmaceutical Research*, 25(4), 781–791.
22. Rathore, A. S., & Winkle, H. (2009). Quality by design for biopharmaceuticals. *Nature Biotechnology*, 27(1), 26–34.
23. Bakeev, K. A. (2010). *Process Analytical Technology: Spectroscopic Tools and Implementation Strategies for the Chemical and Pharmaceutical Industry* (2nd ed.). Wiley.
24. Finnveden, G., Hauschild, M., Ekvall, T., Guinée, J., Heijungs, R., Hellweg, S., … & Suh, S. (2009). Recent developments in Life Cycle Assessment. *Journal of Environmental Management*, 91(1), 1–21.
25. Zabaniotou, A., & Kamaterou, P. (2019). Food waste valorization advocating circular bioeconomy—a critical review of potentialities and perspectives of a non-conventional bioresource. *Fermentation*, 5(1), 4.
26. Demichelis, F., Fiore, S., Confalonieri, R., & Barbero, S. (2013). Anaerobic digestion of the organic fraction of municipal solid waste: A review. *Applied Energy*, 104, 170–184.
27. Koyande, A. K., Show, P. L., Guo, R., Tang, B., Ogino, C., & Chang, J. S. (2020). Bio-processing of algal biomass into biofuels and value-added products – A review. *Bioresource Technology*, 302, 122803.
28. Dinh, A., & Stroncek, D. F. (2024). Healthcare center-based cell therapy laboratories supporting off-site manufactured cell therapies: The experiences of a single academic cell therapy laboratory. *Transfusion*, *64*(2), 357–366.
29. Schubert, J., Köppl, T., Tröndle, K., Greiner, J., Mann, A., Schuller, P., … & Krausslich, H. G. (2022). Mobile modular manufacturing of nucleic acid vaccines. *npj Vaccines*, *7*(1), 1–9.
30. European Medicines Agency. (2021). Good manufacturing practice (GMP) guidelines. *EMA*. Retrieved from https://www.ema.europa.eu/; U.S. Food and Drug Administration. (2020). Guidance for industry: Considerations for decentralized manufacturing of biological products. *FDA*.
31. Jensen, C., Jackson, B., Klymkowsky, M., & Kramer, E. (2023). Vaccines as a global public good: Rethinking the infrastructure of vaccine manufacturing. *Frontiers in Public Health*, 11, 1238535.
32. Farlow, A., Mortensen, J. J., & Singh, S. (2023). More-than-national and less-than-global: The biochemical infrastructure of vaccine manufacturing. *Frontiers in Research Metrics and Analytics*, 8, 1033795.
33. Mathur, A., Sharma, P., Dey, A., & Joshi, A. (2025). Volumetric additive manufacturing for cell printing: Bridging industry adaptation and regulatory frameworks. *Biomaterials*, 302, 122578.
34. Hort, N., Kocken, H., & Radisch, V. (2022). Metallic materials for biodegradable implants. *Acta Biomaterialia*, 123, 1–22. https://doi.org/10.1016/j.actbio.2021.11.001

35. Hort, S., Herbst, L., Bäckel, N., Erkens, F., Niessing, B., Frye, M., König, N., Papantoniou, I., Hudecek, M., Jacobs, J. J. L., & Schmitt, R. H. (2022). Toward rapid, widely available autologous CAR-T cell therapy—Artificial intelligence and automation enabling the smart manufacturing hospital. *Frontiers in Medicine*, 9, 913287. https://doi.org/10.3389/fmed.2022.913287
36. Shah, N. N., Vasu, S., et al. (2023). Point-of-care production of CAR-T products (perspective). *Frontiers in Transplantation*, 2, 1238535. https://doi.org/10.3389/frtra.2023.1238535
37. Lam, C., Meinert, E., Yang, A., & Cui, Z. (2021). Comparison between centralized and decentralized supply chains of autologous CAR-T therapies: A UK case study based on discrete-event simulation. *Cytotherapy*, 23(5), 433–451. https://doi.org/10.1016/j.jcyt.2020.08.007
38. Kedmi, M., Shouval, R., Fried, S., Bomze, D., Fein, J., Cohen, Z., et al. (2022). Point-of-care anti-CD19 CAR-T cells for relapsed and refractory aggressive B-cell lymphoma. *Transplantation and Cellular Therapy*, 28(5), 251–257. https://doi.org/10.1016/j.jtct.2022.02.017
39. Ozbolat, I. T., Peng, W., & Ozbolat, V. (2016). Application areas of 3D bioprinting. *Biofabrication*, 8(3), 032001.
40. Mathur, V., Dsouza, V., Srinivasan, V., & Vasanthan, K. S. (2025). Volumetric additive manufacturing for cell printing: Bridging industry adaptation and regulatory frontiers. *ACS Biomaterials Science & Engineering*, 11, 156–181.
41. George, A. S. (2024). The impact of IT/OT Convergence on digital transformation in manufacturing. *Partners Universal International Innovation Journal*, 2(2), 18–38.
42. Shilenge, M. C., & Telukdarie, A. (2022, June). Optimization of operational and information technology integration towards Healthcare Industry 4.0. In *2022 IEEE 31st international symposium on industrial electronics (ISIE)* (pp. 1076–1081). IEEE.
43. Garimella, P. K. (2018, October). IT-OT integration challenges in utilities. In *2018 IEEE 3rd International Conference on Computing, Communication and Security (ICCCS)* (pp. 199–204). IEEE.
44. Ehie, I. C., & Chilton, M. A. (2020). Understanding the influence of IT/OT Convergence on the adoption of Internet of Things (IoT) in manufacturing organizations: An empirical investigation. *Computers in Industry*, 115, 103166.
45. Gartner (2023b). *IT/OT Integration*. https://www.gartner.com/en/information-technology/glossary/it-ot-integration
46. Meyer, M., Zarnekow, R., & Kolbe, L. M. (2003). IT-Governance - Begriff, Status quo und Bedeutung. *Wirtschaftsinformatik*, 45(4), 445–448.
47. De Haes, S., & Van Grembergen, W. (2005, January). IT governance structures, processes and relational mechanisms: Achieving IT/business alignment in a major Belgian financial group. In *Proceedings of the 38th Annual Hawaii International Conference on System Sciences* (pp. 237b–237b). IEEE.
48. Rajagopal, R. (2014). *Sustainable value creation in the fine and speciality chemicals industry*. John Wiley & Sons.
49. Wagh, D. G., Shahi, S., Ingle, D. M. T., Khadabadi, S., & Karva, R. G. G. (2015). Quality by design in product development: A review. *Journal of Pharmaceutical Research*, 5(05).
50. Boutayeb, A. (2010). The impact of infectious diseases on the development of Africa. *Handbook of Disease Burdens and Quality of Life Measures*, 1171.
51. Sia, S. K., & Kricka, L. J. (2008). Microfluidics and point-of-care testing. *Lab on a Chip*, *8*(12), 1982–1983.
52. Kozel, T.R.; Burnham-Marusich, A.R. Point-of-care testing for infectious diseases: Past, present, and future. *Journal of Clinical Microbiology* 2017, 55, 2313–2320
53. Pashchenko, O.; Shelby, T.; Banerjee, T.; Santra, S.A. Comparison of optical, electrochemical, magnetic, and colorimetric point-of-care biosensors for infectious disease diagnosis. *ACS Infectious Diseases* 2018, 4, 1162–1178.
54. Lowdon, J.W.; Eersels, K.; Rogosic, R.; Boonen, T.; Heidt, B.; Diliën, H.; Van Grinsven, B.; Cleij, T.J. Surface grafted molecularly imprinted polymeric receptor layers for thermal detection of the new psychoactive substance 2-methoxphenidine. *Sensors and Actuators* 2019, 295, 586–595.
55. Vandenryt, T.; Van Grinsven, B.; Eersels, K.; Cornelis, P.; Kholwadia, S.; Cleij, T.J.; Thoelen, R.; De Ceuninck, W.; Peeters, M.; Wagner, P. Single-shot detection of neurotransmitters in whole-blood samples by means of the heat-transfer method in combination with synthetic receptors. *Sensors* 2017, 17, 2701.
56. Lowdon, J.W.; Eersels, K.; Rogosic, R.; Heidt, B.; Diliën, H.; Redeker, E.S.; Peeters, M.; Van Grinsven, B.; Cleij, T.J. Substrate displacement colorimetry for the detection of diarylethylamines. *Sensors and Actuators* 2019, 282, 137–144.
57. Xiong, L.-H.; Cui, R.; Zhang, Z.-L.; Yu, X.; Xie, Z.; Shi, Y.-B.; Pang, D.-W. Uniform fluorescent nanobioprobes for pathogen detection. *ACS Nano* 2014, 8, 5116–5124.
58. Campuzano, S.; Yáñez-Sedeño, P.; Pingarrón, J.M. Molecular biosensors for electrochemical detection of infectious pathogens in liquid biopsies: Current trends and challenges. *Sensors* 2017, 17, 2533.
59. Park, J.Y.; Kricka, L.J. Prospects for nano-and microtechnologies in clinical point-of-care testing. *Lab on a Chip* 2007, 7, 547–549.
60. Pai, N.P.; Vadnais, C.; Denkinger, C.; Engel, N.; Pai, M. Point-of-care testing for infectious diseases: Diversity, complexity, and barriers in low-and middle-income countries. *PLoS Medicine* 2012, 9, e1001306.
61. Haenssgen, M.J.; Charoenboon, N.; Althaus, T.; Greer, R.; Intralawan, D.; Lubell, Y. The social role of C-reactive protein point-of-care testing to guide antibiotic prescription in Northern Thailand. *Social Science & Medicine* 2018, 202, 1–12.
62. Wang, S.; Lifson, M.A.; Inci, F.; Liang, L.-G.; Sheng, Y.-F.; Demirci, U. Advances in addressing technical challenges of point-of-care diagnostics in resource-limited settings. *Expert Review of Molecular Diagnostics* 2016, 16, 449–459.
63. Kimani, F.W.; Mwangi, S.M.; Kwasa, B.; Kusow, A.M.; Ngugi, B.; Chen, J.; Liu, X.; Cademartiri, R.; Thuo, M.M. Rethinking the design of low-cost point-of-care diagnostic devices. *Micromachines* 2017, 8, 317
64. Lyon, L. Transparency: The emerging third dimension of open science and open data. *LIBER Quarterly* 2016, 25, 153–171.
65. Nosek, B.A.; Alter, G.; Banks, G.C.; Borsboom, D.; Bowman, S.D.; Breckler, S.J.; Buck, S.; Chambers, C.D.; Chin, G.; Christensen, G.; et al. Promoting an open research culture. *Science* 2015, 348, 1422–1425
66. Pineda, D. Sharing publication-related data and materials: National Research Council publishes report. *Sci. Edict*
67. Shibayama, S.; Walsh, J.P.; Baba, Y. Academic entrepreneurship and exchange of scientific resources: Material transfer in life and materials sciences in Japanese universities. *American Sociological Review* 2012, 77, 804–830.
68. Hou, C.Y.; Mayernik, M.S. Formalizing an attribution framework for scientific data/software products and collections. *International Journal of Digital Curation* 2016, 11, 87–103.

69. Vilkhovoy, M., Adhikari, A., Vadhin, S., & Varner, J. D. (2020). The evolution of cell free biomanufacturing. *Processes*, 8(6), 675.
70. Timm, A. C., Shankles, P. G., Foster, C. M., Doktycz, M. J., & Retterer, S. T. (2016). Toward microfluidic reactors for cell-free protein synthesis at the point-of-care. *Small*, 12(6), 810–817.
71. Nelson, J. A. D., Barnett, R. J., Hunt, J. P., Foutz, I., Welton, M., & Bundy, B. C. (2021). Hydrofoam and oxygen headspace bioreactors improve cell-free therapeutic protein production yields through enhanced oxygen transport. *Biotechnology Progress*, 37(2), e3079.
72. Hartman, R.L.; McMullen, J.P.; Jensen, K.F. Deciding whether to go with the flow: Evaluating the merits of flow reactors for synthesis. *Angewandte Chemie, International Edition* 2011, 50, 7502–7519.
73. McQuade, D.T.; Seeberger, P.H. Applying flow chemistry: Methods, materials, and multistep synthesis. *The Journal of Organic Chemistry* 2013, 78, 6384–6389
74. Brödel, A. K., Raymond, J. A., Duman, J. G., Bier, F. F., & Kubick, S. (2013). Functional evaluation of candidate ice structuring proteins using cell-free expression systems. *Journal of Biotechnology*, 163(3), 301–310.
75. Zawada, J.F.; Yin, G.; Steiner, A.R.; Yang, J.; Naresh, A.; Roy, S.M.; Gold, D.S.; Heinsohn, H.G.; Murray, C.J. Microscale to manufacturing scale-up of cell-free cytokine production–a new approach for shortening protein production development timelines. *Biotechnology and Bioengineering* 2011, 108, 1570–1578.
76. Yin, G.; Garces, E.D.; Yang, J.; Zhang, J.; Tran, C.; Steiner, A.R.; Roos, C.; Bajad, S.; Hudak, S.; Penta, K.; et al. Aglycosylated antibodies and antibody fragments produced in a scalable in vitro transcription-translation system. *MAbs* 2012, 4, 217–225.
77. Voloshin, A.M.; Swartz, J.R. Efficient and scalable method for scaling up cell-free protein synthesis in batch mode. *Biotechnology and Bioengineering* 2005, 91, 516–521.
78. Swartz, J. Developing cell-free biology for industrial applications. *Journal of Industrial Microbiology & Biotechnology* 2006, 33, 476–485.
79. Ogonah, O.W.; Polizzi, K.M.; Bracewell, D.G. Cell-free protein synthesis: A viable option for stratified medicines manufacturing? *Current Opinion in Chemical Engineering* 2017, 18, 77–83
80. Behrendt, A., Duron, A., Kelly, R., Mussler, S., Rettig, R., & Stoffregen, S. (2018). *Digital manufacturing: Escaping pilot purgatory*. McKinsey & Company.
81. Vamathevan, J., Clark, D., Czodrowski, P., Dunham, I., Ferran, E., Lee, G., ... & Zhao, S. (2019). Applications of machine learning in drug discovery and development. *Nature Reviews Drug Discovery*, 18(6), 463–477.
82. Rogers, A., & Ierapetritou, M. (2015). Challenges and opportunities in modeling pharmaceutical manufacturing processes. *Computers & Chemical Engineering*, 81, 32–39.
83. Benedetti, A., Khoo, J., Sharma, S., Facco, P., Barolo, M., & Zomer, S. (2019). Data analytics on raw material properties to accelerate pharmaceutical drug development. *International Journal of Pharmaceutics*, 563, 122–134.
84. Chatterjee, S., Moore, C. M., & Nasr, M. M. (2017). An overview of the role of mathematical models in implementation of quality by design paradigm for drug development and manufacture. *Comprehensive Quality by Design for Pharmaceutical Product Development and Manufacture*, 9–24.
85. Parasrampuria, D. A., Benet, L. Z., & Sharma, A. (2018). Why drugs fail in late stages of development: case study analyses from the last decade and recommendations. *The AAPS Journal*, 20, 1–16.
86. Glass, H. E., Kolassa, E. M., & Muniz, E. (2016). Drug development through modeling and simulation-The business case. *Applied Clinical Trials*, 25.
87. Appl, C., Moser, A., Baganz, F., & Hass, V. C. (2021). Digital Twins for bioprocess control strategy development and realisation. *Digital Twins: Applications to the Design and Optimization of Bioprocesses*, 63–94.
88. Narayanan, H., Luna, M. F., von Stosch, M., Cruz Bournazou, M. N., Polotti, G., Morbidelli, M., ... & Sokolov, M. (2020). Bioprocessing in the digital age: the role of process models. *Biotechnology Journal*, 15(1), 1900172.
89. Kis, Z., Kontoravdi, C., Dey, A. K., Shattock, R., & Shah, N. (2020). Rapid development and deployment of high-volume vaccines for pandemic response. *Journal of Advanced Manufacturing and Processing*, 2(3), e10060.
90. Portela, R. M., Varsakelis, C., Richelle, A., Giannelos, N., Pence, J., Dessoy, S., & von Stosch, M. (2021). When is an in silico representation a digital twin? A biopharmaceutical industry approach to the digital twin concept. *Digital Twins: Tools and Concepts for Smart Biomanufacturing*, 35–55.
91. Hutter, C., von Stosch, M., Cruz Bournazou, M. N., & Butté, A. (2021). Knowledge transfer across cell lines using hybrid Gaussian process models with entity embedding vectors. *Biotechnology and Bioengineering*, 118(11), 4389–4401.
92. Talib, M. A., Majzoub, S., Nasir, Q., & Jamal, D. (2021). A systematic literature review on hardware implementation of artificial intelligence algorithms. *The Journal of Supercomputing*, 77(2), 1897–1938.
93. Macdonald, G. J. (2020). Biomanufacturing makes sense of the Healthcare Industry 4.0 concept: the digital transformation of biomanufacturing poses short-term difficulties but promises long-term benefits through real-time monitoring and dynamic process optimization. *Genetic Engineering & Biotechnology News*, 40(S3), S7–S10.
94. Mak, K. K., & Pichika, M. R. (2019). Artificial intelligence in drug development: present status and future prospects. *Drug Discovery Today*, 24(3), 773–780.
95. Tripathi, A., Misra, K., Dhanuka, R., & Singh, J. P. (2023). Artificial intelligence in accelerating drug discovery and development. *Recent Patents on Biotechnology*, *17*(1), 9–23.

7 Biomanufacturing Tools for Sustainable Bioenergy Production

Kulvinder Bajwa
University School for Graduate Studies, Chaudhary Devi Lal University, Sirsa, Haryana, India

Rakesh Choudhary and Ajay Kumar
Department of Civil Engineering, National Institute of Technology Delhi, Delhi 110036, India

Muhammad Yousuf Jat Baloch
School of Environmental Science and Engineering, Shandong University, Qingdao, Shandong, PR China

Bhanu Pratap Singh
Department of Civil Engineering, Faculty of Engineering and Technology, SRM Institute of Science and Technology, Ghaziabad, Uttar Pradesh, India

Neha Chaudhary
Research and Development Cell, Swami Vivekanand Subharti University, Meerut, Uttar Pradesh, India

Sowkhya Naidu
Waste to Resources Laboratory, Department of Biotechnology and Chemical Engineering, Manipal University Jaipur, Jaipur, Rajasthan, India

Omid Reza Baghchesaraei
Centre for Infrastructure Engineering, Sydney, Australia

7.1 INTRODUCTION

Biomanufacturing (BM) is a manufacturing process that uses biological systems, such as living microorganisms, cells, plants, animals, tissues, enzymes, or synthetic enzymatic systems, to produce valuable biomolecules for commercial purposes. Biomolecules play a crucial role across various sectors, including agriculture, food production, energy, materials, and pharmaceuticals (Schmaltz et al. 2023). In the bioenergy context, BM enables the conversion of renewable feedstocks into fuels and energy carriers through controllable, scalable bioprocesses. The products of BM are sourced from natural origins, such as blood, microbial cultures, animal cells, or plant cells, which thrive in controlled environments or specialized equipment. The cells, tissues, or enzymes utilized can either be naturally occurring or enhanced through advanced techniques such as genetic engineering, metabolic engineering, synthetic biology, and protein engineering. BM has contributed notably to the previous three industrial revolutions; however, in the 21st century, Biomanufacturing 4.0 (BM 4.0), also referred to as advanced BM, is expected to serve as a foundational element of the ongoing sustainability revolution (Isoko et al. 2024). This transition is characterized by the convergence of high-throughput biology, real-time sensing, and computational decision–support to improve yield, robustness, and reproducibility at scale. The three previous industrial revolutions include: (1) the first, which began in

DOI: 10.1201/9781003637158-7

Phase	Needs and Products	Methods	Key Figures and Time	
Pre-modern BM	Fermented products	Mixed cultures + solid-state fermentation	Louis Pasteur (1860s)	
BM 1.0	Primary metabolites	Mono-culture + anaerobic liquid fermentation	Chaim Weizmann (1910s)	
BM 2.0	Secondary metabolites	Mono-culture + aerobic liquid fermentation	Alexander Fleming (1940s)	
BM 3.0	Proteins, enzymes	rDNA technology + cell cultures	Paul Berg (1980s)	
BM 4.0	Renewable energy, new food, better life	Emerging technologies (e.g., IPS, ME&SB, PE, ivSB)	Future Innovations	

FIGURE 7.1 Historical evolution of BM phases, from Pre-modern BM to BM 4.0, with associated products, methodologies, and pioneering figures (Zhang, Sun, and Ma 2017).

late 18th-century Britain and was marked by the mechanization of the textile industry, steam power, and coal mining; (2) the second, which emerged in the early 20th century in the United States and featured the widespread use of internal combustion engines, fossil fuels, electricity, and mass production through assembly lines; and (3) the third, characterized by the advent and extensive use of computers and the internet. In the current Industry 4.0 era, digitalization, automation, and data-centric control are increasingly embedded into manufacturing systems, enabling higher precision and adaptive operation. In comparison, BM 4.0 is anticipated to serve as a transformative platform, enabling the development of entirely new products or significantly improving existing ones through superior technologies (Macdonald 2020).

7.2 FUNDAMENTALS OF BIOMANUFACTURING IN BIOENERGY

In this context, the history of BM can be broadly classified based on product types, biocatalysts, and technological advancements. Generally, improvements in technology and the development of new products occur in tandem, mutually reinforcing each other and propelling the evolution of biomanufacturing forward. For bioenergy, this co-evolution is reflected in how advances in strain design, reactor control, and downstream processing progressively expanded the range of viable fuels and feedstocks.

To gain a clearer understanding of the evolution of BM, we can categorize it into five distinct phases: Pre-modern BM (often treated as foundational practice prior to industrial BM), Pre-modern BM 1.0, BM 2.0, BM 3.0, and BM 4.0. Significant technological milestones, methodologies, and societal needs define each phase. These phases highlight the transition from early fermentation processes to sophisticated recombinant systems, ultimately leading to Industry 5.0-integrated biosystems, as illustrated in Figure 7.1. In the bioenergy domain, the same progression enables improved conversion efficiency, inhibitor tolerance, and process stability under industrial conditions.

Generally, market demands and potential profits are the primary motivators, while technological innovations facilitate product commercialization, and in turn, market needs stimulate further technological progress. This feedback loop is particularly evident in bioenergy, where policy targets and fuel markets influence both feedstock selection and the pace of bioprocess innovation.

7.2.1 Pre-Modern Biomanufacturing

Pre-modern BM has roots that extend back thousands of years, when humans unknowingly harnessed natural mixed microbial cultures for fermentation. This process transformed basic ingredients such as sugars, fruits, and grains into essential products, such as wine, beer, and bread. Civilizations, including the ancient Chinese, Sumerians, Babylonians, and Egyptians, employed these techniques long before they had a scientific understanding of the underlying processes (Metwaly et al. 2021). These spontaneous, predominantly anaerobic fermentations served not only for food and beverages but also for the production of items, such as dyes and leather. This era is defined by the use of solid-state fermentation and naturally occurring microbes. Although not designed for energy recovery, these early fermentation practices established the biochemical basis for modern anaerobic conversion routes relevant to bioenergy. Even today, traditional practices such as winemaking continue to evolve, with contemporary efforts focused on improving grain saccharification, yeast interactions, and fermentation efficiency (Sant'Ana et al. 2024).

7.2.2 Biomanufacturing 1.0

BM 1.0 began in the 1910s with Chaim Weizmann's development of acetone butanol ethanol (ABE) fermentation. This phase introduced two key innovations: the use of purified single microbial strains instead of mixed cultures and the application of large-scale anaerobic liquid fermentation. Its main products were primary metabolite compounds, such as ethanol, butanol, acetone, amino acids, and organic acids that are essential for microbial growth and function (Sanchez and Demain 2008). From a bioenergy perspective, ABE fermentation represents one of the earliest industrial demonstrations of solvent and fuel-relevant bioconversion.

The ABE fermentation history highlights two significant insights: (1) product focus shifts over time based on demand, and (2) market needs rapidly drive technological progress. Initially, interest in butanol was primarily focused on the production of synthetic rubber. However, following the collapse of the rubber market, Weizmann shifted his focus to acetone, which was critically needed during World War I for the manufacture of explosives. This process was scaled up to industrial levels, making *n*-butanol a valuable byproduct used as a solvent and in various other materials. In recent years, there has been renewed interest in *n*-butanol as a biofuel. Today, Weizmann is honored as the pioneer of industrial fermentation (Nguyen et al. 2018).

During the 20th century, BM 1.0 advanced, enabling the production of various primary metabolites, notably organic acids and amino acids. For instance, a shortage of calcium citrate during World War I, previously imported from Italy, led Pfizer to develop a fermentation-based process using fungi to produce citric acid from sugars, which was commercialized in 1919. In the amino acid sector, monosodium glutamate (MSG) became the most widely produced, reaching around multi-million-tonne annual production levels. Initially derived through protein hydrolysis and later chemical synthesis, production eventually transitioned entirely to microbial fermentation due to its cost-effectiveness, reduced environmental impact, and improved food safety. This shift prompted all MSG manufacturers to adopt fermentation methods. Today, the amino acid industry generates over tens of billions of United States Dollars (USD) in annual revenue. Lysine, the second-most-produced amino acid, is also manufactured through fermentation. Key producers are located in China, Japan, South Korea, the United States, and Europe. These industrial precedents are relevant to bioenergy because they established scale-up principles, contamination control, and cost-sensitive fermentation operation that later informed fuel-focused bioprocesses.

Additionally, the rising demand for biodegradable plastics, like bio-polyesters, is creating expanding markets for organic acids such as D -lactic acid, L-lactic acid, succinic acid, fumaric acid, and glucaric acid (Magalhães Júnior et al. 2021). This diversification signals an early shift toward integrated biorefineries where fuels and chemicals co-exist as co-products.

7.2.3 Biomanufacturing 2.0

BM 2.0 represents a transformative era in industrial biotechnology, emerging in the mid-to-late 20th century. This phase marked a shift from producing basic primary metabolites to more complex and valuable secondary metabolites and recombinant proteins, driven by advancements in molecular biology and genetic engineering. The introduction of recombinant DNA technology in the 1970s enabled scientists to modify microorganisms at the genetic level, leading to the production of non-native or enhanced biomolecules. One of the most notable achievements of this era was the commercial production of recombinant human insulin in *Escherichia coli* by Genentech and Eli Lilly in the 1980s, revolutionizing diabetes treatment by replacing animal-derived insulin. Additionally, microbial strains were optimized for the efficient production of antibiotics and other secondary metabolites, improving drug manufacturing processes (Riggs 2021). BM 2.0 also paved the way for biopharmaceuticals, including therapeutic proteins, vaccines, and monoclonal antibodies (mAbs), with engineered yeasts, bacteria, and mammalian cells serving as production platforms. Alongside these advancements, improvements in bioreactor design, downstream processing, and fermentation conditions enhanced scalability, yield, and purity. Another key development was the rise of industrial enzymes, which replaced conventional chemical processes in industries such as detergents, food production, and textiles. For bioenergy, the most direct relevance of BM 2.0 lies in the maturation of enzyme production platforms and recombinant strain engineering, which later enabled improved biomass deconstruction and targeted pathway optimization. Collectively, this phase set the foundation for modern biopharmaceuticals and enzyme industries, driving the transition from bulk commodity chemicals to precision biologics.

7.2.4 Biomanufacturing 3.0

BM 3.0 began in the 1980s, marking a transition toward the production of complex, large-sized proteins—particularly therapeutic proteins and enzymes rather than primary or secondary metabolites. This new era was made possible by two major scientific breakthroughs: the development of recombinant DNA technology and advancements in cell culture techniques. The foundation of this platform traces back to 1973, when Stanley Cohen and Herbert Boyer created the first recombinant DNA plasmid in vitro. Their innovation sparked the formation of commercial biotech ventures focused on harnessing this technology. Since large proteins are too complex and expensive to synthesize chemically or isolate from natural sources, early efforts focused on producing high-value therapeutic proteins. In 1976, Genentech was founded by Robert Swanson and Herbert Boyer to develop recombinant human insulin and growth hormone. Among the numerous recombinant therapeutic proteins, non-glycosylated proteins are typically produced in *E. coli* or yeast, accounting for approximately 40% of the therapeutic protein market (Huang, Lin, and Yang 2012).

In 1980, Amgen was founded to produce recombinant human erythropoietin (EPO), a glycoprotein hormone that stimulates red blood cell production. Because EPO's biological function depends on its carbohydrate attachments, it must be produced in mammalian cell cultures, rather than in microbes. EPO has become one of the most commercially successful protein therapeutics, with sales reaching $13 billion in 2007 (Chia et al. 2023). However, protein glycosylation can vary depending on the host cells used, thereby affecting their activity. To overcome the challenges of producing high-quality, consistent glycoproteins at lower costs in mammalian systems, Dr. Daniel IC Wang, known as the "son of biochemical engineering," made key contributions. Another significant development in BM 3.0 is the rise of antibody drug conjugates (ADCs) for cancer treatment, such as Adcetris® and Kadcyla®. Unlike traditional therapies that harm healthy tissue, ADCs use mAbs that are accepted to target cancer cells precisely, delivering toxic agents directly to tumors. This targeted approach, which links mAbs to potent chemotherapy drugs via stable linkers, has led to the emergence of a highly effective and rapidly expanding class of anti-cancer biologics (Riccardi et al. 2023).

7.2.5 Emerging Biomanufacturing 4.0

In the early 21st century, rising demands for innovations, such as renewable energy, cultured food, and regenerative therapies, combined with cutting-edge scientific tools, including induced pluripotent stem cells (iPSCs), metabolic and systems biology, synthetic biology, and cascade biocatalysis, have triggered a new era in BM. This advanced approach is designed to generate novel products, such as engineered tissues and new pharmaceuticals, or enhance the production of existing ones with significantly greater efficiency. Key improvements are anticipated across yield, productivity, product concentration, scalability, and environmental impact (Maldini, Klepp, and Laitala 2025). For bioenergy BM, BM 4.0 is characterized by data-rich Design–Build–Test–Learn cycles, automated experimentation, and AI-assisted optimization that shorten development timelines while improving process robustness.

7.3 FEEDSTOCK SELECTION AND PRETREATMENT STRATEGIES FOR BIOENERGY BIOMANUFACTURING

Bioenergy refers to the energy derived from organic matter, or biomass, primarily sourced from plants. During photosynthesis, plants absorb carbon from the atmosphere to produce biomass. This form of energy is highly versatile and utilized in various sectors, including solid biomass, biogases, and liquid biofuels, for transportation modes such as cars, ships, and airplanes (Elazab et al. 2025). From a BM perspective, the chemical composition and heterogeneity of biomass strongly influence microbial conversion efficiency and process design. A significant advantage of bioenergy lies in its compatibility with existing infrastructure. For instance, current natural gas pipelines and equipment transport biomethane. Additionally, most drop-in liquid biofuels are compatible with conventional vehicles and can be seamlessly integrated into existing oil distribution systems with minimal modifications. The processing technology for biomass is essential in producing bioenergy, presenting a promising avenue for reducing carbon emissions and tackling climate change. Bioenergy production depends on a diverse range of feedstocks, typically chosen based on factors such as availability, composition, cost, and sustainability. Among these, food and agro-industrial residues have received increasing attention as versatile feedstocks capable of supporting multiple biofuel production routes through biochemical and thermochemical conversions (Choudhary, Tigga, et al. 2026). Such residues align well with circular bioeconomy principles by valorizing waste while minimizing competition with food resources. This growing evidence base highlights their strategic importance in sustainable bioenergy BM. These feedstocks are generally categorized as follows:

7.3.1 Solid Biomass

Solid biomass includes lignocellulosic materials such as wood, agricultural residues (e.g., straw, husks, and corn stover), forestry byproducts, and dedicated energy crops. These materials are primarily composed of cellulose, hemicellulose, and lignin, which together define their recalcitrance and conversion potential. In bioenergy BM, solid biomass serves as a key feedstock for bioethanol, biogas, and thermochemical fuels, but requires appropriate pretreatment to overcome structural resistance. Factors such as moisture content, ash composition, bulk density, and seasonal availability influence both logistics and process efficiency. Consequently, solid biomass utilization often relies on region-specific supply chains and integrated pretreatment conversion strategies.

7.3.2 Liquid Biofuel Sources (Sugar and Starch-Based Crops)

A promising alternative to petroleum-based fossil fuels is liquid biofuel. Among potential sources, lignocellulosic biomass stands out for its non-competition with food supplies. This type of biomass can be sourced from agricultural and forestry residues, grasses, and dedicated energy crops. Bioethanol, a widely used form of bioenergy, is primarily produced from crops high in starch content, such as sugarcane and maize. Biodiesel, conversely, consists mainly of long-chain alkyl esters and serves as a substitute for conventional diesel. It is typically produced by reacting vegetable oils or animal fats with alcohols. The liquid biofuel (first-generation biofuels) markets are well-established, dynamic, and complex, with strong ties to agricultural commodity markets. Large volumes of feedstocks, such as sugar, starch, and oil-producing crops, are frequently traded and processed into biofuels, such as bioethanol and biodiesel, also often within the importing countries (Roy & Abedin 2022). While technologically mature, these systems face sustainability concerns related to land use and competition for food and fuel, motivating a transition toward waste- and residue-based feedstocks.

7.3.3 Oil-Based Crops and Waste Oils in Bioenergy Biomanufacturing

Oil-rich crops and waste oils are important resources for bioenergy production, especially biodiesel. Common oil-producing crops, such as soybeans, rapeseed (also known as canola), sunflowers, and oil palms, are grown for their ability to yield substantial oil, which can then be processed into fuel. These plant oils are renewable and are typically converted into biodiesel through a chemical process called transesterification, in which oils react with alcohol. Alongside these crops, waste oils-including used cooking oil, fats from food processing, and animal-based greases-offer an environmentally friendly and economical alternative. The use of waste oils reduces feedstock costs and lifecycle emissions while improving overall process sustainability. Using such waste not only helps reduce pollution but also supports sustainable energy practices by repurposing byproducts into fuel. Combining cultivated oils and waste oils in biofuel production helps reduce dependence on fossil fuels, cuts greenhouse gas (GHG) emissions, and enhances overall energy sustainability (Monika & Pathak 2023).

7.3.4 Algae for Bioenergy Biomanufacturing

Algae have emerged as a highly promising feedstock for bioenergy BM due to their fast growth rate, high lipid content, and ability to thrive in a variety of environments. Unlike traditional crops, algae do not require arable land or fresh water, making them a sustainable alternative that avoids competition with food production. Microalgae, in particular, are capable of producing large quantities of oils that can be converted into biodiesel through transesterification. In addition to biodiesel, algae can be processed to generate other forms of bioenergy, such as bioethanol, biogas, and even bio-jet fuel, depending on the processing method used. Moreover, algae cultivation can contribute to environmental benefits by capturing carbon dioxide during photosynthesis and aiding in wastewater treatment. These multifunctional roles position algae as a bridge between energy production and environmental remediation. This dual role in energy production and environmental remediation makes algae a valuable resource in the development of cleaner and more sustainable bioenergy systems (Rawat et al. 2011).

7.3.5 Organic Waste and Industrial Byproducts in Bioenergy Biomanufacturing

Organic waste and byproducts from various industries serve as sustainable, efficient sources of bioenergy. In addition to their availability, several studies have demonstrated that food and agro-industrial residues can be systematically converted into liquid and gaseous biofuels through well-established BM pathways, providing practical evidence for waste-to-energy integration (Choudhary, Harne, et al. 2026). Such case-based demonstrations strengthen the feasibility of integrating organic waste valorization within industrial bioenergy production systems. These include materials such as food waste, crop residues, animal manure, and leftovers from industrial processes, including food production, paper manufacturing, and wastewater treatment. Instead of allowing these wastes to contribute to environmental issues, they can be transformed into renewable energy through methods such as anaerobic digestion, which breaks down organic matter to produce biogas, a fuel rich in methane. Other processes, such as fermentation, can convert suitable waste into bioethanol, while solid residues can be used in combustion or converted into biochar for energy and soil improvement. Beyond energy carriers alone, recent BM case studies have shown that food and agro-industrial wastes can also be transformed into bio-based polymers, thereby expanding the product spectrum from fuels to sustainable materials (Choudhary, Kumar, et al. 2026). This dual pathway of energy and material recovery reinforces the role of BM in advancing a circular bioeconomy. Utilizing these waste streams helps reduce landfill use, lower GHG emissions, and promote energy recovery. Such sustainability gains are largely enabled by biological transformation routes, where microbial and enzymatic processes convert complex organic substrates into value-added bioenergy products under comparatively mild operating conditions (Kumar et al. 2026). These biological pathways form the biochemical foundation upon which modern BM systems are designed. This approach supports environmental sustainability and fosters a more circular economy by turning waste into valuable energy resources (Aslan et al. 2025).

7.3.6 Aquatic Biomass for Bioenergy and Biomanufacturing

Aquatic biomass, derived from water-based organisms such as algae (microalgae and macroalgae), seaweed, and aquatic plants, holds significant promise for sustainable bioenergy production and BM. Unlike terrestrial crops, aquatic biomass can grow rapidly, requires less arable land, and can utilize non-potable water sources (Omokaro et al. 2025). Its biochemical diversity enables multiple conversion pathways, including fermentation, anaerobic digestion, and lipid extraction, although harvesting and seasonal variability remain key challenges for large-scale deployment.

7.3.7 Pretreatment Strategies

The combination of appropriate feedstock selection and tailored pretreatment strategies is pivotal to the viability of bioenergy BM. Innovations in pretreatment technologies and the use of non-conventional feedstocks are driving improvements in process efficiency, economic feasibility, and sustainability. Hence, lignocellulose is considered a carbon-neutral bioenergy feedstock with vast availability, making it a sustainable alternative to fossil fuels. Global lignocellulose production is estimated to reach around 200 billion metric tonne annually, primarily from agricultural and forestry residues, horticultural waste, and commercial crops. The future lies in integrated biorefineries that utilize multiple feedstock sources with optimized pretreatment pipelines to maximize resource utilization and minimize waste. However, effective pretreatment is essential to unlock this potential by overcoming biomass recalcitrance.

Researchers consider pretreatment crucial for disrupting the complex biomass structure, enhancing enzyme accessibility, and increasing fermentable sugar yields. They base the choice of pretreatment method on the type of feedstock and the desired downstream product (Jain and Kumar 2024). Experts categorize lignocellulose pretreatment methods into two main categories: conventional and advanced technologies. Traditional approaches encompass physical, chemical, and biological techniques. The latest advancements in these pretreatment methods are outlined below.

7.3.8 Physical Treatment

Physical pretreatment involves mechanical or thermal processes that alter the structural and physical characteristics of lignocellulosic biomass to enhance its accessibility to enzymes or microbes during subsequent processing stages. These methods (milling, grinding, shredding, and chipping) are typically the first step in the bioconversion process and are often combined with chemical or biological treatments to improve efficiency.

7.3.9 Chemical Treatment

Chemical pretreatment of lignocellulosic feedstock involves the use of various chemicals to break down the complex plant cell wall structure, primarily targeting the removal of lignin and hemicellulose to improve the accessibility of cellulose for enzymatic hydrolysis. Standard methods include acid pretreatment, which hydrolyzes hemicellulose but may produce fermentation inhibitors; alkaline pretreatment, effective in lignin removal with fewer inhibitors; oxidative pretreatment using agents like hydrogen peroxide for lignin breakdown; and organic solvent pretreatment, which employs organic solvents to solubilize lignin while recovering high-purity lignin as a byproduct Emerging methods such as ionic liquids and deep eutectic solvents offer promising, recyclable alternatives for selective biomass deconstruction. However, they remain costly and are not yet widely commercialized. Each method varies in effectiveness, cost, and environmental impact, and is often chosen based on the specific feedstock and desired downstream application (Chozhavendhan et al. 2020).

7.3.10 Biological Treatment

Biological pretreatment of lignocellulosic biomass involves the use of microorganisms, mainly specific fungi and bacteria or their enzymes, to break down the complex structure of the biomass, particularly lignin. White-rot, brown-rot, and soft-rot fungi are commonly used due to their ability to secrete lignin-degrading enzymes, such as laccase, manganese peroxidase, and lignin peroxidase, which help selectively remove lignin while preserving cellulose and hemicellulose. Certain bacteria, including actinomycetes, also play a role in the degradation of lignocellulose. However, their action tends to be slower compared to fungi. This method is considered environmentally sustainable since it operates under mild conditions and avoids the use of harsh chemicals. Despite its eco-friendliness and low energy requirements, biological pretreatment is often limited by slow processing times, making it less favorable for large-scale industrial use unless integrated with other pretreatment methods. Nonetheless, it remains a valuable approach for enhancing cellulose accessibility and minimizing the formation of inhibitory compounds in downstream bioconversion processes (Jain and Kumar 2024).

7.4 METABOLIC ENGINEERING AND SYNTHETIC BIOLOGY APPLICATIONS

Metabolic engineering and **synthetic biology** are crucial for advancing bioprocessing, as they enable targeted modifications of microorganisms to enhance the production of bio-based products. Through metabolic engineering, cellular pathways can be restructured to increase the yield of desired compounds, minimize the formation of unwanted byproducts, and improve microbial tolerance to industrial stress conditions. Meanwhile, synthetic biology allows for the construction of entirely new metabolic routes and regulatory elements, equipping scientists with tools to design microbes with tailored functionalities. These approaches broaden the range of usable raw materials, including low-cost and waste-derived feedstocks, and support the optimization of performance under large-scale production conditions. Together, they enhance multiple stages of bioprocessing, from strain engineering to fermentation and product recovery, contributing to the development of efficient, cost-effective, and sustainable systems for the production of biofuels, chemicals, and other valuable bioproducts (Konwar, Banik, and Geed 2025).

7.5 BIOPROCESS OPTIMIZATION AND SCALE-UP

Bioprocess optimization and scale-up are vital to advancing bench-scale bioenergy developments into large-scale industrial applications. A prime aspect of this is bioreactor arrangement and design. The Stirred Tank Reactor (STR) is often the primary apparatus due to its excellent control over essential metrics such as pH, temperature, dissolved oxygen, and agitation, particularly when seeking consistent yields across varying scales. For particular shear-sensitive cultures or high-density systems such as algae or filamentous fungi used for biodiesel or biohydrogen production, a bubble column or airlift reactor can offer superior mass transfer while reducing shear and improving viability and productivity. Fermentation parameter optimization requires precise levels of control over inoculum density, oxygen, and substrate feed. These controls are interdependent and non-linear, making precise optimization of fermentation parameters a complex proposition. Multivariate optimization requires and can involve the use of Response Surface Methodology, Taguchi design, and genetic algorithms. Real-time monitoring tools, such as dielectric spectroscopy and Raman spectroscopy, are useful for measuring metabolic flux at the moment and stress levels in the system before a significant yield reduction. Predictive control and automated correction are important to relieve yield reduction (Jeyapaul, Shahabudeen, and Krishnaiah 2006). The scale-up precision of laboratory to industrial plateau remains one of the biggest challenges for bioenergy processes. Small-scale bioreactor tasks for mixing and environmental conditions will usually produce much more uniformly than industrial-scale STRs, which can yield significant gradients in pH, oxygen level, and nutrients, leading to reduced cell count and yield reduction. Advanced Computational Fluid Dynamics (CFD) models aid this work by predicting impeller design and sparging modifications (Nickel et al. 2024; Ebrahimi et al. 2019). Smart manufacturing in the Industry 5.0 paradigm uses cyber-physical systems (CPS) with highly sensorized reactors, and the Internet of Things (IoT) to capture data in real time for adaptive control strategies (Isoko et al. 2024). Virtual bioprocesses, known as digital twins, can use live data and AI to enhance operations (Udugama et al. 2021; Isoko et al. 2024). Together, these approaches provide predictive diagnostics and can operate with more uptime to increase throughput (Nickel et al. 2024). For example, in bioenergy systems such as biogas or biodiesel, technical pathways or strategies, such as Continuous Stirred Tank Reactors (CSTRs) and membrane bioreactors, are used to enhance volumetric productivity. CSTRs and anaerobic fixed-film reactors have been shown to be tolerant to fluctuations in feedstock. Membrane bioreactors can facilitate *in situ* product recovery and cell retention, thereby increasing yields while reducing energy input (Gunes 2021). Photobioreactors are increasingly popular; however, they are capital intensive. They enable sustainable CO_2 fixation and lipid accumulation in microalgae (Camarero et al. 2014). For scaling-up processes, downstream integration, harvesting, recovery, and purification have been identified as critical components of adequate scale-up (or higher productivity). For instance, Consolidated Bioprocessing (CBP) is a novel approach for lignocellulosic ethanol production in which enzyme production, hydrolysis, and fermentation occur within a single process unit. This integration reduces costs and energy associated with the unit operations, while increasing output (Olguin-Maciel et al. 2020).

7.6 LIFECYCLE ASSESSMENT AND ENVIRONMENTAL IMPACT

Lifecycle assessment (LCA) is a systematic and quantitative way of measuring the environmental impacts of bioenergy systems, presented in a standardized framework. Following the International Organization for Standardization (ISO) 14040 framework as a guiding point, LCA has four stages: goal and scope, inventory, impact assessment, and interpretation. The key impact categories that can be assessed, as determined by the LCA methodology, are global warming potential (GWP), eutrophication, water footprint, fossil energy demand, and land-use (Volkart et al. 2013). The differences in water and land-use between first-generation biofuels, such as corn-based ethanol, and third-generation biofuels, such as those derived from algae, are substantial. Third-generation biofuels have less land-use needs but require more energy to grow. When comparing the LCA of biofuels from starches and sugars, such as corn and cane, the processing energy for converting biomass to produce ethanol increases the GHG emissions, although they are a little bit lower compared to lignocellulosic ethanol, where the theoretical strategy is to cultivate the biomass but keep the energy inputs low when converting the biomass into sugars and then to ethanol. Lignocellulosic ethanol also has environmental impacts associated with land use, but thus far it yields a lower GWP than ethanol derived from starches or sugars, and has higher processing energy requirements for those biomass resources (Dhanya et al. 2020; Morales et al. 2015). Biogas derived from organic waste has a low environmental burden, especially in LCA metrics, because the organic waste is not an added carbon source to the environment, and the environmental consequences of the organic waste can be valued as a decrease in methane emissions.

Table 7.1 provides a comparative snapshot of lifecycle metrics, including GWP, water footprint, fossil energy demand, and land use, across major bioenergy pathways, highlighting the environmental efficiency of biogas and lignocellulosic ethanol compared to first-generation fuels. Indirect consequences are being evaluated with consequential and attributional LCA models, which are being used to evaluate indirect consequences, such as land-use change and displacement behavior of bioenergy production from food systems. For instance, in one study research

TABLE 7.1
Comparative Environmental Metrics (GWP, Water Footprint, Energy Demand, and Land Use) for Key Bioenergy Pathways per Unit Energy Produced

Bioenergy Pathway	GWP (kg CO_2 eq/MJ)	Water Footprint (L/MJ)	Fossil Energy Demand (MJ/MJ)	Land Use (m^2/MJ)	Reference
Corn-based bioethanol	0.52–0.59	10.8–13.2	0.78–0.85	0.72–0.80	Morales et al. (2015)
Algae-based biodiesel	0.28–0.35	4.5–6.1	0.90–0.98	0.12–0.18	Camarero et al. (2014)
Lignocellulosic bioethanol	0.22–0.27	3.8–4.6	0.57–0.64	0.18–0.22	Volkart et al. (2013)
Biogas from organic waste	0.13–0.17	1.0–1.5	0.18–0.23	0.04–0.06	Dhanya et al. (2020)

converting food waste to produce biohydrogen on an LCA framework claimed indirect environmental impacts, thus providing more of an environmental benefit to avoid storing food waste in a landfill and produce biohydrogen. Multifunctional analysis of biochar, for example, can provide more understanding about sustainability when defining the impact and performing an LCA. Therefore, when biomass resources are harvested for biochar applications, GHG emissions to the environment can be appropriately categorized and linked to biofuels to assess environmental impacts, thereby enabling a multifunctional analysis. Burning biomass for energy in a power conversion setting can also offer similar advantages. Electricity produced can provide ignition for burning woody biomass, which has been harvested intentionally to reduce GHG emissions. Linking LCA values and metrics with techno-economic assessments (TEA) in the realm of bioenergy production can capture the economic feasibility, environmental burdens, mitigation strategies, and functional applications of a biomass resource. For example, algal biodiesel may show strong GHG savings but weak economic benefits (Camarero et al. 2014). Hybrid LCA-TEA models can be used to identify pathways with favorable sustainability-cost trade-offs in sustainable processes. As a more direct example of how LCA and TEA methodologies are used to assess the sustainability impacts of biofuels, the European Union Renewable Energy Directive (EU RED) II policy framework requires a minimum GHG savings of 65% compared to fossil fuels for biofuels to qualify for government incentives. The GHG savings required by the government at the EU level rely heavily upon metrics derived from LCA. Similarly, the recently developed Environmental Product Declarations and corporate carbon accounting rely on both LCA and TEA metrics ingrained in the data. To do LCA with data for economic inventory and regional-based factors and maturity, databases, such as SimPro, Ecoinvent, GaBi, GREET, and OpenLCA, contain relevant data from LCA studies to help establish benchmarks when using the software applications for dematerialization and decarbonization metrics. Additionally, spatial LCA studies and methods can consider location-based issues such as water scarcity and biodiversity impacts, when assessing the sustainability of site-specific bioenergy (Choudhary, Kumar, Narayan, et al. 2025). LCA also faces challenges, such as methodological limitations related to data gaps, boundary setting, and uncertainty regarding temporal and spatial relevance, which are particularly concerning for emerging synthetic biology systems (Chavez et al. 2025). To address such uncertainty and multi-criteria trade-offs, hybrid assessment frameworks that integrate LCA with fuzzy logic and decision–support tools have been proposed to improve robustness in sustainability evaluations. These approaches are especially relevant for emerging bioenergy systems where experimental and inventory data remain incomplete. However, it would seem that closing the biology vs. biofuels or non-biofuels gap and improving scientific knowledge in that arena will require transparency toward data and ecosystem cross-sector cooperation, for sustainability.

7.7 ARTIFICIAL INTELLIGENCE AND MACHINE LEARNING IN BIOMANUFACTURING

AI and machine learning (ML) contribute to BM as one of the most rapidly adopted new technologies and allow for data-driven optimization, autonomous control, and agility in decision-making. Figure 7.2 illustrates a representative AI workflow, outlining the sequential stages from raw data acquisition to deployment of digital twins with real-time feedback loops, supporting intelligent decision-making across the BM lifecycle.

In light of the increasing complexity of bioenergy systems, including engineered strains and customized inputs, and advanced AI and ML-enabled hardware, computational approaches provide efficient means to process high-dimensional and non-linear biological data. In upstream bioenergy applications, ML models such as random forests, support vector machines (SVMs), and neural networks predict strain performance (i.e., tolerance to varied conditions and substrate uptake) based on specific data genotypes and/or omics data, resulting in a reduction in the amount of time spent on experimental screening. Advances in deep learning are also utilized with imaging and high-throughput data to identify candidates faster. In addition, reinforcement learning offers a unique approach for multi-objective optimization in metabolic pathway engineering and related bioprocess applications. In addition, real-time process control

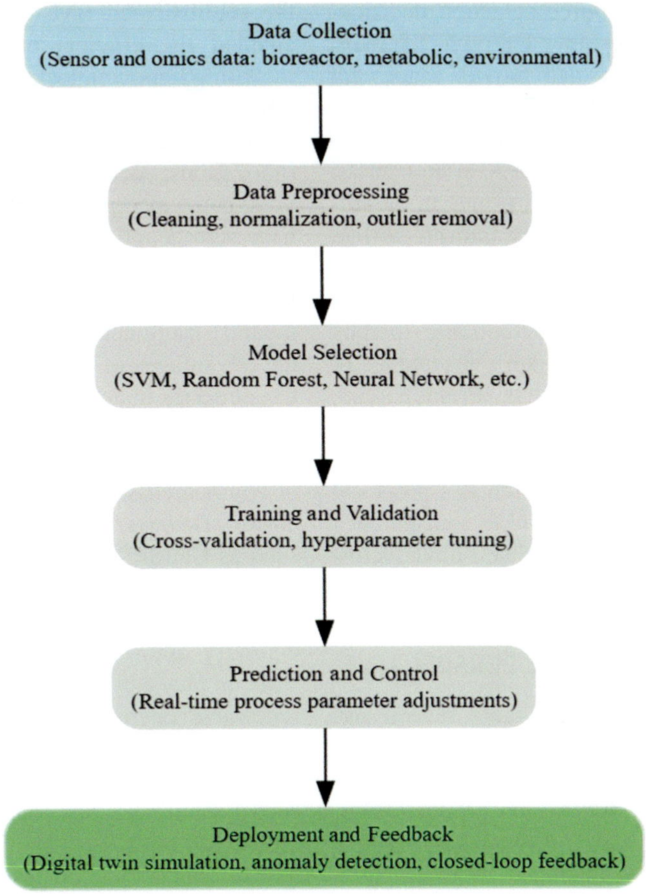

FIGURE 7.2 Machine learning (ML) pipeline for intelligent bioprocess management.

applications transitioning from traditional PID-based decision-making to ML-based predictive models have shown superior performance in integrating dynamic biological feedback. ML-based predictive models forecast essential biometrics (biomass and metabolite levels, as applicable) and feed the model data into Bayesian optimization and model predictive control (MPC) systems for autonomous regulation of the bioprocess. As depicted in Figure 7.3, AI/ML platforms can be integrated with bioreactor systems through real-time sensor networks (e.g., pH, dissolved oxygen [DO], and temperature) and cloud-based analytics to enable dynamic feedback control. This enables for continuous monitoring and autonomous adjustment of critical process parameters, ensuring optimal performance in real time.

In addition to ML-based process control optimization, digital twins combine mechanistic bioprocess models with real-time sensors to simulate reactor performance and optimize it through in silico experimentation. Digital twins can also be leveraged for remote monitoring of facilities and are particularly well suited for decentralized bioenergy applications. AI can also provide a route for anomaly detection through clustering and PCA identifying and predicting deviations before a sample is batch-failed, minimizing operational waste. Hybrid mechanistic ML models can also be leveraged for life cycle optimization and LCA, along with TEA, by providing rapid surrogate-based LCA and

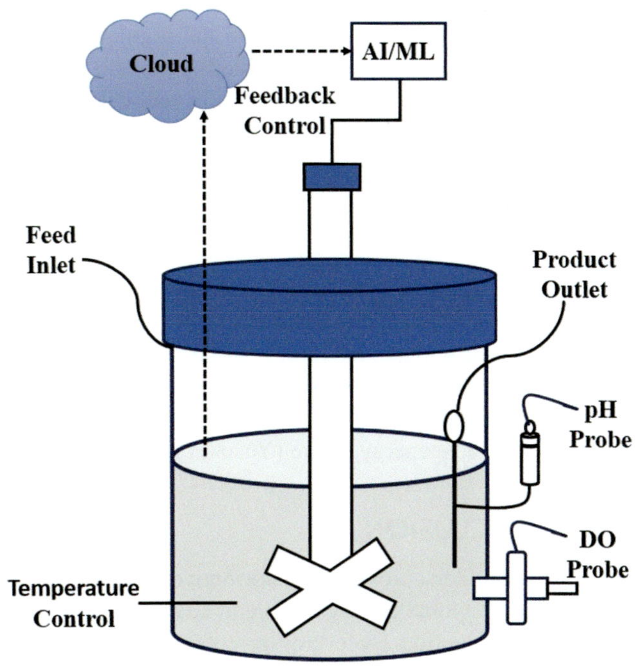

FIGURE 7.3 Schematic of an AI-integrated bioreactor system with sensor-driven feedback control via cloud computing.

TEA simulations to support resource selection decision-making. Toolkits, such as TensorFlow, PyTorch, KNIME, COBRApy, and BioMet Toolbox, can provide stacks for integrating AI into bioprocessing data. There are also commercial platforms that facilitate the introduction of AI into bioprocess industry applications to enhance AI effectiveness. However, there are still challenges in AI usage, in bioprocessing applications, including data quality, interpretability, and regulatory compliance. Recent advancements in explainable AI demonstrate how model transparency can be achieved by quantifying feature contributions, thereby improving trust, auditability, and regulatory acceptance of data-driven systems (Choudhary, Kumar, Priyadharsini, et al. 2025). Such explainability principles are increasingly critical for deploying AI-enabled control strategies in regulated BM environments. Emergent research has proposed frameworks for explainable AI to respond to transparency and trust issues related to AI use in regulated manufacturing systems.

7.8 CHALLENGES AND FUTURE PERSPECTIVE

Bioprocessing for bioenergy production faces several significant challenges that limit its large-scale commercial viability. One of the primary obstacles is the high production cost, driven by expensive feedstocks, low product yields, and complex downstream processing (Osman et al. 2021). Additionally, limited microbial tolerance to toxic intermediates and end-products, such as ethanol or butanol, hampers productivity in fermentation systems. The recalcitrance of lignocellulosic biomass to degradation also poses a barrier, requiring energy-intensive pretreatment methods. Furthermore, genetic instability in engineered strains and variability in process performance under industrial conditions reduce long-term reliability. On the regulatory front, the deployment of genetically modified organisms (GMOs) in open environments raises biosafety and public acceptance concerns. Despite these challenges, future perspectives remain promising (Hamdan and Tan 2025). Advances in systems biology, CRISPR-based genome editing, synthetic consortia, and AI-driven metabolic modeling are expected to revolutionize strain engineering and process optimization. Additionally, innovations in cell-free BM, continuous bioprocessing, and carbon capture integration offer new pathways for sustainable, scalable bioenergy production. By addressing these bottlenecks through interdisciplinary approaches, bioprocessing can play a crucial role in achieving a low-carbon, renewable energy future (Yusof et al. 2025).

7.9 CONCLUSION

The chapter underscores the transformative potential of BM to address the global energy crisis and advance sustainable bioenergy production. By leveraging cutting-edge technologies such as metabolic engineering, synthetic biology, AI, and advanced bioprocessing techniques, BM has evolved through distinct phases from Pre-modern fermentation to the sophisticated BM 4.0 era. These advancements have enabled the efficient conversion of diverse feedstocks, including lignocellulosic biomass, algae, and organic waste, into biofuels, such as bioethanol, biodiesel, and biogas, while simultaneously reducing environmental impacts. Key innovations, such as tailored pretreatment strategies, metabolic pathway optimization, and AI-driven process control, have significantly enhanced the scalability, efficiency, and economic viability of bioenergy systems. LCA has further provided a robust framework for evaluating the environmental sustainability of bioenergy pathways, highlighting the advantages of third-generation biofuels and integrated biorefineries in achieving carbon neutrality and fostering a circular economy. Despite these advancements, challenges persist, including high production costs, genetic instability of engineered strains, and regulatory hurdles surrounding GMOs. However, the integration of emerging technologies, such as CRISPR-based genome editing, cell-free BM, and digital twins, offers promising solutions to overcome these barriers. By addressing these challenges through interdisciplinary approaches and fostering global collaboration, BM can play a pivotal role in transitioning to a low-carbon, renewable energy future. Ultimately, this chapter highlights the critical role of BM in revolutionizing the bioenergy industry, paving the way for a sustainable and resilient energy landscape. The continued evolution of BM technologies holds immense potential to mitigate climate change, reduce dependence on fossil fuels, and achieve long-term environmental and economic sustainability.

REFERENCES

Aslan, Nisa, Onur Salman, Rabia Konakci, and Aytac Perihan Akan. 2025. "Circular Economy Approach to Promote Sustainable Energy from Waste." *Sustainable Futures* 10 (December): 101238. doi:10.1016/j.sftr.2025.101238

Camarero, S., M.J. Martínez, and A.T. Martínez. 2014. Understanding Lignin Biodegradation for the Improved Utilization of Plant Biomass in Modern Biorefineries. *Biofuels, Bioproducts and Biorefining* 8: 615–25. https://doi.org/10.1002/bbb.1467

Chavez, Diego Larrahondo, Catherine Azzaro-Pantel, Florent Montignac, and Alain Ruby. 2025. "Integrating Life Cycle Assessment in Multi-Objective Optimization of Green Hydrogen Systems: A Review of Literature and Methodological Challenges." *Renewable and Sustainable Energy Reviews* 217 (July): 115689. doi:10.1016/j.rser.2025.115689

Chia, Sean, Shi Jie Tay, Zhiwei Song, Yuansheng Yang, Ian Walsh, and Kuin Tian Pang. 2023. "Enhancing Pharmacokinetic and Pharmacodynamic Properties of Recombinant Therapeutic Proteins by Manipulation of Sialic Acid Content." *Biomedicine & Pharmacotherapy* 163 (July): 114757. doi:10.1016/j.biopha.2023.114757

Choudhary, Rakesh, Kailash Harne, Subhankar Dhar, Monaem Hamad Elmnifi, Anoop Gupta, Ajay Kumar, Fahad M. Alqahtani, et al. 2026. "Case Study on Biofuels Derived from Food and Agro-Industrial Wastes." In *Food and Agro-Industrial Wastes*, 455–70. Elsevier. doi:10.1016/B978-0-443-31654-8.00030-8

Choudhary, Rakesh, Ajay Kumar, Vikash Kumar, Monaem Hamad Elmnifi, Subhankar Dhar, Fahad M. Alqahtani, Deepak Sharma, et al. 2026. "Case Study on Biopolymers Derived from Food and Agro-Industrial Wastes." In *Food and Agro-Industrial Wastes*, 439–54. Elsevier. doi:10.1016/B978-0-443-31654-8.00026-6

Choudhary, Rakesh, Ajay Kumar, Saurav Narayan, Igor Mishchenko, Moharana Choudhury, Laxmi Kant Bhardwaj, Aqueel Ahmad, Mykola Andrienko, Petro Gaman, and Oksana Boiko. 2025. "Integrating Fuzzy Logic into Life Cycle Assessment for Sustainable Municipal Solid Waste Management in Metro Cities." *Ecological Engineering & Environmental Technology* 26 (7): 316–31. doi:10.12912/27197050/205500

Choudhary, Rakesh, Ajay Kumar, C. Priyadharsini, Mude Murali Naik, Moharana Choudhury, and Nadeem A. Khan. 2025. "Predicting Water Quality Index Using Stacked Ensemble Regression and SHAP Based Explainable Artificial Intelligence." *Scientific Reports* 15 (1): 31139. doi:10.1038/s41598-025-09463-4

Choudhary, Rakesh, Salina Tigga, Laxmi Kant Bhardwaj, Ajay Kumar, Subhankar Dhar, Fahad M. Alqahtani, Deepak Sharma, Aqueel Ahmad, Viola Vambol, and Satyam Kumar Chaudhari. 2026. "Biofuels from Food and Agro-Industrial Wastes." In *Food and Agro-Industrial Wastes*, 201–14. Elsevier. doi:10.1016/B978-0-443-31654-8.00004-7

Chozhavendhan, S., M. Rajamehala, G. Karthigadevi, R. Praveenkumar, and B. Bharathiraja. 2020. "A Review on Feedstock, Pretreatment Methods, Influencing Factors, Production and Purification Processes of Bio-Hydrogen Production." *Case Studies in Chemical and Environmental Engineering* 2 (September): 100038. doi:10.1016/j.cscee.2020.100038

Dhanya, B. S., A. Mishra, A. K. Chandel, and M. L. Verma. 2020. Development of Sustainable Approaches for Converting the Organic Waste to Bioenergy. *Science of the Total Environment* 723: 138109. https://doi.org/10.1016/j.scitotenv.2020.138109

Ebrahimi, M., M. Tamer, R. M. Villegas, A. Chiappetta, and F. Ein-Mozaffari. 2019. Application of CFD to Analyze the Hydrodynamic Behaviour of a Bioreactor with a Double Impeller. *Processes* 7 (10): 694. https://doi.org/10.3390/pr7100694

Elazab, M.A., Abdelrahman T. Elgohr, Mohamed Bassyouni, Abd Elnaby Kabeel, Mohammed El Hadi Attia, Mohamed Kamel Elshaarawy, Abdelrahman Kamal Hamed, and Hassan A.H. Alzahrani. 2025. "Green Hydrogen: Unleashing the Potential for Sustainable Energy Generation." *Results in Engineering* 27 (September): 106031. doi:10.1016/j.rineng.2025.106031

Gunes, Burcu. 2021. "A Critical Review on Biofilm-Based Reactor Systems for Enhanced Syngas Fermentation Processes." *Renewable and Sustainable Energy Reviews* 143 (June): 110950. doi:10.1016/j.rser.2021.110950

Hamdan, Mohd Fadhli, and Boon Chin Tan. 2025. "Genetic Modification Techniques in Plant Breeding: A Comparative Review of CRISPR/Cas and GM Technologies." *Horticultural Plant Journal* 11 (5): 1807–29. doi:10.1016/j.hpj.2024.02.012

Huang, Chung-Jr, Henry Lin, and Xiaoming Yang. 2012. "Industrial Production of Recombinant Therapeutics in *Escherichia Coli* and Its Recent Advancements." *Journal of Industrial Microbiology and Biotechnology* 39 (3): 383–99. doi:10.1007/s10295-011-1082-9

Isoko, Kesler, Joan L. Cordiner, Zoltan Kis, and Peyman Z. Moghadam. 2024. "Bioprocessing 4.0: A Pragmatic Review and Future Perspectives." *Digital Discovery* 3 (9): 1662–81. doi:10.1039/D4DD00127C

Jain, Sanyam, and Shushil Kumar. 2024. "Advances and Challenges in Pretreatment Technologies for Bioethanol Production: A Comprehensive Review." *Sustainable Chemistry for Climate Action* 5 (December): 100053. doi:10.1016/j.scca.2024.100053

Jeyapaul, R., P. Shahabudeen, and K. Krishnaiah. 2006. "Simultaneous Optimization of Multi-Response Problems in the Taguchi Method Using Genetic Algorithm." *The International Journal of Advanced Manufacturing Technology* 30 (9–10): 870–78. doi:10.1007/s00170-005-0095-9

Konwar, Gyanab, Riya Banik, and Sachin Rameshrao Geed. 2025. "Bioprocess Strategies for Bioethanol Production from Algae-Derived Carbohydrates: Scalable Solutions and Industry Prospects in a Circular Bioeconomy." *Chemical Engineering Journal* 515 (July): 163456. doi:10.1016/j.cej.2025.163456

Kumar, Pritha, Sanee Chauhan, Godfred Addai, Rakesh Choudhary, Prapti Sudan, Moharana Choudhury, Ajay Kumar, et al. 2026. "Valorization of Food and Agroindustrial Wastes—Biological Transformation." In *Food and Agro-Industrial Wastes*, 113–28. Elsevier. doi:10.1016/B978-0-443-31654-8.00008-4

Macdonald, Gareth John. 2020. "Biomanufacturing Makes Sense of the Industry 4.0 Concept." *Genetic Engineering & Biotechnology News* 40 (S3): S7–10. doi:10.1089/gen.40.S3.03

Magalhães Júnior, Antonio Irineudo, Carlos Ricardo Soccol, Marcela Candido Camara, Denisse Tatiana Molina Aulestia, Luciana Porto de Souza Vandenberghe, and Júlio Cesar de Carvalho. 2021. "Challenges in the Production of Second-Generation Organic Acids (Potential Monomers for Application in Biopolymers)." *Biomass and Bioenergy* 149 (June): 106092. doi:10.1016/j.biombioe.2021.106092

Maldini, Irene, Ingun Grimstad Klepp, and Kirsi Laitala. 2025. "The Environmental Impact of Product Lifetime Extension: A Literature Review and Research Agenda." *Sustainable Production and Consumption* 56 (June): 561–78. doi:10.1016/j.spc.2025.04.020

Metwaly, Ahmed M., Mohammed M. Ghoneim, Ibrahim H. Eissa, Islam A. Elsehemy, Ahmad E. Mostafa, Mostafa M. Hegazy, Wael M. Afifi, and Deqiang Dou. 2021. "Traditional Ancient Egyptian Medicine: A Review." *Saudi Journal of Biological Sciences* 28 (10): 5823–32. doi:10.1016/j.sjbs.2021.06.044

Monika, Sangita Banga, and Vinayak V. Pathak. 2023. "Biodiesel Production from Waste Cooking Oil: A Comprehensive Review on the Application of Heterogenous Catalysts." *Energy Nexus* 10 (June): 100209. doi:10.1016/j.nexus.2023.100209

Morales, M., J. Quintero, R. Conejeros, and G. Aroca. 2015. Life Cycle Assessment of Lignocellulosic Bioethanol: Environmental Impacts and Energy Balance. *Renewable and Sustainable Energy Reviews* 42: 1349–61. https://doi.org/10.1016/j.rser.2014.10.097

Nguyen, Ngoc-Phuong-Thao, Céline Raynaud, Isabelle Meynial-Salles, and Philippe Soucaille. 2018. "Reviving the Weizmann Process for Commercial N-Butanol Production." *Nature Communications* 9 (1): 3682. doi:10.1038/s41467-018-05661-z

Nickel, N., J. Fitschen, I. Haase, M. Kuschel, T. W. Schulz, T. Wucherpfennig, and M. Schlüter. 2024. Novel Sparging Strategies to Enhance Dissolved Carbon Dioxide Stripping in Industrial Scale Stirred Tank Reactors. *Frontiers in Chemical Engineering* 6: 1470991. doi:10.3389/fceng.2024.1470991

Olguin-Maciel, Edgar, Anusuiya Singh, Rubi Chable-Villacis, Raul Tapia-Tussell, and Héctor A. Ruiz. 2020. "Consolidated Bioprocessing, an Innovative Strategy towards Sustainability for Biofuels Production from Crop Residues: An Overview." *Agronomy* 10 (11): 1834. doi:10.3390/agronomy10111834

Omokaro, Godspower Oke, Zipporah Simiyu Nafula, Nwankwo Evalistus Iloabuchi, Abdulmajeed Allan Chikukula, Osazemen Godswill Osayogie, and Eucharia Chisom Nnoli. 2025. "Microalgae as Biofactories for Sustainable Applications: Advancing Carbon Sequestration, Bioenergy, and Environmental Remediation." *Sustainable Chemistry for Climate Action* 7 (December): 100098. doi:10.1016/j.scca.2025.100098

Osman, A. I., U. Qasim, F. Jamil, A. H. Al-Muhtaseb, A. Abu Jrai, M. Al-Riyami, S. Al-Maawali, L. Al-Haj, A. Al-Hinai, M. Al-Abri, A. Inayat, A. Waris, C. Farrell, M. I. A. Abdel Maksoud, and D. W. Rooney. 2021. Bioethanol and Biodiesel: Bibliometric Mapping, Policies and Future Needs. *Renewable and Sustainable Energy Reviews*, 152: 111677. https://doi.org/10.1016/j.rser.2021.111677

Rawat, I., R. Ranjith Kumar, T. Mutanda, and F. Bux. 2011. "Dual Role of Microalgae: Phycoremediation of Domestic Wastewater and Biomass Production for Sustainable Biofuels Production." *Applied Energy* 88 (10): 3411–24. doi:10.1016/j.apenergy.2010.11.025

Riccardi, Federico, Michele Dal Bo, Paolo Macor, and Giuseppe Toffoli. 2023. "A Comprehensive Overview on Antibody-Drug Conjugates: From the Conceptualization to Cancer Therapy." *Frontiers in Pharmacology* 14 (September). doi:10.3389/fphar.2023.1274088

Riggs, Arthur D. 2021. "Making, Cloning, and the Expression of Human Insulin Genes in Bacteria: The Path to Humulin." *Endocrine Reviews* 42 (3): 374–80. doi:10.1210/endrev/bnaa029

Roy, Dalim Kumer, and Mohammad Zoynal Abedin. 2022. "Potentiality of Biodiesel and Bioethanol Production from Feedstock in Bangladesh: A Review." *Heliyon* 8 (11): e11213. doi:10.1016/j.heliyon.2022.e11213

Sanchez, Sergio, and Arnold L. Demain. 2008. "Metabolic Regulation and Overproduction of Primary Metabolites." *Microbial Biotechnology* 1 (4): 283–319. doi:10.1111/j.1751-7915.2007.00015.x

Sant'Ana, Anderson S., and Wilson J.F. Lemos Junior. 2024. "Microbial Synergies and Their Impact on Economic and Quality Innovation in Sustainable Winemaking: Yeast and Lactic Acid Bacteria Interconnections." *Food Bioscience* 62 (December): 105238. doi:10.1016/j.fbio.2024.105238

Schmaltz, Silvana, Marco Antônio Silva, Renata Gulart Ninaus, Jerson Vanderlei Carus Guedes, Giovani Leone Zabot, Marcus Vinícius Tres, and Marcio Antonio Mazutti. 2023. "Biomolecules in Modern and Sustainable Agriculture." *3 Biotech* 13 (2): 70. doi:10.1007/s13205-023-03486-2

Udugama, I. A., P. C. Lopez, C. L. Gargalo, X. Li, C. Bayer, and K. V. Gernaey. 2021. "Digital Twin in Biomanufacturing: Challenges and Opportunities Towards Its Implementation". *Systems Microbiology and Biomanufacturing* 1 (3): 257–74. https://doi.org/10.1007/s43393-021-00024-0

Volkart, K., C. Bauer, and C. Boulet. 2013. Life Cycle Assessment of Carbon Capture and Storage in Power Generation and Industry in Europe. *International Journal of Greenhouse Gas Control* 16: 91–106. https://doi.org/10.1016/j.ijggc.2013.03.003

Yusof, Zuhaili, Yen Wah Tong, Kumar Selvarajoo, Sheetal Kishor Parakh, and Su Chern Foo. 2025. "Overcoming Challenges in Microalgal Bioprocessing through Data-Driven and Computational Approaches." *Current Opinion in Food Science* 61 (February): 101253. doi:10.1016/j.cofs.2024.101253

Zhang, Yi-Heng Percival, Jibin Sun, and Yanhe Ma. 2017. "Biomanufacturing: History and Perspective." *Journal of Industrial Microbiology and Biotechnology* 44 (4–5): 773–84. doi:10.1007/s10295-016-1863-2

8 Bridging Biology and Industry

Next-Generation Biomanufacturing Strategies and Process Intensification for Sustainable Biochemicals and Enzymes

Rachna Shashikant Jadhav
Faculty of Applied Sciences and Biotechnology, Shoolini University, Bajhol, Himachal Pradesh, India

Saurabh Kulshrestha
Faculty of Applied Sciences and Biotechnology, Shoolini University, Bajhol, Himachal Pradesh, India
Bennet University, Greater Noida, Uttar Pradesh, India

Himanshu Sharma
Amity School of Biological Sciences, Amity University Punjab, Mohali, Punjab, India

Santanu Mukherjee
School of Agriculture, Shoolini University, Bajhol, Himachal Pradesh, India
Department of Environmental and Civil Engineering, Toyama Prefectural University, Imizu, Japan

8.1 INTRODUCTION

The need for alternative production methods is rising owing to the chemical industry's growing environmental impact, reliance on fossil fuels, and the quick depletion of natural resources. Long-term environmental and economic sustainability is threatened due to conventional petrochemical-based manufacturing methods, which produce hazardous wastes, use non-renewable resources, and contribute significantly to greenhouse gas emissions. A recent method known as "biomanufacturing" employs biological systems, such as bacteria, enzymes, and plants, to generate chemicals, fuels, and materials in a sustainable way, and has become a ground-breaking solution within the last 20 years. The ability of microbial and plant biofactories has been greatly increased by the recent developments and advances in the field of synthetic biology, metabolic engineering, Clustered Regularly Interspaced Short Palindromic Repeats (CRISPR)-based genome editing, and artificial intelligence (AI)-assisted bioprocess optimization methods.

The goal of developing cell-free systems and integrated biorefineries is to transform CO_2 and renewable biomass into high-value products with minimal environmental impact, thereby aligning with circular bioeconomy principles. Although these advances have proven to be beneficial so far, opinions about the systematic integration of platform technologies, scalable bioprocessing methods, and enzyme engineering approaches across different biomanufacturing systems remain divided. Most of the literature and studies focus on isolated innovations, such as upstream strain engineering or downstream process (DSP) improvements, without offering a comprehensive understanding of how these components interact to produce sustainable biochemicals. A thorough and precise understanding is necessary to bridge the gap between laboratory-scale innovations and industrial-scale biomanufacturing methods. As industries move toward decarbonization and green manufacturing, there is an increasing need to integrate microbial, plant, and algal platform advancements, enzyme designs, and process intensification techniques to enable efficient, cost-effective, and environmentally friendly bioproduction methods.

This chapter provides an extensive review of next-generation biomanufacturing platforms for environmentally

DOI: 10.1201/9781003637158-8

friendly biochemical and enzyme production techniques. Key microbial, plant, algal, and cell-free systems are highlighted, along with enzyme engineering methods, including directed evolution and rational design (RD), immobilization enabled by nanotechnology, downstream processing improvements, and scale-up tactics. It also examines economic, environmental, and regulatory factors and identifies emerging trends that will influence biomanufacturing in the future. This chapter thus attempts to offer a coherent framework to encourage the shift toward sustainable industrial bioprocessing by integrating various viewpoints.

8.1.1 Overview and Importance of Biomanufacturing

Biomanufacturing is a process that employs renewable resources to provide ecologically sustainable alternatives to conventional manufacturing methods, which rely entirely on fossil fuels and generate substantial waste (Nielsen and Keasling 2016). It is a cutting-edge industrial production method that combines biological systems such as enzymes and microbes with synthetic biology to manufacture sustainable materials, chemicals, and biofuels. When it comes to the development of the nation's bioeconomy and a global movement to reduce industrial waste and carbon emissions, etc., biomanufacturing plays a vital role and promotes circular economies. The use of biochemicals, biocatalysts, and enzymes is one of the key pillars in biomanufacturing, enabling efficient and sustainable processes across multiple industries. Instead of chemicals, biochemicals serve as green alternatives to petroleum-derived products, thereby reducing environmental impact and ensuring economic viability (Clomburg et al. 2017). Enzymes are an example of biocatalysts that accelerate industrial reactions with high specificity, reduce energy consumption, and minimize toxic byproducts as well (Sheldon and Woodley 2018). Enzyme-based biomanufacturing may transform biomass into valuable goods such as biofuels, biodegradable polymers, and medicinal intermediates (Fasim et al. 2021).

As industries transition from traditional fossil-based manufacturing to next-generation biomanufacturing, advancements in synthetic biology, metabolic engineering, and automation are revolutionizing production techniques. The creation of high-yield bioprocesses, which optimize microbial strains to increase the efficiency of biochemical synthesis, has been enabled by advances in genome editing and precision fermentation. Furthermore, the widespread adoption of sustainable manufacturing techniques is being aided by biorefineries, which are integrated facilities that transform biomass into fuels, chemicals, and materials (Lee and Kim 2024).

Biomanufacturing has a lot of potential, but it also has a lot of problems, such as high production costs, problems with regulatory compliance, and limited feedstock availability (Jullesson et al. 2015). However, new potential for scalable and affordable bio-based manufacturing has been made possible by recent developments in microbial metabolic modeling, enzyme engineering, and AI-driven bioprocess optimization (Table 8.1). These problems must be addressed if biomanufacturing is to fully contribute to the creation of a sustainable, environmentally conscious future (Ndochinwa et al. 2024a).

8.2 BIOMANUFACTURING PLATFORMS FOR SUSTAINABLE BIOCHEMICAL PRODUCTION

By using microbial cell factories to create bio-based chemicals, fuels, and medications in an environmentally friendly way, biomanufacturing has become a preferable alternative to conventional chemical synthesis (Yadav et al. 2022; Cho et al. 2022). The advancement of microbial strains has been transformed by advancements in metabolic engineering and synthetic biology, which have increased the productivity of biochemical production (Jullesson et al. 2015). High-yield synthesis of biofuels, bioplastics, and specialized chemicals has been made possible by engineering microbial hosts, such as *Escherichia coli* or yeast *Saccharomyces cerevisiae*, filamentous fungi, and cyanobacteria (Ding and Ye 2023).

8.2.1 Microbial Cell Factories for Biomanufacturing

Microbial cell factories are genetically engineered microbes optimized to produce valuable biochemicals from renewable feedstocks. Using metabolic engineering and synthetic biology, filamentous bacteria, fungi, and yeasts can be converted into efficient microbial cell factories through genetic, tolerance, routes, or microbial engineering. These systems offer high specificity and efficiency in biochemical production, renewable carbon sources (e.g., sugars, lignocellulose, and CO_2), and scalability and economic feasibility for industrial applications (Hussain et al. 2022).

8.2.1.1 Synthetic Biology and Metabolic Engineering Approaches

Precision alterations in microbial genomes have been made possible by advances in metabolic or genetic engineering and synthetic biology, increasing sustainability and productivity. The three most popular synthetic biology-based techniques are route modularization, synthetic circuits and dynamic regulation, and genome editing using Clustered Regularly Interspaced Short Palindromic Repeats and CRISPR-associated protein 9 (CRISPR-Cas9), also known as Transcription Activator-Like Effector Nucleases (TALENs). According to Nemudryi

TABLE 8.1
Key Challenges and Emerging Opportunities in Biomanufacturing for Sustainable Development

S. No.	Category	Challenges	Opportunities	Reference
1.	Production costs and scalability	• High costs of enzyme production and microbial fermentation. • Difficulty scaling up bio-based processes. • Expensive bioreactor infrastructure.	• Employing metabolic engineering and synthetic biology to enhance microbial yield. • Process optimization by AI-driven technology for cost reduction. • Using engineered enzymes with higher efficiency.	Jullesson et al. (2015)
2.	Feedstock availability and sustainability	• Competitive food supply due to increased dependence on agricultural biomass. • Seasonal variability affecting the quality of the feedstock. • Increased cost of logistics for biomass transport.	• Development of non-food feedstocks (e.g., algae, agricultural residues, and waste-derived biomass). • Use of microbial electrosynthesis for feedstock conversion.	Yadav et al. (2022); Jourdin and Burdyny (2021)
3.	Regulatory hurdles andmarket adoption	• Approval issues for genetically modified organisms (GMOs) and bio-based products. • Stringent environmental and biosafety regulations. • Limited consumer acceptance of bio-based alternatives.	• Global standardization of biomanufacturing regulations. • Incentives for bio-based product adoption. • Increased public awareness and policy support for green technologies.	Tang et al. (2022)
4.	Process efficiency and enzyme stability	- Low stability of industrial enzymes in extreme conditions. - Short enzyme half-life increases operational costs. - Limited substrate specificity of enzymes.	- Protein engineering to improve enzyme stability and robustness. - Enzyme immobilization techniques to enhance the reuse of enzymes. - Development of multi-functional biocatalysts.	Sheldon and Woodley (2018)
5.	Biomanufacturing in healthcare and biopharmaceuticals	- High cost of biopharmaceutical production. - Scalability challenges in vaccine and bio-based drug production. - Complex regulatory approvals for cell-based therapies.	- Cell-free biomanufacturing for faster and cost-effective drug synthesis. - Optimization of biosynthetic pathways using AI-driven technology. - Expansion of biomanufacturing in precision medicine and gene therapy.	Zheng et al. (2022); Lee and Kim (2024)
6.	Expansion of biorefineries and circular economy	- Inefficient biomass conversion processes. - High energy consumption in large-scale biorefineries. - Limited integration of waste valorization strategies.	- Development of integrated lignocellulosic biorefineries. - Microbial conversion of industrial waste into valuable biochemicals. - Zero-waste manufacturing models for sustainable production.	Yogalakshmi et al. (2023); Kumar et al. (2025)

et al. (2014), genome editing is utilized to improve metabolic pathways, incorporate heterologous pathways into microorganisms, and eliminate inhibitory genes. Dynamic regulations and synthetic circuits are constructed control mechanisms that can minimize the formation of harmful byproducts and optimize the flow balance (Venayak et al. 2015; Chen et al. 2025). However, to improve biosynthetic efficiency, route modularization will be accomplished by dividing the metabolic pathways into modular parts (Jiang et al. 2017).

Metabolic engineering strategies include direct evolution and adaptive laboratory evolution (ALE), heterologous pathway expression, and metabolic flux analysis and computer modeling (Figure 8.1). In ALE, enzyme activity is enhanced, and microbes are allowed to adapt in a way that they are tolerant to industrial conditions (Dragosits and Mattanovich 2013). Heterologous pathway expression involves transferring biosynthetic pathways from non-model organisms to industrial strains (Markina et al. 2020). Whereas metabolic flux analysis and computational

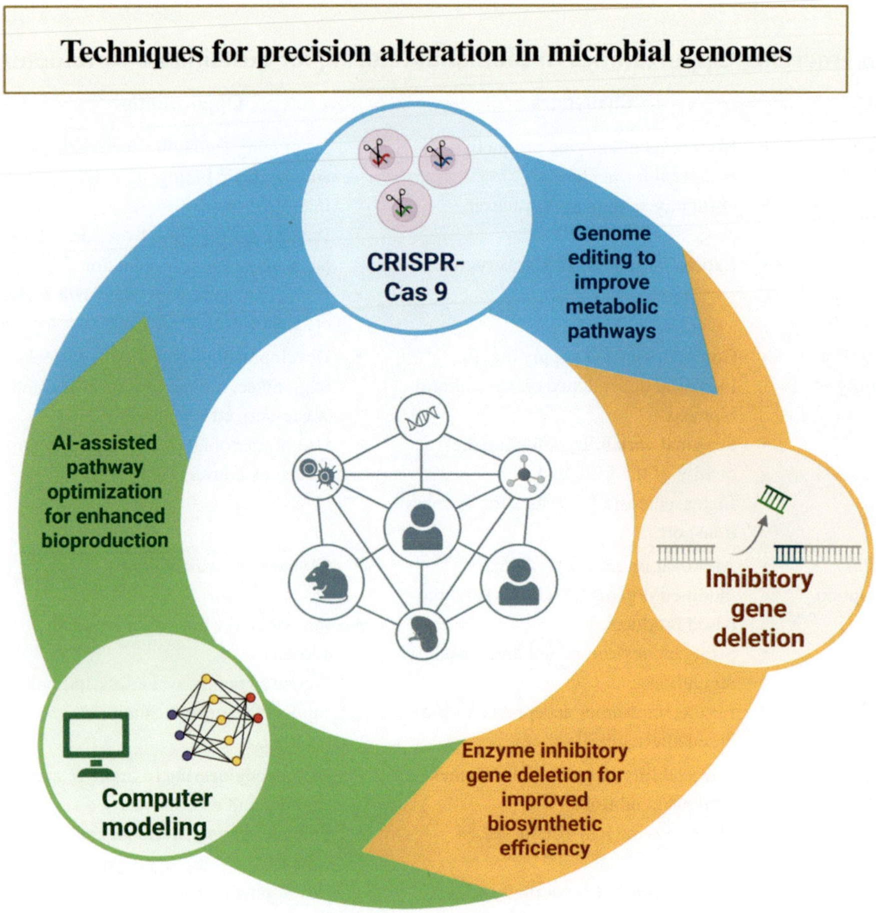

FIGURE 8.1 Techniques for precision alterations in microbial genomes.

modeling will employ AI-assisted pathway optimization for enhanced bioproduction (Aurich and Thiele 2016).

8.2.1.2 Engineering Microbial Hosts for Sustainable Biochemical Production

The most often used microbial hosts for sustainable biochemical production are *Aspergillus* species, *Trichoderma* species, filamentous fungi, bacteria such as *E. coli, S. cerevisiae*, and cyanobacteria. *E. coli*'s rapid growth, well-comprehended genetics, and ease of genetic modification make it one of the most popular platform bacteria. Biofuels, organic acids, such as lactic and succinic acid, and even biopolymers such as PHAs have all been produced using *E. coli* (Koppolu and Vasigala 2016; Chiang et al. 2021; Liu et al. 2011; Fidler and Dennis 1992). Microbes can also be altered by deleting competing routes and adding synthetic biosynthetic pathways to boost the efficiency of their production (Savino et al. 2022).

8.2.1.3 *Saccharomyces cerevisiae* (Yeast) as a Biofactory

Yeast is frequently used as a microbe for the production of bioethanol, terpenoids, and pharmaceutical compounds due to its ability to withstand fermentation. Among its many applications are flavonoids, high-value biochemicals, and lipids for manufacturing biodiesel (Sánchez-Solís et al. 2024; Yang et al. 2024; Dai et al. 2015). Synthetic biology has also been used with CRISPR to enhance lipid biosynthesis and ethanol tolerance by deletion or modification of genes such as glycerol 3-phosphate dehydrogenase-2 (GPD2) and alcohol dehydrogenase-2 (ADH2) (Li et al. 2021; Yang et al. 2022).

8.2.1.4 Industrial Bioprocessing by Filamentous Fungi

A strong filamentous fungus that secretes hydrolytic enzymes, *Aspergillus oryzae*, is regarded as a biosafe strain for fermentation and bioproduction technologies (Sun et al. 2024). Additionally, *Trichoderma reesei* has been employed as a low-cost, scalable strain that produces important enzymes for industry (de Lima et al. 2022). *Aspergillus awamori* and *T. reesei* have been used to produce cellulase enzyme using submerged fermentation methods (Naher et al. 2021); whereas filamentous fungi, such as *Penicillium fellutanum, Penicillium expansum, Penicillium brevicompactum, Trichothecium roseum, A. oryzae, Aspergillus flavus, Aspergillus fumigatus, Aspergillus niger, A. awamori, Aspergillus terreus, Aspergillus carbonarius, Thermomyces lanuginosus, Rhizopus delemar*, and *Rhizopus oryzae* have been used for manufacturing alpha amylase using solid-state fermentation (Mishra et al. 2020). Gene knockout strategies and protein secretion pathway modifications have also been conducted for enhancing enzyme yield by using filamentous fungi as hosts (Liu et al. 2013). Utilizing glucose as a carbon source, Rantasalo et al. (2019) have reported a novel engineering of *T. reesei* to produce highly enriched lipase B from *Candida antarctica* (calB) at high levels. Using this recently developed synthetic expression system (SES) in conjunction with multiplexed CRISPR/Cas9 enabled the rapid creation of strains that generated large quantities of extremely pure calB. Xue et al. (2024) reported using progressive genetic modification techniques to increase the secretion of an acid protease (AopepA) from *A. oryzae* in *Komagataella phaffii*. These included co-expression of the hemoglobin gene, disruption of the hypoxia heme-dependent repressor gene, overexpression of initiation proteins in eukaryotic translation, deletion of the β-1,3-glucanosyltransferase gene, and co-expression of secretion-associated factors of the endoplasmic reticulum.

8.2.1.5 Cyanobacteria for Carbon Dioxide-Based Biomanufacturing

Cyanobacteria have been proven to be an efficient carbon dioxide-based biomanufacturing platform that utilizes photosynthesis to generate biochemicals. It has been used for biosynthesizing bioplastics, biohydrogen, isoprenoids, biofuels, etc. (Nawaz et al. 2024; Singh et al. 2024). Zhong et al. (2024), has focused on engineering cyanobacteria to produce taxol, an effective anticancer agent, directly from carbon dioxide. To achieve this, *Synechocystis* sp. Pasteur Culture Collection (PCC) 6803, with modular pathways of metabolism that include the methylerythritol phosphate pathway enzymes and taxol biosynthesis enzymes, was developed. The potential contribution of cyanobacterial biotechnology to an environment-conscious bioeconomy through negative emissions technologies (NETs),such as carbon sequestration or bioproduction, has also been studied (Victoria et al. 2024). Wu et al. (2023) successfully induced the synthesis of isomaltulose and accumulation in the recombinant *Synechococcus elongatus* PCC 7942 strains by introducing several sucrose isomerases. In this research, the final titer was greatly increased, and isomaltulose was secreted efficiently when sucrose isomerases and an *E. coli*-sourced sucrose permease, CscB, were co-expressed. Consequently, a new pathway for the biosynthesis of isomaltulose using carbon dioxide as a substrate was demonstrated, and new insights into the adaptability of the cyanobacterial photosynthetic metabolism network were gained.

In conclusion, biomanufacturing systems that employ modified microbial cell factories provide sustainable options for biochemical synthesis. Enhancements in metabolic engineering, synthetic biology, and microbial host optimization have significantly raised yields and economic feasibility. Future research integrating AI-driven strain optimization, advanced fermentation technologies, and non-model host engineering will further expand the promise of next-generation biomanufacturing for a circular bioeconomy.

8.2.2 Plant-Based Biomanufacturing

Plant-based biomanufacturing (PBB) is another bio-based manufacturing method that uses plants as biofactories. It is a sustainable way to produce important biochemicals, pharmaceuticals, and industrial enzymes. Advances in genetic engineering have made it possible to modify plants to boost their capacity to manufacture particular biomolecules effectively and commercially (Zarrelli et al. 2022). The significant advancements, advantages, and difficulties surrounding the usage of genetically modified (GM) plants in the synthesis of biochemicals are discussed below.

8.2.2.1 Role of Genetically Modified Plants in Biochemical Production

Several biochemicals, including vaccines, medicinal proteins, and industrial enzymes, can be produced by genetically engineered/GM plants. According to Kowalczyk et al. (2022), these changes involve the introduction of certain genes that code for desired products and are fueled by strong promoters to ensure high expression levels. The manufacturing of medications is among the most important uses of GM plants. Using plant-based expression systems, monoclonal antibodies (mAb), vaccines, therapeutic proteins, and antiviral compounds have all been created (England et al. 2023). Compared to plant viral vectors, transgenic plants have slower transient mAb expression. As a result, single-chain antibodies (scFv) tailored for cancer patients with distinct epitopes can be produced quickly by inoculating the viral vectors (Moussavou et al. 2015). McCormick et al. (1999) developed a modified tobamoviral vector that encodes 38C13 mouse B cell lymphoma's idiotype-specific single-chain Fv fragment (scFv) immunoglobulin. The extracellular compartment of infected *Nicotiana benthamiana* plants was shown to have elevated levels of secreted scFv protein. A range of antibodies was produced using the refined geminiviral vectors, with special focus on the creation and production of a chimeric anti-flavivirus antibody that is broadly neutralizing. The variable parts of this murine antibody, 2A10G6, were fused to a human IgG1 using codon optimization. The chimeric antibody was discovered to be effectively expressed in plants, neutralize the Zika virus potently, retain its capacity to detect the virus's envelope protein, and be purified to almost homogeneity using a straightforward single purification procedure (Diamos et al. 2020).

Enzymes for industrial usage have also been produced by GM plants. Cellulases and xylanases, which are essential for biofuel generation, have been produced by modifying tobacco and maize (Melo-Duran et al. 2024). Furthermore, plants, such as oilseed rape and Arabidopsis, have been modified to make biodegradable plastics like polyhydroxyalkanoates (PHAs) (Surendran et al. 2020; Dobrogojski et al. 2018). As they require less money for fermentation and purification than microbial or mammalian cell systems, plants provide an affordable platform for large-scale production. Large-scale cultivation of GM plants yields significant biomolecule yields; moreover, unlike microbial systems, plants do not host human diseases, reducing the risk of contamination (Sethi et al. 2021).

8.2.3 Algal Biorefineries

As sustainable feedstocks for biochemicals, microalgae and macroalgae have captured interest due to their high productivity, diverse biochemical composition, and ability to flourish in barren regions without competing with food crops (Medipally et al. 2015). Lipids, proteins, carbohydrates, and bioactive molecules are among the useful chemicals that algae biorefineries extract for use in the food, pharmaceutical, cosmetic, and biofuel industries by combining a number of processing techniques (Branco-Vieira et al. 2020).

8.2.3.1 Microalgae Usage in Algal Biorefineries

The unicellular photosynthetic organisms known as microalgae have the capability to collect valuable chemicals under controlled conditions. Their rapid growth rates and capacity to adapt to many environments make them suitable for large-scale cultivation. Important biochemicals derived from microalgae include acids and fats, proteins, amino acids, polysaccharides, carotenoids, and other colors, and bioactive substances. Certain species, such as *Chlorella vulgaris* and *Nannochloropsis*, accumulate high lipid content, making them valuable for manufacturing biodiesel (Ma et al. 2016). Moradi and Saidi (2022), examined the electrochemical and thermal process potential for generating biodiesel from hydrated oil by comparing and observing the esterification reaction of microalgal-derived oil at low and middle temperatures. Examples of species that are appropriate for use in food and feed due to their excess protein content include *Spirulina* and *Chlorella* (Alagawany et al. 2021; Abreu et al. 2023). Sulfated polysaccharides of Porphyridium possess antiviral and immunomodulatory properties (Panggabean et al. 2022).

8.2.3.2 Macroalgae for Algal Biorefineries

Seaweeds, also known as macroalgae, are multicellular marine algae that fall into three groups: brown (Phaeophyceae), green (Chlorophyta), and red (Rhodophyta). According to Leandro et al. (2019), they are sustainable sources of bioactive metabolites, polysaccharides, and polyphenols. Important biochemical uses include the synthesis of phenolic compounds, thickeners made from hydrocolloids such as red and brown algae, and even biofuels. Red algae produce the polysaccharide carrageenan, which has long been used as a food additive. Alongside the development of innovative food technology, extensive research has been done on the potential uses of carrageenan, such as encapsulation, plant-based analogs, edible films and coatings, and 3D/4D printing (Udo et al. 2023). Hydrocolloids are frequently used in food recipes to enhance shelf life and quality characteristics. They are mostly used as thickening and gelling agents in thickening sauces, gravies, soups, toppings, and salad dressings, as well as serving as gelling agents in goods such as marmalade, jam, jelly, restructured meals, and low-sugar/calorie gels (Saha and Bhattacharya 2010). Brown algae, such as *Ascophyllum nodosum*, contain phlorotannins with antioxidant, antidiabetic, and antimicrobial properties (Gisbert et al. 2023). The presence of substances, such as lipids and carbohydrates, along with the absence or insufficiency of lignin, makes macroalgae an ideal feedstock for the production of biofuels. Thus, they have also been explored for bioethanol and biogas production due to their high carbohydrate content (Roesijadi et al. 2010). Green macroalgae hydrolysate has been used for biofuel production using *Ulva rigida* (Sayed et al. 2024).

8.2.4 Cell-Free Biomanufacturing

Cell-free biomanufacturing has emerged as a powerful platform for the manufacture of biochemicals by using enzymatic cascades to catalyze complex biosynthetic processes without the use of living cells. This technique enhances scalability, yield optimization, and reaction control, making it a promising alternative to traditional microbial fermentation (Bundy et al. 2018). Cell-free systems assemble specific enzyme cascades to enable the effective and selective production of biochemicals, biofuels, and medicines (Brookwell et al. 2021).

Sequential enzyme-catalyzed reactions known as "enzymatic cascades" convert basic precursors into valuable biochemicals. It is possible to design these cascades for the biosynthesis of amino acids and organic acids, which are examples of linear cascades that are sequential reactions in which intermediates are transferred from one enzyme to another (Wohlgemuth 2024). The lower concentrations of flavonoids and their modified forms found in biomass, which are essential for the production of secondary metabolites, limit the traditional supply of flavonoids from plants. Ex planta flavonoid synthesis now has new and intriguing potential thanks to developments in metabolic engineering, synthetic biology, protein engineering, and machine learning (Sajid et al. 2021).

Since microbial fermentation technologies currently exhibit low conversion yields and host cell toxicity, cell-free industrial chemical and biofuel syntheses are gaining significant attention as a viable alternative. Cellular components, such as DNA, RNA, proteins, and metabolites, can be reassembled into various target molecules, much as an electronic device is disassembled, and its components reassembled for a specific function. The cell-free synthesis system's open nature allows for flexible and mix-and-match-type reaction conditions for target protein generation and optimization, which is not possible with expression techniques based on cell-based genes. This facilitates the quick creation of proteins, such as essential enzymes for the synthesis of biofuel, that are challenging to express in living cells (Lim and Kim 2022).

8.3 BIOCATALYST ENGINEERING FOR ENHANCED PERFORMANCE

In industrial and biological applications, biocatalysts—mainly enzymes—are essential because of their high selectivity, efficiency, and environmental friendliness. To improve stability, activity, and selectivity under the intended conditions, native enzymes are often optimized. Enzyme optimization has advanced significantly through two key approaches in enzyme engineering: directed evolution (DE) and RD. As they can model protein structures, forecast advantageous mutations, and direct experimental designs, computational methods are essential to these tactics.

8.3.1 Directed Evolution and Rational Design

DE is a well-liked method for designing enzymes for a variety of applications. Through iteratively creating enzyme variations and screening for better qualities, DE mimics natural selection. This procedure includes high-throughput screening, recombination, and random mutagenesis (Figure 8.2). By boosting the likelihood that mutated sequences have the desired features, structural data, and knowledge of how proteins react to recombination and mutation are utilizing these approaches to create more targeted evolution approaches (Bloom et al. 2005). One sub-strategy of DE is mutagenesis, which includes DNA shuffling, site saturation mutagenesis, and error-prone polymerase chain reaction (epPCR). Another sub-strategy is screening and selection, which uses high-throughput screening such as fluorescence-based assays or microfluidic droplet-based screening. The endothelial Protein C Receptor (ePCR) introduced random mutations during DNA amplification (Gruet et al. 2012). It is possible to increase catalytic efficiency, protein–protein/DNA interactions, or inhibitor resistance by a thousand times. DNA shuffling recombines a subspace of analogous sequences by operating inside that subspace. People are chosen when they reach a predetermined threshold after being checked for one or two specific traits (Werkman et al. 2011). According to research by Parikh and Matsumura (2005), SSM is considered the most effective mutagenesis strategy for the DE of β-fucosidase from β-galactosidase compared to DNA shuffling. DE offers a chance to improvise hydrolase promiscuous activity in high-throughput screening and variety generation cycles. Proteases and possibly other hydrolases (such as lipases or esterases) should have their perhydrolytic activity increased. A screening platform was developed and validated by Despotovic et al. (2012). Vallejo et al. (2020)created a microfluidic-based method that evolves novel polymerase activities by sorting fluorescent water-in-oil (w/o) microdroplets at ultrahigh throughput. Numerous polymerase variants are expressed in *E. coli*, microfluidic droplets containing one or fewer *E. coli* are created, bacteria are lysed to release the polymerase and by encoding plasmid into the neighboring droplet, variants with the desired activity are identified using a fluorescence-based activity assay, fluorescent droplets are isolated using a fluorescence activated droplet sorting (FADS) device, and the functional variants are identified by plasmid recovery with DNA sequencing. Any artificial nucleic acid and/or polymerase function, such as DNA-templated synthesis, reverse transcription, and replication, can be used with this method.

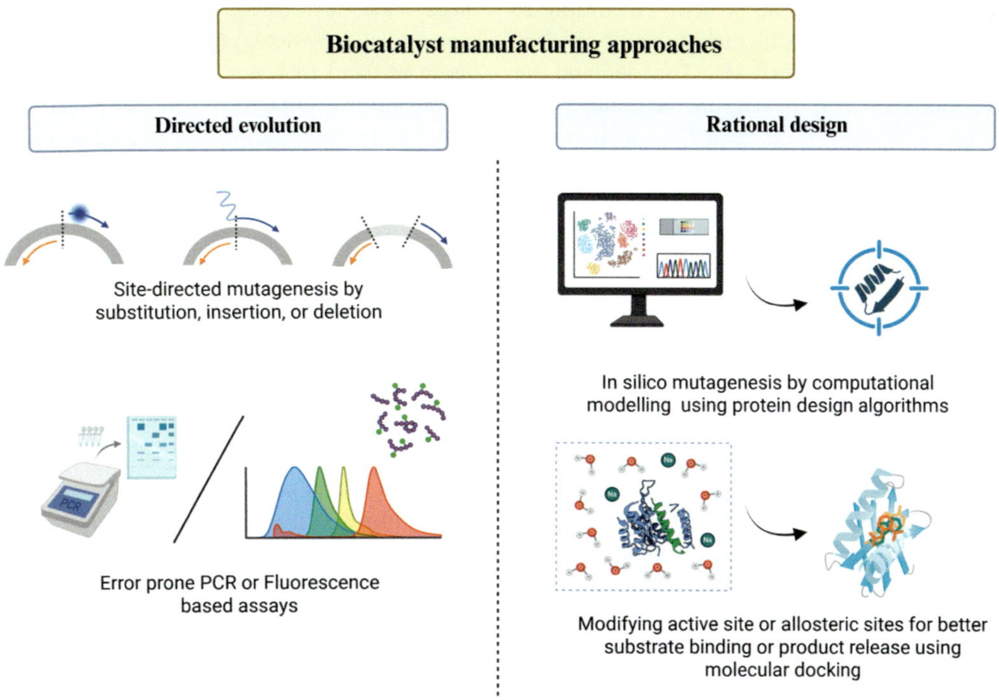

FIGURE 8.2 Biocatalyst manufacturing approaches.

To introduce particular mutations that improve the characteristics of enzymes, RD makes use of structural and mechanistic understanding. This method makes extensive use of bioinformatics tools and computational modeling. Using universal and relatively fast RD techniques, algorithms that can quantitatively forecast the performance of the given sequences can be developed (Song et al. 2023). With so many possible uses, the RD of enzymes presents a significant challenge to our knowledge of physical chemistry and protein structure. Proteins that fold, catalyze more quickly, signal, and adopt preferred structural states have all been designed or engineered using protein design algorithms (Tiwari et al. 2012). One of the main requirements for protein function is thermodynamic stability, and improving stability is one of the objectives of protein engineering, especially for biotechnology applications. Zeiske et al. (2016) estimated the in vitro thermostability of endonucleases belonging to the bacterial Ribonuclease HI (RNase H) family using molecular dynamics simulations. A method for in silico mutational screens to increase the thermostability of important biological and commercial enzymes was presented in this work.

The use of a range of techniques requiring multidisciplinary research and extensive knowledge is required because site-directed mutagenesis is a traditional method based on the principles of protein folding. These techniques include the biological computing of complex data from multiple sequencing projects and the forecasting of future protein structures that may be altered chemically or genetically. The desired protein can be selected using advanced technology from a random collection of many massive libraries of mutant proteins that are produced by reordering the genes encoding different domains or by unplanned mutations in the gene encoding the protein. According to Khan and Waheed Akhtar (2011), this technique is referred to as DE or non-rational mutagenesis. The demand for highly effective enzymes in industrial biotechnology creates new difficulties, especially the requirement to greatly boost catalytic turnover (k cat) or efficiency (k cat/K m). However, a crucial component of enzyme engineering is substrate selectivity. It describes an enzyme's ability to attach and catalyze a certain substrate more selectively than others. The enzyme's distinct three-dimensional structure, which is made to bind particular substrates, determines its specificity. By ensuring that designed enzymes perform their intended reactions effectively and selectively, this specificity reduces undesirable side reactions in an industrial context. For instance, the pharmaceutical industry uses enzymes with increased substrate specificity to make medications, which reduces the amount of contaminants. In a similar vein, given the harsh conditions of industrial processes, resistance to heat and chemical solvents is very crucial. Enhancing catalytic capabilities has required DE, and efforts have been made to create enzymes with distinct catalytic activity for processes that are not seen in nature. According to Ndochinwa et al. (2024b), current research projects have merged DE, computational methodologies, and RD to achieve this goal.

Both DE and RD are crucial for improving biocatalysts for application in industrial and medical settings with the use of computational tools. The combination of AI and machine learning is further increasing the efficiency of enzyme engineering. Advances in computer simulations, high-throughput screening, and synthetic biology will continue to improve biocatalyst design for a range of applications.

8.3.2 Protein Engineering for Industrial Applications

Through the optimization of crucial characteristics, such as stability, specificity, and efficiency, biocatalyst engineering seeks to improve enzyme performance for industrial applications. Enzyme functionality has been greatly enhanced by protein engineering techniques such as DE, computational modeling, and RD, which have made them more appropriate for a variety of industrial processes.

8.3.3 Immobilization Strategies for Biocatalysts

Enhancing enzyme stability, activity, and reusability for industrial and biotechnological applications is the goal of biocatalyst engineering. Immobilization, which entails securing enzymes to solid supports or inside matrices to improve their stability and functionality, is among the best strategies. Immobilization's primary benefit is that it protects the enzymes from harmful environmental conditions, such as high temperatures and extremely high pH levels. The immobilized enzymes can be used in various large-scale businesses, such as the food, detergent, pharmaceutical, medical, and textile sectors, in addition to being utilized in water treatment facilities. Depending on the required application, the right carrier materials and enzyme immobilization technique are chosen. Techniques for immobilizing enzymes include covalent binding, encapsulation, entrapment, adsorption, cross-linking, and other procedures.

In covalent bonding, the stability of the enzyme is increased by a strong chemical bond between the substrate and the enzyme. In encapsulation, polymeric or nanostructured carriers carry enzymes. Entrapment is when enzymes are trapped inside porous matrices, such as hydrogels, silica gels, or alginate. Adsorption is achieved through physical interactions, such as hydrogen bonds or van der Waals forces, in which enzymes bind to solid supports. Whereas cross-linking is used to aggregate enzymes or crystals that have been cross-linked to increase the activity and durability of the enzymes Maghraby et al. (2023). The development of sustainable enzymes has become essential to advancing environmentally conscious industrial processes and green chemistry. As extremely effective and selective biocatalysts, enzymes have many benefits over conventional chemical catalysts, such as lower energy costs and fewer harmful byproducts. However, their broad use is constrained by their intrinsic instability in demanding industrial settings. Immobilization methods have been created to enhance the stability, reusability, and performance of enzymes to overcome this difficulty. By providing innovative nanoscale materials and platforms that increase enzyme activity, preserve structural integrity, and prolong operational lifespan, nanotechnology has significantly contributed to the recent revolution in enzyme immobilization. In addition to aligning with sustainability principles, this combination of nanotechnology and biocatalysis opens new avenues for enzyme-driven innovations in waste management, biofuels, food processing, and pharmaceutical sectors.

8.3.3.1 Role of Nanotechnology in Enzyme Stabilization

Some of the most promising biomaterials have been produced by combining cutting-edge nanobiotechnology with enzymes in a synergistic manner, yielding nano-biocatalysts. To create nanoparticles (NPs) that are rarely hazardous to the environment, enzymes are crucial to nanotechnology. Nonetheless, functional nanostructures (nanomaterials) are novel, intriguing, and extraordinary matrices for the development of nano-biocatalysts due to their distinct/specific physicochemical properties and supramolecular nature. These can significantly improve the stability, kinetics, function, efficiency, susceptibility to diffusional constraints, and engineering performance of enzymes in bioprocessing. As a result, the created nano-biocatalysts possess remarkable properties and a wide range of potential uses (Oke et al. 2023).

Metallic NPs, such as gold, silver, and ferrihydrite, offer high enzyme stability and easy recovery. Das et al. (2022) used fruit extract from *Prunus nepalensis* to create gold NPs with catalytic activity similar to peroxidase. The Au NPs' catalytic activity was corroborated by their capacity to catalyze the oxidation of the substrate TMB in the presence of hydrogen peroxide. Silver NPs derived from cucumber and aloe vera extracts, the latter of which acted as a capping agent, reduced methyl orange hue and para-nitro-phenol (Riaz et al. 2022). This demonstrated the NPs' capacity to break down the nitro-phenols found in industrial waste. Peroxidase-like activity was also shown by ferrihydrite NPs from *Trichoderma guizhouense*, which are produced when the fungus interacts with hematite and creates NPs by absorbing minerals (Yu et al. 2020).

Enzyme immobilization is also promoted by dendritic mesoporous Si nanoparticles (DMSNs), a novel family of solid porous materials, due to their exceptional biocompatibility and distinctive open central–radial topologies with extensive pore channels (Zhang et al. 2024).

Numerous polymers, including cellulose, alginate, chitosan, and their derivatives, are frequently used in enzyme immobilization. A viable choice for enzyme support, chitosan is available in a number of forms for immobilization applications, including microspheres, gels, and beads. Yi et al. (2009) have developed chitosan beads modified with amino acids to encapsulate the lipase enzyme from *Candida rugosa*. Many nanostructured materials are now more reasonably priced for a wider range of applications thanks to recent advancements in nanotechnology. Numerous nanostructures have been investigated as hosts for the encapsulation of enzymes in the application of these materials for biocatalysis. These techniques include enzyme adsorption, covalent attachment, immobilization, and complex combinations of approaches (Kim et al. 2006). When metal ions or clusters are combined with organic linkers, one- or more-dimensional structures known as metal–organic frameworks (MOFs) are created. The large pore size, crystalline structure, tunability, and enhanced surface area of many MOFs have made them interesting candidates for several applications, such as gas, energy transport and storage, catalysis, and biosensing (Yusuf et al. 2022).

Some advantages of employing nanotechnology in biocatalyst immobilization include the following:

1. Improved Stability: Protection from denaturation and severe weather.
2. Enhanced Enzyme Activity: In optimal microenvironments, a native structure is maintained.
3. Reusability and Cost Reduction: Facilitates multiple use cycles and the recovery of enzymes.
4. Selective Binding and Improved Kinetics: Nanomaterials allow for improved substrate accessibility and precise enzyme orientation.

Despite its many advantages, the application of NPs for enzyme immobilization also presents some challenges. Cost and scalability: manufacturing nanocarriers at a reasonable cost is essential for industrial uses are few of these challenges (Table 8.2).

8.4 SUSTAINABLE ENZYME DEVELOPMENT AND INDUSTRIAL APPLICATIONS

As mentioned earlier, since enzymes offer efficient and sustainable alternatives to conventional chemical methods, they are crucial to a wide range of industrial operations. The primary objective of sustainable enzyme development is to improve enzyme stability, activity, and recyclability while minimizing environmental impact. Advances in biotechnology, such as protein engineering and nanotechnology, have enabled the production of powerful, highly specialized enzymes for application in food processing, wastewater treatment, medicine, and biofuels. By adopting sustainable practices, industries can reduce energy consumption, waste output, and reliance on hazardous chemicals, paving the way for a more productive and ecologically friendly future.

8.4.1 Next-Generation Enzyme Discovery

Next-generation enzyme discovery has advanced in response to the need for sustainable biocatalysts, especially through

TABLE 8.2
Protein Engineering for Industrial Applications, Focusing on Stability, Specificity, and Efficiency Improvements

S. No	Parameter	Challenges	Engineering Strategies	Examples	References
1.	Stability, including resistance to pH, temperature, and solvents	Enzyme denaturation due to extreme industrial environmental conditions.	• Directed evolution using random mutagenesis and selection. • RD using site-directed mutagenesis for structural stability. • Computational modeling using predicting stabilizing mutations. • Chemical modifications using polyethylene glycol (PEG)ylation and immobilization.	• Lipases and extremophile lipases for detergents (alkaline stability) • Cellulases for biofuels (thermostability)	Chandra et al. (2020); Vivek et al. (2022); Ejaz et al. (2021)
2.	Specificity for targeting desired substrates	Reduced efficiency and the production of side products are caused by low selectivity.	• Active site modifications by altering substrate binding pockets. • Computational docking by modeling enzyme-substrate interactions. • Chimeric enzymes by hybridization for new functionalities.	• Proteases for cheese production (casein specificity) • Amylases for starch processing (high substrate selectivity)	Leulmi et al. (2023); de Souza and Magalhães (2010)
3.	Efficiency (catalytic activity and kinetics)	Low turnover rates increase enzyme dosage and costs.	• Enzyme kinetics optimization by modifying catalytic residues. • Allosteric regulation engineering by improving distant site interactions. • Fusion proteins by combining domains for enhanced function.	• Laccases for wastewater treatment (faster pollutant degradation) • Glucose oxidase for biosensors (higher activity)	Rodríguez-Couto (2024); Tong et al. (2024)

metagenomics and enzyme mining. Through the identification of novel enzymes and the investigation of uncultured microbial populations, metagenomics offers access to a vast genetic pool. Researchers can find strong and varied enzymes with improved stability, activity, and adaptability to industrial settings by eschewing conventional culturing techniques. High-throughput screening and bioinformatics enable enzyme mining, which makes it easier to find enzymes with desired characteristics, such as solvent tolerance, heat resistance, and catalytic efficiency. These methods have greatly increased the range of enzymes that can be used in bioremediation, green chemistry, biofuels, and pharmaceuticals. Next-generation enzyme discovery, when combined with protein engineering and nanotechnology, is transforming industrial biocatalysis and opening the door to more economical, ecologically friendly, and efficient processes (Figure 8.3).

To enable the targeted identification of pectinolytic bacteria and enzymes, Zhou et al. (2017) combined an enrichment culture technique with a metagenomics sequence-guided approach. Apple pomace-adapted compost (APAC) was created as a habitat to improve the enrichment of pectinolytic microbes. APAC was created as a habitat to improve the enrichment of pectinolytic microbes. According to this study's phylogenetic analysis, the apple pomace-adapted compost microbial community(APACMC) had a significant number of pectinolytic bacteria that had not yet been discovered, as depicted in the image below. Massive volumes of metagenomic sequencing data have provided insight into the metabolic potential of the greatest number of uncultivated microorganisms. Although genome-resolved information on microbial communities from nearly every environment on Earth is available, accurately estimating biocatalytic activities directly from sequencing

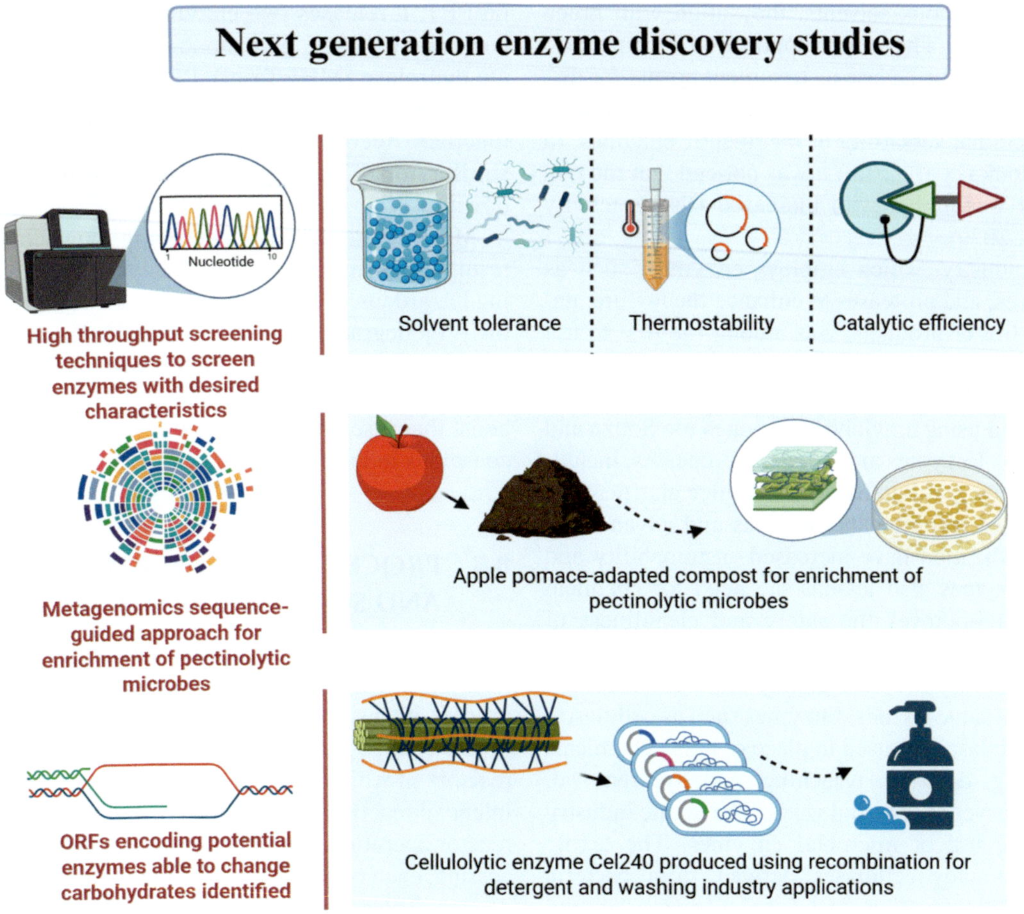

FIGURE 8.3 Next-generation enzyme discovery studies.

data remains challenging. In contrast to core metabolic pathways, secondary metabolism enzymes are valuable resources for the discovery of new enzymology, as they often catalyze specific reactions with a variety of substrates. Until now, the primary technique for deriving functional insights from research on environmental DNA (eDNA) has been PCR, or activity-based screening of eDNA fragments cloned in fosmid or cosmid libraries (Robinson et al. 2021).

Oliva et al. (2024)examined a metagenome from cold and alkaline ikaite columns in Greenland and found 282 open reading frames (ORFs) encoding potential enzymes that change carbohydrates. These enzymes could be used in detergents and other processes that require operation at high pH and low temperature. Two host organisms, *E. coli* and *Aliivibrio wodanis*, were used to generate and express17selected ORFs, representing 8 enzyme families. Compared to *E. coli*, *A. wodanis* showed expression of a broader range of enzyme classes, highlighting the importance of different expression methods for enzymes from extremophilic microbes. In this study, a cellulolytic enzyme produced by recombination from the metagenome was selected as the method for additional characterization, as depicted in the image below. Cel240 was thus observed to be a viable candidate for detergent and washing industry applications since it demonstrated activity at lower temperatures, higher enzyme stability at 25°C, activity at alkaline pH, and improved enzyme activity in the presence of $CaCl_2$.

8.4.2 Industries Benefiting from Enzyme Innovations

Many large industrial sectors have been significantly impacted by the introduction of sustainable enzyme development. Traditional chemical processes can be replaced with affordable, efficient, and eco-friendly enzyme technology. Enzyme advances are being used by a variety of industries, including waste management, bioremediation, textiles, food and beverage, pharmaceuticals, and biofuels, to achieve sustainable growth. Enzymes have revolutionized drug development and biocatalysis in pharmaceutical manufacturing. They are used in the production of chiral intermediates, which are necessary for the synthesis of enantiomerically pure drugs. Chiral intermediates for medications were produced using biocatalytic methods. Patel (2000) showed that *C. antarctica* lipase B could enzymatically resolve racemic 2-pentanol and 2-heptanol. One essential chiral intermediate needed to produce anti-Alzheimer's medications was S-(+)-2-pentanol. As enzyme-based processes have a high specificity, require fewer reaction steps, and use fewer hazardous solvents, they align with green chemistry principles. The direct application of enzymes, such as asparaginase or DNase as treatment agents for diseases, such as cancer and cystic fibrosis, further demonstrates the functional versatility of therapeutic enzymes. In a study by Golonka et al. (2018) it was noticed that the risk of liver cancer was higher with increased DNase activity (Lauková et al. 2020).

The food industry, which employs enzymes, such as lipases, amylases, and proteases to enhance the texture, flavor, and shelf life of products, was among the first to use enzyme technologies. Glucose syrups, high fructose corn syrups, crystalline glucose, maltose syrups, and other products are produced using amylolytic enzymes (de Souza and Magalhães 2010). Enzymes aid in several processes, including brewing, cheesemaking, baking, and juice clarification. The production of recombinant enzymes and advances in enzyme immobilization have increased sustainability and yield. Using enzymes also lessens the need for chemical additives, which improves the safety and cleanliness of food processing.

Innovations in enzymes have greatly improved the manufacture of sustainable textiles. Enzymes such as cellulases, laccases, and catalases are used in place of harsh chemicals for bio-polishing, desizing, bleaching, and dye removal, which also conserves energy and water. The textile industry makes extensive use of microbial enzymes. The textile industry may employ cellulases derived from bacteria (*Bacillus* and *Paenibacillus*) and fungi (*Aspergillus* and *Trichoderma*). From textile effluents, a facultative anaerobic catalase-producing *Bacillus maroccanus* was identified by Sukumaran et al. (2005); Gomaa and Momtaz (2006), which was resistant to hydrogen peroxide. Enzymatic treatments offer a more biodegradable alternative to conventional textile finishing methods while also reducing pollutants and enhancing fabric quality. As they transform lignocellulosic biomass into fermentable sugars—a critical step in the synthesis of second-generation bioethanol—enzymes are essential to the biofuel sector. Cellulases and hemicellulases efficiently break down plant cell walls, providing a more affordable and sustainable renewable energy source. Improvements in substrate selectivity and thermal stability brought about by enzyme engineering have increased process efficiency and reduced carbon footprints.

Enzymes are essential for waste treatment and bioremediation. They aid in the conversion of industrial waste, polymers, and organic pollutants into innocuous byproducts. Lipases, proteases, and ligninases are frequently employed in composting and wastewater treatment. A circular bioeconomy is also being promoted by the creation of enzyme-based solutions for plastic biodegradation, such as polyethylene terephthalate hydrolase (PETase) for the breakdown of polyethylene terephthalate (PET). PET film is a significant source of energy and carbon. It has been reported that when *Ideonellasakaiensis* is cultured on PET, it releases two enzymes called PET-hydrolyzing enzymes (PETase and mono-2-hydroxyethyl terephthalate hydrolase [MHETase]). PET polymers are hydrolyzed into monomeric forms by these enzymes working together. According to Liu et al. (2018), cutinase is most similar to PETase, an enzyme required for the breakdown of PET.

Many industries are undergoing long-term change as a result of enzyme innovations. Enzymes are ideal for replacing hazardous, energy-intensive industrial processes due to their biodegradability, catalytic efficiency, and bioengineered adaptability. Continued investigation into the stability of enzymes, cost reduction, and substrate diversity will boost their use and aid in creating a more environmentally conscious industrial future.

8.5 PROCESS INTENSIFICATION AND SCALE-UP STRATEGIES

Transferring biotechnological innovations from the laboratory to the industrial environment requires strategies for process intensification and scale-up. A crucial component of this change is the development and improvement of bioreactors in which microorganisms or enzymes catalyze the intended reactions. One of the most crucial decisions in bioreactor operation is whether to use batch or continuous processing, each of which has advantages and disadvantages of its own (Table 8.3).

8.5.1 Downstream Processing Innovations

The process of extracting and purifying biological products, such as enzymes, from complicated mixtures is known as DSP in bioprocessing. Nowadays, downstream costs can make up as much as 50–80% of total production costs, as upstream technologies increase enzyme yields. Therefore, both process intensification and commercial feasibility depend on the development of affordable, scalable purification techniques. High-purity enzymes are not necessary for many industrial uses, including waste treatment and detergents. Instead, raw or partially purified enzyme preparations are used directly, causing a drastic reduction in the number of processing steps, energy usage, and material costs.

Upstream and downstream bioprocessing are significant, interconnected processes in the synthesis of enzymes for a variety of biotechnological applications. The key objective of upstream bioprocessing, which encompasses strain selection, media optimization, and bioreactor design, is to maximize enzyme production (John et al. 2020). The selected microorganism or enzyme source should be maintained under ideal conditions to maximize enzyme yield and productivity (Adhyaru et al. 2015). Downstream bioprocessing includes the recovery, purification, and formulation of the produced enzyme. Its goals are to remove

TABLE 8.3
Comparison of Batch and Continuous Bioprocessing

S. No.	Feature	Batch Processing	Continuous Processing	Reference
1.	Operation mode	All reactants loaded at once; run in discrete cycles	Continuous input of substrates and output of products	Narses (2003)
2.	Productivity	Lower overall due to downtime between batches	Higher due to steady-state operation	Bukhtiyarova et al. (2023)
3.	Process control	Easier to manage and validate; suited for sensitive systems	Requires precise, real-time monitoring and control	Macchietti et al. (2024)
4.	Flexibility	High; suitable for multi-product and variable production	Low; optimized for consistent, high-volume output	Yang et al. (2021)
5.	Scale-up complexity	More challenges with gradients and mixing in larger volumes	Scalable with proper engineering and design	Li et al. (2015)
6.	Equipment requirements	Larger vessels for batch size; frequent cleaning and sterilization are needed	Smaller vessels can be used continuously; integration with PAT systems	Goršek (1997)
7.	Contamination risk	Lower risk as it is easier to contain and sterilize between runs	Higher risk as longer run time makes detection and containment harder	Kumar et al. (2020)
8.	Regulatory compliance	Well-established in the pharma and food industries; easier traceability	Increasingly accepted but more complex validation	Domokos et al. (2021)
9.	Sustainability	Higher resource usage due to downtime and cleaning cycles	More sustainable with reduced energy and water usage	Ince et al. (2025)
10.	Application suitability	Ideal for pharmaceuticals, Research and Development (R&D), and small to mid-scale production	Best for large-scale enzyme production, biofuels, and continuous manufacturing lines	Yang et al. (2021)

impurities, prepare it for use, and extract the enzyme from the complex mixture that comes from upstream processing. Effective and cost-effective methods are essential because DSP often contributes significantly to the overall cost of bioprocessing (González and Guajardo 2023). The aqueous two-phase system (ATPS), a liquid-liquid fractionation technique, has garnered interest from both industry and academia due to its vast potential for the extraction, purification, separation, and enrichment of proteins, membranes, viruses, enzymes, nucleic acids, and other biomolecules. However, the partition behavior of the technique is complex and uncertain. According to research by Iqbal et al. 2016, it has also been effectively utilized for a number of other tasks, including the treatment of sewage, the separation of valuable metals, and the identification of veterinary medication residues in food. The ATPS can offer a high recovery yield and is readily scalable. It is also an extremely economical and ecologically responsible method. The influenza vaccines currently on the market are only effective against analogous viruses; therefore, a seasonal update based on circulating subtypes is necessary. As the vaccine product's physicochemical properties vary, this continuous reformulation creates additional manufacturing challenges, especially in purification. It is crucial to have a universal platform strategy that can manage this kind of variation. Carvalho et al. (2019) investigated the purification of influenza virus-like particles using a filtration-based method. The retention variations seen for various influenza strains can be overcome by moving from adsorptive separation to size-based purification. Modern protein purification frequently uses immobilized metal affinity chromatography (IMAC). IMAC differs from most other affinity chromatography technologies in that it can work on a native, unmodified protein without the need fora particular affinity "tag" to encourage binding. This can be specifically important when a protein of interest needs to be separated from a complex mixture, such as serum or an environmental isolate. IMAC can offer strong discrimination between minute variations in protein sequence and structure by relying on the interactions of particular target protein surface amino acids with chelated metal ions. Furthermore, it has been shown that IMAC works well as a cation exchanger, enabling two purification methods to be used with a single column (Charlton and Zachariou 2008). *In situ* product recovery (ISPR) integrates product separation directly into bioprocesses, offering significant advantages by removing the required product during fermentation or reaction (Santos et al. 2021). This approach is particularly helpful when the product is poisonous or inhibitory since it reduces inhibition, alters the equilibrium, and boosts overall productivity (Woodley et al. 2008). ISPR employs techniques, such as extraction, adsorption, and membrane separation, to boost product yield, simplify DSP, and enable continuous production—despite ongoing challenges with maintaining sterility, selectivity, and compatibility with biological systems (Santos et al. 2021). Successful industrial application depends on a number of factors, such as process structure, robustness, choice of separation operation, as well as product market features (Woodley et al. 2008).

8.5.2 Integration with Circular Bioeconomy

Industries can reduce waste, improve sustainability, and cut costs by integrating process intensification with the circular bioeconomy. Fermentation leftovers, such as discarded biomass, can be recycled into bioactive compounds, animal feed, or fertilizers via enzymatic treatment. Agro-industrial waste, such as the peels of fruits and grain bran, can be employed as low-cost substrates for enzyme production, thereby lowering the price of raw materials(Yuliana et al. 2023). Circular scale-up strategies that improve overall efficiency include energy recovery, process water recycling, and enzyme reuse. Using circular models, biorefineries create bio-based compounds or enzymes from wastes, such as glycerol (Protte-Freitag et al. 2024).

8.6 REGULATORY, ECONOMIC, AND ENVIRONMENTAL CONSIDERATIONS

Strict regulatory, financial, and environmental requirements must be met for contemporary biomanufacturing to be secure, commercially viable, and sustainable. The following considerations are essential to be taken care of:

1. The Regulation Framework: Organizations, such as the US Food and Drug Administration (FDA) and European Medicines Agency (EMA), enforce strict rules on the safety, efficacy, and purity of enzymes, especially for use in food and medicine (Ilic et al. 2012). Public trust and product approval depend on compliance with environmental safety laws and Good Manufacturing Practices (GMP) (Gouveia et al. 2015).
2. Technology and Economics Analysis (TEA): TEA considers yield, scale-up potential, and capital and operating costs when evaluating the economic feasibility of bioprocesses. It is critical to identify bottlenecks and increase the cost-competitiveness of biomanufactured goods relative to chemical alternatives. The model predicts considerable reductions in capital investment and a >50% reduction in the cost of goods compared to published figures for comparable production sizes, relative to standard biomanufacturing platforms using mammalian cells cultured in bioreactors. Others can edit or adapt the Nandi et al.(2016) simulation model to evaluate the profitability of different designs, apply various process assumptions, and aid in the creation and optimization of processes (Nandi et al. 2016).
3. Life Cycle Assessment (LCA) and Sustainability Measures: LCA quantifies the environmental impact of a bioprocess from the procurement of basic materials to the disposal of the finished product. Metrics, such as waste production, water use, and energy use, are used to evaluate and enhance sustainability (Oluyemi et al. 2024).
4. The Carbon Footprint and Environmental Benefits: Biomanufacturing technologies often provide lower carbon emissions than petrochemical pathways. Enzyme-based technologies can boost biodegradability, reduce negative effects, and ease the shift to low-carbon, circular production models (Scown 2022).

8.7 FUTURE PERSPECTIVES AND EMERGING TRENDS IN BIOMANUFACTURING

Rapid technological advancement and the increasing demand for sustainable solutions are driving the future of biomanufacturing. Precise genetic modifications are becoming possible thanks to the advancements in synthetic biology and CRISPR, which enable the customized synthesis of enzymes and bioproducts with improved yields and utility. AI and automation are transforming industrial workflows by enabling real-time monitoring, predictive analytics, and automated control systems, all of which boost productivity and reduce operating costs. Furthermore, portable, modular bioreactors and other decentralized bioproduction systems are enabling localized and on-demand manufacturing, which is especially advantageous for specialized or remote applications. These technological developments are further cemented by supportive regulatory frameworks and commercial incentives that promote circular, bio-based economies and reduce carbon footprints. When taken as a whole, these advancements are establishing biomanufacturing as a key component of sustainable industrial change.

8.8 CONCLUSIONS

At the nexus of biotechnology, sustainability, and industrial innovation, biomanufacturing has become a game-changing platform. It is clear from this analysis that developments in DSP, enzyme engineering, and process intensification are not only increasing production efficiency but also changing how industries manage resource use and environmental impacts. Bioprocesses have the potential to be in line with the circular bioeconomy's tenets when cost-effective, scalable technologies are combined with waste valorization techniques.

More sophisticated, adaptable bioprocesses will be the center of attention of future R&D. This means looking into AI-driven optimization tools, developing modular, decentralized production systems that can function well at different scales and places, and using synthetic biology and CRISPR to construct tailored microbial platforms. Furthermore, advancements in sustainability metrics and

LCA will be employed to quantify and direct the environmentally friendly advantages of these advances.

Beyond technological advancements, biomanufacturing plays a vital and timely role in addressing global sustainability issues. It provides workable solutions to reduce greenhouse gas emissions, replace petroleum-based products, and promote environmentally friendly supply chains. Biomanufacturing is expected to play an important role in enabling a resilient, low-carbon, and resource-efficient future, driven by increasing regulatory support, market demand for environmentally acceptable alternatives, and interdisciplinary collaboration.

REFERENCES

Abreu, A. P., R. Martins, and J. Nunes. (2023). Emerging Applications of *Chlorella* sp. and *Spirulina* (*Arthrospira*) sp. *Bioengineering*, 10(8), 955.

Adhyaru, D.N., N.S. Bhatt, and H.A. Modi. (2015). Optimization of Upstream and Downstream Process Parameters for Cellulase-Poor-Thermo-Solvent-Stable Xylanase Production and Extraction by *Aspergillus tubingensis* FDHN1. *Bioresources and Bioprocessing* 2, 3.

Alagawany, Mahmoud, Ayman E. Taha, Ahmed Noreldin, Khaled A. El-Tarabily, Mohamed E. Abd El-Hack. (2021). Nutritional Applications of Species of Spirulina and Chlorella in Farmed Fish: A Review, *Aquaculture*, 542, 736841, ISSN 0044-8486.

Aurich, M. K., and I. Thiele. 2016. "Computational Modeling of Human Metabolism and Its Application to Systems Biomedicine." *Methods in Molecular Biology*1386: 253–281. https://doi.org/10.1007/978-1-4939-3283-2_12

Bloom, J. D., M. M. Meyer, P. Meinhold, C. R. Otey, D. MacMillan, and F. H. Arnold. (2005). Evolving Strategies for Enzyme Engineering. *Current Opinion in Structural Biology*, 15(4), 447–452.

Branco-Vieira, Monique, Sergio San Martin, Cristian Agurto, Marcos A.V. Freitas, António A. Martins, Teresa M. Mata, Nídia S. Caetano. 2020. Biotechnological Potential of Phaeodactylum tricornutum for Biorefinery Processes. *Fuel*, 268, 117357, ISSN 0016-2361.

Brookwell, A., J. P. Oza, and F. Caschera. (2021). Biotechnology Applications of Cell-Free Expression Systems. *Life (Basel, Switzerland)*, 11(12), 1367.

Bukhtiyarova, M. V., A. L. Nuzhdin, and G. A. Bukhtiyarova. (2023). Comparative Study of Batch and Continuous Flow Reactors in Selective Hydrogenation of Functional Groups in Organic Compounds: What Is More Effective? *International Journal of Molecular Sciences*, 24(18), 14136. https://doi.org/10.3390/ijms241814136

Bundy, B. C., J. Porter Hunt, Michael C. Jewett, James R. Swartz, David W. Wood, Douglas D. Frey, Govind Rao. (2018). Cell-Free Biomanufacturing. *Current Opinion in Chemical Engineering*, 22, 177–183. https://doi.org/10.1016/j.coche.2018.10.003

Carvalho, S. B., R. J. S. Silva, M. G. Moleirinho, B. Cunha, A. S. Moreira, A. Xenopoulos, P. M. Alves, M. J. T. Carrondo, and Peixoto, C. (2019). Membrane-Based Approach for the Downstream Processing of Influenza Virus-Like Particles. *Biotechnology Journal*, 14(8), e1800570.

Chandra, P., Enespa, R. Singh, et al. (2020). Microbial Lipases and Their Industrial Applications: A Comprehensive Review. *Microbial Cell Factories* 19, 169. https://doi.org/10.1186/s12934-020-01428-8

Charlton, A., and M. Zachariou. (2008). Immobilized Metal Ion Affinity Chromatography of Native Proteins. *Methods in Molecular Biology* 421, 25–35.

Chen, R., S. Ren, S. Li, H. Zhou, X. Jia, D. Han, and Z. Gao. 2025. "Synthetic Biology for the Food Industry: Advances and Challenges." *Critical Reviews in Biotechnology* 45: 23–47. https://doi.org/10.1080/07388551.2024.2340530

Chiang, C. J., R. Hu, Z. Huang, and Y. Chao. 2021. "Production of Succinic Acid from Amino Acids in Escherichia coli." *Journal of Agricultural and Food Chemistry* 69: 8172–8178. https://doi.org/10.1021/acs.jafc.1c02958

Cho, J. S., G. B. Kim, H. Eun, C. W. Moon, and S. Y. Lee. 2022. "Designing Microbial Cell Factories for the Production of Chemicals." *JACS Au* 2: 1781–1799. https://doi.org/10.1021/jacsau.2c00344

Clomburg, J. M., A. M. Crumbley, and R. Gonzalez. 2017. "Industrial Biomanufacturing: The Future of Chemical Production." *Science* 355: aag0804. https://doi.org/10.1126/science.aag0804

Dai, Z., Y. Liu, J. Guo, L. Huang, and X. Zhang. 2015. "Yeast Synthetic Biology for High-Value Metabolites." *FEMS Yeast Research* 15: 1–11. https://doi.org/10.1111/1567-1364.12187

Das, B., J. Lou-Franco, B. Gilbride, M. G. Ellis, L. D. Stewart, I. R. Grant, et al. (2022). Peroxidase-Mimicking Activity of Biogenic Gold Nanoparticles Produced from *Prunus nepalensis* Fruit Extract: Characterizations and Application for the Detection of *Mycobacterium bovis*. *ACS Appl. Bio Mater.* 5 (6), 2712–2725. https://doi.org/10.1021/acsabm.2c00180

de Lima, E. A., F. Mandelli, D. Kolling, J. Matsusato Souza, C. A. de Oliveira Filho, M. Ribeiro da Silva, I. Lobo de Mesquita Sampaio, et al.2022. "Development of an Economically Competitive *Trichoderma*-Based Platform for Enzyme Production: Bioprocess Optimization, Pilot Plant Scale-Up, Techno-Economic Analysis and Life Cycle Assessment." *Bioresource Technology*364: 128019. https://doi.org/10.1016/j.biortech.2022.128019

de Souza, P. M., P. O. Magalhães. (2010). Application of Microbial α-Amylase in Industry - A Review. *Brazilian Journal of Microbiology*, 41(4):850–861.

Despotovic, D., L. Vojcic, R. Prodanovic, R. Martinez, K. H. Maurer, and U. Schwaneberg. (2012). Fluorescent Assay for Directed Evolution of Perhydrolases. *Journal of Biomolecular Screening*, 17(6), 796–805.

Diamos, A. G., M. D. Pardhe, H. Sun, J. G. L. Hunter, T. Mor, L. Meador, J. Kilbourne, Q. Chen, and H. S. Mason. (2020). "Codelivery of Improved Immune Complex and Virus-Like Particle Vaccines Containing Zika Virus Envelope Domain III Synergistically Enhances Immunogenicity. *Vaccine*, 38(18), 3455–3463.

Ding, Q., and C. Ye. 2023. "Microbial Cell Factories Based on Filamentous Bacteria, Yeasts, and Fungi." *Microbial Cell Factories*22: 20. https://doi.org/10.1186/s12934-023-02025-1

Dobrogojski, J., M. Spychalski, R. Luciński, et al. (2018). Transgenic Plants as a Source of Polyhydroxyalkanoates. *Acta Physiologiae Plantarum*40, 162.

Domokos, A., B. Nagy, B. Szilágyi, G. Marosi, and Z. K. Nagy. (2021). Integrated Continuous Pharmaceutical Technologies—A Review. *Organic Process Research & Development*, 25(4), 721–739. doi:10.1021/acs.oprd.0c00504

Dragosits, M., and D. Mattanovich 2013. "Adaptive Laboratory Evolution: Principles and Applications for Biotechnology." *Microbial Cell Factories*12: 64. https://doi.org/10.1186/1475-2859-12-64

Ejaz, U., M. Sohail, and A. Ghanemi. (2021). Cellulases: From Bioactivity to a Variety of Industrial Applications. *Biomimetics (Basel, Switzerland)*, 6(3), 44. https://doi.org/10.3390/biomimetics6030044

England, R., S. Kumar, and N. Sharma. 2023. "Plant-Based Expression Systems for Monoclonal Antibodies and Vaccines." *Current Pharmaceutical Biotechnology* 24: 59–72.

Fasim, A., V. S. More, and S. S. More. 2021. "Large-Scale Production of Enzymes for Biotechnology Uses." *Current Opinion in Biotechnology* 69: 68–76. https://doi.org/10.1016/j.copbio.2020.12.002

Fidler, S., and D. Dennis. 1992. "Polyhydroxyalkanoate Production in Recombinant *Escherichia coli*." *FEMS Microbiology Reviews* 9: 231–235. https://doi.org/10.1016/0378-1097(92)90314-e

Gisbert, M., D. Franco, J. Sineiro, and R. Moreira. (2023). Antioxidant and Antidiabetic Properties of Phlorotannins from *Ascophyllum nodosum* Seaweed Extracts. *Molecules (Basel, Switzerland)*, 28(13), 4937.

Golonka,R. M., B.S. Yeoh, J.L. Petrick, S.J. Weinstein, D. Albanes, A.T. Gewirtz, K.A. McGlynn, and M. Vijay-Kumar. (2018). Deoxyribonuclease I Activity, Cell-Free DNA, and Risk of Liver Cancer in a Prospective Cohort. *JNCI Cancer Spectrum* 2: pky083. doi: 10.1093/jncics/pky083

Gomaa, O. M., and A. M. Momtaz. (2006). Characterization of the Hydrogen Peroxide Tolerating 'Bacillus maroccanus' Type Strain Isolated from Textile Wastewater. *Arab Journal of Biotechnology* 9: 83–94.

González, J., and N. Guajardo. (2023). Flow Enzymatic Esterification of 5-Hydroxymethylfurfural and Liquid–Liquid Extraction of 5-Acetyl-Hydroxymethylfurfural Using Deep Eutectic Solvents in Semicontinuous Mode. *Process Biochemistry* 130: 409–418. doi: 10.1016/j.procbio.2023.04.026

Goršek, P. G. (1997). Design of Batch Versus Continuous Processes: Part I: Single-Purpose Equipment. *Chemical Engineering Research and Design* 75 (7), 709–717. https://doi.org/10.1205/026387697524209

Gouveia, B. G., P. Rijo, T. S. Gonçalo, and C. P. Reis. (2015). Good Manufacturing Practices for Medicinal Products for Human Use. *Journal of Pharmacy & Bioallied Sciences*, 7(2), 87–96.

Gruet, A., S. Longhi, and C. Bignon. (2012). One-Step Generation of Error-Prone PCR Libraries Using Gateway® Technology. *Microbial Cell Factories*11, 14.

Hussain, M. H., M. Z. Mohsin, W. Q. Zaman, J. Yu, X. Zhao, Y. Wei, Y. Zhuang, A. Mohsin, and M. Guo. 2022. "Multiscale Engineering of Microbial Cell Factories: A Step Forward towards Sustainable Natural Products Industry." *Synthetic and Systems Biotechnology*7: 586–601. https://doi.org/10.1016/j.synbio.2021.12.012

Ilic, N., S. Savic, E. Siegel, K. Atkinson, and L. Tasic. (2012). Examination of the Regulatory Frameworks Applicable to Biologic Drugs (Including Stem Cells and Their Progeny) in Europe, the U.S., and Australia: Part I--A Method of Manual Documentary Analysis. *Stem Cells Translational Medicine* 1(12), 898–908.

Ince, M. C., B. Benyahia, and G. Vilé. (2025). Sustainability and Techno-Economic Assessment of Batch and Flow Chemistry in Seven Industrial Pharmaceutical Processes. *ACS Sustainable Chemistry & Engineering*, 13(7), 2864–2874. https://doi.org/10.1021/acssuschemeng.4c09289

Iqbal, M., Y. Tao, S. Xie et al. (2016). Aqueous Two-Phase System (ATPS): An Overview and Advances in Its Applications. *Biological Procedures Online* 18, 18.

Jiang, W., J. B. Qiao, G. J. Bentley, et al. 2017. "Modular Pathway Engineering for the Microbial Production of Branched-Chain Fatty Alcohols." *Biotechnology for Biofuels* 10: 244. https://doi.org/10.1186/s13068-017-0936-4

John, J., K.K.S. Kaimal, M.L. Smith, P.K.S.M. Rahman, P.V. Chellam. (2020). Advances in Upstream and Downstream Strategies of Pectinase Bioprocessing: A Review. *International Journal of Biological Macromolecules*162: 1086–1099. doi: 10.1016/j.ijbiomac.2020.06.224

Jourdin, L., and T. Burdyny. (2021). Microbial Electrosynthesis: Where Do We Go from Here? *Trends in Biotechnology*, 39(4), 359–369. https://doi.org/10.1016/j.tibtech.2020.10.014

Jullesson, D., F. David, B. Pfleger, and J. Nielsen. 2015. "Impact of Synthetic Biology and Metabolic Engineering on Industrial Production of Fine Chemicals." *Biotechnology Advances* 33: 1395–1402. https://doi.org/10.1016/j.biotechadv.2015.02.011

Khan, I., and M. Waheed Akhtar. (2011). Different Approaches For Protein Engineering In Industrial Biotechnology. *Nature Preceedings* 6.

Kim, Jungbae, Jay W. Grate, Ping Wang. (2006). Nanostructures for Enzyme Stabilization, *Chemical Engineering Science*61 (3), 1017–1026, ISSN 0009-2509.

Koppolu, V., and V. K. Vasigala. 2016. "Role of Escherichia coli in Biofuel Production." *Microbiology Insights*9: 29–35. https://doi.org/10.4137/MBI.S10878

Kowalczyk, A., M. Kowalski, and J. Kowalewska. 2022. "Genetically Modified Plants as Biofactories: Advances in Molecular Farming." *Plant Biotechnology Journal* 20: 1824–1840.

Kumar, A., I. A. Udugama, C. L. Gargalo, and K. V. Gernaey. (2020). Why Is Batch Processing Still Dominating the Biologics Landscape? Towards an Integrated Continuous Bioprocessing Alternative. *Processes* 8(12): 1641. https://doi.org/10.3390/pr8121641

Kumar, R., G. Gupta, A. Hussain, et al. (2025). Pioneering Zero-Waste Technologies Utilization and Its Framework on Sustainable Management: International, National and State Level. *Discover Applied Sciences* 7, 224. https://doi.org/10.1007/s42452-025-06693-z

Lauková, L., B. Konečná, Ľ. Janovičová, B. Vlková, and P. Celec. (2020). Deoxyribonucleases and Their Applications in Biomedicine. *Biomolecules*, 10(7), 1036.

Leandro, A., L. Pereira, and A. M. M. Gonçalves. (2019). Diverse Applications of Marine Macroalgae. *Marine Drugs*, 18(1), 17.

Lee, S. J., and Kim, D. M. (2024). Cell-Free Synthesis: Expediting Biomanufacturing of Chemical and Biological Molecules. *Molecules (Basel, Switzerland)*, 29(8), 1878.

Leulmi, I., M. N. Zidoune, K. Hafid, F. Djeghim, H. Bourekoua, D. Dziki, and R. Różyło. (2023). New Coagulant Proteases for Cheesemaking from Leaves and Latex of the Spontaneous Plant *Pergularia tomentosa*: Biochemical Characterization of Coagulants and Sensorial Evaluation of Cheese. *Foods*, 12(13), 2467. https://doi.org/10.3390/foods12132467

Li, J., J. Jaitzig, P. Lu, et al. (2015). Scale-up Bioprocess Development for Production of the Antibiotic Valinomycin in *Escherichia coli* Based on Consistent Fed-Batch Cultivations. *Microbial Cell Factories* 14, 83. https://doi.org/10.1186/s12934-015-0272-y

Li, M., X. Lang, M. Moran Cabrera, et al.2021. "CRISPR-Mediated Multigene Integration Enables Shikimate Pathway Refactoring for Enhanced 2-Phenylethanol Biosynthesis in *Kluyveromyces marxianus*." *Biotechnology for Biofuels* 14: 3. https://doi.org/10.1186/s13068-020-01852-3

Lim, Hye Jin, and Dong-Myung Kim. (2022). Cell-Free Synthesis of Industrial Chemicals and Biofuels from Carbon Feedstocks, *Current Opinion in Biotechnology*73, 158–163, ISSN 0958-1669, https://doi.org/10.1016/j.copbio.2021.08.002

Liu, B., L. He, L. Wang, T. Li, C. Li, H. Liu, et al. (2018). Protein Crystallography and Site-Direct Mutagenesis Analysis of the Poly (Ethylene Terephthalate) Hydrolase PETase from *Ideonella sakaiensis*. *ChemBioChem* 19, 1471–1475. doi: 10.1002/cbic.201800097

Liu, H., J. Kang, Q. Qi, and G. Chen. 2011. "Production of Lactate in *Escherichia coli* by Redox Regulation Genetically and Physiologically." *Applied Biochemistry and Biotechnology* 164: 162–169. https://doi.org/10.1007/s12010-010-9123-9

Liu, L., H. Yang, H. D. Shin, R. R. Chen, J. Li, G. Du, and J. Chen. 2013. "How to Achieve High-Level Expression of Microbial Enzymes: Strategies and Perspectives." *Bioengineered* 4: 212–223. https://doi.org/10.4161/bioe.24761

Ma, Xiaochen, Hongli Zheng, Min Addy, Erik Anderson, Yuhuan Liu, Paul Chen, Roger Ruan. 2016. Cultivation of Chlorella vulgaris in Wastewater with Waste Glycerol: Strategies for Improving Nutrients Removal and Enhancing Lipid Production. *Bioresource Technology*, 207, 252–261, ISSN 0960-8524.

Macchietti, L., D. Melucci, L. Menarini, F. Consoli, and A. Zappi. (2024). Analytical Comparison Between Batch and Continuous Direct Compression Processes for Pharmaceutical Manufacturing Using an Innovative UV-Vis Reflectance Method and Chemometrics. *International Journal of Pharmaceutics*, 656, 124090. https://doi.org/10.1016/j.ijpharm.2024.124090

Maghraby, Y. R., R. M. El-Shabasy, A. H. Ibrahim, and H. M. E. Azzazy. (2023). Enzyme Immobilization Technologies and Industrial Applications. *ACS Omega*, 8(6), 5184–5196.

Markina, N. M., A. A. Kotlobay, and A. S. Tsarkova. 2020. "Heterologous Metabolic Pathways: Strategies for Optimal Expression in Eukaryotic Hosts." *Acta Naturae* 12: 28–39. https://doi.org/10.32607/actanaturae.10966

McCormick, A., J. Palmer, and D. Chapman. 1999. "Modified Tobamoviral Vector for Antibody Expression in *Nicotiana benthamiana*." *Plant Molecular Biology* 41: 295–302.

Medipally, Srikanth Reddy, Fatimah Md Yusoff, Sanjoy Banerjee, M. Shariff. (2015). Microalgae as Sustainable Renewable Energy Feedstock for Biofuel Production. *BioMed Research International*, 2015, 519513.

Melo-Duran, Diego, Gemma González-Ortiz, Sandra Villagomez-Estrada, Michael R. Bedford, Mercè Farré, José F. Pérez, and David Solà-Oriol. 2024. Using in Feed Xylanase or Stimbiotic to Reduce the Variability in Corn Nutritive Value for Broiler Chickens, *Poultry Science*, 103 (3), 103401.

Mishra, S. K., P. R. Kumar, R. K. Singh, and A. K. Singh. 2020. "Agro-Industrial Residues as Solid Substrate for α-Amylase Production Using Solid-State Fermentation by Filamentous Fungi: A Review." *Bioscience Biotechnology Research Communications*13: 2

Moradi, Pantea, and Majid Saidi. 2022. Biodiesel Production from Chlorella vulgaris Microalgal-Derived Oil via Electrochemical and Thermal Processes, *Fuel Processing Technology* 228, 107158, ISSN 0378-3820.

Moussavou, G., K. Munyendo, D. Jouda, and C. Ndinteh. 2015. "Single-Chain Antibodies Produced by Plant Viral Vectors." *Applied Microbiology and Biotechnology*99: 8399–8412.

Naher, L., S. N. Fatin, M. A. H. Sheikh, L. A. Azeez, S. Siddiquee, N. M. Zain, and S. M. R. Karim. 2021. "Cellulase Enzyme Production from Filamentous Fungi *Trichoderma reesei* and *Aspergillus awamori* in Submerged Fermentation with Rice Straw." *Journal of Fungi*7: 868. https://doi.org/10.3390/jof7100868

Nandi, S., A. T. Kwong, B. R. Holtz, R. L. Erwin, S. Marcel, and K. A. McDonald. (2016). Techno-Economic Analysis of a Transient Plant-Based Platform for Monoclonal Antibody Production. *MAbs*8(8), 1456–1466.

Narses, B. (2003). *Batch Processing, Encyclopedia of Physical Science and Technology* (Third Edition), Academic Press, pp. 41–56. https://doi.org/10.1016/B0-12-227410-5/00049-1

Nawaz, T., S. Fahad, L. Gu, S. Saud, and R. Zhou. 2024. "Cyanobacteria: Role in Sustainable Biomanufacturing and Nitrogen Fixation." *Biofuels, Bioproducts and Biorefining* 18: 2132–2155. https://doi.org/10.1002/bbb.2674

Ndochinwa, O. G., Q. Y. Wang, O. C. Amadi, T. N. Nwagu, C. I. Nnamchi, E. S. Okeke, and A. N. Moneke. 2024a. "Current Status and Emerging Frontiers in Enzyme Engineering: An Industrial Perspective." *Heliyon*10: e32673. https://doi.org/10.1016/j.heliyon.2024.e32673

Ndochinwa, G. O., Q. Y. Wang, N. O. Okoro, O. C. Amadi, T. N. Nwagu, C. I. Nnamchi, A. N. Moneke, and A. S. Odiba. (2024b). New Advances in Protein Engineering for Industrial Applications: Key Takeaways. *Open Life Sciences*, 19(1), 20220856. https://doi.org/10.1515/biol-2022-0856

Nemudryi, A. A., K. R. Valetdinova, S. P. Medvedev, and S. M. Zakian. 2014. "TALEN and CRISPR/Cas Genome Editing Systems: Tools of Discovery." *Acta Naturae*6: 19–40.

Nielsen, J., and J. D. Keasling. 2016. "Engineering Cellular Metabolism." *Cell*164: 1185–1197. https://doi.org/10.1016/j.cell.2016.02.004

Oke, M. A., S. A. Ojo, S. A. Fasiku, and E. A. Adebayo. (2023). Nanotechnology and Enzyme Immobilization: A Review. *Nanotechnology*, 34(38). https://doi.org/10.1088/1361-6528/acda35

Oliva, B., A. Zervas, P. Stougaard, P. Westh, and M. S. Thøgersen. (2024). Metagenomic Exploration of Cold-Active Enzymes for Detergent Applications: Characterization of a Novel, Cold-Active and Alkali-Stable GH8 Endoglucanase from Ikaite Columns in SW Greenland. *Microbial Biotechnology*, 17(6), e14466.

Oluyemi, G. F., R. O. Afolabi, S. C. Zamora, Y. Li, and D. McElroy. (2024). Environmental Impact Assessment of a Plant Cell-Based Bio-Manufacturing Process for Producing Plant Natural Product Ingredients. *Sustainability* 16(19), 8515.

Panggabean, J. A., S. P. Adiguna, S. I. Rahmawati, P. Ahmadi, E. N. Zainuddin, A. Bayu, and M. Y. Putra. (2022). Antiviral Activities of Algal-Based Sulfated Polysaccharides. *Molecules (Basel, Switzerland)*, 27(4), 1178.

Parikh, M. R., and I. Matsumura. (2005). Site-Saturation Mutagenesis is More Efficient Than DNA Shuffling for the Directed Evolution of Beta-Fucosidase from Beta-Galactosidase. *Journal of Molecular Biology*, 352(3), 621–628.

Patel, R. N. (2000). Microbial/Enzymatic Synthesis of Chiral Drug Intermediates. *Advances in Applied Microbiology* 47, 33–78.

Protte-Freitag, K., S. Gotzig, H. Rothe, O. Schwarz, N. Silber, and R. Miehe (2024). Enzyme-Assisted Circular Additive Manufacturing as an Enabling Technology for a Circular

Bioeconomy—A Conceptual Review. *Sustainability*, 16(5), 2167.

Rantasalo, A., M. Vitikainen, T. Paasikallio, et al. 2019. "Novel Genetic Tools That Enable Highly Pure Protein Production in *Trichoderma reesei*." *Scientific Reports*9: 5032. https://doi.org/10.1038/s41598-019-41573-8

Riaz, M., U. Sharafat, N. Zahid, M. Ismail, J. Park, B. Ahmad, et al. (2022). Synthesis of Biogenic Silver Nanocatalyst and Their Antibacterial and Organic Pollutants Reduction Ability. *ACS Omega*7 (17), 14723–14734. https://doi.org/10.1021/acsomega.1c07365

Robinson, S. L., J. Piel, and S. Sunagawa (2021). A Roadmap for Metagenomic Enzyme Discovery. *Natural Product Reports*, 38(11), 1994–2023.

Rodríguez-Couto, S. (2024). Immobilized-Laccase Bioreactors for Wastewater Treatment. *Biotechnology Journal*, 19, e2300354. https://doi.org/10.1002/biot.202300354

Roesijadi, G., S. B. B. Jones, L. J. Snowden-Swan, et al. 2010. *Macroalgae as a Biomass Feedstock: A Preliminary Analysis*. Dep Energy Contract DE-AC05-76RL01830 Pac Northwest. Natl Lab; 1–50.

Saha, D., and S. Bhattacharya. (2010). Hydrocolloids as Thickening and Gelling Agents in Food: A Critical Review. *Journal of Food Science and Technology*, 47(6), 587–597.

Sajid, M., C. N. Channakesavula, S. R. Stone, and P. Kaur. (2021). Synthetic Biology towards Improved Flavonoid Pharmacokinetics. *Biomolecules*, 11(5), 754.

Sánchez-Solís, A., O. Lobato-Calleros, R. Moreno-Terrazas, P. Lappe-Oliveras, and E. Neri-Torres. 2024. "Biodiesel Production Processes with Yeast: A Sustainable Approach." *Energies*17: 302. https://doi.org/10.3390/en17020302

Santos, A.G., T.L. de Albuquerque, B.D. Ribeiro, and M.A.Z. Coelho. (2021). In Situ Product Recovery Techniques Aiming to Obtain Biotechnological Products: A Glance to Current Knowledge. *Biotechnology and Applied Biochemistry*, 68: 1044–1057.

Savino, S., T. Desmet, and J. Franceus. 2022. "Insertions and Deletions in Protein Evolution and Engineering." *Biotechnology Advances* 60: 108010. https://doi.org/10.1016/j.biotechadv.2022.108010

Sayed, W., A. Cabrol, A. Salma, A. Amrane, M. Benoit, R. Pierre, H. Djelal. (2024). Green Macroalgae Hydrolysate for Biofuel Production: Potential of *Ulva rigida*. *Applied Microbiology* 4(2): 563–581.

Scown, Corinne D.. (2022). Prospects for Carbon-Negative Biomanufacturing, *Trends in Biotechnology* 40 (12), 1415–1424, ISSN 0167-7799. https://doi.org/10.1016/j.tibtech.2022.09.004

Sethi, L., K. Kumari, and N. Dey. (2021). Engineering of Plants for Efficient Production of Therapeutics. *Molecular Biotechnology*63, 1125–1137.

Sheldon, R. A., and J. M. Woodley. 2018. "Role of Biocatalysis in Sustainable Chemistry." *Chemical Reviews* 118: 801–838. https://doi.org/10.1021/acs.chemrev.7b00203

Singh, V. K., S. Jha, P. Rana, R. Soni, R. Lalnunpuii, P. K. Singh, R. P. Sinha, and G. Singh. 2024. "Cyanobacteria as a Biocatalyst for Sustainable Production of Biofuels and Chemicals." *Energies* 17: 408. https://doi.org/10.3390/en17020408

Song, Z., Q. Zhang, W. Wu, Z. Pu, and H. Yu. (2023). Rational Design of Enzyme Activity and Enantioselectivity. *Frontiers in Bioengineering and Biotechnology* 11, 1129149.

Sukumaran, R. K., R.R. Singhania, A. Pandey. (2005). Microbial cellulases - Production, Applications and Challenges. *Journal of Scientific and Industrial Research*64: 832–844.

Sun, Z., Y. Wu, S. Long, S. Feng, X. Jia, Y. Hu, M. Ma, J. Liu, and B. Zeng. 2024. "*Aspergillus oryzae* as a Cell Factory: Research and Applications in Industrial Production." *Journal of Fungi*10: 248. https://doi.org/10.3390/jof10040248

Surendran, A., M. Lakshmanan, J. Y. Chee, A. M. Sulaiman, D. V. Thuoc, and K. Sudesh. (2020). Can Polyhydroxyalkanoates Be Produced Efficiently From Waste Plant and Animal Oils?. *Frontiers in Bioengineering and Biotechnology*, 8, 169.

Tang, C., L. Wang, L. Zang, Q. Wang, D. Qi, and Z. Dai. (2022). On-Demand Biomanufacturingthrough Synthetic Biology Approach. *Materials Today: Bio*, 18, 100518. https://doi.org/10.1016/j.mtbio.2022.100518

Tiwari, M. K., R. Singh, R. K. Singh, I. W. Kim, and J. K. Lee. (2012). Computational Approaches for Rational Design of Proteins with Novel Functionalities. *Computational and Structural Biotechnology Journal*, 2, e201209002.

Tong, X., L. Jiang, Q. Ao, X. Lv, Y. Song, and J. Tang. (2024). Highly Stable Glucose Oxidase Polynanogel@MXene/Chitosan Electrochemical Biosensor Based on a Multi-Stable Interface Structure for Glucose Detection. *Biosensors & Bioelectronics*, 248, 115942. https://doi.org/10.1016/j.bios.2023.115942

Udo, T., G. Mummaleti, A. Mohan, R. K. Singh, and F. Kong. (2023). Current and Emerging Applications of Carrageenan in the Food Industry. *Food Research International*, 173(Pt 2), 113369.

Vallejo, D., A. Nikoomanzar, and J. C. Chaput. (2020). Directed Evolution of Custom Polymerases Using Droplet Microfluidics. *Methods in Enzymology*, 644, 227–253.

Venayak, N., N. Anesiadis, W. R. Cluett, and R. Mahadevan. 2015. "Engineering Metabolism through Dynamic Control." *Current Opinion in Biotechnology* 34: 142–152. https://doi.org/10.1016/j.copbio.2014.12.022

Victoria, S., M. J. Astbury, and A. J. McCormick. (2024). "Engineering Highly Productive Cyanobacteria Towards Carbon Negative Emissions Technologies." *Current Opinion in Biotechnology*, 87, 103141. https://doi.org/10.1016/j.copbio.2024.103141

Vivek, K., G. S. Sandhia, and S. Subramaniyan. (2022). Extremophilic Lipases for Industrial Applications: A General Review. *Biotechnology Advances*, 60, 108002. https://doi.org/10.1016/j.biotechadv.2022.108002

Werkman, J. R., S. Pattanaik, and L. Yuan. (2011). Directed Evolution through DNA Shuffling for the Improvement and Understanding of Genes and Promoters. *Methods in Molecular Biology* 754, 325–342.

Wohlgemuth, R. (2024). Enzyme Catalysis for Sustainable Value Creation Using Renewable Biobased Resources. *Molecules* 29(23), 5772.

Woodley, J.M., M. Bisschops, A.J.J. Straathof, and M. Ottens (2008), Future Directions for *In-Situ* Product Removal (ISPR)†. *Journal of Chemical Technology and Biotechnology*, 83: 121–123.

Wu, W., J. Chen, Z. Zhou, et al. 2023. "Efficient Isomaltulose Biosynthesis in Recombinant *Synechococcus elongatus* PCC 7942 through Sucrose Isomerase and Sucrose Permease Co-Expression." *Metabolic Engineering*74: 140–152.

Xue, Y., Q. Yan, X. Tian, D. Han, and Z. Jiang. 2024. "High-Level Secretory Expression and Characterization of an Acid Protease in *Komagataella phaffii* and Its Application in Soybean Meal Protein Degradation." *International Journal of Biological Macromolecules* 282: 137011. https://doi.org/10.1016/j.ijbiomac.2024.137011

Yadav, R., B. Chattopadhyay, R. Kiran, A. Yadav, A. K. Bachhawat, and S. A. Patil. 2022. "Microbial Electrosynthesis from Carbon Dioxide Feedstock Linked to Yeast Growth for the Production of High-Value Isoprenoids." *Bioresource Technology*363: 127906. https://doi.org/10.1016/j.biortech.2022.127906

Yang, P., S. Jiang, S. Jiang, S. Lu, Z. Zheng, J. Chen, W. Wu, and S. Jiang. 2022. "CRISPR-Cas9 Approach Constructed Engineered *Saccharomyces cerevisiae* with the Deletion of GPD2, FPS1, and ADH2 to Enhance the Production of Ethanol." *Journal of Fungi* 8: 703. https://doi.org/10.3390/jof8070703

Yang, S., R. Chen, X. Cao, et al. 2024. "De Novo Biosynthesis of the Hops Bioactive Flavonoid Xanthohumol in Yeast." *Nature Communications* 15: 253. https://doi.org/10.1038/s41467-023-44654-5

Yang, W., W. Qian, Z. Yuan, and Chen, B. (2021). Perspectives on the Flexibility Analysis for Continuous Pharmaceutical Manufacturing Processes. *Chinese Journal of Chemical Engineering* 41, 29–41.

Yi, S.-S., J.-M. Noh, and Y.-S. Lee. (2009). Amino Acid Modified Chitosan Beads: Improved Polymer Supports for Immobilization of Lipase from Candida rugosa. *Journal of Molecular Catalysis B: Enzymatic*, 57, 123–129.

Yogalakshmi, K.N., T.M. Mohamed Usman, S. Kavitha, S. Sachdeva, S. Thakur, S. Adish Kumar, and J. Rajesh Banu. (2023). Lignocellulosic Biorefinery Technologies: A Perception into Recent Advances in Biomass Fractionation, Biorefineries, Economic Hurdles and Market Outlook. *Fermentation*, 9(3), 238. https://doi.org/10.3390/fermentation9030238

Yu, G.-H., Z.-L. Chi, A. Kappler, F.-S. Sun, C.-Q. Liu, H. H. Teng, et al.(2020). Fungal Nanophase Particles Catalyze Iron Transformation for Oxidative Stress Removal and Iron Acquisition. *Current Biology* 30 (15), 2943–2950. https://doi.org/10.1016/j.cub.2020.05.058

Yuliana, Cadavid-Mora, Márquez-Cardozo Carlos Julio, Cadena-Chamorro Edith, Londoño-Hernandez Liliana, Yepes-Betancur Diana. (2023). Chapter 2 - Enzymes Production From Fruit and Vegetable Waste and Their Industrial Applications, *Value-Addition in Agri-food Industry Waste Through Enzyme Technology*, Academic Press, pp. 17–36. https://doi.org/10.1016/B978-0-323-89928-4.00014-6

Yusuf, V. F., N. I. Malek, and S. K. Kailasa (2022). Review on Metal-Organic Framework Classification, Synthetic Approaches, and Influencing Factors: Applications in Energy, Drug Delivery, and Wastewater Treatment. *ACS Omega* 7(49), 44507–44531.

Zarrelli, A., G. Rossi, and M. De Luca. 2022. "Plant-Based Biomanufacturing for Biochemicals and Pharmaceuticals." *Phytochemistry Reviews* 21: 873–892.

Zeiske, T., K. A. Stafford, and A. G. Palmer, 3rd (2016). Thermostability of Enzymes from Molecular Dynamics Simulations. *Journal of Chemical Theory and Computation*, 12(6), 2489–2492.

Zhang, Shuling, Jing Bai, Weixi Kong, Haolei Song, Yunting Liu, Guanhua Liu, Li Ma, Liya Zhou, Yanjun Jiang. (2024). Dendritic Mesoporous Silica Nanoparticles for Enzyme Immobilization, *Green Chemical Engineering*5 (2), 173–186, ISSN 2666-9528.

Zheng, S., T. Zeng, C. Li, et al. (2022). Deep Learning Driven Biosynthetic Pathways Navigation for Natural Products with BioNavi-NP. *Nature Communications* 13, 3342. https://doi.org/10.1038/s41467-022-30970-9

Zhong, J., Y. Zhou, L. Wang, et al. 2024. "Engineering Cyanobacteria for Production of Taxol Directly from Carbon Dioxide." *Metabolic Engineering* 80: 56–68. https://doi.org/10.1186/s13068-024-02555-9

Zhou, M., P. Guo, T. Wang et al. (2017).Metagenomic Mining Pectinolytic Microbes and Enzymes from an Apple Pomace-Adapted Compost Microbial Community. *Biotechnology for Biofuels* 10, 198.

9 Sustainable Microalgal Biomanufacturing in Agriculture and Food Production

Keerti Singh, Rhea Gupta, Rinku Kaushik, Abhishek Saxena, and Archana Tiwari
Diatom Research Laboratory, Amity Institute of Biotechnology, Amity University, Noida, Uttar Pradesh, India

9.1 INTRODUCTION

Traditional agriculture tends to be a linear process, involving the extraction of resources, cultivation, harvest, and waste generation at various stages. This has resulted in soil degradation, reduced biodiversity, and significant greenhouse gas (GHG) emissions (Adedibu, 2023). A growing population with a limited resource pool makes it evident that this unsustainable model cannot be maintained for a long time in the future. The circular economy offers a paradigm shift and a new framework that aims to redefine growth by focusing on sustainable development and providing positive and society-wide benefits. Agricultural systems can be redesigned positively and sustainably to establish a food system that is resilient to people and the planet (Kirchherr et al., 2018). A circular economy is designed to minimize waste and maximize the use of resources, as opposed to the conventional 'take, make, dispose' model. (Toop et al., 2017). The transition toward an agricultural circular economy is a regenerative and resource-efficient model that opens up new possibilities for innovation, particularly in the area of biological systems such as microalgae. Microalgae are characterized by high growth rates, photosynthetic efficiency, and the ability to utilize waste streams, including nutrient-rich wastewater and industrial CO_2 emissions. They can provide a viable substitute for traditional agricultural inputs. Microalgae can play multiple roles, including serving as a source of bio-stimulants, biopesticides, biofertilizers, and soil conditioners, which improve crop resilience and productivity. The biomass of microalgae grown in marginal lands or closed-loop systems is rich in nutrients and thus recovers nutrients besides minimizing the use of synthetic agrochemicals. Microalgal systems are also capable of assisting in carbon sequestration as well as the reduction of greenhouse gas emissions (Ma et al., 2022).

The diatoms have become a useful tool in sustainable biomanufacturing because of their special characteristics that make them suitable for the agricultural and food production sectors. Diatomaceous Earth (DE) is made of fossils of diatoms and is a natural pesticide. Other than pest control, it enhances soil properties, such as water retention, aeration, and nutrient availability (Korunic, 2013). Diatom extracts contain silicic acid, which serves as a bio-stimulant to enhance plant growth and resistance. It is also used in the purification of water and in food supplements for livestock to increase digestion and uptake of nutrients (Monlau et al., 2015). Diatoms are minute but ubiquitous in aquatic ecosystems and are highly important in the paradigm of the circular economy. Their ability to efficiently sequester heavy metals and pollutants through bioremediation shows they have environmental restoration capabilities (Elizabeth George & Wan, 2023). Besides being a prolific producer of biomass, diatoms open up avenues for sustainable biofuel and bioplastic production (Basbous et al., 2024). They increase the amount of carbon sequestration by efficient fixation of carbon and deposition of organic materials in sediments (Lobus & Kulikovskiy, 2023). From here, it can be understood that the concept of the circular economy is resource-efficient, environmentally sustainable, and economically efficient, helping reduce pollution and energy consumption, create jobs, and foster innovation (Geissdoerfer et al., 2017).

For the implementation of the circular economy within agriculture, it is important to focus on full utilization of resources while minimizing consumption and eliminating waste (Monlau et al., 2015). Second, it should look at long-term ecological, economic, and social sustainability (Burgo Bencomo et al., 2019). Finally, it must promote regenerative practices that will close nutrient cycles and reduce resource losses (Morseletto, 2020) Circular agriculture is also aimed at minimizing waste through the reuse or recycling of farm products. These crop residues are thus converted into biofuels or used as animal feed, while manure is made into fertilizers or biogas. It lays emphasis on the necessity to maintain biodiversity and maintain healthy ecosystems, recognizing the linkage between the agricultural practices and nature (Gaba et al., 2015). Crop rotation and agroforestry not only improve soil fertility but also have a positive impact on the regulation of pests and pollination, and resilience is also

DOI: 10.1201/9781003637158-9

enhanced. Circular farming has a positive effect on climate change as it lowers the emission of greenhouse gases and increases the absorption of carbon. The circular principles enhance the yield, reduce the input cost, and create additional revenue sources for the farmers. The transition to the principles of the agricultural circular economy is not a smooth one. There are technological and infrastructural hinders in the adoption of circular practices, which require major investments (Lamb et al., 2016). Knowledge gap and behavior change among farmers are some of the challenges that demonstrate the importance of education, training, and extension services. It also needs economic viability, favorable policies, and good supply chain coordination (Burgo Bencomo et al., 2019).

This chapter discusses the multiple applications of microalgae, especially diatoms, as circular biomanufacturing agents, their role in biogeochemical cycling, sustainable agricultural production, enhancement of soil and crop productivity, biofertilization, pest control, integration of agricultural systems, and broader biomanufacturing applications, as well as the challenges and opportunities of using these extraordinary microorganisms in sustainable agriculture.

9.2 MICROALGAE AS AGENTS OF CIRCULAR BIOMANUFACTURING

Microalgae present a paradigm shift in circular biomanufacturing, offering a sustainable and circular economy by converting waste streams into valuable bioproducts while utilizing by-products with minimal environmental impact. Microalgae are particularly interesting as they serve as a renewable feedstock for several commercially important compounds, and by using nutrients in wastewater, they offer an ideal avenue for waste valorization and resource recovery (Lee et al., 2024). Integrating photosynthetic microorganisms into circular biomanufacturing systems can help address multiple sustainability issues at the same time, where bioremediation is particularly aligned to the sustainable circular economy, as the bioremediated mass can be valorized into different industrial uses to transform the traditional linear "take-make-dispose" into circular systems, where waste streams can function as feedstock for valuable product generation (Santos et al., 2025).

Modern microalgal cultivation systems exhibit remarkable flexibility by utilizing a variety of waste streams, including food waste, agricultural residues, and wastewater, as nutrient sources. This integration of waste valorization and biomass production aligns closely with circular economy principles. The recovered biomass can subsequently be converted into biofuels and other value-added products, demonstrating the potential of microalgae for sustainable resource recovery (Giwa et al., 2022). The biorefinery concept proves central in maximizing economic and environmental benefits as valorization of waste into microalgae bioprocessing represents a greener and cost-effective circular bioeconomy approach that enables the modern microalgal biorefineries to produce multiple high-value products, including biofuels, pharmaceuticals, nutraceuticals, food additives, and specialty chemicals from identical biomass feedstock (Leong et al., 2021). Microalgae are considered to be a promising group of organisms for fuel production, waste processing, pharmaceutical applications, and as a source of food, with this multi-product approach enhancing economic viability while reducing waste generation throughout production processes (Rozenberg et al., 2024).

Recent biotechnological advances have significantly enhanced microalgae's potential in circular biomanufacturing, where faster growth rates and higher protein and lipid content make microalgae a popular chassis for many industrial applications, with genetic engineering and synthetic biology approaches being employed to optimize metabolic pathways for enhanced production of target compounds (Dhokane et al., 2023). Microalgae, with their complex metabolic capacities, are continuously explored for nutraceuticals, pharmaceuticals, and other industrially important bioactive compounds. CRISPR-based bioengineering has emerged as a powerful tool for developing strains with enhanced productivity and novel biosynthetic capabilities (Kumar et al., 2020). Among the most significant advantages lies their ability to simultaneously treat waste streams while producing valuable products, as the biomass produced during microalgae-based wastewater treatment can be valorized via pyrolysis to generate multiple valuable products, such as biochar, bio-oil, and pyrolytic gas (de Morais et al., 2023), addressing both environmental remediation and resource recovery objectives through dual functionality. Despite considerable potential, there exist several challenges in scaling microalgae-based circular biomanufacturing systems, where microalgae cultivation represents a critical research area as laboratory-scale achievements often encounter obstacles during industrial implementation, necessitating continued research focusing on strain optimization, process integration, and scale-up technologies to realize the full potential of microalgae in sustainable biomanufacturing systems that support the transition (Guieysse & Plouviez, 2024) (Figure 9.1).

9.3 ROLE OF DIATOMS IN BIOGEOCHEMICAL CYCLING

Diatoms, being a vital part of the aquatic ecosystems, contribute to primary production and biogeochemical cycling of carbon, nitrogen, phosphorus, and silicon, with their crucial ecological function being recycling of nutrients and maintaining productivity in both marine and freshwater ecosystems. Nutrient cycling by these organisms follows multiple pathways that link photosynthesis and nutrient uptake to growth and cell division, the exudation of organic compounds, sinking and decomposition, and the silica cycle. They are photosynthetic organisms that follow the process of photosynthesis to produce energy, utilizing

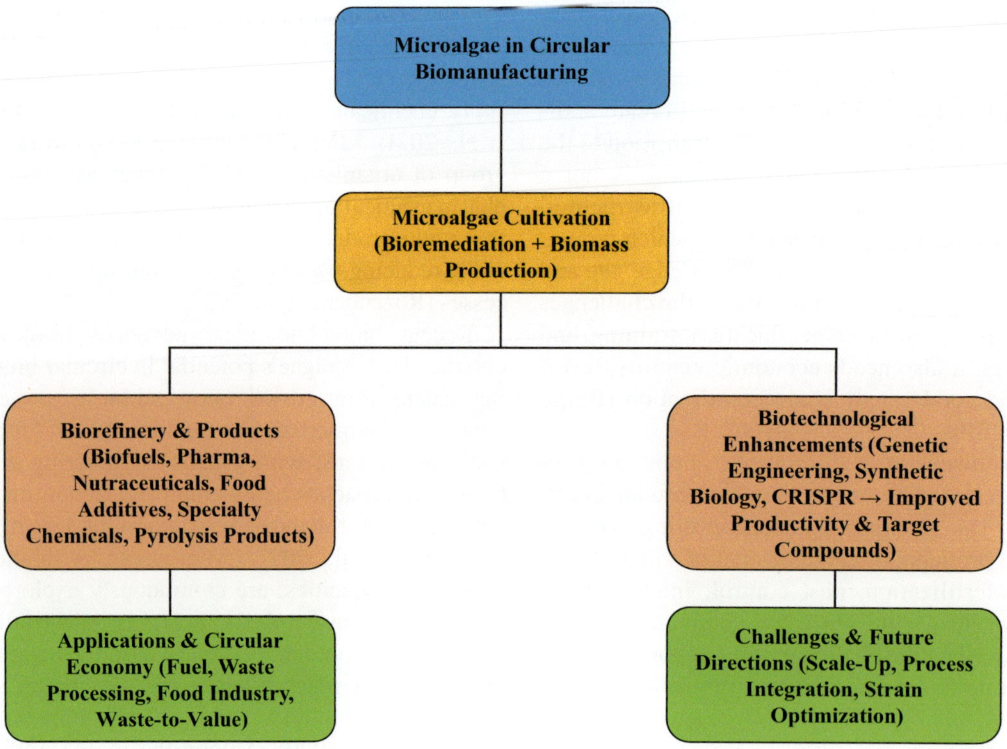

FIGURE 9.1 Flowchart explaining the role of microalgae in circular biomanufacturing.

sunlight to convert carbon dioxide into organic matter. This process that forms the basic fundamental basis of energy production also drives the uptake of essential nutrients from their aquatic environment where the key nutrients absorbed in diatoms are nitrogen (N), phosphorus (P), and silicon (Si), all these nutrients have essential functions in physiological processes where nitrogen serves to form proteins, sustaining cellular growth, and phosphorus is a key component in energy transfer and the formation of nucleic acids, silicon turns out to be essential for diatoms since it is required in the construction of the intricately silica-based cell walls or frustules (Falkowski et al., 1998). The rate of nutrient uptake by diatoms is determined by the concentration of these nutrients in the water. As nutrient elements are sufficiently available, they can quickly uptake them and store them to maintain their growth and reproduction when the nutrient supply is low. This capability is of great significance in dynamic water environments, where the level of available nutrients can change significantly (Falkowski et al., 1998). These nutrients are subsequently absorbed and integrated into various metabolic processes, thereby promoting the growth and division of diatom cells, which enables their survival and success across diverse aquatic habitats (Armbrust, 2009).

The development and cell division of diatoms occur in relation to how these organisms assimilate and utilize the nutritional elements present in the stationary life forms or in aquatic habitats. These organisms grow by increasing cell size or by dividing into new cells through mitosis. To be able to divide, diatoms must continuously remove some of the essential elements for growth from the surrounding water. Nitrogen and phosphorus are found in every protein and nucleic acid, as well as in the majority of other cell components. These nutrients determine the growth rate and division of diatoms. The formation of high-density populations, referred to as blooms, is made possible by proliferating rapidly in the presence of an abundant nutrient supply. These blooms are crucial to primary production and have a strong impact on the ecosystem. Diatoms are capable of dramatically slowing down their metabolic process under unfavorable conditions, thus reducing activity enables survival until the conditions become more favorable. Diatoms may endure nutrient shortages through mechanisms of intracellular nutrient storage. This ability allows for diatoms to rapidly resume their development and division after recovery to favorable conditions (Armbrust, 2009). Diatoms also have mechanisms for intracellular storage of nutrients, which help them endure periods of nutrient scarcity. The capacity for nutrient storage enables diatoms to develop and divide rapidly when favorable conditions return. Diatoms undergo cell division as soon as the frustule replicates, a complex and energetically costly process that requires coordinated, timely nutrient uptake and metabolic activity (Falkowski et al., 1998).

Diatoms are known for exuding and releasing various organic compounds into their surroundings in their aquatic environment. In one such process, they secrete dissolved organic carbon (DOC), which includes multiple organic molecules, primarily polysaccharides, amino acids, and fatty acids. The exudates are essential as they provide substrates for heterotrophic bacteria and other microorganisms in the aquatic ecosystem. Interactions with microorganisms are necessary for diatoms in terms of nutrient recycling (Armbrust, 2009). The release of organic compounds by diatoms occurs continuously and is influenced by several factors, including nutrient availability, light intensity, and diatom cell physiology. It appears that diatoms release a greater quantity of organic compounds into the water during periods of peak growth and photosynthesis. Such exudates play a crucial role in the microbial loop, a process that involves the conversion of dissolved organic matter into inorganic nutrients by bacteria, making these nutrients available again for *de novo* synthesis by primary producers, such as diatoms (Azam & Malfatti, 2007). The organic material from diatoms contributes to the formation of marine snow, a mass of organic aggregates that sinks to the sea floor, facilitating the vertical transport of carbon and other nutrients while sustaining deep-sea ecosystems and providing a food source for benthic organisms. Exudation of organic material can alter the chemical composition and physical properties of the surrounding water, thereby influencing processes such as nutrient cycling and the development of biofilms (Falkowski et al., 1998).

With the death of diatoms, their silica-based cell walls cause them to sink to the ocean or lake floor, a process that serves as an essential part of the vertical transport of organic and inorganic matter in aquatic ecosystems. This sinking is a significant process in the biological pump, which is a mechanism transporting carbon from the surface to the deep ocean, helping in sequestering carbon and influencing global carbon cycles and climate regulation (Armbrust, 2009). Sinking to the bottom of a lake or an ocean initiates destruction of skeletal remains through microbial activity, primarily by bacteria and fungi, and decomposition of the organic material into ammonium, phosphate, and silicic acid happens into nutrients that support new diatom growth and maintain the silica cycle (Azam & Malfatti, 2007). Silicic acid generated from the breakdown of frustules is reabsorbed by living diatoms to build new frustules (Armbrust, 2009).

Diatoms are used as fertilizers in agriculture as they contain organic material, which greatly enriches soil fertility due to their high silica content. Diatoms on decomposition not only provide nitrogen, phosphorus, and potassium, but also an enriched soil that supports microbial and plant growth (Lundholm et al., 2019). With increased soil structure and porosity, enhanced water and nutrient infiltration also aid root development. With increased silica, reduced stress on the plant, along with improved productivity, is seen. Diatoms contribute to sustainable practices by expanding the availability of organic matter and silicates, which in turn enhance soil nutrient content and promote crop growth.

Diatoms play essential roles in nitrogen dynamics within aquatic ecosystems, including nitrogen fixation and assimilation. Nitrogen fixation is the process by which atmospheric nitrogen gas (N_2) is converted to its biologically available forms, such as ammonia (NH_3) or nitrate (NO^{3-}). Although diatoms are not typically regarded as nitrogen fixers, such as certain bacteria, recent studies have shown that some species of diatoms may have the potential to fix nitrogen through symbiosis with diazotrophic bacteria (Zehr & Capone, 2021). These bacteria can fix atmospheric nitrogen into ammonium, which can then be assimilated by the diatoms for growth and metabolism. Additionally, diatoms assimilate nitrogen from the environment in the form of dissolved inorganic nitrogen (ammonium and nitrate) and organic nitrogen, such as amino acids and urea. Diatoms are equipped with transporters and enzymes that catalyze the transport and assimilation of nitrogen compounds. Nitrogen is assimilated inside the cell into organic molecules, such as proteins, nucleic acids, and chlorophyll, which are essential for the growth, metabolism, and photosynthesis of diatoms. Diatoms could thus be introduced into soil nitrogen cycling through various means, including their organic matter content, potential nitrogen fixation, and silica. Upon decomposition, diatoms release nitrogen-rich compounds that become a nutrient source for microbes in the soil, aiding in soil nitrogen cycling (Zehr & Capone, 2021). Some species of diatoms have been found to harbor nitrogen-fixing bacteria, and their introduction into the soil would enhance nitrogen availability through biological nitrogen fixation.

Diatoms are essential in phosphorus dynamics in aquatic ecosystems in terms of cycling (Sunda & Hardison, 2010). These microbes have very specialized transporters and enzymes that facilitate efficient phosphate uptake from their surroundings. Inside the diatom cell, phosphate is utilized for various metabolic processes and converted into organic molecules to support diatom growth and productivity (Sunda & Hardison, 2010). During diatom cell division and growth, phosphate is consumed and incorporated into their biomass, making it unavailable in the water column. After these cells die or are grazed by predators, diatom biomass is re-mineralized, which causes the release of phosphate back into the surrounding water as inorganic phosphate. Sometimes referred to as the "biological pump," this process enables the conversion of phosphates from organic to inorganic forms, thereby returning phosphorus and making it available for the productivity of ecosystems in aquatic environments. There are several promising approaches in which diatoms can be utilized to influence phosphorus availability in aquatic ecosystems through various strategies. One approach is the use of diatoms in bioremediation efforts aimed at removing phosphorus pollution from water bodies. The phosphate uptake capacity can be increased by the addition of nutrients or modification of environmental conditions, such as the optimization of light conditions for algal growth that promotes the development of diatoms, which in turn decreases phosphate concentrations in the water column by that amount of phosphate taken up (Kiran et al., 2017).

Diatoms can be used as components of biofertilizers in both aquaculture and agriculture. Application of the

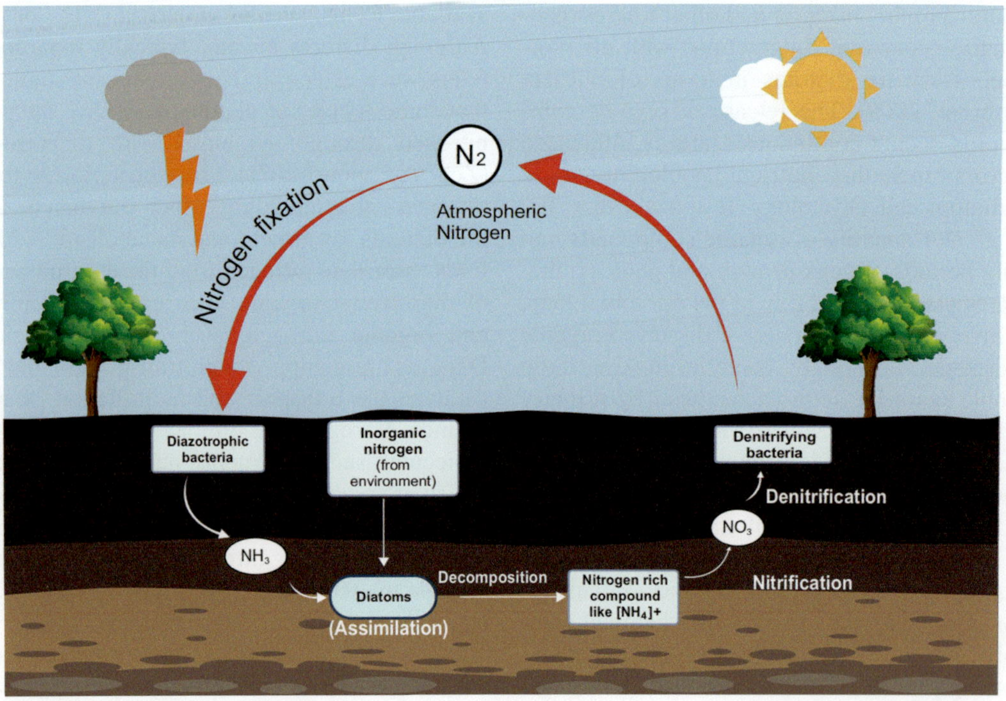

FIGURE 9.2 Illustration explaining the incorporation of diatoms in the soil nitrogen cycle. (BioRender.com.)

diatom-based biofertilizers increases the fertility of the soil or sediment through the addition of organic matter rich in phosphorus that is released during decomposition, and constantly makes phosphorus available for uptake by plants (Zehr & Capone, 2021). The frustules of diatoms are capable of adsorbing phosphate and may serve as a phosphorus reservoir in many aquatic environments. Such silica-rich material that is formed by the accumulation of the fossilized remains of diatoms when introduced into a water body as DE can help in the sequestration of phosphate, making it unavailable to algal blooms and enhancing the possibility of improvement in water quality (Zehr & Capone, 2021). These methods utilize the biological and chemical characteristics of diatoms to influence the dynamics of phosphorus, thereby mitigating phosphorus-associated environmental impacts on aquatic systems (Figure 9.2)

9.4 DIATOMS IN SOIL AND CROP PRODUCTIVITY ENHANCEMENT

Diatoms affect soil health and sustainability, performing vital functions supporting fundamental functions of the soil ecosystem: soil aggregation, porosity, water retention, moisture levels, and relation to soil microbes

Diatoms drive soil aggregation, a fundamental process essential for maintaining soil stability and fertility (Lundholm et al., 2019). Silica-enriched frustules and the Extracellular Polymeric Substances (EPS) that they release during growth and reproduction help to bind soil, as diatoms mediate soil aggregation. Diatom silicate frustules serve as physical binders, bonding with soil particles to form stable microaggregates (Lundholm et al., 2019). Moreover, the frustules provide a framework for soil particles to adhere to each other, contributing to the formation of larger macroaggregates. The EPS released by diatoms functions as an organic binder, connecting particles and forming a sticky matrix that facilitates aggregate stability (Lundholm et al., 2019). This organic matrix contributes to increased macroaggregation by binding soil particles together, thereby improving soil stability and structure. Additionally, diatoms interact reciprocally with soil microbes in aggregate formation. The EPS diatoms release provides carbon and energy to soil organisms, thereby increasing microbial activity and promoting the decomposition of organic matter. Increased microbial activity and the by-products of microbial activity (glues, sugars, and other organic compounds) create aggregate stability by improving the "stickiness" of soil particles for aggregate formation (Lundholm et al., 2019). Diatoms can enhance aggregate stability, which has important implications for soil health and functioning, as they promote soil structural benefits such as decreased erosion and compaction, stabilization of soil porosity, and improved air and water movement (Lundholm et al., 2019).

Enhancing soil structure offers benefits, including deeper plant rooting, greater nutrient availability in the soil for plants, and enhanced microbial activity. Ultimately, this contributes to increased soil fertility and a better understanding of the functions of ecosystems, which impact soil porosity through various mechanisms. Additionally, the nanostructured diatom frustule, with its unique morphology, also contributes to the formation of microscale pore spaces within soil aggregates. The tiny pore spaces provide habitat for the soil microbial community and facilitate gas diffusion of oxygen and carbon dioxide for microbial activity and root respiration. The diatom frustules also provide aeration in aerobic soils, which enhances the growth of soil microorganisms and the development of the plant root system (Palomino et al., 2011). Diatoms also create macropores in the soil, which are defined as large, well-connected pore spaces in soil that are critical for water movement and drainage (Lundholm et al., 2019). The complex structure of the diatom frustule and the way it fosters aggregation yield stable soil aggregates that develop soil porosity. Soil aggregates contribute to a complex soil matrix that enhances water infiltration into the soil, reduces soil runoff and erosion, and provides a stable moisture regime that supports plant growth.

Diatoms have a considerable influence on water retention and moisture due to the usefulness of their structure and the porosity of their silica frustules, which work together to retain moisture in the soil. The frustules provide physical adsorption along with capillary action to enhance the soil's water-holding properties. The silanol groups (-Si-OH) located on the diatom surface make them hydrophilic and are responsible for forming hydrogen bonds with water molecules, which helps retain moisture in the soil. Furthermore, diatoms have a high surface area to volume ratio due to the elaborate structure of their frustules, allowing a sufficient bonding site with water molecules to help increase the overall water-holding capacity of the soil (Palomino et al., 2011).

As diatoms are the primary producers in terrestrial environments, they have been shown to influence the soil microbial community through various processes (Battin et al., 2009). One significant mechanism is the provision of organic matter. Through photosynthesis, diatoms contribute to the overall input of organic carbon in the soil ecosystem. Upon death and subsequent decomposition, they release organic compounds into the soil, serving as a nutrient source for soil microorganisms (Battin et al., 2009). Such an influx of organic matter could alter microbial growth and activity as well as microbial community composition and dynamics. Their siliceous frustules provide a surface for microbial attachment and colonization, thereby creating microhabitats within the soil matrix. These microhabitats offer protection and niches to microbial communities. Some diatom species also produce EPS that enhance soil aggregation and microbial colonization. These aggregates serve as hotspots of microbial function, which, in turn, have a secondary effect on structuring microbial communities. Diatoms also produce allelochemicals/secondary metabolites that can affect microbial communities in soils. Few diatoms release bioactive compounds with allelopathic effects on other microorganisms that can be either inhibitory or promote growth and activity (Leibold et al., 2004). These allelochemicals can serve as mediators of competitive interactions in soil microbial communities, influencing their structure and functioning.

9.5 DIATOM AS BIOFERTILIZERS

Diatoms are gaining recognition as promising candidates for sustainable agriculture due to their nutrient-rich biomass. This section presents diatom biomass and its nutritional composition, followed by a discussion of the bioavailability of those nutrients for plant uptake and use, especially in the context of their potential to enhance soil fertility and crop yield (Lobus et al., 2021; Slocombe et al., 2015).

9.5.1 Nutritional Composition of Diatom Biomass

Diatom biomass is a micronutrient-rich resource that can be considered a nutrient for plant growth (Slocombe et al., 2015). Numerous studies have shown that diatom biomass provide sufficient levels of the macronutrients nitrogen, phosphorus, and potassium, as well as micronutrient trace minerals, including iron, manganese, and zinc (Lobus et al., 2021). The frustule (the siliceous cell wall of diatoms) can serve as a nutrient store because it contains all the requisite nutrient elements, and it slowly release them into the surrounding environment. Nutritional characteristics of biomass also differ based on species diversity, environmental conditions, and ecologies (Table 9.1) (Lobus et al., 2021).

9.5.2 Bioavailability of Nutrients for Plants

The potential of diatoms as biofertilizers depends on the bioavailability of the nutrients they contain (Singh et al., 2023). The silica frustules are porous, allowing nutrients to leach out of the diatom cells into the soil solution for plant roots to uptake. In addition, diatom cells can provide organic matter to microbes, thereby contributing to the mineralization and cycling of nutrients in the rhizosphere. The mutualism between diatoms and microbes, and ultimately with plants, accelerates nutrient uptake and utilization, enhancing crop productivity (Alvarez et al., 2021).

TABLE 9.1
Nutritional Composition of Diatom Biomass

Nutrient	*Skeletonema costatum*	*Chaetoceros calcitrans*	References
Silicate	2.4 mM	1.2 mM	Bastos et al. (2022)
Nitrate	4 mM	4 mM	Bastos et al. (2022)
Phosphate	100 μM	100 μM	Bastos et al. (2022)
Iron	20 μM	80 μM	Bastos et al. (2022)
Micronutrients	0.5 mL L^{-1}	80 μM	Bastos et al. (2022)
Eicosapentaenoic acid (EPA)	2.6-fold	2.3-fold	Bastos et al. (2022)
Docosahexaenoic acid (DHA)	2.6-fold	Not specified	Bastos et al. (2022)

9.5.3 Application Techniques

9.5.3.1 Incorporation into Soil

Incorporating diatoms into soil is one of the simplest ways to take advantage of their biofertilization potential (Singh et al., 2023). DE has been used in agriculture for many years because of its ability to condition soils. DE has been shown to improve soil quality by improving soil structure by retaining water, aggregate formation, and enhancing nutrient acquisition and root penetration, which builds the overall fertility of crops. The porous structure of DE creates an air-exchange zone that encourages gas exchange and microbial activities that are ideal for plant growth (Adewara et al., 2024).

Studies show that diatom-rich soil improves soil fertility and increases crop productivity. Diatomaceous materials serve as a source of bioavailable silica and may contribute small amounts of nutrients, including nitrogen, phosphorus, and potassium, depending on their origin and composition. In addition to nutrients, diatoms enhance beneficial soil microbes, promote the growth of beneficial soil microorganisms, support the activities of mycorrhizal fungi, and support nitrogen-fixing bacteria, all of which are essential for nutrient recycling and for establishing symbiotic relationships with plants. Finally, all of these forms of microbial activity are ultimately favorable for enhancing soil health and better agricultural plant growth outcomes (Alvarez et al., 2021).

They could be integrated into soil by mixing diatomaceous powders or suspending them in soil at land preparation time, or by introducing them into composting processes to enhance the decomposition of organic matter and the nutrient recycling process (Alvarez et al., 2021). Also, introduction and application rates and times need to be at optimal diatom levels for soil fertility without affecting the equilibrium of soil pH and microbes (Adewara et al., 2024).

9.5.3.2 Foliar Sprays and Seed Coatings

Diatom-based foliar sprays improve overall plant growth, vigor, and resistance to most biotic and abiotic stressors. Diatom silica strengthens plant cell and tissue structure, enhancing defense against pests and diseases, as well as other environmental stressors such as drought and salinity. Foliar applications of diatoms may also stimulate the production of phytohormones and antioxidants, which are known to promote plant growth and vigor (Soubeih et al., 2017). Another approach to this is seed coating with diatomaceous materials to tap the potential of diatoms as biofertilizers. Applying a diatom-based formulation to seeds prior to sowing provides an initial nutrient-rich microenvironment for germinating seeds. It supports the gradual release of nutrients during germination and early seedling growth. Additionally, diatom-coated seeds can help mitigate pathogens and pests, thereby enhancing vigor and seedling survival (Ma, 2019).

9.5.4 Comparative Analysis with Conventional Fertilizers

9.5.4.1 Performance Compared to Chemical Fertilizers

Recently, diatomaceous biofertilizers have attracted interest because they can compete with, and in some cases outperform, chemical fertilizers for crop productivity in agricultural systems. Research has shown that diatomaceous biofertilizers can be as effective and offer the same benefits as chemical fertilizers to stimulate plant growth and yields (Das et al., 2019). Diatomaceous biofertilizers are reported to improve nutrient uptake, especially nitrogen and silica in plants. The more steady and controlled release of plant nutrients from diatomaceous material provides a constant supply of nutrients essential for growth into the future, providing improved resistance against environmental limitations, such as drought and salinity of soil (Table 9.2) (Kovács et al., 2022).

TABLE 9.2
Comparative Analysis with Conventional Fertilizers

Fertilizer Type	Crop Yield Impact	Soil Health Effects	Economic Considerations	References
Diatomaceous Earth (DE) + 50% nitrogen (N), phosphorus (P), and potassium (NPK)	Increased potato yield by 38.7% over control and 12.9% over full-dose NPK alone	Enhances soil porosity, aggregation, and water retention; reduces disease	Lower synthetic input costs; higher returns from yield gain	Al-Suhaibani et al. (2020); Artyszak, (2018)
Urea (100%)	High initial yields; may decline over time due to soil degradation	Can lead to soil acidification and loss of microbial diversity	Low upfront cost; may incur long-term remediation expenses	Hammad et al. (2020)
Neem-coated urea (NCU)	Yield increase in soybean by ~13% over normal urea	Improves nitrogen use efficiency; reduces leaching	Higher input cost than urea, but better net returns	Ramappa (2020)
NPK (full dose)	Standard yield benchmarks; not always sustainable long-term	Potential for nutrient runoff; little effect on organic matter	Consistent costs; may face environmental compliance issues	Al-Suhaibani et al. (2020); Artyszak (2018)
DE alone	Moderate yield improvements; best when combined with other fertilizers	Improves structure and water retention; acts as a natural pest deterrent	Low-cost; saves on pesticides and irrigation	Artyszak (2018)

In addition, ongoing research is evaluating the potential for biofertilizers based on diatoms to influence long-term fertility and soil structure (Alvarez et al., 2021). Chemical fertilizers can degrade soil by depleting it of nutrients. In contrast, diatom-derived products will work to improve fertility by enhancing microbial activity and decomposing organic matter (Das et al., 2019). Use of this integrated approach to maintain soil health will not only increase future crop productivity, but it will also help to sustain a viable model of agriculture for future generations (Kovács et al., 2022).

9.5.4.2 Environmental and Economic Considerations

The environmental benefits of using diatom-based biofertilizers vs. the chemical equivalents are significant. For example, while well known for their potential to pollute due to nutrient leaching—excess nitrogen or phosphorus that the plant cannot take up—chemical fertilizers are infamous for polluting water and soil. Nutrients exceeding the plant's uptake capacity can leach into adjacent water bodies, leading to eutrophication and degradation of aquatic ecosystems. In contrast, diatom-based fertilizers typically have a slower release nutrient profile, which reduces the risk of environmental impacts (Das et al., 2019). They can biodegrade and have a much lower risk of environmental pollution. Additionally, other ecological benefits—carbon sequestration, wastewater treatment, and diatom production for biofertilization—enhanced the sustainability of biofertilization (Das et al., 2019).

The economic implications will depend on the analysis that remains to be completed on how the production process affects economic factors and how it ultimately scales, as well as on market demand. The cost of production and the up-front capital investment may be greater than that of conventional fertilizer facilities, but as noted above, it leads to savings in the long term for farmers by reducing reliance on factors, such as human health and external inputs (Das et al., 2019). Moreover, as consumers are becoming increasingly interested in sustainably grown agricultural products, this could lead to some market advantages of crops grown with biofertilizers that could further economically justify a diatom-based fertilization strategy (Das et al., 2019; Singh et al., 2023).

9.6 DIATOMS IN PEST MANAGEMENT

9.6.1 Biocontrol Properties

DE is gaining recognition for its properties and potential for use in pest management; its mode of action involves both physical and chemical effects that disorient and manage a range of arthropod pests when used properly. This section pulls together pertinent findings from three critical studies that demonstrate the efficacy of DE on various pest species, including the challenging and hard-to-manage bed bug (*Cimex lectularius*) and other stored product pests (Lilly et al., 2016; Zeni et al., 2021). The studies represented here provide valuable illustrations of the unique structural and functional characteristics of DE and their relevance to application as a sustainable biocontrol.

9.6.2 Mechanism of Pest Control

9.6.2.1 Physical Mechanism of Desiccation

The primary method of action of DE is through physical desiccation of its intended targets. DE possesses a high level of abrasiveness and is micro-sharp enough to penetrate and damage arthropods' protective cuticle upon contact. The cuticle is a significant barrier for an insect against dehydration. The altered condition leads to excessive water loss from the insect's body, which ultimately results in its mortality. Evidence shows that pyrethroid-resistant strains of Cimex lectularius exhibited some tolerance; however, they ultimately succumbed to the drying effects of DE, albeit at a slower rate than the pyrethroid-susceptible strains (Lilly et al., 2016).

9.6.2.2 Lipid Absorption and Enhanced Dehydration

Moreover, DE has an absorptive property, which facilitates water loss. The absorptive mechanism refers to DE particles removing the waxy-layer lipids from the insect cuticle, thereby interfering with the cuticle's ability to retain moisture and accelerating moisture loss. It is established that both the mechanism of mechanical abrasion and the removal of the lipids are the reasons DE is an effective insecticide against several types of arthropod pests, including some of those that infest stored products (Zeni et al., 2021).

9.6.3 Impact on Stored Product Insects

Knowledge of how DE works was expanded through learning about its use with plant extracts for more effective pest control. In fact, there is some synergy in pest control when mechanical injury and botanical toxicity are combined. The study revealed that DE-based formulations were more potent against major stored-product pests, such as *Tribolium castaneum* and *Sitophilus oryzae*, than other natural pesticides used in bioassay experiments, resulting in very high mortality in the treated insect population within a few days. Natural formulations also showed good potential for reducing pest populations without synthetic chemicals, hence being a sustainable option for pest management in stored products.

9.6.4 Environmental and Safety Considerations

Another significant advantage of DE is that it represents no toxicity to humans and animals. Unlike synthetic chemical insecticides, DE does not produce hazardous residues; hence, it can be used freely at home, in food stores, and in fields without posing any danger. The environmentally friendly nature and safety features of DE make it an essential material for green pest control (Zeni et al., 2021).

9.6.5 Examples of Diatom-Based Biopesticides

Diatomix produces biopesticides with diatoms as the active ingredient. The company brands the products under the trademark name "DiatomGuard." The products are designed to be environmentally friendly pest management tools, killing pests through the physical mechanism of desiccation (Diatomix - Garden Pond Algae Control, n.d.). Safer Brand, selling organic gardening and pest control products, under which comes the DE insecticides brand name "Safer Brand Diatomaceous Earth." This product line is designed to kill a myriad of insects without chemicals and is hence safe for the environment (Home Diatomaceous Earth, n.d.). Perma-Guard specializes in natural pest control solutions. It has a line of products with DE as the active ingredient. Its products, branded as "Perma-Guard Crawling Insect Control," mechanically damage the exoskeleton of crawling insects, resulting in their dehydration (Perma-Guard Diatomaceous Earth, n.d.).

9.6.6 Application Methods

9.6.6.1 Spray Applications

Spray applications of DE are accomplished by suspending the powder in water- or oil-based carriers for more uniform distribution and better targeting. This method is very effective for reaching pests in hard-to-reach or hidden areas. The formulations are changed to either a water-dispersible granule or a wettable powder, which are mixed with water before application. (Stejskal et al., 2021).

The effectiveness of DE sprays is affected by a myriad of factors, including suspension concentration, carrier type, and environmental conditions such as humidity. According to Zeni et al., 2021, a high concentration of DE in spray formulations may increase pest mortality, but this would further exacerbate clogging in spray equipment. This factor should therefore be balanced with the application equipment to achieve optimal outcomes (Table 9.3).

9.6.6.2 Incorporation into Soil Amendments

DE could be applied directly to soil in its solid, powdered form or as a suspended material in a slurry system. It has the advantage of ensuring that the DE, when used appropriately, is distributed throughout the soil, thereby increasing contact with pests traveling through or living in the soil. DE not only helps control pest populations but also creates healthier soils by providing improved aeration, water retention, and nutrient availability. In addition, DE promotes healthier plant growth by allowing water to infiltrate more readily and by enabling roots to move through its porous structure. Another critical point about including DE in soil amendments is that it effectively controls a broad spectrum of soil-dwelling pests. The action of DE is mechanical, and there is no risk of the pest developing resistance, unlike with the use of chemical pesticides. DE has demonstrated effective control against ants, termites, and beetles (Soubeih et al., 2017).

9.7 INTEGRATION WITH INTEGRATED PEST MANAGEMENT (IPM) STRATEGIES

9.7.1 Compatibility with Integrated Pest Management

DE has been widely used in IPM strategies. IPM is a concept based on the use of multiple complementary approaches to pest control, minimizing the use of chemical pesticides, and reducing environmental impacts. The mechanism by which DE acts—abrading and desiccating the exoskeleton of insects—aligns well with IPM concepts, providing a non-toxic, environmentally friendly alternative.

Among the significant advantages of DE in IPM is its compatibility with biological control agents. DE does not harm beneficial insects, such as predators and parasitoids, unlike chemical pesticides. The selectivity of DE thus maintains and supports the natural process of pest control,

TABLE 9.3
Pest Control Efficiency

Mechanism	Target Pests	Efficiency	Application Method	Synergies	References
Physical desiccation	*Cimex lectularius*	Effective, slower in resistant strains	Direct contact with DE particles	Chemical Control: Enhances susceptibility to chemical agents	Lilly et al. (2016)
Lipid absorption	Various arthropod pests	Highly effective, dual action of abrasion and absorption	Direct contact with DE particles	Biological Control: Compatible with predatory insects and fungi	Zeni et al. (2021)
Botanical synergy	*Tribolium castaneum*, *Sitophilus oryzae*	High mortality rates, sustainable	DE with botanical extract	Cultural Practices: Enhanced by crop rotation and sanitation	Zeni et al. (2021)
Spray applications	Soil-dwelling pests	Varies with concentration and humidity	Suspended in water/oil carriers	Physical Controls: Complements barriers, such as screens or row covers	Stejskal et al. (2021)
Soil amendments	Ants, termites, beetles	Effective, improves soil health	Mixed into soil or applied as a slurry	-	Soubeih et al. (2017)

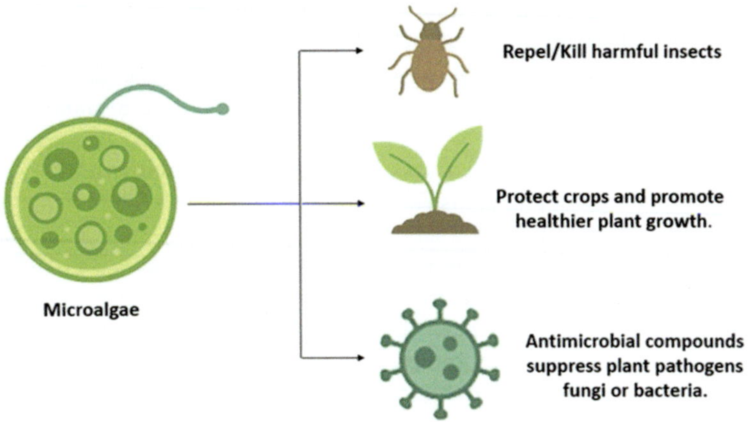

FIGURE 9.3 Importance of algae as biocontrol agents.

thereby enhancing the efficiency of IPM programs. It has been proven that DE can be applied simultaneously with entomopathogenic fungi and nematodes without impairing their activities, thus expanding the range of pest control options under application (Subramanyam & Roesli, 2000).

The effectiveness of DE is affected by environmental conditions such as humidity and temperature, as it relies on desiccation, and careful consideration of local climate and pest biology is required for its integration into IPM. DE is most effective in dry conditions and is especially beneficial for protecting stored grain and for use in arid environments. The integration of other non-chemical approaches, including sanitation and habitat manipulation, can further facilitate pest suppression and delay the development of pest resistance (Athanassiou et al., 2005).

DE is often viewed as the less cost-effective choice in IPM programs, except in organic farming systems that prohibit synthetic chemical pesticides. Nevertheless, DE remains a low-cost option with extended residual action; thus, it remains a feasible ingredient in a sustainable pest management strategy. Its compatibility with other biocontrol methods, long residual efficacy, relatively low cost, and environmental safety make it relevant for environmentally sustainable agriculture (Figure 9.3).

9.7.2 Potential Synergies with Other Control Methods

Integration of DE into IPM programs is highly promising because it presents a non-toxic mode of action and is compatible with a wide array of control measures. A combination of DE with other methods greatly enhances the effectiveness and sustainability of pest control programs.

9.7.2.1 Chemical Control Synergisms

DE can be applied with lower doses of chemical insecticides to delay the onset of resistance. The physical mode of action, with respect to the integrity of the cuticle of the insects, can increase susceptibility to chemical agents (Athanassiou et al., 2005).

9.7.2.2 Biological Control Enhancement

Other biological control agents, such as predatory insects or entomopathogenic fungi, in combination with DE, may show synergistic interactions beyond the action of pathogens. In these organisms, no deleterious effects of the diatoms were observed, and at times the reduction in pest populations was low enough to allow natural predators to keep them below the economic injury level.

9.7.2.3 Cultural Practices

DE's efficacy can be improved by cultural practices that reduce pest habitats. Crop rotation and sanitation practices, for instance, have few pests, hence increasing the impact of DE on the reduced initial population (Subramanyam & Roesli, 2000).

9.7.2.4 Combination with Other Physical Controls

By combining DE with other physical suppression methods, such as screens or row covers, two layers of defense are created. This setup would physically deter target pests from initial access and further infestation, while DE acts on those that make it past the barriers (Agrafioti et al., 2023).

9.8 BIOMANUFACTURING APPLICATIONS BEYOND AGRICULTURE

Microalgae biomanufacturing applications go well beyond agriculture, spanning a range of other high-value sectors, including pharmaceuticals, nanotechnology, biomaterials, and energy. Microalgae are a rich source of lipids, proteins, carbohydrates, and pigments, with significant nutritional and health benefits, and are increasingly being added to functional foods and feeds, as well as cosmetics and agricultural products (Su et al., 2023). Microalgae have emerged as promising sources of bioactive compounds with potential health benefits, having medicinal, anticancerous, and pharmaceutical applications. The pharmaceutical sector is a significant application domain in which the microalgal biotechnology industry is expanding rapidly due to the availability of highly valuable products, such as pigments, pharmaceuticals, and nutraceuticals (Mutanda et al., 2020). Microalgae offer additional potential as light-driven cell factories for the production of novel metabolites. Plant secondary products are one of the most critical forms of target compounds.

Beyond pharmaceutical applications, biogenic silica and diatoms prove attractive for biotechnological and industrial applications, especially in biomedicine, industrial/synthetic manufacturing processes, and biomedical/pharmaceutical sciences. The deposition of silica by diatoms allows them to create microscale or nanoscale structures, which may be utilized in nanomedicine and drug/gene delivery (Rabiee et al., 2021). Diatom frustules consist of an intricate, ordered porous silica matrix. This has recently found several applications in optoelectronics, sensing, solar light harvesting, filtering, and drug delivery. The micro- to nanoscale properties of diatom frustules enable applications in nanotechnology and biotechnology, including bioimaging/biosensing, biosensors, drug/gene delivery, photodynamic therapy, microfluidics, biophotonics, solar cells, and molecular filtrations. (Rabiee et al., 2021).

Biomaterials provide a new source of inspiration for the design and fabrication of advanced nanostructured materials. Diatoms can synthesize inorganic structures into complex architectures with ordered features from the microscale to the nanoscale, through the exceptional variety of patterned silica structures. In recent decades, diatoms have been proposed for use in drug delivery, microfluidics, batteries, templates, biosensors, and for energy conversion and storage. Diatoms, used as a platform for nanotechnology, offer crucial advantages. Diatom frustules function as natural biomaterials that feature regular shape and intricate hierarchical micro/nano structures, showing excellent performance in biosensing (Wang et al., 2021)

The potential of microalgae as expression systems for biomanufacturing highly valuable proteins remains significant. Nuclear and chloroplast genomes of some microalgae are being engineered to express recombinant proteins. These diverse applications collectively demonstrate that microalgal biomanufacturing represents a rapidly expanding biotechnology sector. Applications span from traditional pharmaceutical manufacturing to cutting-edge nanotechnology platforms, positioning microalgae as essential components of the emerging circular bioeconomy and sustainable industrial biotechnology landscape (Udayan et al., 2021)

9.9 PROSPECTS FOR HARNESSING DIATOMS IN AGRICULTURE

With advancements in biotechnology and nanotechnology, the potential for utilizing diatoms in agriculture is increasing. Innovations in cultivation techniques, such as optimized growing conditions and genetic engineering, are increasing diatom productivity. Collaborative research and knowledge sharing are crucial for developing innovations, including biofertilizers and pest control agents to support sustainable agricultural practices. Genetic manipulation in the diatom genome can be used to modify any desirable trait for agricultural production; increased lipid production makes it worthwhile for biofertilizers and bio-stimulants. Genetic transformation of diatoms can overexpress specific genes, thereby enhancing lipid accumulation and improving tolerance to abiotic stresses such as drought and salinity, resulting in more stable productivity. Gene-editing methods have dramatically altered diatoms' tolerance to stressors. Genetic engineering also enhances the growth rate of the diatoms, which increases the biomass obtained from culturing and promotes cultivation on larger scales. Methods of editing genes using CRISPR/Cas9 edit diatom genomes with precision in pathways related to their growth (Huang & Daboussi, 2017). Nanotechnology utilizes diatom biosilica in new agricultural applications, where DE is used in the development of nano-fertilizers that provide plants with controlled and targeted release of nutrients (Yadav et al., 2024). Diatom-based nanoparticles act as carriers for slow-release pesticides, providing prolonged pest control and reducing chemical use and, consequently, environmental impact (Puranik et al., 2024). Functionalized diatom biosilica nanoparticles can also be used in water purification systems and to improve soil fertility, thus providing adequate conditions for the proper growth of plants.

New cultivation techniques are discovering better ways to harness diatom potential in agriculture, including photobioreactor systems and their integration into aquaponics/hydroponics farming systems. Photobioreactors provide a controlled environment for growing diatoms under optimal light, temperature, and nutrient conditions, enabling maximum biomass production. They can be configured in various designs, for example, flat-panel, tubular, and column reactors, each having the added advantage of efficiency or scalability. In photobioreactors, the closed setup limits contamination and allows strict monitoring of culture parameters, enabling high, stable yields of diatomic biomass. These controlled conditions, besides yielding high productivity, allow tailoring the production to specific diatom strains, for example, in the agricultural sector as a source of biofertilizers and bio-stimulants. An instance of the microalgae cultivation method showing innovation is their integration into hydroponic and aquaponic systems (Wijffels et al., 2010). As far as aquaponics goes, diatoms may, for instance, be farmed in conjunction with fish by using the fish tank water, full of nutrients (wastewater), as a growth medium. This type of symbiosis not only "strengthens" the processes involved in nutrient recycling but also improves water quality by reducing waste through the addition of a natural biofilter. The diatoms, in turn, generate oxygen and provide other essential nutrients for both the aquatic and plant parts of the system. In the same vein, the presence of diatoms in hydroponic nutrient solutions would enrich them; thus, plants would grow faster and, at the same time, more healthily. Consequently, their incorporation into hydroponic systems will simultaneously improve nutrient uptake by plants and provide them with some protection against pathogens through various bioactive compounds (Subashchandrabose et al., 2011). These combined systems are designed to be sustainable in terms of resource optimization, which implies that such systems have been designed to use all available resources, yet keep the impact on the environment at a minimum level, and therefore, the concept of diatoms is that of a great potential in the systems as mentioned above to improve the modern agricultural techniques. Collaboration among experts from various fields, such as biotechnology, nanotechnology, and agriculture, strengthens the innovation process for the use of diatom properties in sustainable agriculture. Knowledge exchange through workshops, conferences, and joint research projects is likely to accelerate the implementation of such innovations. Capacity building not only empowers scientists but also equips practitioners with the knowledge and skills needed to implement diatom-based solutions to enhance sustainable agricultural practices, a prerequisite for addressing global challenges such as environmental sustainability and alleviating hunger.

9.10 CHALLENGES IN INTEGRATING DIATOMS INTO AGRICULTURAL SYSTEMS

9.10.1 Scalability Issues

There is enormous potential for diatoms' use in agroecosystems to enhance soil health, foster sustainable agriculture, and increase crop yields. Despite these promising benefits, the large-scale and commercially viable implementation of diatom-based biofertilizers faces several challenges (Alvarez et al., 2021). This section discusses the major scalability challenges associated with production constraints, process bottlenecks, and the complexities involved in commercial-scale implementation.

9.10.2 Production Challenges and Constraints

One of the significant challenges of large-scale production of diatoms is that they require special biological and environmental conditions to thrive, which include specific light, temperature, and nutrients that are costly and challenging (Sharma et al., 2021). In addition, cultivating diatoms must be free of microbial contamination, which is challenging and adds complexity and cost. Another aspect of diatom culture production that poses considerable difficulty is the need for cultures to yield diatoms of consistent quality and growth rates. Small changes in environmental conditions can lead to variations in growth rates and biomass yield, leading to difficulties with the standardization and commercialization of products based on diatoms (Sharma et al., 2021).

9.10.3 Scaling Up for Commercial Application

There are various technical and economic challenges and drawbacks associated with scaling diatom production from research scale to commercial scale. The overall system will include photobioreactors and an open-pond system for large-scale production, which can entail high costs. Although less expensive, open systems will limit control over environmental variables and increase the risk of environmental contamination. Although closed systems (such as photobioreactors) offer greater control over growth conditions and reduced risk of environmental contamination, they will be significantly more costly to build and operate (Heidi & Jakob, 2016). Establishing ideal environments for diatom growth will entail innovative designs for the growth system and nutrient and CO_2 recycling. Harvesting live microalgae requires a departure from standard practices such as centrifugation or filtration, which are feasible for pilot systems but unlikely to be practical, effective, or economical at commercial scales (Sun et al., 2018). The commercial production of diatom-based biofertilizer will require reliable production systems, streamlined distribution mechanisms, appropriate storage systems for solid or liquid products, and suitable mechanisms to maintain product quality and consistency. A sustainable supply chain and continuity across producers, processors, distributors, and end-users must be established; businesses all have a functional interdependent role. On the other end of the production and marketing process, regulatory approval is paramount; 34 respondents in the survey cited regulatory constraints on safety and efficacy as a deterrent to the innovation and commercialization of diatom-based biofertilizers (Heidi & Jakob, 2016; Sun et al., 2018).

9.10.4 Technological Limitations

Key challenges in the effective utilization of diatoms in agriculture lie primarily in technological limitations. Focus needs to be diverted toward bioprocessing, extraction technologies, product formulation, and stability (Heidi & Jakob, 2016).

9.10.5 Bioprocessing and Extraction Technologies

Significant challenges include efficiently growing and harvesting diatoms, as traditional methods, such as open ponds and photobioreactors, have limitations in scalability and cost efficiency. Techniques for harvesting, such as centrifugation and filtration, are too expensive and energy-intensive. Electro-flocculation and membrane technology are techniques that have potential but still need some development (Heidi & Jakob, 2016). Extraction of valuable compounds, such as lipids, proteins, and silica, from diatoms is also complex. The classic extraction methods have low efficiency, whereas advanced techniques, such as supercritical fluid and ultrasonic-assisted extraction, improve yields and preserve bioactive compounds but require adaptation for industrial-scale use (Mahindrakar & Rathod, 2022).

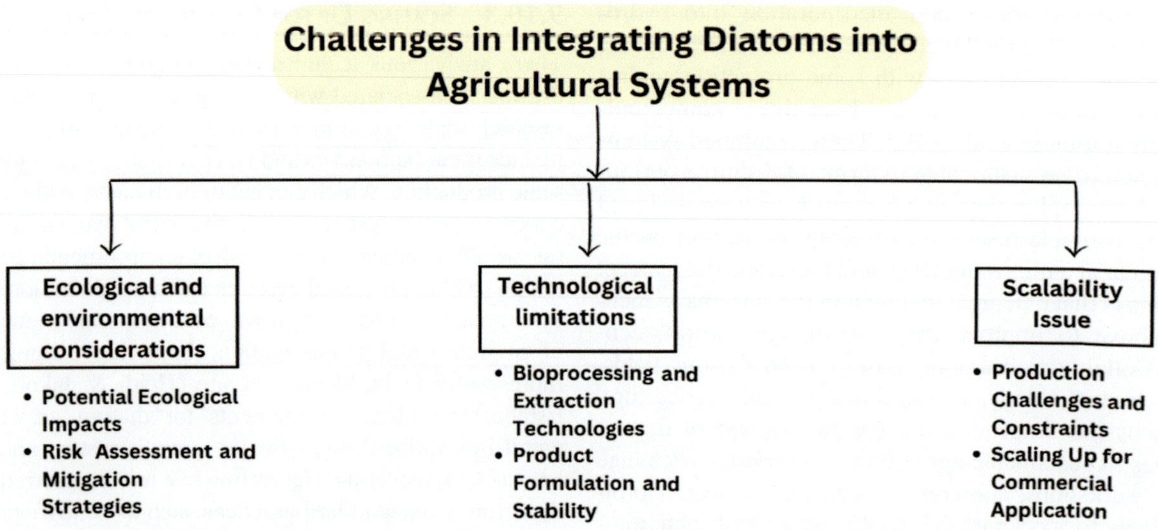

FIGURE 9.4 Challenges associated with the incorporation of diatoms in agriculture.

9.10.6 Product Formulation and Stability

Another challenge is stabilizing diatom-based biofertilizers and biopesticides. The efficacy of these products must be maintained across different environmental conditions. Among the factors that may degrade the bioactive compounds are temperature, humidity, and light (Chabili et al., 2024). Hence, encapsulation techniques, such as micro- or nanoencapsulation, have been studied as strategies to protect bioactive compounds and control their release, thereby enhancing stability and efficacy (Vieira et al., 2020). Such products should be compatible with existing agricultural practices. In other words, diatom-based products must be easily integrated into current farming systems, which requires formulations that are easy to handle, apply, and distribute. Research continues to find a balance between effectiveness, ease of use, and cost (Heidi & Jakob, 2016). It undergoes rigorous safety and efficacy testing, and the regulatory approval process is time-consuming and costly, thereby slowing the deployment of innovation. The process can be streamlined by collaboration among researchers, industry players, and regulators (Heidi & Jakob, 2016).

9.10.7 Ecological and Environmental Considerations

The application of diatoms in agricultural systems opens new vistas for green farming, but also raises ecological challenges that need to be addressed to ensure safe adoption (Sharma et al., 2021).

9.10.7.1 Potential Ecological Impacts

The introduction of diatoms may disrupt nutrient cycling within an ecosystem, depleting or altering the balance of soil nutrients that promote plant growth. The influence of these organisms on the environment would alter soil structure, as changes in soil microbial communities affect decomposition and nutrient mineralization. In addition to these potential impacts, runoff of blooms of diatoms into the aquatic environment may negatively influence both marine and terrestrial organisms through hypoxic events, and the release of toxins (Glibert & Burford, 2017).

9.10.7.2 Risk Assessment and Mitigation Strategies

Risk assessments will have to be detailed and specific to species interaction reports, or environmental effects assessments that refer to the potential for diatoms to proliferate, as well as the possible impacts to the ecosystem (Brown et al., 2020).

Mitigation measures that can be adopted to address risks to the environment may include the application process itself to ensure that application does not over-proliferate diatom populations, diversity of species to help lessen biodiversity impacts, salvage buffers so that there are some distance between the fields and the aquatic environment, and ongoing monitoring of the health status of both soil and water to provide the earliest warning of possible adverse effects. Adaptive management approaches concerning application rates and bioremediation also remain essential components of mitigating risks as they emerge (Brown et al., 2020; Glibert & Burford, 2017) (Figure 9.4).

9.11 CONCLUSION

Diatoms and their uses are a significant element of sustainable development and the development of the circular economy conceptions. This chapter identifies various functions of diatoms in nutrient cycling, improving soil health, pest control, and biofertilization in food production. Recycling essential nutrients, such as nitrogen and phosphorus, using diatoms can support the development of circular agriculture, improve soil structure and water retention, and increase microbial community diversity. In the agricultural sector, diatom biocontrol capabilities may offer more environmentally friendly alternatives to pest control and be more effective when combined with other systems in IPM strategies. When comparing the functionality of ecosystem services and resource utilization, the nutrient-rich biomass of diatoms can also serve as an environmentally sound alternative to traditional fertilizers. Despite their considerable potential, several barriers remain to the large-scale adoption of diatom-based agricultural practices, including production, technological, and ecological constraints. The combination of more traditional farming methods, the latest developments in biotechnology, the emergence of nanotechnology, and the latest innovations in cultivation may lead to the desired conservation and agricultural use of diatoms. Cross-disciplinary collaboration and knowledge sharing (through the diatoms) can also facilitate this, which, in turn, can support the acceleration and adoption of diatom-based agricultural practices. Agricultural practices that are based on diatoms provide a chance to establish sustainability and efficiency in food production systems and ultimately a more sustainable future. To conclude, the diatoms' future in agriculture is quite bright, and they will be able to advance biomimetic, more environmentally friendly production processes. They say that innovation, cross-disciplinary, and cross-sector collaboration offer a unique opportunity to regenerate the agricultural landscape, particularly through exposure to diatom applications.

REFERENCES

Adedibu, P. A. (2023). Ecological problems of agriculture: Impacts and sustainable solutions. *scienceopen.com*. https://doi.org/10.14293/PR2199.000145.v1

Adewara, O. A., Adebayo-Olajide, T. C., Ayedun, J. S., Kotun, B. C., Adeleke, A. J., Brown, A. D., Alabi, O. J., & Ogunbanwo, S. T. (2024). Soil formation, soil health and soil biodiversity In *Prospects for Soil Regeneration and Its Impact on Environmental Protection* (pp. 95–121). Cham: Springer Nature Switzerland. https://doi.org/10.1007/978-3-031-53270-2_5

Agrafioti, P., Vrontaki, M., Rigopoulou, M., Lampiri, E., Grigoriadou, K., Ioannidis, P. M., Rumbos, C. I., & Athanassiou, C. G. (2023). Insecticidal Effect of Diatomaceous Earth Formulations for the Control of a Wide Range of Stored-Product Beetle Species. *Insects*, *14*(7), 656. https://doi.org/10.3390/insects14070656

Al-Suhaibani, N., Selim, M., Alderfasi, A., & El-Hendawy, S. (2020). Comparative Performance of Integrated Nutrient Management between Composted Agricultural Wastes, Chemical Fertilizers, and Biofertilizers in Improving Soil Quantitative and Qualitative Properties and Crop Yields under Arid Conditions. *Agronomy*, *10*(10), 1503. https://doi.org/10.3390/agronomy10101503

Alvarez, A. L., Weyers, S. L., Goemann, H. M., Peyton, B. M., & Gardner, R. D. (2021). Microalgae, soil and plants: A critical review of microalgae as renewable resources for agriculture. *Algal Research*, *54*, 102200. https://doi.org/10.1016/j.algal.2021.102200

Armbrust, E. V. (2009). The life of diatoms in the world's oceans. *Nature*, *459*(7244), 185–192. https://doi.org/10.1038/nature08057

Artyszak, A. (2018). Effect of silicon fertilization on crop yield quantity and quality—A literature review in Europe. *Plants*, *7*(3), 54. https://doi.org/10.3390/plants7030054

Athanassiou, C. G., Vayias, B. J., Dimizas, C. B., Kavallieratos, N. G., Papagregoriou, A. S., & Buchelos, C. T. (2005). Insecticidal efficacy of diatomaceous earth against Sitophilus oryzae (L.) (Coleoptera: Curculionidae) and Tribolium confusum du Val (Coleoptera: Tenebrionidae) on stored wheat: influence of dose rate, temperature and exposure interval. *Journal of Stored Products Research*, *41*(1), 47–55. https://doi.org/10.1016/j.jspr.2003.12.001

Azam, F., & Malfatti, F. (2007). Microbial structuring of marine ecosystems. *Nature Reviews Microbiology*, *5*(10), 782–791.

Basbous, M., El-Mounsef, J. R., El Hokayem, J., Kanaan, H. Y. (2024). A comprehensive review of biofuel and bioplastic production from microalgae. *In 2024 IEEE 22nd Mediterranean Electrotechnical Conference (MELECON)* (pp. 435–442). IEEE.

Bastos, C. R. V., Maia, I. B., Pereira, H., Navalho, J., & Varela, J. C. S. (2022). Optimisation of biomass production and nutritional value of two marine diatoms (Bacillariophyceae), Skeletonema costatum and Chaetoceros calcitrans. *Biology*, *11*(4). https://doi.org/10.3390/biology11040594

Battin, T. J., Luyssaert, S., Kaplan, L. A., Aufdenkampe, A. K., Richter, A., & Tranvik, L. J. (2009). The boundless carbon cycle. *Nature Geoscience*, *2*. https://doi.org/10.1038/ngeo618

Brown, A. R., Lilley, M., Shutler, J., Lowe, C., Artioli, Y., Torres, R., Berdalet, E., & Tyler, C. R. (2020). Assessing risks and mitigating impacts of harmful algal blooms on mariculture and marine fisheries. *Reviews in Aquaculture*, *12*(3), 1663–1688. https://doi.org/10.1111/raq.12403

Burgo Bencomo, O. B., León González, J. L., Cáceres Mesa, M. L., Pérez Maya, C. J., Espinoza Freire, E. E., Burgo Bencomo, O. B., León González, J. L., Cáceres Mesa, M. L., Pérez Maya, C. J., & Espinoza Freire, E. E. (2019). Algunas reflexiones sobre investigación e intervención educativa. *Revista Cubana de Medicina Militar*, *48*(sup), 316. http://scielo.sld.cu/scielo.php?script=sci_arttext&pid=S0138-65572019000500003&lng=es&nrm=iso&tlng=en

Chabili, A., Minaoui, F., Hakkoum, Z., Douma, M., Meddich, A., & Loudiki, M. (2024). A comprehensive review of microalgae and cyanobacteria-based biostimulants for agriculture uses. *Plants*, *13*(2), 159. https://doi.org/10.3390/plants13020159

Das, P., Khan, S., Chaudhary, A. K., AbdulQuadir, M., Thaher, M. I., & Al-Jabri, H. (2019). Potential applications of algae-based bio-fertilizer. In *Biofertilizers for Sustainable Agriculture and Environment* (pp. 41–65). Cham: Springer International Publishing. https://doi.org/10.1007/978-3-030-18933-4_3

de Morais, E. G., da Silveira, J. T., Schüler, L. M., de Freitas, B. C. B., Costa, J. A. V., de Morais, M. G., Ferrer, I., & Barreira, L. (2023). Biomass valorization via pyrolysis in microalgae-based wastewater treatment: Challenges and opportunities for a circular bioeconomy. *Journal of Applied Phycology*, *35*(6), 2689–2708. https://doi.org/10.1007/s10811-023-03104-x

Dhokane, D., Shaikh, A., Yadav, A., Giri, N., Bandyopadhyay, A., Dasgupta, S., & Bhadra, B. (2023). CRISPR-based bioengineering in microalgae for production of industrially important biomolecules. *Frontiers in Bioengineering and Biotechnology*, *11*. https://doi.org/10.3389/fbioe.2023.1267826

Diatomix - Garden Pond Algae Control. (n.d.). Retrieved May 20, 2024, from https://diatomix.com.au/?v=c86ee0d9d7ed

Elizabeth George, S., & Wan, Y. (2023). Microbial functionalities and immobilization of environmental lead: Biogeochemical and molecular mechanisms and implications for bioremediation. *Journal of Hazardous Materials*, *457*, 131738. https://doi.org/10.1016/j.jhazmat.2023.131738

Falkowski, P. G., Barber, R. T., & Smetacek, V. (1998). Biogeochemical Controls and Feedbacks on Ocean Primary Production. *Science*, *281*(5374), 200–206. https://doi.org/10.1126/SCIENCE.281.5374.200

Gaba, S., Lescourret, F., Boudsocq, S., Enjalbert, J., Hinsinger, P., Journet, E. P., Navas, M. L., Wery, J., Louarn, G., Malézieux, E., Pelzer, E., Prudent, M., & Ozier-Lafontaine, H. (2015). Multiple cropping systems as drivers for providing multiple ecosystem services: from concepts to design. *Agronomy for Sustainable Development*, *35*(2), 607–623.

Geissdoerfer, M., Savaget, P., Bocken, N. M. P., & Hultink, E. J. (2017). The Circular Economy – A new sustainability paradigm? *Journal of Cleaner Production*, *143*, 757–768. https://doi.org/10.1016/J.JCLEPRO.2016.12.048

Giwa, A., Abuhantash, F., Chalermthai, B., & Taher, H. (2022). Bio-based circular economy and polygeneration in microalgal production from food wastes: A concise review. *Sustainability*, *14*(17), 10759. https://doi.org/10.3390/su141710759

Glibert, P. M., & Burford, M. A. (2017). Globally changing nutrient loads and harmful algal blooms: Recent advances, new paradigms, and continuing challenges. *Oceanography*. 30 (1), 58–69.

Guieysse, B., & Plouviez, M. (2024). Microalgae cultivation: Closing the yield gap from laboratory to field scale. *Frontiers in Bioengineering and Biotechnology*, *12*. https://doi.org/10.3389/fbioe.2024.1359755

Hammad, H. M., Khaliq, A., Abbas, F., Farhad, W., Fahad, S., Aslam, M., Shah, G. M., Nasim, W., Mubeen, M., & Bakhat, H. F. (2020). Comparative effects of organic and inorganic fertilizers on soil organic carbon and wheat productivity under arid region. *Communications in Soil Science and Plant Analysis*, *51*(10), 1406–1422. https://doi.org/10.1080/00103624.2020.1763385

Heidi, G., & Jakob, N. (2016). *Scale-up cultivation of Australian algae New approaches to isolation, mid-scale cultivation and harvesting of Australian wild type algal strains*. PhD thesis. The University of Queensland, Australia.

Home Diatomaceous Earth. (n.d.). Retrieved May 20, 2024, from https://www.saferbrand.com/store/home/diatomaceous-earth

Huang, W., & Daboussi, F. (2017). Genetic and metabolic engineering in diatoms. *Philosophical Transactions of the Royal Society B: Biological Sciences*, *372*(1728), 20160411. https://doi.org/10.1098/rstb.2016.0411

Kiran, B., Pathak, K., Kumar, R., & Deshmukh, D. (2017). Phycoremediation: An eco-friendly approach to solve water pollution problems. In *Microbial Applications* Vol. 1 (pp. 3–28). Springer International Publishing. https://doi.org/10.1007/978-3-319-52666-9_1

Kirchherr, J., Piscicelli, L., Bour, R., Kostense-Smit, E., Muller, J., Huibrechtse-Truijens, A., & Hekkert, M. (2018). Barriers to the circular economy: Evidence from the European Union (EU). *Ecological Economics*, *150*, 264–272. https://doi.org/10.1016/j.ecolecon.2018.04.028

Korunic, Z. (2013). Diatomaceous earths: Natural insecticides. *Pesticidi i Fitomedicina*, *28*(2), 77–95.

Kovács, S., Kutasy, E., & Csajbók, J. (2022). The Multiple Role of Silicon Nutrition in Alleviating Environmental Stresses in Sustainable Crop Production. *Plants*, *11*(9), 1223. https://doi.org/10.3390/plants11091223

Kumar, G., Shekh, A., Jakhu, S., Sharma, Y., Kapoor, R., & Sharma, T. R. (2020). Bioengineering of microalgae: Recent advances, perspectives, and regulatory challenges for industrial application. *Frontiers in Bioengineering and Biotechnology*, *8*, 914. https://doi.org/10.3389/fbioe.2020.00914

Lamb, A., Green, R., Bateman, I., Broadmeadow, M., Bruce, T., Burney, J., Carey, P., Chadwick, D., Crane, E., Field, R., Goulding, K., Griffiths, H., Hastings, A., Kasoar, T., Kindred, D., Phalan, B., Pickett, J., Smith, P., Wall, E., … Balmford, A. (2016). The potential for land sparing to offset greenhouse gas emissions from agriculture. *Nature Climate Change*, *6*(5), 488–492. https://doi.org/10.1038/nclimate2910

Lee, A., Lan, J. C.-W., Jambrak, A. R., Chang, J.-S., Lim, J. W., & Khoo, K. S. (2024). Upcycling fruit waste into microalgae biotechnology: Perspective views and way forward. *Food Chemistry: Molecular Sciences*, *8*, 100203. https://doi.org/10.1016/j.fochms.2024.100203

Leibold, M. A., Holyoak, M., Mouquet, N., Amarasekare, P., Chase, J. M., Hoopes, M. F., Holt, R. D., Shurin, J. B., Law, R., Tilman, D., Loreau, M., & Gonzalez, A. (2004). The metacommunity concept: A framework for multi-scale community ecology. *Ecology Letters*, *7*(7), 601–613. https://doi.org/10.1111/j.1461-0248.2004.00608.x

Leong, H. Y., Chang, C.-K., Khoo, K. S., Chew, K. W., Chia, S. R., Lim, J. W., Chang, J.-S., & Show, P. L. (2021). Waste biorefinery towards a sustainable circular bioeconomy: a solution to global issues. *Biotechnology for Biofuels*, *14*(1), 87. https://doi.org/10.1186/s13068-021-01939-5

Lilly, D., Webb, C., & Doggett, S. (2016). Evidence of tolerance to silica-based desiccant dusts in a pyrethroid-resistant strain of Cimex lectularius (Hemiptera: Cimicidae). *Insects*, *7*(4), 74. https://doi.org/10.3390/insects7040074

Lobus, N. V., & Kulikovskiy, M. S. (2023). The co-evolution aspects of the biogeochemical role of phytoplankton in aquatic ecosystems: A review. *Biology*, *12*(1), 92. https://doi.org/10.3390/biology12010092

Lobus, N. V., Kulikovskiy, M. S., & Maltsev, Y. I. (2021). Multi-element composition of diatom Chaetoceros spp. from natural phytoplankton assemblages of the Russian Arctic Seas. *Biology*, *10*(10), 1009. https://doi.org/10.3390/biology10101009

Lundholm, N., McCarthy, A., & Baer, P. (2019). Diatoms: A bioremediation tool for phosphorus pollution in aquatic environments. *Algal Research*, *44*, 101705.

Ma, Y. (2019). Seed coating with beneficial microorganisms for precision agriculture. *Biotechnology Advances*, *37*(7), 107423. https://doi.org/10.1016/j.biotechadv.2019.107423

Ma, Z., Cheah, W. Y., Ng, I.-S., Chang, J.-S., Zhao, M., & Show, P. L. (2022). Microalgae-based biotechnological sequestration of carbon dioxide for net zero emissions. *Trends in Biotechnology*, *40*(12), 1439–1453. https://doi.org/10.1016/j.tibtech.2022.09.002

Mahindrakar, K. V., & Rathod, V. K. (2022). Ultrasound-assisted extraction of lipids, carotenoids, and other compounds from marine resources. In *Innovative and Emerging Technologies in the Bio-marine Food Sector* (pp. 81–128). Elsevier. https://doi.org/10.1016/B978-0-12-820096-4.00001-8

Monlau, F., Sambusiti, C., Antoniou, N., Barakat, A., & Zabaniotou, A. (2015). A new concept for enhancing energy recovery from agricultural residues by coupling anaerobic digestion and pyrolysis process. *Applied Energy*. https://www.sciencedirect.com/science/article/pii/S0306261915003086

Morseletto, P. (2020). Targets for a circular economy. *Resources, Conservation and Recycling*, *153*, 104553. https://doi.org/10.1016/j.resconrec.2019.104553

Mutanda, T., Naidoo, D., Bwapwa, J. K., & Anandraj, A. (2020). Biotechnological applications of microalgal oleaginous compounds: current trends on microalgal bioprocessing of products. *Frontiers in Energy Research*, *8*. https://doi.org/10.3389/fenrg.2020.598803

Palomino, A. M., Kim, S., Summitt, A., & Fratta, D. (2011). Impact of diatoms on fabric and chemical stability of diatom–kaolin mixtures. *Applied Clay Science*, *51*(3), 287–294. https://doi.org/10.1016/j.clay.2010.12.002

Perma-Guard Diatomaceous Earth. (n.d.). Retrieved May 20, 2024, from https://www.perma-guard.com/

Puranik, S., Sruthy, K. S., Manoj, M., Vikram, K. V., Karijadar, P., Singh, S. K., & Shukla, L. (2024). Microbial inoculants and their potential application in bioremediation. In *Microbes Based Approaches for the Management of Hazardous Contaminants* (pp. 118–145). Wiley. https://doi.org/10.1002/9781119851158.ch8

Rabiee, N., Khatami, M., Jamalipour Soufi, G., Fatahi, Y., Iravani, S., & Varma, R. S. (2021). Diatoms with invaluable applications in nanotechnology, biotechnology, and biomedicine: Recent advances. *ACS Biomaterials Science & Engineering*, *7*(7), 3053–3068. https://doi.org/10.1021/acsbiomaterials.1c00475

Ramappa, K. (2020). Comparative economics of neem coated urea vis-a-vis normal urea: Evidence from a field-based study in the Indian context. *Economic Affairs*, *65*(2). https://doi.org/10.46852/0424-2513.2.2020.21

Rozenberg, J. M., Sorokin, B. A., Mukhambetova, A. N., Emelianova, A. A., Kuzmin, V. V., Belogurova-Ovchinnikova, O. Y., & Kuzmin, D. V. (2024). Recent advances and fundamentals of microalgae cultivation technology. *Biotechnology Journal*, *19*(3). https://doi.org/10.1002/biot.202300725

Santos, B., Freitas, F., Sobral, A. J. F. N., & Encarnação, T. (2025). Microalgae and circular economy: unlocking waste to resource pathways for sustainable development. *International Journal of Sustainable Engineering*, *18*(1). https://doi.org/10.1080/19397038.2025.2501488

Sharma, N., Simon, D. P., Diaz-Garza, A. M., Fantino, E., Messaabi, A., Meddeb-Mouelhi, F., Germain, H., & Desgagné-Penix, I. (2021). Diatoms biotechnology: Various industrial applications for a greener tomorrow. *Frontiers in Marine Science*, *8*. https://doi.org/10.3389/fmars.2021.636613

Singh, P. K., Saxena, A., Tyagi, R., Sindhu, R., Binod, P., & Tiwari, A. (2023). Biomass valorization of agriculture wastewater grown freshwater diatom Nitzschia sp. for metabolites, antibacterial activity, and biofertilizer. *Bioresource Technology*, *377*, 128976. https://doi.org/10.1016/j.biortech.2023.128976

Slocombe, S. P., Zhang, Q., Ross, M., Anderson, A., Thomas, N. J., Lapresa, Á., Rad-Menéndez, C., Campbell, C. N., Black, K. D., Stanley, M. S., & Day, J. G. (2015). Unlocking nature's treasure-chest: Screening for oleaginous algae. *Scientific Reports*, *5*. https://doi.org/10.1038/srep09844

Soubeih, K., Ali, E., & El-Hadidy, A. (2017). Effect of kaolin and diatoms on growth, productivity and pests of potato under north Sinai conditions. *Egyptian Journal of Desert Research*, *67*(1), 83–114.

Stejskal, V., Vendl, T., Aulicky, R., & Athanassiou, C. (2021). Synthetic and natural insecticides: Gas, liquid, gel and solid formulations for stored-product and food-industry pest control. *Insects*, *12*(7), 590. https://doi.org/10.3390/insects12070590

Su, M., Bastiaens, L., Verspreet, J., & Hayes, M. (2023). Applications of microalgae in foods, pharma and feeds and their use as fertilizers and biostimulants: Legislation and regulatory aspects for consideration. *Foods*, *12*(20), 3878. https://doi.org/10.3390/foods12203878

Subashchandrabose, S. R., Ramakrishnan, B., Megharaj, M., Venkateswarlu, K., & Naidu, R. (2011). Consortia of cyanobacteria/microalgae and bacteria: Biotechnological potential. *Biotechnology Advances*, *29*(6), 896–907. https://doi.org/10.1016/j.biotechadv.2011.07.009

Subramanyam, B., & Roesli, R. (2000). Inert dusts. In *Alternatives to Pesticides in Stored-Product IPM* (pp. 321–380). Springer US. https://doi.org/10.1007/978-1-4615-4353-4_12

Sun, H., Zhao, W., Mao, X., Li, Y., Wu, T., & Chen, F. (2018). High-value biomass from microalgae production platforms: strategies and progress based on carbon metabolism and energy conversion. *Biotechnology for Biofuels*, *11*(1), 227. https://doi.org/10.1186/s13068-018-1225-6

Sunda, W. G., & Hardison, D. R. (2010). Evolutionary tradeoffs among nutrient acquisition, cell size, and grazing defense in marine phytoplankton promote ecosystem stability. *Marine Ecology Progress Series*, *401*, 63–76.

Toop, T., Ward, S., Oldfield, T., Hull, M., & Procedia, M. K.-E. (2017). *AgroCycle–Developing a Circular Economy in Agriculture*. Elsevier. https://www.sciencedirect.com/science/article/pii/S1876610217328436

Udayan, A., Pandey, A. K., Sharma, P., Sreekumar, N., & Kumar, S. (2021). Emerging industrial applications of microalgae: challenges and future perspectives. *Systems Microbiology and Biomanufacturing*, *1*(4), 411–431. https://doi.org/10.1007/s43393-021-00038-8

Vieira, M. V., Pastrana, L. M., & Fuciños, P. (2020). Microalgae encapsulation systems for food, pharmaceutical and cosmetics applications. *Marine Drugs*, *18*(12), 644. https://doi.org/10.3390/md18120644

Wang, Z., Gong, D., & Cai, J. (2021). Diatom frustule array for flow-through enhancement of fluorescent signal in a microfluidic chip. *Micromachines*, *12*(9), 1017. https://doi.org/10.3390/mi12091017

Wijffels, R. H., Barbosa, M. J., & Eppink, M. H. M. (2010). Microalgae for the production of bulk chemicals and biofuels. *Biofuels, Bioproducts and Biorefining*, *4*(3), 287–295. https://doi.org/10.1002/BBB.215

Yadav, M., Gupta, S., Sharma, S., & George, N. (2024). Biosilica nanoparticles from agricultural waste and their applications. In *Biogenic Silica* (pp. 79–115). Royal Society of Chemistry. https://doi.org/10.1039/9781839169717-00079

Zehr, J. P., & Capone, D. G. (2021). *Marine Nitrogen Fixation*. Springer International Publishing. https://doi.org/10.1007/978-3-030-67746-6

Zeni, V., Baliota, G. V., Benelli, G., Canale, A., & Athanassiou, C. G. (2021). Diatomaceous Earth for arthropod pest control: Back to the future. *Molecules*, *26*(24), 7487. https://doi.org/10.3390/molecules26247487

10 Biomanufacturing of Pharmaceutical and Personal Care Products (PPCPs) and Cosmetics to Foster a Sustainable Future

Priyanka Gupta, Aditi Mishra and Vaibhav Srivastava
Naithani Plant Genetics and Ecotoxicology Laboratory, Department of Botany, University of Allahabad, Prayagraj, India

Suvendu Manna
School of Health Sciences & Technology, UPES, Dehradun, India

Vishal Tripathi
Department of Biotechnology, Graphic Era (Deemed to be University), Dehradun, India

Rajeev Pratap Singh
Waste Management, Resource Recovery & Ecotoxicology Laboratory, Institute of Environment and Sustainable Development, Banaras Hindu University, Varanasi, India

10.1 INTRODUCTION

Studies conducted in recent years have revealed the presence of emerging micro-pollutants that currently lack legislative regulation (Sarkar et al., 2022). These substances are found in everyday products, such as personal care items, medications, plasticizers, insecticides, and surfactants, and are difficult to remove quickly (Wilkinson et al., 2017). To preserve the ecological status of the aquatic environment, the European Union (EU) has designated 33 chemical substances found in surface water (including freshwater and marine water bodies) as priority pollutants that must be eliminated within 20 years (Hena et al., 2021). Diclofenac, Phthalates, carbamazepine, iopamidol, triclosan, clofibric acid, ibuprofen, musks, and other pharmaceutical and personal care products (PPCPs) were added to the list of priority pollutants in 2007. PPCPs are a broad class of chemicals with everyday uses, mainly in personal hygiene, cosmetics, and medicine (Büning et al., 2021).

The pharmaceutical industry has significantly improved human well-being, but it is one of the most resource-, energy-, and pollution-intensive industries (Parvatker et al., 2019). According to the World Health Organization, global per capita pharmaceutical consumption, measured in defined daily doses (DDD), has increased steadily over the past decade and is expected to continue to grow due to the aging population and increased global access to healthcare (Fermini & Bell, 2022). Approximately 3,000 medications are regularly used on the EU market, and their use is continually increasing worldwide (Patel et al., 2019). PPCPs have attracted researchers' attention as they have become increasingly abundant in the environment and pose significant ecological risks, including endocrine disruption, carcinogenicity, bioaccumulation, antibiotic resistance, and clinical toxicity (Priya et al., 2022).

Despite pharmaceutical companies being the foremost contributors to the global economy with collective revenue of 1.27 trillion US dollars, this growth increases environmental pressure resulting from various stages of PPCP's life cycle, namely, sourcing, synthesis, packaging, distribution, consumer, and post-consumer use and disposal (Bom et al., 2019). The pharmaceutical industry also contributes to the carbon footprint, emitting about 1.9 million tons of CO_2 annually (Bandichhor, 2018). The pharmaceutical industry generates substantial amounts of pollutants and waste by-products, including depleted reagents, air pollutants, and contaminated solvents, which are significant contributors to energy consumption, resource depletion, and emissions (World Economic Forum, 2019). As a result,

DOI: 10.1201/9781003637158-10

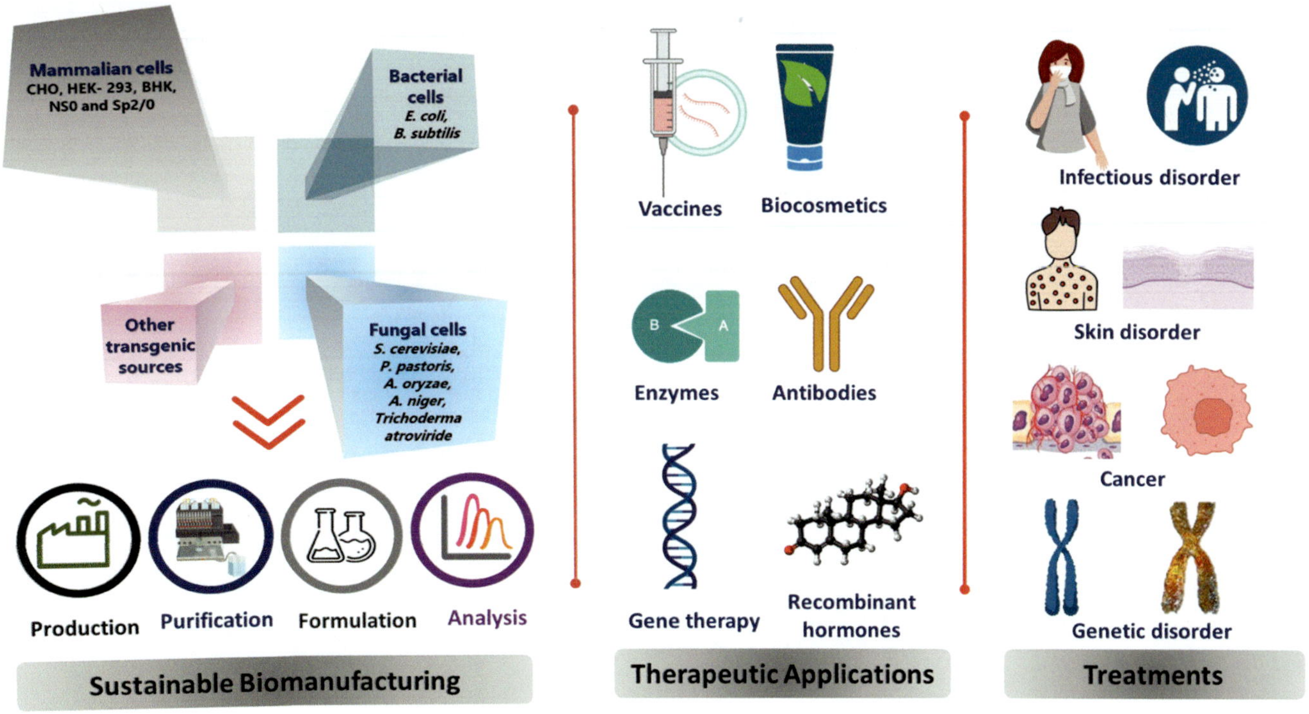

FIGURE 10.1 Sustainable biomanufacturing of PPCPs and their therapeutic applications: a step toward a linear to circular economy.

sustainability is a significant concern for the pharmaceutical sector, necessitating greater attention.

The pharmaceutical sector has transitioned from fossil-based to bio-based ingredients to align it with sustainable development goals (SDGs), move from a linear to a circular economy, and promote environmental sustainability (Figure 10.1). The current market size for biopharmaceuticals is huge and grew from 139 billion US dollars in 2012 to 346 billion US dollars in 2021, with a compound annual growth rate (CAGR) of 10.7% is expected to reach 975 billion US dollars by 2030, with a CAGR of 12.2%. Along with this, the proportion of biopharmaceuticals in the total pharmaceutical market also increased from 24% in 2014 to 33% in 2020 and is anticipated to reach 41% in 2028 (Etit et al., 2024). Considering the global emphasis on sustainability and the rising demand for natural personal care products and cosmetics, this chapter focuses on advanced biomanufacturing approaches that enable sustainable production of PPCPs and cosmetics, in line with the concepts of green chemistry and the circular bioeconomy.

10.2 SUSTAINABLE SYNTHESIS OF PPCPS

Biopharmaceuticals and biocosmetics are derived from biological sources and play a vital role in modern medicine. Biopharmaceutical manufacturing is a complex process that requires rigorous control over production parameters to guarantee consistent quality, safety, efficacy, and yield. The antibacterial qualities of marine algae are a noteworthy example, having been demonstrated to be effective against a range of bacterial and fungal species, including *Salmonella typhimurium*, *Escherichia coli*, *Enterococcus faecalis*, and *Staphylococcus aureus* (Alsenani et al., 2020). The antibacterial properties of marine algae offer potential prospects for developing novel biopharmaceutical treatments, provided they are carefully monitored and controlled during production. Microalgae and cyanobacteria are abundant sources of bioactive chemicals with various properties. These organisms are expected to continue playing a significant role in the commercial synthesis of valuable bioactive substances. They can synthesize various chemicals, making them a desirable source for biopharmaceutical production, cosmetics ingredients, and other uses (Table 10.1).

ALS, ammonium lauryl sulfate; APGs, alkyl polyglucosides; DMDM, dimethylo dimethyl; EDDS, Ethylene diamine-N,N′-disuccinic acid; EDTA, Ethylene diamine tetraacetic acid; PEG, Propylene glycol; QACs, quaternary ammonium compounds; SLES, sodium laureth sulfate; SLS, sodium lauryl sulfate; UV, ultraviolet.

Biopharmaceutical companies are investing in innovative technologies, such as biotechnology and single-use technology (SUT), to enhance operational efficiency, reduce facility costs, and mitigate the environmental impact of traditional systems and chemical processes. Biotechnology harnesses the natural ability of living cells to produce desired compounds, resulting in improved output while minimizing waste directly. It could be easily expanded to satisfy the increasing demand for pharmaceutical raw materials (Shokoohi and Attar, 2024). SUT eliminates the need for substantial quantities of water and energy for cleaning, compared to traditional systems, resulting in a more sustainable production model that aligns with the increasing importance of sustainability in the biopharmaceutical sector.

TABLE 10.1
Cosmetic Ingredients, Sustainability Issues, and Sustainable Alternatives (Bom et al., 2019)

Cosmetic Ingredient	Type	Traditional Examples	Sources	Sustainability Issues	Sustainable Alternatives
Surfactants	Anionic; Cationic; and non-ionic	Alkylsulfates (SLS, ALS, and SLES), etc. QACs PEG	Synthetic or naturally derived synthetic	Petrochemical origin, non-biodegradable, and harmful to marine life	Naturally derived surfactants: Lecithin (extracted from soybeans or egg yolk), plant saponins obtained from *Yucca* and *Quillaia*, APG, and glucose (from corn or potatoes)
Moisturizer	Hydrocarbon based fatty alcohols; Ester-based silicones and derivatives	Squalane cetylalcohol; Octyldodecanol Isopropyl myristate; Olive oil caprylic/capric triglyceride; Dimethicone cyclomethicone	Naturally derived from squalene (animal-based or plant-based); Spermaceti is found in sperm whales or dolphins; Synthetic or naturally derived from animal or plant sources; Synthetic	Animal exploitation or deforestation; Petrochemical origin pollution due to the use of heavy metal catalysts; Animal exploitation or deforestation; and Bio-accumulative	Amaranth oil, butters, and waxes or naturally derived esters. Fatty alcohols derived from Sustainably sourced plants. Application of biocatalytic processes. Blend of coconut alkane and coco-caprylate/caprate, sourced from vegetable oils
Exfoliators	Physical; Chemical	Plastic microbeads, alpha hydroxy acids, beta hydroxy acids	Synthetic naturally derived	Non-biodegradable; Water insoluble; Bio-accumulative; Harmful to the aquatic environment	Mineral microbeads (e.g., silica, bentonite, pumice, mica, montmorillonite, salt, and quartz sand), Biodegradable microbeads (from cellulose),exfoliator particles(obtained from nuts or barks/shells) organic (sugar) –
Solvents	–	Propylene glycol; Butylene glycol glycerin	Synthetic	Skin irritation and sensitization; weaken protein and cellular structures	Biofermentation-derived glycols, 1,3-propanediol (natural propylene glycol, lack irritation and sensitization potential, even at high concentration)
Preservatives	–	Parabens, DMDM hydantoin, imidazolidinyl urea, and glutaraldehyde; Phenol derivatives	Synthetic	Negative health impacts	natural preservatives, mainly plant extracts such as rosemary and *Santolina chamaecyparissus* extract
Colorants	Inorganic pigments; Organic pigments; Organic dyes	Iron, titanium, zinc, and chromium oxides (mica); Curcumin, betalain, red lycopene, capsanthin, and capsorubin; and Carmine, azo dyes, quinoline, and xanthone	Natural or naturally derived; Plant or animal-based synthetic	Deforestation; Animal exploitation and Petrochemicals origin	Utilizing plants grown under sustainable conditions for colorants; Microbial-derived colorants, either natural or produced using genetically engineered strains grown in renewable carbon and hydrogen sources

Fragrances	Natural synthetic	Derived from plant essential oils, nature-identical, chemically modified, naturally derived compounds	Natural; synthetic	Massive deforestation and animal exploitation concerns; Skin sensitization	Obtained from essential oils from plants grown under sustainable conditions
Chelating agents	–	Citrates, phosphonates, and EDTA	Synthetic	Deleterious to human health, non-biodegradable	Phytic acid (nuts, seeds, and grains), EDDS (biodegradable structural isomer of EDTA)
UV filters	Physical; Chemical	ZnO, TiO_2, octylcrylene, avobenzone, octinoxate, octisalate, oxybenzone, and homosalate	Synthetic	UV filters reach coastal waters, bioaccumulation and biomagnification affect the marine ecosystem	Phytochemicals (polyphenols- *Rosa damascene* flower extracts, aromatic compounds, and aromatic and flavonoid compounds from saffron, *Crocus sativus*)
Antioxidants	–	Butylated hydroxytoluene, butylated hydroxyanisole	Synthetic	Skin allergy, carcinogenicity, and endocrine disruption	Vitamins (tocopherol, retinyl palmitate, and ascorbic acid) obtained from plant source, *Camellia assamica* seed oil, flavonoids, and pomegranate polyphenols
pH adjustment	–	Alkaline salts	Synthetic	Skin irritation and sensitization	Sustainably derived alpha hydroxyacids, such as lactic or citric acid, from plant and animal tissues

10.2.1 Biotechnological Approaches

The main drawbacks of traditional techniques are their poor efficacy, purity, and efficiency, which hinder the development of novel medications and therapies. Additionally, conventional methods are susceptible to fluctuations in availability and pricing because they primarily rely on limited natural resources. Biotechnology enables the manipulation of living systems and harnessing of biological functions, making it possible to produce complex chemicals that are difficult to synthesize using conventional methods. Likewise, biotechnology promotes a more resilient and sustainable pharmaceutical raw material supply chain by reducing the demand for limited natural resources. With the potential to modify the manufacturing of pharmaceutical raw materials, the rise of biotechnology represents a paradigm shift as the limitations of conventional methods become more apparent (Shokoohi and Attar, 2024).

10.2.1.1 rDNA Technology

The development of recombinant DNA (rDNA) technology has revolutionized pharmaceutical production by assisting in the design of customized DNA molecules. The rDNA technique involves manipulating organisms at the gene level to produce specific proteins, such as insulin, growth hormones, and antibodies, on a large scale (Khan et al., 2016). Additionally, this technology has significantly impacted the manufacturing of several other pharmaceutical products, including vaccines, interferons, interleukins (ILs), and erythropoietin, while mitigating the risk of allergic reactions associated with animal-derived sources. It enables the development of specific cell lines that can produce antibodies, allowing for their consistent quality production on a large scale (Parasuraman et al., 2024).

Using rDNA technology, microorganisms such as *E. coli, Bacillus*, Yeast, and *Pichia* are exploited as production platforms for recombinant proteins for biopharmaceuticals. The first step is to choose the expression system for protein amplification, considering important factors such as protein quality, functionality, production speed, and yield. An ideal expression system is crucial for high-level therapeutic production, allowing for highly regulated and effective transcription. Bacteria have been the most effective producers of heterologous proteins due to several factors: i) Well-developed molecular tools for genetic manipulation, ii) Annotated genomes and metabolic pathways, iii) High cell density cultivation capacity and growth rate, and iv) High yield of recombinant proteins. The earliest platform for producing recombinant proteins was *E. coli* (Terpe, 2006), which remains in use. *E. coli* K12 and its derivatives are the main strains utilized to produce recombinant therapeutics. The *E. coli* expression system is cost-effective, and the ease of bacterial culture makes it a preferable choice for producing therapeutic proteins on a laboratory scale and in industry. The expression of valuable heterologous proteins at high expression rates is facilitated by rDNA technology. In *E. coli*, the overexpression of proteins often leads to the formation of inclusion bodies (IBs), which can be used to produce therapeutics such as interferons, ILs, and Fc-fusion proteins (Matejczyk, 2014). Likewise, *Ralstonia eutropha* has been used to develop an improved Gram-negative host for recombinant protein production (Barnard et al., 2004). Other hosts that have been exploited for recombinant protein production include *Pseudomonas putida* for antibody fragments and the Pfenex system, which uses *Pseudomonas fluorescens* for trimeric TNF-alpha (Squires and Lucy, 2008).

Yeast strains have been widely used as hosts for the production biopharmaceuticals, enabling higher protein titers (>1 g/L) in a few fermentation steps. The biopharmaceuticals produced by *Saccharomyces cerevisiae* include insulin and its analogs, human serum albumin, hepatitis vaccines, and virus-like particles (Nielsen, 2013). Similarly, specific human-like glycoforms of proteins have been produced successfully by engineering *Pichia* species (Martínez et al., 2012).

10.2.1.2 Cell Culture Technology

Cell culture technology involves cultivating animal or plant cells under controlled conditions to generate a variety of complex molecules of medical importance. The production and development of life-saving products, including vaccines, monoclonal antibodies, and cell therapy, depend substantially on this approach (Li et al., 2010). Large-scale plant cell culture under submerged conditions for manufacturing bioactive phytochemicals, active cosmetic additives, food supplements, and nutraceuticals is known as plant cell culture technology (PCCT). Mass propagation of plant cells, tissues, organs, or genetically modified plants in a controlled environment through PCCT is being up-scaled for the commercialization of a broad spectrum of biopharmaceuticals, such as plant-made pharmaceuticals (PMPs, including secondary metabolites and recombinant therapeutic proteins), plant-made vaccines (PMVs), and plant-made industrials (PMIs, including technical enzymes, antibodies, cosmetic active ingredients, food supplements, and nutraceuticals) (Sil & Jha, 2014). PMPs available in the market include anti-inflammatory drug shikonin, anticancer drug paclitaxel, anticholinergic alkaloid scopolamine, anti-inflammatory protoberberines, anti-inflammatory phenolic rosmarinic acid, anticancer compound geraniol, anticancer and antiviral agent podophyllotoxin, cardioprotective triterpene saponins mix Ginseng, tyrosinase inhibitor arbutin, and immunostimulant polysaccharides from *Lithospermum erythrorhizon*, *Taxus* spp., *Duboisia* spp., *Coptis japonica*, *Coleus blumei*, *Geramineae* spp., *Podophyllum* spp., *Panax ginseng, Catharanthus roseus, and Echinaceae* spp., respectively (Sharma et al., 2014).

Between 1987 and 2020, the United States and the EU approved over 207 mammalian biopharmaceutical products. The application of mammalian cell expression systems to produce biopharmaceuticals began in 1987 with the approval of tissue plasminogen activator (Activase®) derived from Chinese hamster ovary (CHO) cells for treating myocardial infarction. In recent decades, a variety of mammalian cell expression systems have been used to create biopharmaceuticals, including epithelial cells of the human embryonic kidney (HEK-293), baby hamster kidney (BHK), and murine myeloma lymphoblastoid type cells, such as NS0 and Sp2/0 (Al-Majmaie et al., 2022).

10.2.1.3 Fermentation

The fermentation process utilizes the metabolic powers of microorganisms, such as yeast and bacteria, to produce various essential medicinal compounds. The culture medium can be composed of biological materials, such as grains, sugar, molasses, yeast, or synthetic nutrients, such as potassium phosphate, ammonium chloride, ammonium nitrate, or petrochemicals, such as paraffin, followed by extraction. Extraction using physical and mechanical methods, such as heat or maceration, as well as chemical methods, such as petroleum solvents, ammonia, strong acids, strong bases, and/or ion exchange, can be applied, ultimately yielding the product in crystallized form (Kusumoto, 2001). Antibiotics, vitamins, amino acids, and enzymes are among the raw materials used as ingredients in cosmetics, pharmaceuticals, and medical products produced using this versatile method. The first antibiotic, penicillin, was derived from the microbial fermentation of the fungus Penicillium notatum, leading to a groundbreaking discovery that revolutionized the treatment of bacterial infections (Hook, 2006). Improved strains of mold fungi produce the necessary antibiotics through fermentation (Baranova et al., 2023). Beta-lactams, including *Penicillins* and *Cephalosporins*, are derived from the fermentation products of *Penicillium chrysogenum* and *Acremonium chrysogenum*, respectively, and are among the most significant fungal antibiotics (Liu et al., 2022).

The use of bacteria, yeasts, or microalgae for the fermentation-based synthesis of vitamins offers several benefits over conventional chemical synthesis techniques. Vitamins produced through biological processes may be preferred for internal and external applications due to their safety, biological activity, and absorption rate. Microbial fermentation, using bacteria such as *E. coli* and *S. typhimurium*, is used to synthesize Vitamin B12, which is vital for nervous system function. The production method of Vitamin C is primarily based on single-bacteria fermentation, eliminating the reliance on associated bacteria (Wang et al., 2021). With the ongoing advances in synthetic biotechnology and metabolic engineering, the industrialization of vitamin fermentation is expected to evolve into a safer, more sustainable manufacturing process. Three primary biotechnological techniques are mainly used in the industrial production of amino acids: i) direct fermentation, ii) semi-fermentation, and iii) enzymatic. Fermentation offers economic and environmental benefits due to its simple procedure and the use of affordable, renewable carbon sources, such as glucose, molasses, and sucrose.

Okafor (2007) reported considerable production of l-glutamine from sugar and ammonia by *Corynebacterium glutamicum*. Subsequently, it was also discovered that a homoserine auxotrophic mutant of *C. glutamicum* produced a significant amount of l-lysine through fermentation. Almost all amino acids can be synthesized by developed mutants of *C. glutamicum* or *E. coli*. *Brevibacterium* spp. (*B. divaricartum* and *B. alanicum*), *Microbacterium* spp. (*M. flavum* var. *glutamicum*), *Arthrobacter* spp. (*A. globiformis; A. aminofaciens*) are also used in amino acid biomanufacturing (Ivanov et al., 2013).

Enzymes used in the pharmaceutical industry are produced by fermenting suitable strains of bacteria and fungi, due to their ease of handling, rapid growth rates, and scalability in fermenters. The biotechnology industries around the world use an array of microorganisms to produce enzymes in very high yields, including bacteria, such as *E. coli*, *Bacillus subtilis*, lactic acid bacteria, and filamentous fungi such as *Aspergillus oryzae*, *Aspergillus niger*, *Trichoderma atroviride*, and *Yeasts*, for example, *S. cerevisiae*, *Pichia pastoris*, etc., that have been enhanced through genetic engineering. Two primary methods used in enzyme manufacturing are submerged fermentation (SmF) and solid-state fermentation (SSF). The nutrient components are either dissolved or remain suspended in an aqueous medium during submerged fermentation. Submerged cultures are used to cultivate aerobic bacteria. The synthesis of enzymes using SSF is enormous and has several perks compared to the SmF process, including a higher product titer, reduced wastewater production, the use of basic fermentation equipment, less trained labor, and more. In SSF processes, agro-industrial residues are frequently used as substrates, including but not limited to sugar cane bagasse, wheat bran, wheat straw, rice bran, rice straw, sawdust, corncobs, cassava waste, banana waste, palm oil mill waste, and oil cakes (Meghwanshi et al., 2020)

10.2.1.4 Biocatalysis

Biocatalysis harnesses an enzyme's catalytic strength and offers an intricate and effective means to transform raw materials into specific pharmaceutical constituents. It upgrades the efficiency and sustainability of several steps within the pharmaceutical manufacturing process (Bell et al., 2021). Biocatalysts, including reductases, oxidases, hydrolases, lyases, isomerases, and transaminases, produce pharmaceutical intermediates and pharmaceuticals. A new isolate of the aerobic, Gram-negative soil bacterium Pseudomonas pseudoalcaligenes XW-40 exhibits reductase properties, producing an intermediate reduced product for the manufacturing process of Sitagliptin, used in the treatment of type 2 diabetes. Oxidases are used in the industrial production of telaprevir and esomeprazole, which are hepatitis C virus protease inhibitors and proton pump inhibitors, respectively. *Penicillium simplicissimum* and bacterial laccases produce vanillyl alcohol oxidase, which has applications in the production of phytoestrogen pinoresinol, demonstrating a protective effect against multiple health disorders. Novel (+)-ɣ-lactamases have been found to produce a chiral intermediate for drugs carbovir and abacavir used for the treatment of HIV infection and swine flu, respectively (Sun et al., 2018) (Table 10.2).

TABLE 10.2
Classes of Biocatalysts, Microbial Strains, Pharmaceutical Intermediates, Pharmaceutical and Their Application

Classes of Biocatalysts	Microbial Strain	Pharmaceutical Intermediate	Pharmaceutical	Application
Reductases	1. *Pseudomonas pseudoalcaligenes XW-40* 2. *Rhizopus microsporus* var. *rhizopodiformis ZJPH1308*	HTPP	Sitagliptin	Treatment of type 2 diabetes
Oxidases-Alcohol oxidase; Amine oxidases	1. *Penicillium simplicissimum* 2. *Aspergillus niger*	Coniferyl alcohol	Phytoestrogen pinoresinol; Solifenacin; and Levocetirizine	Protective effect against multiple health disorders
Hydrolases	*Bradyrhizobium japonicum* USDA 6	(+)-ɣ-Lactamases	Carbovir; Abacavir	Treats HIV infection Swine flu
Lyases	*Erwinia herbicola*	_	(L-DOPA),	Treatment of Parkinson's disease
Isomerases cellobiose 2-epimerase (CsCE)	*Caldicellulosiruptor saccharolyticus*	_	Lactulose	Treatment of constipation
Transaminases	*Vibrio fluvialis*	Chiral amines	Spiroindolone	Treatment of malaria

Source: Adapted by Sun et al., 2018.
HTPP-(S)-3-Hydroxy-1-(3-(trifluoromethyl)-5,6-dihydro[1,2,4]triazolo[4,3-a]pyrazine-7(8H)-yl)-4-(2,4,5trifluorophenyl);
L-DOPA-(L-3,4-dihydroxyphenylalanin.

10.2.2 Single-Use Technology

SUT refers to disposable elements used in the biomanufacturing process that are designed for single use before being discarded, including bioreactors, mixers, tubing, filters, and bags, intended to replace conventional stainless-steel equipment. In the early 2000s, SUTs emerged as the biopharmaceutical industry recognized the need for more flexible and rapid manufacturing solutions. SUTs have been widely adopted in the biopharmaceutical industry for biologics, vaccines, and personalized pharmaceuticals due to advancements in materials science, production techniques, and regulatory acceptability. Biopharmaceutical businesses of all sizes, from large multinational enterprises to small biotech start-ups, have adopted SUT to overcome the limitations of conventional stainless-steel systems, including high capital costs, extensive setup times, and intricate cleaning and validation procedures. SUT reduces cleaning, validation, and maintenance costs, as there is no need for large-scale clean-in-place (CIP) and sterilize-in-place (SIP) systems, which in turn minimize water, chemical, and energy consumption. High levels of cleanliness and purity are ensured by the pre-sterilized nature of the components, which reduces the risk of contamination and is essential for the integrity of biopharmaceutical products. With SUT, resource-intensive processes are no longer necessary, resulting in a more sustainable production model that aligns with the biopharmaceutical industry's growing emphasis on sustainability (Soni & Patel, 2024).

Scalability, or the ease of increasing or decreasing production according to demand, is another significant advantage of single-use systems, which is especially helpful to biopharmaceutical companies that need to manufacture batches of varying sizes. The rapid development and distribution of COVID-19 vaccines underscored the importance of SUT in accelerating the biomanufacturing process. SUT's adaptability and speed during the epidemic enabled vaccine producers to increase output to satisfy demand worldwide swiftly.

SUT reduces the need for energy-intensive cleaning and sterilization procedures, but at the expense of using more disposable materials that require careful handling. This balance is essential for SUT to be used sustainably in biomanufacturing (Manda et al., 2018). To address key issues, such as improved compatibility with biologics and reduced leachability associated with SUTs, and to enable their use in biomanufacturing, considerable material improvements and integration with digital technologies must be prioritized. Furthermore, advances in bioplastics and biodegradable materials are receiving increased attention to lessen the adverse environmental effects of single-use systems by offering more sustainable substitutes.

10.2.3 Revamping PPCPs and Cosmetics via Bio-Based Ingredients

Various pharmaceutical industries have transitioned toward bio-based ingredients, driven by a vision to achieve environmental sustainability and meet consumer demand for natural products. They include formulations derived from plants, algae, and fungi. These natural alternatives are rich in pigments, antimicrobial properties, various antioxidants, and different types of polysaccharides, which contribute to the success and effectiveness of any product, thereby promoting environmental sustainability (Cavinato et al., 2017).

FIGURE 10.2 Comparison of sustainable biomanufacturing and conventional approaches for PPCPs production.

These natural alternatives are skin-compatible, biodegradable, and serve as a cleaner and greener substitute for the environment and human health. This section highlights the importance of several plant, algae, and fungi-derived compounds. Each part explores the diversity among several bio-based groups with different functional roles. According to Mukherjee et al., (2011), bio-based groups possess antimicrobial, anti-aging properties, and can serve as substitutes in hair care products and cleansing agents (Figure 10.2).

10.2.3.1 Plant-Derived PPCPs

Plants have been utilized for centuries as a bioformulation in various PPCPs, fulfilling diverse functional and medicinal needs. They are rich in diverse bioactive compounds, highlighting the importance of plant-derived PPCPs and positioning them as a focal point for the sustainable development of the cosmetics and pharmaceutical industries. Some of the categories are discussed below.

10.2.3.1.1 Antioxidants and Anti-Aging Agents

Plants are rich in several antioxidants that help combat reactive oxygen species (ROS) and oxidative stress, which are known to contribute to aging. One such plant is aloe vera, which is enriched with anti-aging properties. The nanoparticles obtained from aloe skin protect from UV-induced damage, enhance skin pliability and growth, and help in tissue repair. Similarly, grape (*Vitis vinifera)* seed extracts help maintain skin integrity by hindering lipid peroxidation. Another example is *Camellia sinensis* leaves supplemented with catechins (polyphenols), which display antioxidant values (Heinrich et al., 2011). These catechins have been attributed to therapeutic and medicinal values as they have been widely used as free radical scavengers. The main catechins are epigallocatechin (EGC), epicatechin (EC), and gallocatechin (GC) (Heinrich et al., 2011).

10.2.3.1.2 Skin-Whitening and Brightening Agents

Nowadays, cosmetology focuses on skin enhancement, adjusting skin tone, and treating skin pigmentation, among other areas. Skin-whitening and brightening treatments are in demand. Compared to primitive synthetic agents, such as hydroquinone, plant-based formulations offer skin-compatible and safer options. These older synthetic depigmenting agents are now banned in various countries due to concerns about their cytotoxicity. These bioactive compounds typically inhibit tyrosinase, the rate-limiting enzyme in melanin production, thereby reducing melanin synthesis (Faccio, 2020).

One such example is *C. sativus* (saffron), enriched with kaempferol, a type of flavonol that has been used to inhibit tyrosinase, thereby minimizing melasma and improving skin health. Clinical and in vitro trials have shown that saffron extracts can reduce melanin synthesis and exhibit antioxidant synergy, thereby amplifying their brightening effects while offering photoprotection (Kubo & Kinst-Hori, 1999). Another potent skin-lightening plant is *Morus nigra* (black mulberry). Its root bark is rich in phenolic compounds such as morachalcone A and oxyresveratrol, which have shown comparable tyrosinase inhibition to synthetic agents (Zheng et al., 2010).

10.2.3.1.3 Moisturizers and Emollients

The key to healthy, glowing skin is proper moisturization and hydration. Three categorical classifications of moisturizers are as follows: 1) Emollients, 2) Humectants, and 3) Occlusives. Emollients are used to fill the spaces between skin cells with lipids, thereby softening the skin surface. Humectants draw and attract moisture into the cornified layer from the surrounding environment, while occlusives prevent moisture loss. For instance, *Moringa oleifera* (drumstick tree) seed oil enriched with tocopherols and oleic acid nourishes the skin. It also contains antioxidants, which mitigate lipid peroxidation in cells, conferring anti-aging properties. The oil in the plant impersonates sebum, thereby nourishing the skin without clogging pores (Anwar & Bhanger, 2003). Additionally, the aloe vera gel acts as a humectant, thereby hydrating and moisturizing the skin. It comprises acemannan and glucomannans (polysaccharides), ideal for all skin types. Moreover, the wound-healing and anti-inflammatory attributes of aloe vera mark its potential use in various skin formulations and products.

10.2.3.1.4 Natural Surfactants and Cleansing Agents

Several personal care products contain synthetic surfactants and cleansing agents, which are non-biodegradable and can act as skin irritants, thereby impeding the health of the environment. Alternatively, saponins, which are secondary metabolites (glycosides) present in various plants, possess surfactant properties. They have already been used as natural cleansing agents in conditioners, shampoos, and skincare products. They are skin-compatible and biodegradable as well. For instance, *Sapindus mukorossi (soapnut), enriched with saponins*, is used for its foaming properties (Bezerra et al., 2021). Plants such as Glycine max (soybean) and Chenopodium quinoa are also used for obtaining biosurfactants due to *their* emulsifying properties. They provide a sustainable alternative to synthetic and non-natural ingredients, paving the way toward a greener world.

10.2.3.1.5 Anti-Inflammatory and Soothing Agents

Inflammation is a common indicator of the underlying problem for nearly all skin disorders, ranging from mild acne to skin allergies and hypersensitivity, to UV-induced damage, eczema, melasma, and various other conditions. Inflammation is a standard indicator of the underlying problem. A solution to this could be plant-derived formulations already in practice that substantiate soothing and anti-inflammatory ability.

The roots and leaves of Taraxacum officinale (dandelion) are full of chlorogenic acid, which is used as an antioxidant and anti-inflammatory agent. Chlorogenic acid reduces ROS-mediated cellular damage and regulates Nuclear Factor kappa B (NF-κB) signaling, an essential pathway in inflammation (Schütz et al., 2006). Dandelion extracts are increasingly used in formulations targeted at acne-prone, inflamed, or UV-stressed skin. Another example is the petals of *Rosa gallica* (French rose), enriched with flavonoids, anthocyanin pigment, quercetin, and essential oils, which act as an anti-inflammatory agent by inhibiting Tumor Necrosis Factor (TNF)-α and IL-6 (Kazemipoor et al., 2012). The topical application of these rose extracts reduces edema and erythema, making them suitable for use in soothing creams and serums.

10.2.3.1.6 Antimicrobial Agents

Certain plant extracts possess antimicrobial properties, making them valuable in formulations for preventing or treating skin infections. Compounds such as caffeic acid, protocatechuic acid, and vanillic acid have demonstrated broad-spectrum antimicrobial activity against bacteria, including *E. coli*, *S. aureus*, and *Bacillus cereus* (Rybczyńska-Tkaczyk et al., 2023). These compounds thus ensure the durability and longevity of cosmetics, making them skin-compatible and a safer option. Another example is the extract of Bacopa monnieri (Brahmi), which exhibits antibacterial properties against *S. aureus*. The extracts have demonstrated their potential in various skin products for combating microbial infections. They also exhibit antifungal properties against *Candida albicans*, which has led to their inclusion in various topical products (Mkk et al., 2019).

10.2.3.1.7 Hair Care Applications and Fragrance/Aromatherapy

Plant-derived formulations are also used in hair care products, fragrances, and perfumes. One such example is *Jasminum grandiflorum* (jasmine). It is enriched with essential oils, including jasmine, benzyl alcohol, and other volatile compounds, which make it suitable for use in perfumes and deodorants (Ahmed et al., 2016). Also, clove (*Syzygium aromaticum*) is used for its antibacterial and analgesic properties in various aromatherapy products. The eugenol and several antioxidants in cloves offer therapeutic benefits, enhancing digestive and nutritional values (Cortés-Rojas et al., 2014). Pomegranate extracts are rich in tannins and polyphenols, which are used extensively in hair dyes and several other hair care formulations to minimize discoloration and bleaching caused by ultraviolet A (UVA) exposure (Sharma et al., 2014).

10.2.3.2 Algae-Derived PPCPs

Algae-based bioformulations have a wide range of applications in cosmetics and PPCPs. Both microalgae and macroalgae produce abundant natural ingredients. Species, such as *Chlorella* and *Spirulina*, which belong to the microalgae group, are rich in proteins, pigments, and vitamins. They are also considered a source of single-cell protein (SCP). Bioactive compounds derived from them are used in various skin care products, providing anti-aging, skin-rejuvenating, antioxidant, and therapeutic properties. Pigments, such as phycobiliproteins, used as natural colorants, offer an alternative to non-natural products in cosmetics (Dini, 2023).

Members of the Chlorophyceae (green algae), Rhodophyceae (red algae), and Phaeophyceae (brown algae) consist of agar, alginates, carrageenan polysaccharides, and several other polysaccharides, which exhibit anti-inflammatory and antioxidant properties, thereby improving skin health. These compounds are used as stabilizers, thickeners, moisturizers, emollients, etc., in cosmetics (Fonseca et al., 2023). Some of the categories where algae-derived bio-actives are used are as follows:

10.2.3.2.1 Moisturizers and Emollients

Species, such as *Sargassum japonica* and *Chondrus crispus, have a calming effect, regulating water circulation and distribution in the skin through the use of carrageenan and fucoidan polysaccharides* (Joshi et al., 2018). According to Dini (2023), these compounds provide hydration and nourishment to the skin, locking in moisture content and thereby giving the skin a smooth texture and longevity. Additionally, agar from *Gracilaria* and alginates from *Fucus* species have surpassed glycerin, exhibiting superior water-holding properties (Jahan et al., 2017). They also exhibit antioxidant properties, thereby preventing aging and skin disorders. Moreover, by impeding lipid peroxidation, these aid in maintaining the sensory properties of cosmetics.

10.2.3.2.2 Antioxidants and Anti-Aging Agents

Several algae serve as a reservoir of antioxidants, including vitamins, carotenoids (such as fucoxanthin), and phlorotannins, which are utilized against ROS and help reduce oxidative stress. As a result, they help delay skin aging and contribute to healthy, glowing skin. Brown algae, such as *Macrocystis pyrifera* and *Ecklonia stolonifera*, are enriched in polyphenols, which exhibit antioxidant properties and can inhibit enzymes responsible for skin aging. Phlorotannins help inhibit the tyrosinase and hyaluronidase enzymes involved in various biochemical pathways associated with aging and skin issues (Ariede et al., 2017).

10.2.3.2.3 Skin-Whitening and Brightening Agents

Melanin is responsible for skin pigmentation, protecting mammalian skin from UV-induced damage and scavenging toxins. In this pursuit, compounds derived from algae, such as phlorotannins and fucoidans, help reduce melanin synthesis and serve as skin-whitening and brightening agents (Costa et al., 2023). For instance, *Ecklonia maxima* and *Fucus vesiculosus* are known to minimize wrinkles and display anti-melanogenesis properties. They can diminish the melanin index, nourish skin, and reduce hyperpigmentation, thus serving as a potent source of skin-whitening agents (Pereira, 2018). Thus, these skin care products fulfill modern-day beauty standards by providing an opalescent, homogeneous complexion.

10.2.3.2.4 UV Protection and Photoprotective Agents

According to Aslam et al. (2021), algae mass produces scytonemin and mycosporine-like amino acids (MAAs), which help absorb UV rays, thereby protecting the skin from sunburn and photodamage. Scytonemin is a pigment in cyanobacteria that absorbs UVA, exhibiting antioxidant properties, thus marking its potential use in makeup products and sunscreens (Ding et al., 2022). Similarly, MAAs, such as mycosporine taurine, porphyra-334, shinorine, palythine, and mycosporine glycine, have cyclohexenimine or cyclohexenone ring structures that help them absorb UV radiation, enabling them to act as photoprotective agents. They dissipate the energy without forming any ROS, serving as natural UV filters (Fuentes-Tristan et al., 2019). Certain red algae, such as *Porphyra haitanensis*, also have these photoprotective properties, enhancing dermal health and protection via sulfated polysaccharides.

10.2.3.2.5 Anti-Inflammatory and Soothing Agents

Certain algae act as anti-inflammatory agents, calming the skin and reducing the swelling, redness, puffiness, and underlying pain. MAAs are known to minimize the expression of genes such as cyclooxygenase (COX-2), interferons (IFN), ILs, and C-reactive protein, which play a key role in inflammation, indicating their potent use as an anti-inflammatory agent (Jesumani et al., 2019). Similarly, *Padina boergesenii (*a type of brown algae) is used as a soothing agent because it contains magnesium, silicon, diterpenes, and fatty alcohols, which help calm irritated skin and have anti-aging and moisturizing effects. The ethyl acetate fraction of this alga serves as a natural UV filter in cosmetics (Kalasariya et al., 2023).

10.2.3.2.6 Hair Care Applications and Fragrance/Pigments

Algal pigment, including carotenoids and chlorophylls, is a safer and greener alternative to synthetic hair dyes and coloring agents. These pigments, rich in antioxidant properties, provide long-lasting impact and vibrant shades, acting as natural colorants (Fonseca et al., 2023). Similarly, *Spirulina* and *Chlorella* are used in various hair care products due to their protein- and nutrient-rich compounds. These are beneficial in promoting hair growth, nourishment, strength, and overall health. The SCP, vitamins, and essential fatty acids strengthen the hair follicles, promoting hair vitality and overall health (Andrade et al., 2018).

10.2.3.3 Fungi-Derived PPCPs

Fungi-derived bioactive compounds from several fungal species are used in different personal care products and cosmetic industries. A diverse array of compounds, including chitin, glucans, ceramides, phenols, lipids, naphthoquinone derivatives, beta-carotene, and various other secondary metabolites, have been utilized in PPCPs to achieve sustainability. For example, ceramides produced from *Aspergillus fumigatus* and *Phellinus pini* help in hydration and act as a skin barrier, thus providing an alternative to animal-based products and validating bio-derived formulations (Gupta et al., 2019). Similarly, polysaccharides, such as chitin and glucans, are known for their antimicrobial properties as well as their moisturizing and soothing effects, which pave the way for their use in oral and skin care products.

Moreover, *Suillus* species, which contain beta carotene and phenols and are considered edible mushrooms, exhibit antimicrobial and antioxidative properties that can aid in the formulation of cosmetic products (Abdulhadi et al., 2023). Naphthoquinone derivatives from *Ophiocordyceps unilateralis* have outsmarted synthetic dyes and coloring agents, making cosmetics a greener and safer natural colorant (Gupta et al., 2019). Some of the categories are discussed below:

10.2.3.3.1 Moisturizers and Emollients

According to Wu et al. (2016), a polysaccharide from *Tremella fuciformis* (snow mushroom), structurally resembles hyaluronic acid, retaining moisture and thereby providing hydration and nourishment. Due to their small molecular size, these polysaccharides can penetrate more deeply into the skin, thereby keeping their moisture content. Similarly, *Pleurotus ostreatus* (oyster mushroom) and *Lentinula edodes* (shiitake), enriched with beta-glucans, serve as skin barriers that enhance hydration and exhibit soothing effects. These hydrophilic biopolymers function as humectants and emollients, nurturing overall dermal health.

10.2.3.3.2 Antioxidants and Anti-Aging Agents

Several fungal species show potent antioxidative properties because of the polysaccharides, phenols, ergothioneine, glycosides, flavonoids, and ascorbate present in them. For instance, *Inonotus obliquus* (Chaga) plays a dual role by enhancing skin durability and resilience, while also protecting against photo-oxidative stress (Shashkina et al., 2006). Another example is *Ganoderma lucidum* (Reishi), which is used in anti-aging topicals and serums and helps protect against ROS and oxidative stress. It contains ganoderic acids and other polysaccharides that help protect skin fibroblast cells against photo-oxidative damage. It also promotes collagen synthesis, providing strength and elasticity to the dermal tissue (Zhou et al., 2018).

10.2.3.3.3 Skin Brightening Agents and UV Protection and Photoprotective Agents

Compounds from *Schizophyllum commune* inhibit tyrosinase enzyme activity, making them suitable for use as skin-whitening and brightening agents. Various fungal bioformulations that inhibit melanin production demonstrate skin-lightening effects. Polysaccharides, such as schizophyllan, serve as a barrier against UV radiation, reducing photo-oxidative damage and skin inflammation, and acting as UV filters (Gupta et al., 2019). Similarly, cordycepin, produced from *Cordyceps militaris*, protects against UVB-induced damage and exhibits anti-melanogenesis properties.

10.2.3.3.4 Anti-Inflammatory and Soothing Agents

Various mushroom extracts offer rich anti-inflammatory benefits, helping calm, soothe, hydrate, and promote healthy skin. For example, several polysaccharides and hispidin obtained from *Phellinus linteus* protect from oxidative stress and reduce inflammation in dermal cells exposed to UV rays and pollution. Another instance is *Hericium erinaceus* (Lion's Mane), which downregulates the inflammatory cascade and cytokines, thus acting as a potent anti-inflammatory agent (Kim et al., 2013).

10.2.3.3.5 Hair Care Applications and Fragrance/Pigments

Fungal-derived ingredients, such as lactic acid produced by *Rhizopus* species, are utilized in hair care products for their exfoliating and moisturizing effects (Gupta et al., 2019). Lactic acid helps maintain scalp health and improve hair texture. Some fungi produce pigments and aromatic compounds that can be used to provide natural color and fragrance in cosmetics. Also, *Penicillium roqueforti* produces geosmin and other volatiles with earthy aromas that could inspire fungal-derived fragrance bases in natural perfumes (Alam et al., 2022). Therefore, integrating plant-, algae-, and fungi-derived ingredients into PPCPs and cosmetics not only enhances product functionality but also supports the industry's shift toward sustainability and natural product development.

10.3 CHALLENGES

Despite its unquestionable potential, the significant initial expenditure impedes the sustainable synthesis of therapeutic raw materials. The establishment and maintenance of biomanufacturing facilities require highly qualified workers, controlled surroundings, and specialized equipment (Broeze, 2006). Technology and equipment such as purification systems, bioreactors, fermentation apparatus, and quality control tools, as well as building a facility and installing the necessary equipment to maintain sterility and adhere to strict regulatory requirements, can result in high upfront costs. Extensive research and development efforts are often required for specialized research and development focused on specific pharmaceutical raw materials, further increasing the initial expense (Gargalo et al., 2020). Regulatory and safety considerations must also be taken into account; biotechnologically derived drugs are subject to stringent regulations and lengthy approval processes. The complexity and novelty inherent in biologics necessitate a thorough assessment of their safety and efficacy. Biomanufacturing involves working with living systems, including genetically modified organisms and potentially hazardous materials, which pose safety issues for workers and the environment. To overcome these obstacles and ensure the safe and responsible development of PPCP biomanufacturing, collaboration among scientists, industry stakeholders, and regulatory agencies is essential.

10.4 SUMMARY AND CONCLUSION

The growing emphasis on the sustainable synthesis of pharmaceutical raw materials marks the beginning of a new era in healthcare innovation. Since the pharmaceutical sector is a major polluter, attempts to synthesize PPCPs sustainably, along with the advancement of biopharmaceutical production technologies, could yield substantial progress. Conventional methods are becoming increasingly limited in their effectiveness, environmental impact, and ability to adjust to changing healthcare demands. The efficient and scalable manufacture of pharmaceutical raw materials with unprecedented accuracy and purity is made possible by biotechnology through methods such as microbial fermentation, cell culture technology, biocatalysis, and rDNA technology. Integrating biotechnology signifies a commitment to a more sustainable, tailored, and efficient healthcare future, as well as a paradigm shift in pharmaceutical production.

Genetic manipulation and SUTs drug discovery, production, and delivery could be revolutionized by utilizing the power of living systems. This will ultimately improve the lives of millions of people worldwide without adversely affecting the environment. Sustainable pharmaceutical production is a testament to human ingenuity and our collective commitment to sustainable development and advancing the frontiers of healthcare.

ACKNOWLEDGEMENT

The authors (Vaibhav Srivastava, Priyanka Gupta, and Aditi Mishra) are thankful to the Head of the Department, Department of Botany, University of Allahabad, for providing institutional support. Priyanka Gupta and Aditi Mishra are grateful to the Department of Biotechnology (DBT, Haryana) (DBT/2025-26/UOA/2729) and the Indian Council of Medical Research (ICMR, New Delhi) (No.3/1/3/BRET-2024/HRD (L1)) for granting Junior Research Fellowships, respectively.

REFERENCES

Abdulhadi, S. Y., Gergees, R. N., & Hasan, G. Q. (2023). Molecular Identification, Antioxidant efficacy of phenolic compounds, and antimicrobial activity of beta-carotene isolated from fruiting bodies of Suillus sp. *arXiv preprint arXiv:2303.08768*.

Ahmed, N., Hanani, Y. A., Ansari, S. Y., & Anwar, S. (2016). Jasmine (Jasminum sambac L., Oleaceae) oils. In *Essential Oils in Food Preservation, Flavor and Safety* (pp. 487–494). Academic Press.

Alam, K., Mazumder, A., Sikdar, S., Zhao, Y. M., Hao, J., Song, C., Wang, Y., Sarkar, R., Islam, S., Zhang, Y & Li, A. (2022). Streptomyces: The biofactory of secondary metabolites. *Frontiers in Microbiology*, *13*, 968053.

Al-Majmaie, R., Kuystermans, D., & Al-Rubeai, M. (2022). Biopharmaceuticals produced from cultivated mammalian cells. In *Cell Culture Engineering and Technology: In appreciation to Professor Mohamed Al-Rubeai* (pp. 3–52). Cham: Springer International Publishing.

Alsenani, F., Tupally, K. R., Chua, E. T., Eltanahy, E., Alsufyani, H., Parekh, H. S., & Schenk, P. M. (2020). Evaluation of microalgae and cyanobacteria as potential sources of antimicrobial compounds. *Saudi Pharmaceutical Journal*, *28*(12), 1834–1841.

Andrade, L. M., Andrade, C. J., Dias, M., Nascimento, C., & Mendes, M. A. (2018). Chlorella and spirulina microalgae as sources of functional foods. *Nutraceuticals, and Food Supplements*, *6*(1), 45–58.

Anwar, F., & Bhanger, M. I. (2003). Analytical characterization of Moringa oleifera seed oil grown in temperate regions of Pakistan. *Journal of Agricultural and Food Chemistry*, *51*(22), 6558–6563.

Ariede, M. B., Candido, T. M., Jacome, A. L. M., Velasco, M. V. R., de Carvalho, J. C. M., & Baby, A. R. (2017). Cosmetic attributes of algae-A review. *Algal Research*, *25*, 483–487.

Aslam, A., Bahadar, A., Liaquat, R., Saleem, M., Waqas, A., & Zwawi, M. (2021). Algae as an attractive source for cosmetics to counter environmental stress. *Science of The Total Environment*, *772*, 144905.

Bandichhor, R. (2018). *Hazardous Reagent Substitution-A Pharmaceutical Perspective*. Royal Society of Chemistry.

Baranova, A. A., Alferova, V. A., Korshun, V. A., & Tyurin, A. P. (2023). Modern trends in natural antibiotic discovery. *Life*, *13*(5), 1073.

Barnard, G. C., Henderson, G. E., Srinivasan, S., & Gerngross, T. U. (2004). High level recombinant protein expression in Ralstonia eutropha using T7 RNA polymerase based amplification. *Protein Expression and Purification*, *38*(2), 264–271.

Bell, E. L., Finnigan, W., France, S. P., Green, A. P., Hayes, M. A., Hepworth, L. J., Lovelock, S.L., Niikura, H., Osuna, S., Romero, E., Ryan, K.S., Turner, N. S. & Flitsch, S. L. (2021). Biocatalysis. *Nature Reviews Methods Primers*, *1*(1), 46.

Bezerra, K. G., Silva, I. G., Almeida, F. C., Rufino, R. D., & Sarubbo, L. A. (2021). Plant-derived biosurfactants: Extraction, characteristics and properties for application in cosmetics. *Biocatalysis and Agricultural Biotechnology*, *34*, 102036.

Bom, S., Jorge, J., Ribeiro, H. M., & Marto, J. O. A. N. A. (2019). A step forward on sustainability in the cosmetics industry: A review. *Journal of Cleaner Production*, *225*, 270–290.

Broeze, R. J. (2006). Key challenges facing bio manufacturing. *Bioprocessing and Biopartnering*, *2006*, 14–16.

Büning, B., Rechtenbach, D., Behrendt, J., & Otterpohl, R. (2021). Removal of emerging micropollutants from wastewater by nanofiltration and biofilm reactor (MicroStop). *Environmental Progress & Sustainable Energy*, *40*(3), e13587.

Cavinato, M., Waltenberger, B., Baraldo, G., Grade, C. V., Stuppner, H., & Jansen-Dürr, P. (2017). Plant extracts and natural compounds used against UVB-induced photoaging. *Biogerontology*, *18*, 499–516.

Cortés-Rojas, D. F., De Souza, C. R. F., & Oliveira, W. P. (2014). Clove (Syzygium aromaticum): A precious spice. *Asian Pacific Journal of Tropical Biomedicine*, *4*(2), 90–96.

Costa, J. P., Custódio, L., & Reis, C. P. (2023). Exploring the potential of using marine-derived ingredients: from the extraction to cutting-edge cosmetics. *Marine Drugs*, *21*(12), 620.

Ding, J., Wu, B., & Chen, L. (2022). Application of marine microbial natural products in cosmetics. *Frontiers in Microbiology*, *13*, 892505.

Dini, I. (2023). The potential of algae in the nutricosmetic sector. *Molecules*, *28*(10), 4032.

Etit, D., Meramo, S., Ögmundarson, Ó., Jensen, M. K., & Sukumara, S. (2024). Can biotechnology lead the way toward a sustainable pharmaceutical industry?. *Current Opinion in Biotechnology*, *87*, 103100.

Faccio, G. (2020). Plant complexity and cosmetic innovation. *IScience*, *23*(8).

Fermini, B., & Bell, D. C. (2022). On the perspective of an aging population and its potential impact on drug attrition and pre-clinical cardiovascular safety assessment. *Journal of Pharmacological and Toxicological Methods*, *117*, 107184.

Fonseca, S., Amaral, M. N., Reis, C. P., & Custódio, L. (2023). Marine natural products as innovative cosmetic ingredients. *Marine Drugs*, *21*(3), 170.

Fuentes-Tristan, S., Parra-Saldivar, R., Iqbal, H. M., & Carrillo-Nieves, D. (2019). Bioinspired biomolecules: Mycosporine-like amino acids and scytonemin from Lyngbya sp. with UV-protection potentialities. *Journal of Photochemistry and Photobiology B: Biology*, *201*, 111684.

Gargalo, C. L., de Las Heras, S. C., Jones, M. N., Udugama, I., Mansouri, S. S., Krühne, U., & Gernaey, K. V. (2020). Towards the development of digital twins for the bio-manufacturing industry. In: Herwig, C., Pörtner, R., Möller, J. (Eds) *Digital Twins. Advances in Biochemical Engineering/ Biotechnology* (vol 176, pp. 1–34). Springer Cham.

Gupta, P. L., Rajput, M., Oza, T., Trivedi, U., & Sanghvi, G. (2019). Eminence of microbial products in cosmetic industry. *Natural Products and Bioprospecting*, *9*, 267–278.

Heinrich, U., Moore, C. E., De Spirt, S., Tronnier, H., & Stahl, W. (2011). Green Tea Polyphenols Provide Photoprotection, Increase Microcirculation, and Modulate Skin Properties of Women12. *The Journal of Nutrition*, *141*(6), 1202–1208.

Hena, S., Gutierrez, L., & Croué, J. P. (2021). Removal of pharmaceutical and personal care products (PPCPs) from wastewater using microalgae: A review. *Journal of Hazardous Materials*, *403*, 124041.

Hook, D. J. (2006). Production of antibiotics by fermentation. *Basic Biotechnology*, 433–453.

Ivanov, K., Stoimenova, A., Obreshkova, D., & Saso, L. (2013). Biotechnology in the production of pharmaceutical industry ingredients: Amino acids. *Biotechnology & Biotechnological Equipment*, *27*(2), 3620–3626.

Jahan, A., Ahmad, I. Z., Fatima, N., Ansari, V. A., & Akhtar, J. (2017). Algal bioactive compounds in the cosmeceutical industry: A review. *Phycologia*, *56*(4), 410–422.

Jesumani, V., Du, H., Aslam, M., Pei, P., & Huang, N. (2019). Potential use of seaweed bioactive compounds in skincare—A review. *Marine Drugs*, *17*(12), 688.

Joshi, S., Kumari, R., & Upasani, V. N. (2018). Applications of algae in cosmetics: An overview. *Int. J. Innov. Res. Sci. Eng. Technol*, *7*(2), 1269.

Kalasariya, H. S., Pereira, L., & Patel, N. B. (2023). Comprehensive phytochemical analysis and bioactivity evaluation of Padina boergesenii: unveiling its prospects as a promising cosmetic component. *Marine Drugs*, *21*(7), 385.

Kazemipoor, M., Mohamed, C. W. J. W., Cordell, G. A., & Yaze, I. (2012). Safety, efficacy and metabolism of traditional medicinal plants in the management of obesity: a review. *International Journal of Chemical Engineering and Applications*, *3*(4), 288.

Khan, S., Ullah, M. W., Siddique, R., Nabi, G., Manan, S., Yousaf, M., & Hou, H. (2016). Role of recombinant DNA technology to improve life. *International Journal of Genomics*, *2016*(1), 2405954.

Kim, S. P., Nam, S. H., & Friedman, M. (2013). Hericium erinaceus (Lion's Mane) mushroom extracts inhibit metastasis of cancer cells to the lung in CT-26 colon cancer-tansplanted mice. *Journal of Agricultural and Food Chemistry*, *61*(20), 4898–4904.

Kubo, I., & Kinst-Hori, I. (1999). Flavonols from saffron flower: tyrosinase inhibitory activity and inhibition mechanism. *Journal of Agricultural and Food Chemistry*, *47*(10), 4121–4125.

Kusumoto, I. (2001). Industrial production of L-glutamine. *The Journal of Nutrition*, *131*(9), 2552S–2555S.

Li, F., Vijayasankaran, N., Shen, A., Kiss, R., & Amanullah, A. (2010). Cell culture processes for monoclonal antibody production. *MAbs* 2 (5), pp. 466–479.

Liu, L., Chen, Z., Liu, W., Ke, X., Tian, X., & Chu, J. (2022). Cephalosporin C biosynthesis and Fermentation in Acremonium chrysogenum. *Applied Microbiology and Biotechnology*, *106*(19), 6413–6426.

Manda, P., Popescu, C., Juluri, A., Janga, K., Kakulamarri, P. R., Narishetty, S., Murthy, S. N. & Repka, M. A. (2018). Micronized zaleplon delivery via orodispersible film and orodispersible tablets. *AAPS PharmSciTech*, *19*, 1358–1366.

Martínez, J. L., Liu, L., Petranovic, D., & Nielsen, J. (2012). Pharmaceutical protein production by yeast: towards production of human blood proteins by microbial Fermentation. *Current Opinion in Biotechnology*, *23*(6), 965–971.

Matejczyk, M. (2014). Environmental and recombinant microorganisms for biopharmaceuticals production. *BudownictwoiI nżynieriaSrodowiska*, *5*(1), 15–21.

Meghwanshi, G. K., Kaur, N., Verma, S., Dabi, N. K., Vashishtha, A., Charan, P. D., Praveen, P., Bhandari, H. S., Bhojak, N & Kumar, R. (2020). Enzymes for pharmaceutical and therapeutic applications. *Biotechnology and Applied Biochemistry*, *67*(4), 586–601.

Mkk, F., Sp, D., & Irfan, M. (2019). Antibacterial and antifungal activity of various extracts of Bacopa monnieri. *arXiv preprint arXiv:1909.01856*.

Mukherjee, P. K., Maity, N., Nema, N. K., & Sarkar, B. K. (2011). Bioactive compounds from natural resources against skin aging. *Phytomedicine*, *19*(1), 64–73.

Nielsen, J. (2013). Production of biopharmaceutical proteins by yeast: advances through metabolic engineering. *Bioengineered*, *4*(4), 207–211.

Okafor N. (2007) *Modern industrial Microbiology and Biotechnology*, Science Publishers Inc., U.S., Enfield, p. 530

Parasuraman, S., Kumar, L. N. D., Thanapakiam, G., Sayem, A. S. M., Chuah, J. J., & Venkateskumar, K. (2024). Biopharmaceutical Production by Recombinant DNA Technology: Future Perspectives. In *Microbial Products for Health and Nutrition* (pp. 285–303). Singapore: Springer Nature Singapore.

Parvatker, A. G., Tunceroglu, H., Sherman, J. D., Coish, P., Anastas, P., Zimmerman, J. B., & Eckelman, M. J. (2019). Cradle-to-gate greenhouse gas emissions for twenty anesthetic active pharmaceutical ingredients based on process scale-up and process design calculations. *ACS Sustainable Chemistry & Engineering*, *7*(7), 6580–6591.

Patel, M., Kumar, R., Kishor, K., Mlsna, T., Pittman Jr, C. U., & Mohan, D. (2019). Pharmaceuticals of emerging concern in aquatic systems: chemistry, occurrence, effects, and removal methods. *Chemical Reviews*, *119*(6), 3510–3673.

Pereira, L. (2018). Seaweeds as source of bioactive substances and skin care therapy—cosmeceuticals, algotheraphy, and thalassotherapy. *Cosmetics*, *5*(4), 68

Priya, A. K., Gnanasekaran, L., Rajendran, S., Qin, J., & Vasseghian, Y. (2022). Occurrences and removal of pharmaceutical and personal care products from aquatic systems using advanced treatment-A review. *Environmental Research*, *204*, 112298.

Rybczyńska-Tkaczyk, K., Grenda, A., Jakubczyk, A., Kiersnowska, K., & Bik-Małodzińska, M. (2023). Natural compounds with antimicrobial properties in cosmetics. *Pathogens*, *12*(2), 320.

Sarkar, B., Dissanayake, P. D., Bolan, N. S., Dar, J. Y., Kumar, M., Haque, M. N., Mukhopadhyay, R., Ramanayaka, S., Biswas, J. K., Tsang, D. C. W., Rinklebe, J & Ok, Y. S. (2022). Challenges and opportunities in sustainable management of microplastics and nanoplastics in the environment. *Environmental Research*, *207*, 112179.

Schütz, K., Carle, R., & Schieber, A. (2006). Taraxacum—a review on its phytochemical and pharmacological profile. *Journal of Ethnopharmacology*, *107*(3), 313–323.

Sharma, P., Sharma, S., Yadav, S., Srivastava, A., Purohit, I., & Shrivastava, N. (2014). Plant derived bioactive molecules: culture vessels to bioreactors. In: Paek, K.Y., Murthy, H., Zhong, J.J. (Eds), *Production of Biomass and Bioactive Compounds Using Bioreactor Technology* (pp. 47–60). Springer, Dordrecht.

Shashkina, M. Y., Shashkin, P. N., & Sergeev, A. V. (2006). Chemical and medicobiological properties of chaga. *Pharmaceutical Chemistry Journal*, *40*(10), 560–568.

Shokoohi, M., & Attar, T. (2024). Biomanufacturing for a sustainable future: unleashing the potential of biotechnology in pharmaceutical raw material production. *International*

Journal of New Findings in Health and Educational Sciences (IJHES), *2*(2), 142–153.

Sil, B., & Jha, S. (2014). Plants: The future pharmaceutical factory. *American Journal of Plant Sciences*, *5*(3), 319.

Soni, S. J., & Patel, A. N. (2024). Single-use technologies in bio manufacturing: benefits and implementation challenges. *Asian Journal of Hospital Pharmacy*, *4*(4), 1–7.

Squires, C. H., & Lucy, P. (2008). Vendor voice: A new paradigm for bacterial strain engineering. *BioProcess Int*, *6*(Suppl 4), 22–27.

Sun, H., Zhang, H., Ang, E. L., & Zhao, H. (2018). Biocatalysis for the synthesis of pharmaceuticals and pharmaceutical intermediates. *Bioorganic & Medicinal Chemistry*, *26*(7), 1275–1284.

Terpe, K. (2006). Overview of bacterial expression systems for heterologous protein production: from molecular and biochemical fundamentals to commercial systems. *Applied Microbiology and Biotechnology*, *72*, 211–222.

Wang, Y., Liu, L., Jin, Z., & Zhang, D. (2021). Microbial cell factories for green production of vitamins. *Frontiers in Bioengineering and Biotechnology*, *9*, 661562.

Wilkinson, J., Hooda, P. S., Barker, J., Barton, S., & Swinden, J. (2017). Occurrence, fate and transformation of emerging contaminants in water: An overarching review of the field. *Environmental Pollution*, *231*, 954–970.

World Economic Forum. (2019). *Making Manufacturing Sustainable by Design*. October.

Wu, Y., Choi, M. H., Li, J., Yang, H., & Shin, H. J. (2016). Mushroom cosmetics: the present and future. *Cosmetics*, *3*(3), 22.

Zheng, Z. P., Cheng, K. W., Zhu, Q., Wang, X. C., Lin, Z. X., & Wang, M. (2010). Tyrosinase inhibitory constituents from the roots of Morus nigra: A structure– activity relationship study. *Journal of Agricultural and Food Chemistry*, *58*(9), 5368–5373.

Zhou, S., Tang, Q., Tang, C., Liu, Y., Ma, F., Zhang, X., & Zhang, J. (2018). Triterpenes and soluble polysaccharide changes in lingzhi or reishi medicinal mushroom, Ganoderma lucidum (Agaricomycetes), during fruiting growth. *International Journal of Medicinal Mushrooms*, *20*(9), 859–871.

11 Biomanufacturing for Pollution Control and Ecosystem Restoration

Rajni Garg
Department of Applied Sciences and Humanities, Galgotias College of Engineering & Technology, Greater Noida, Uttar Pradesh, India

Rishav Garg
Department of Civil Engineering, Galgotias College of Engineering & Technology, Greater Noida, Uttar Pradesh, India

Nnabuk Okon Eddy
Department of Applied Science, Walter Sisulu University, East London, Republic of South Africa
Department of Nuclear Science, University of Nigeria, Nsukka, Nigeria

11.1 INTRODUCTION

Pollution and ecosystem degradation constitute the dual crises confronting the world today. Rapid industrialization, urbanization, and intensified agriculture have led to extensive deterioration of global terrestrial and marine resources (Schirpke et al. 2023). Land degradation presently impacts around 1.5 billion people worldwide, undermining essential ecosystem services, such as food production, biodiversity preservation, nutrient cycling, and climate regulation (Thomas et al. 2023). The issue is intensified by significant contamination from a diverse array of pollutants, including persistent organic pollutants (POPs), heavy metals, and plastics, which directly endanger ecological and human health (Garg et al. 2025). Worldwide, numerous severely contaminated sites necessitate treatment, with annual cleaning costs estimated in the billions of dollars and expected to rise significantly (Cecchin et al. 2017).

Traditional remediation solutions, including physicochemical treatments and excavation followed by landfilling, can cause ecological disruption, require substantial energy, and are economically unfeasible for large-scale implementation (Anjum et al. 2024). This context requires the development of sustainable, economical, and non-invasive technologies that can rehabilitate degraded ecosystems to a functional condition (Abhilash et al. 2016).

Biomanufacturing, characterized by the utilization of biological systems (such as microorganisms, plant cells, and enzymes) to generate valuable chemicals, signifies a transformative approach to addressing environmental issues (Asin-Garcia et al. 2025). The historical emphasis on valuable products for the pharmaceutical and food sectors is evolving, as biomanufacturing now encompasses the commodity-scale production of chemicals and materials as renewable substitutes for petroleum-derived counterparts (Scown 2022). This chapter emphasizes that biomanufacturing's utility encompasses not only the creation of sustainable products but also its direct role in pollution mitigation and ecosystem rehabilitation. This entails utilizing biological catalysis to provide targeted, implementable solutions for environmental remediation, including the production of particular enzymes to break down xenobiotics or the synthesis of biosurfactants to enhance pollutant bioavailability.

This chapter aims to deliver a thorough introduction to this burgeoning topic. It delineates the principal biomanufacturing processes used to produce instruments for pollution mitigation, encompassing biocatalysts, biosurfactants, and biopolymers. This chapter explores the incorporation of these tools into engineered biological systems, ranging from tailored microbial consortia to sophisticated phytoremediation techniques, for in-situ ecosystem restoration. The implementation of these integrated systems to address challenges such as hydrocarbon and plastic pollution, has been demonstrated through analytical case studies. The chapter ultimately discusses the significant obstacles to extensive implementation and delineates future outlooks, establishing biomanufacturing as a fundamental technology for a circular bioeconomy that promotes sustainable growth via environmental remediation.

DOI: 10.1201/9781003637158-11

11.2 BIOMANUFACTURING PROCESSES FOR POLLUTION CONTROL

The application of biomanufacturing to pollution control is predicated on the targeted, industrial-scale production of biological agents and materials capable of transforming or sequestering environmental contaminants. This approach moves beyond reliance on indigenous microbial activity toward a proactive strategy of deploying precisely formulated "bioremediation tools" (Halecký and Kozliak 2020). Advances in metabolic engineering and fermentation technology now permit the efficient production of enzymes (Ramírez et al. 2024), biosurfactants (Kalvandi et al. 2022), and biopolymers (Sharma et al. 2024) with properties tailored for environmental applications, offering higher specificity and lower ecological impact than conventional chemical treatments. Table 11.1 summarizes the primary "tools" produced via biomanufacturing and their specific roles in remediation.

11.2.1 Production of Biocatalysts and Enzymes

A primary application of biomanufacturing is the production of specific biocatalysts for the degradation of recalcitrant pollutants. Many persistent organic pollutants (POPs) are resistant to abiotic degradation due to their stable chemical structures, yet various microorganisms have evolved enzymatic pathways for their catabolism. Biomanufacturing can harness these pathways and scale for environmental application (Elias, Ranque, and Malleret 2025). Biomanufacturing enables the identification of these enzymes, their heterologous expression in robust industrial hosts (e.g., *Escherichia coli*, *Aspergillus*, *penicillium*, and *Phlebia*), and their production at scales sufficient for environmental application. Key enzymes include oxidoreductases such as laccases (Marshall et al. 2021), peroxidases (Dadi, Celik, and Ocsoy 2020), and cytochrome P450 monooxygenases (Behrendorff 2021), which catalyze the initial, often rate-limiting, step in pollutant degradation. Fungal laccases, for example, have demonstrated high efficacy in transforming PAHs, such as pyrene, phenanthrene, fluoranthene, and benzo(a)pyrene (Crittenden, Raudabaugh, and Gunsch 2025).

A significant technical challenge is the limited stability of free enzymes in field conditions due to factors such as pH, temperature, and proteolysis (Alaidaroos 2023). Biotechnological solutions focus on enzyme immobilization to enhance durability. Techniques include covalent attachment or encapsulation onto stable carriers. Immobilizing manganese peroxidase (MnP) on chitosan beads enhances its catalytic stability and effectiveness in eliminating industrial dyes (Bilal et al. 2016). Advanced systems employ encapsulation within materials such as alginate beads or integration into electrospun fibrous membranes, which not only stabilize the enzyme but also concentrate hydrophobic pollutants at the catalytic site, thereby enhancing reaction efficiency (Weng et al. 2023). The biomanufacturing of stabilized enzyme formulations is essential for producing effective, field-deployable remedial treatments.

11.2.2 Biosurfactants for Improved Pollutant Bioavailability

A significant impediment in the bioremediation of hydrophobic organic compounds (HOCs), including those present in crude oil, is their limited bioavailability due to substantial sorption to soil organic matter (Bielská et al. 2018). Biosurfactants, amphiphilic molecules generated by microorganisms, mitigate this constraint by diminishing surface and interfacial tension, facilitating the emulsification and solubilization of HOCs. In contrast to synthetic

TABLE 11.1
Important Biomanufactured Products for Pollution Control

Biomanufactured Product	Examples	Mechanism of Action	Target Pollutants	References
Biocatalysts (enzymes)	Laccases, peroxidases, and oxygenases	Directly catalyze the chemical breakdown (oxidation or hydrolysis) of complex organic molecules into simpler, and less toxic compounds.	Persistent organic pollutants (POPs), polycyclic aromatic hydrocarbons (PAHs), pesticides, and industrial dyes	Marshall et al. (2021); Dadi, Celik, and Ocsoy (2020); Ramírez et al. (2024)
Biosurfactants	Rhamnolipids, sophorolipids, and lipopeptides	Reduce surface and interfacial tension, emulsify hydrophobic pollutants and increase their solubility, and bioavailability for microbial degradation.	Crude oil, diesel fuel, motor oil, and other hydrocarbon-based contaminants	Dobler et al. (2020); Jadhav, Pratap, and Kale (2019); Kalvandi et al. (2022)
Biopolymers	Polyhydroxyalkanoates (PHAs), polylactic acid (PLA), and microbial cellulose	Sequestration: Bind heavy metal ions through functional groups. Filtration: Form membranes to physically remove particulates. Matrix: Acts as a carrier for immobilizing enzymes.	Heavy metals (lead, cadmium, and copper), dissolved organic compounds, and suspended solids	Sharma et al. (2024)

chemical alternatives, microbial surfactants are biodegradable and demonstrate reduced toxicity, rendering them environmentally acceptable (Satpute, Płaza, and Banpurkar 2017). Biomanufacturing facilitates the production of compounds via the fermentation of microbial strains, such as *Pseudomonas aeruginosa* (which produces rhamnolipids) and *Bacillus subtilis* (which produces lipopeptides), recognized for their high-yield synthesis. The utilization of a lipopeptide biosurfactant derived from *B. subtilis* has demonstrated an enhancement of over 85% in the degradation of used motor oil from contaminated soil (Umar et al. 2021). Additionally, biosurfactants may offer ancillary advantages, such as the enhancement of beneficial soil fungal proliferation, thereby aiding in overall ecosystem restoration (Kumar et al. 2021).

11.2.3 Biopolymers for Sequestration and Water Purification

Biomanufacturing polymers, such as polylactic acid (PLA) and polyhydroxyalkanoates (PHAs) is crucial for efforts designed to alleviate plastic pollution by substituting petrochemical plastics with biodegradable alternatives (Naser, Deiab, and Darras 2021). In addition to their intended biodegradability, the material characteristics of these biopolymers render them appropriate for active pollution management. The functional groups (e.g., carboxyl and hydroxyl) in the polymer structure can act as binding sites for heavy metal cations, facilitating their application as biosorbents for the sequestration of metals such as lead, cadmium, and copper from contaminated water (Raji et al. 2023). By fabricating these biopolymers into films or beads, they can be utilized in passive remediation systems to immobilize hazardous metals. Additionally, biopolymers are increasingly utilized in the fabrication of advanced filtration membranes for water treatment. Materials, such as PLA can be formed into micro- or ultrafiltration membranes that can be functionalized with biocatalysts, creating integrated systems that simultaneously filter particulates and degrade dissolved organic pollutants. This synergy between a biomanufactured material and a biomanufactured agent exemplifies the potential to create multifunctional pollution-control technologies.

11.3 BIOMANUFACTURING IN ECOSYSTEM RESTORATION

While the biomanufacturing of individual agents, such as enzymes and biosurfactants, provides a powerful toolkit for pollution control, the true potential for ecosystem restoration lies in their integration into sophisticated, engineered biological systems. Real-world contaminated sites are rarely characterized by a single, well-defined pollutant; they are complex, heterogeneous environments containing mixtures of contaminants and stressed native microbiomes (Schirpke et al. 2023). Addressing this complexity requires moving beyond single-agent solutions toward holistic systems that combine multiple biological components to create a robust, synergistic cleanup strategy. This involves engineering interactions between different microorganisms, between microbes and plants, and even between biological agents and advanced materials to create highly effective, site-specific restoration technologies. This section explores three key frontiers in the development of these engineered biological systems: customized microbial consortia (Pascual-García, Bonhoeffer, and Bell 2020), bioaugmentation by designer plants and their microbial partners (Xun et al. 2015), and the integration of nanobiotechnology (Chugh, Siddique, and Solaiman 2021). Table 11.2 outlines

TABLE 11.2
Engineered Systems for Ecosystem Restoration

Engineered System	Key Components	Primary Application and Function	Advantages/Limitations	References
Customized microbial consortia	A synergistic mix of multiple bacterial and/or fungal species (natural or genetically engineered).	Degradation of mixed waste. Different microbes work together to break down a complex cocktail of pollutants that a single strain cannot.	Advantage: Broad substrate range, metabolic synergy. Limitation: Potential for interspecies competition, instability in field conditions.	Pascual-García, Bonhoeffer, and Bell (2020)
Bioaugmented phytoremediation	A combination of pollutant-tolerant plants (e.g., poplar, and mustard) and beneficial microbial inoculants (PGPR, Endophytes, AMF).	In-situ soil and groundwater cleanup. Plants stabilize soil and extract/degrade pollutants, while microbes boost plant health and pollutant breakdown.	Advantage: Low cost, solar-powered, and improves soil health. Limitation: Slow process, limited by root depth and plant tolerance.	Xun et al. (2015)
Nanobiotechnology systems	Integration of nanomaterials (e.g., nZVI) with biological components (e.g., enzymes and whole microbes).	Enhanced and targeted remediation. Nanoparticles can deliver enzymes, directly catalyze reactions, or immobilize toxins, boosting the efficiency of biological processes.	Advantage: High reactivity, increased stability of biocatalysts. Limitation: Potential for nanoparticle toxicity, high cost, and regulatory concerns.	Chugh, Siddique, and Solaiman (2021); Dong et al. (2019)

PGPR, plant growth-promoting rhizobacteria; AMF, arbuscular mycorrhizal fungi; nZVI, nanoscale zero-valent iron the integrated, multi-component systems, highlighting their composition and primary applications.

11.3.1 Customized Microbial Consortia

In nature, microbial life rarely exists in isolation. Instead, microorganisms form complex communities, or consortia, where different species occupy distinct metabolic niches and often work together to break down complex substrates (Pascual-García, Bonhoeffer, and Bell 2020). This natural principle is being harnessed for bioremediation. Since most contaminated sites contain a diverse array of pollutants, a single microbial strain, even one with potent degradative abilities, is unlikely to possess the full enzymatic repertoire needed for complete cleanup. Microbial consortia, by contrast, can pool their metabolic capabilities to simultaneously tackle a wider range of chemical structures. When manufactured into standardized formulations, these consortia can be deployed as reliable remediation products, demonstrating how biomanufacturing transforms ecological interactions into field-ready technologies.

The development of effective microbial consortia is increasingly a core focus of environmental biotechnology. An early approach involves enriching and selecting for naturally occurring consortia from contaminated sites that have already adapted to the pollutants present (Li et al. 2020). For instance, a consortium of five different fungal species isolated from a site heavily contaminated with PAHs demonstrated the ability to remove significant amounts of phenanthrene (74%), pyrene (63%), and benzopyrene (81%) (Zafra et al. 2015). The efficacy of such consortia can be further augmented. In the same study, the addition of two genetically engineered *Aspergillus niger* strains, biomanufactured to express specific peroxidase enzymes, boosted the degradation of these PAHs even further. This highlights a powerful strategy: combining the resilience of a native consortium with the specialized power of bioaugmented, factory-grown microbes. Such examples illustrate that biomanufacturing does not merely replicate natural processes but actively enhances them, producing hybrid systems of native and engineered strains that outperform either component alone.

However, simply mixing different microbial strains can be challenging. Interspecies competition for resources, the production of inhibitory compounds, and difficulties in maintaining the optimal spatial distribution and population ratios can lead to instability and suboptimal performance (Hibbing et al. 2010). Modern biomanufacturing and synthetic biology are providing innovative solutions to these problems. Advances in "omics" technologies, such as next-generation sequencing and metagenomics, allow researchers to rapidly profile the microbial communities at contaminated sites, identifying not only the key players but also their functional genes, even for strains that cannot be grown in the laboratory (the viable but non-culturable [VBNC], state) (Desai, Pathak, and Madamwar 2010). This provides a blueprint for rationally designing a consortium of compatible, synergistic members. Beyond selection, biomanufacturing techniques can be used to create synthetic ecosystems that provide an ideal microenvironment for the consortium. For example, researchers have created systems where different bacterial species, *Pseudomonas* sp. National Collection of Industrial Bacteria (NCIB) 9816 and *Synechococcus elongatus* Pasteur Culture Collection (PCC) 7942 were co-encapsulated in silica gel. This physical structuring controls their interactions and microenvironment, enabling a synergistic pathway where *S. elongatus* PCC 7942 provided the necessary oxygen for the breakdown of pollutant (naphthalene) and *Pseudomonas* sp. NCIB 9816 completed the transformation to CO_2 (Mutlu et al. 2016). This level of control allows for the biomanufacturing of stable, high-performance consortia as reliable "off-the-shelf" products for remediation. By framing these engineered consortia explicitly as biomanufactured goods, they become not only biological systems but also deployable technologies that bridge laboratory design with field-scale application.

11.3.2 Designer Plants and Their Microbial Partners (Phytoremediation)

Phytoremediation, the use of plants to remove, degrade, or stabilize pollutants, is an aesthetically pleasing, solar-powered, and low-cost approach to restoring large areas of contaminated land (Jahan 2023). Plants can absorb certain pollutants, such as heavy metals, and accumulate them in their tissues (phytoextraction), or they can stimulate the degradation of organic pollutants in the soil around their roots, a zone known as the rhizosphere (rhizodegradation). This rhizosphere effect is critical; plants are not working alone. They actively recruit and nurture beneficial microorganisms by releasing a variety of compounds, known as root exudates (Ravelo-Ortega, Raya-González, and López-Bucio 2023). These exudates can serve as a food source for microbes or as signaling molecules that attract specific bacteria and fungi with pollutant-degrading capabilities. For example, maize seedlings are known to exude a specific compound 2,4-dihydroxy-7-methoxy-2*H*-1,4-benzoxazin-3(4*H*)-one (DIMBOA) that actively recruits the plant-growth-promoting bacterium *Pseudomonas putida*, which can help protect the plant from stresses (Neal et al. 2012). Such plant–microbe partnerships illustrate how natural rhizosphere interactions can be deliberately enhanced through biotechnology and biomanufacturing.

The concept of "designer plants" aims to optimize this natural process. Through selective breeding or genetic engineering, plants can be developed that are more tolerant to pollutants, have enhanced uptake capabilities, or are better at recruiting beneficial microbes (Trivedi et al. 2021). However, an equally powerful and often more readily deployable strategy is to focus on the other half of the partnership: the microbes. Biomanufacturing allows for the large-scale production of specific, high-performance microbial inoculants that can be paired with tolerant plant species to create a powerful phytoremediation system. Recent studies show that pairing stress-tolerant plants with tailored microbial inoculants can significantly improve pollutant uptake efficiency and plant biomass, making the combination more effective than either component alone. These inoculants can include the following:

- **Plant Growth-Promoting Rhizobacteria (PGPR):** These bacteria can enhance plant health and biomass production in contaminated soils by fixing nitrogen, solubilizing phosphorus, or producing hormones that stimulate growth, thus indirectly boosting the plant's remediation capacity (Cochard et al. 2022).). In oil-polluted soils, PGPR have been shown to increase plant survival rates significantly, directly supporting remediation efforts.
- **Endophytes:** These are microorganisms that live inside the plant's tissues without causing harm. They can play a direct role in remediation by degrading pollutants that the plant has taken up, effectively detoxifying the contaminant within the plant itself (Trivedi et al. 2021). Co-inoculating plants with specific endophytes and soil bacteria has been shown to improve plant growth and enhance the dissipation of pollutants, such as hexachlorocyclohexane (HCH) from the soil (Alvarez et al. 2022). This indicates that endophytes are not only passive residents but active biomanufactured partners for pollutant detoxification.
- **Arbuscular Mycorrhizal Fungi (AMF):** These fungi belong to the phylum *Glomeromycota* and form a symbiotic relationship with the roots of most plants, extending a vast network of hyphae into the soil. This network dramatically increases the surface area for water and nutrient uptake and can also enhance the degradation of organic pollutants. AMF hyphae can alter the soil microbial community in a way that favors the degradation of stubborn contaminants, such as polychlorinated biphenyls (PCBs) (Rajtor and Piotrowska-Seget 2016). Biomanufactured AMF inocula have been successfully applied to contaminated soils, reducing pollutant levels while simultaneously improving nutrient cycling.

By biomanufacturing these tailored microbial partners, we can assemble a complete phytoremediation "package," a tolerant plant species plus a scientifically selected microbial inoculant, designed to maximize restoration success at a specific site. This packaging concept of pairing engineered or resilient plants with factory-produced microbial formulations illustrates how phytoremediation is evolving from a passive, natural process into a structured, deployable biomanufacturing solution.

11.3.3 Nanobiotechnology in Remediation

Nanobiotechnology emerges at the intersection of materials science and biology, offering a new dimension of control and efficiency in environmental remediation. This field involves the use of materials at the nanoscale (typically 1–100 nanometers) integrated with biological components or systems to create novel remediation agents (Osuntokun, Onwudiwe, and Ebenso 2018). Biomanufacturing plays a crucial role here, either by producing the biological components that are integrated with nanomaterials or by using biological processes to create the nanoparticles themselves.

One of the primary applications of nanobiotechnology is to enhance the performance and stability of the biocatalysts discussed earlier. For example, enzymes can be packaged within nanoscale biological structures called "vaults," which protect the enzyme from degradation in the harsh field environment and allow for more targeted delivery (Lothe et al. 2020). Vault nanoparticles packaged with MnP have shown significantly higher stability and threefold greater phenol biodegradation compared to the free enzyme. This represents a sophisticated form of biomanufactured enzyme delivery system.

Another powerful strategy is the creation of nano-bio-integrated systems, where synthetic nanoparticles are combined with living microbes to accelerate remediation. For example, the degradation of the pesticide chlorpyrifos was significantly enhanced by a system combining palladium on zinc oxide nanoparticles (ZnONPs) with the aerobic bacterium *Pseudomonas* sp. CF7b (Sreekanth et al. 2024). In this symbiotic relationship, the nanoparticles initiate the chemical breakdown of the pollutant, creating intermediates that are more easily consumed and fully mineralized by the bacteria. This integrated approach was found to be more than twice as efficient as using either the nanoparticles or the bacteria alone. Such examples underline how biomanufactured microbial strains, when combined with nanoscale catalysts, achieve efficiencies impossible by either component in isolation.

Furthermore, biotechnology is being used to manufacture nanoparticles through "green synthesis" routes. Certain microbes, such as *Geobacter metallireducens*, can be cultured under specific conditions to produce biogenic magnetite nanoparticles. These biologically synthesized nanoparticles have been shown to be effective in degrading pollutants, such as Cr(III/VI) (Kim and Roh 2019). This approach is not only more sustainable than conventional chemical synthesis of nanoparticles but also produces nanoparticles with unique surface properties that can enhance their reactivity. Framing these as biomanufactured nanoparticles emphasizes their dual advantage: environmental sustainability in production and superior reactivity in application. While the potential of nanobiotechnology is immense, it is crucial to proceed with caution. The very properties that make nanoparticles effective, their small size and high reactivity, also raise concerns about their potential long-term fate, transport, and toxicity in ecosystems (Garg et al. 2021). Therefore, a critical area of ongoing research is to thoroughly assess the environmental impact of these nanomaterials to ensure that the solution does not create a new problem.

11.4 INTEGRATING BIOMANUFACTURING AND RESTORATION

The true measure of the environmental utility of biomanufacturing lies not in the performance of its individual products in a laboratory, but in their synergistic application in the field. To restore complex, polluted ecosystems, a multi-faceted approach is required, integrating different biotechnologies in a logical sequence. This section presents three illustrative case studies, restoring oil-contaminated soils, tackling plastic pollution, and revitalizing industrial brownfields, to demonstrate how biomanufactured agents and engineered biological systems can be combined to form comprehensive and effective restoration strategies. These cases highlight the transition from theoretical potential to practical application, showcasing how biomanufacturing can be deployed to solve some of the most pressing environmental problems.

11.4.1 Remediation of Hydrocarbon-Contaminated Soils

Terrestrial sites contaminated with petroleum hydrocarbons (PHCs), such as crude oil or diesel fuel, present a significant remediation challenge. The hydrophobic nature of these contaminants leads to strong sorption onto soil matrices, limiting their bioavailability for microbial degradation and creating toxic conditions that suppress native ecological functions (Phulpoto et al. 2020). An effective restoration strategy requires a multi-phasic approach integrating several biomanufactured technologies in a logical sequence, as discussed here:

1. **Phase 1: Enhancing pollutant bioavailability via biosurfactants**. The primary rate-limiting step in the bioremediation of PHCs is their low aqueous solubility. The initial intervention would therefore involve the application of microbially produced surfactants. The biomanufacturing of agents such as lipopeptides from *B. subtilis* (Umar et al. 2021) or rhamnolipids from *P. aeruginosa* (Dobler et al. 2020) provides a scalable source of these compounds. Application through soil tilling or irrigation facilitates the emulsification of hydrocarbon sludge and desorption from soil particles, thereby increasing the contaminant flux into the aqueous phase and rendering it accessible for enzymatic attack. This can be augmented by the application of biomanufactured, cell-free enzymes, such as lipases and oxygenases, to initiate the cleavage of long-chain alkanes and aromatic structures, reducing acute phytotoxicity and preparing the matrix for subsequent microbial colonization.
2. **Phase 2: Accelerated degradation via bioaugmentation with designed consortia**. Following the increase in bioavailability, the site would be bioaugmented with a high-density microbial consortium produced via industrial fermentation. The composition of this consortium would be rationally designed, informed by metagenomic analyses of known hydrocarbon-degrading environments, to ensure a synergistic collection of metabolic pathways. Such a consortium would necessarily include specialist organisms, such as *Rhodococcus* and *Sphingomonas* species, for the degradation of complex polycyclic aromatic hydrocarbons (PAHs), alongside generalist hydrocarbon degraders (Alizadeh-Sani et al. 2018). This strategy of applying a pre-grown, high-concentration inoculum is designed to overcome the initial lag phase and outcompete less efficient native microbes, establishing a robust catabolic potential for the mineralization of the diverse hydrocarbon mixture.
3. **Phase 3: Phytoremediation for final polishing and ecosystem recovery**. Once bulk contaminant concentrations are reduced to sub-toxic levels, the final phase shifts to establishing a vegetative cover for long-term soil health and degradation of residual pollutants (Feng et al. 2017). This phase employs phytoremediation, utilizing hardy, fast-growing plant species (e.g., *Poaceae* or *Fabaceae*) selected for their tolerance to residual PHCs and their dense root systems. Crucially, the seeds or seedlings would be inoculated with a biomanufactured formulation of PGPR and AMF (Xun et al. 2015). The PGPR enhances plant establishment and stress tolerance, while the extensive mycelial network of AMF explores the soil matrix, improving soil structure and facilitating the final breakdown of recalcitrant contaminant fractions through enhanced rhizodegradation. This final step is critical for transitioning the site from a remediated matrix to a self-sustaining, ecologically functional system.

11.4.2 Tackling Plastic Pollution

The global crisis of plastic pollution requires a dual approach: remediating the vast quantities of plastic waste already choking our ecosystems and preventing future accumulation by fundamentally changing how plastics are made and managed.

Integrated Biomanufacturing Strategy:

1. **Remediation via enzymatic recycling:** for legacy plastic waste, particularly polyethylene terephthalate (PET), which is common in bottles and packaging, biomanufacturing offers a path toward a circular economy through enzymatic recycling. This involves the large-scale production of specialized

plastic-degrading enzymes, such as PET hydrolase (PETase) and mono(2-hydroxyethyl) terephthalate hydrolase (MHETase), originally discovered in the bacterium *Ideonella sakaiensis* (Yoshida et al. 2021). In this model, collected PET waste is processed in a bioreactor containing these biomanufactured enzymes. The enzymes depolymerize the PET back to its core chemical building blocks: terephthalic acid (TPA) and ethylene glycol (EG) (Soong, Sobkowicz, and Xie 2022). These monomers can then be purified and used to synthesize new, virgin-quality PET plastic, creating a closed-loop system that reduces the need for fossil-fuel-based plastic production and lessens the burden on landfills and incinerators (Figure 11.1). The success of this strategy depends on the ability of biomanufacturing to produce these enzymes cost-effectively and at an industrial scale.

2. **Prevention via bio-based, biodegradable polymers:** The long-term solution lies in replacing persistent petrochemical plastics with sustainable alternatives. Biomanufacturing is central to producing biodegradable polymers such as PLA and PHAs

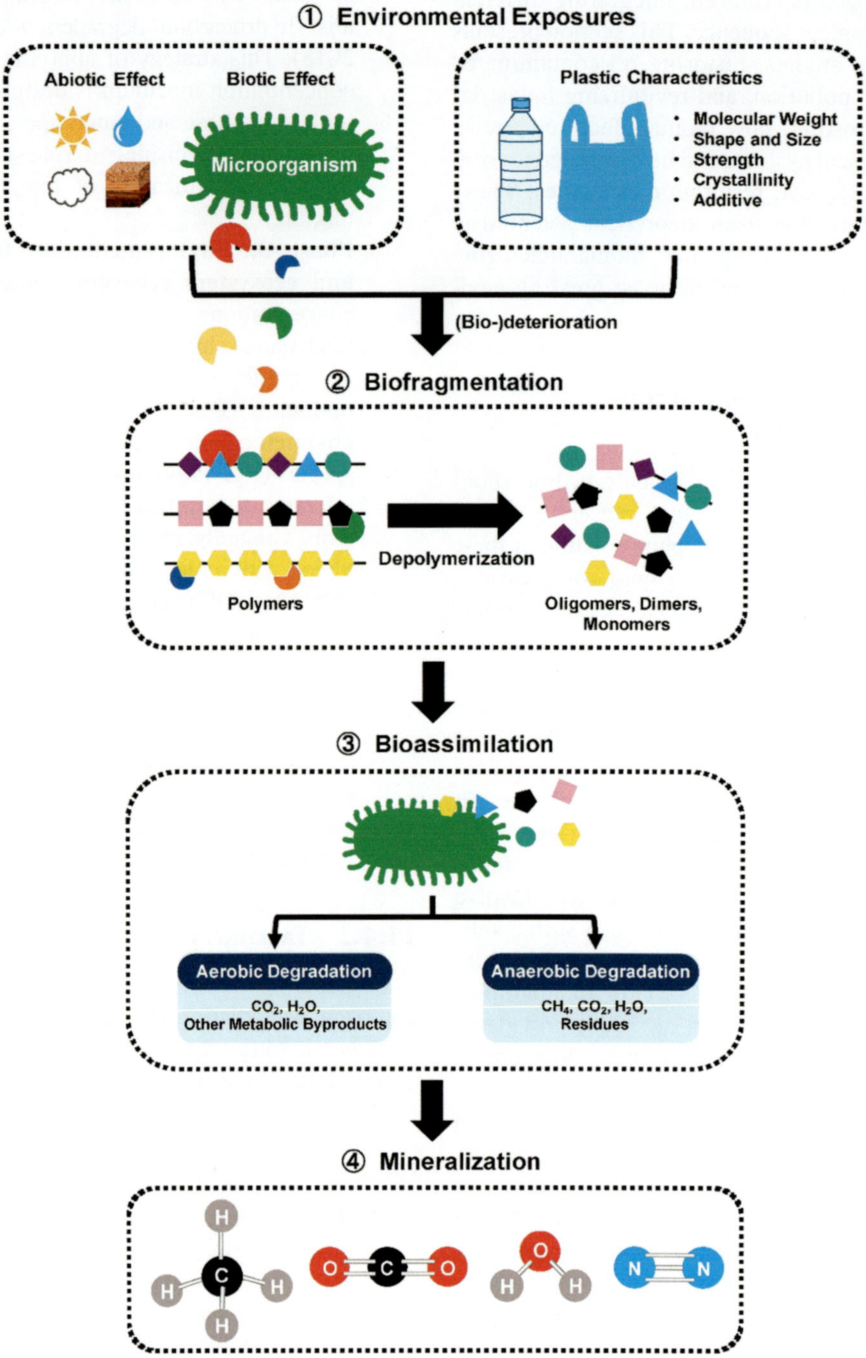

FIGURE 11.1 Mechanism for microbial degradation of plastics (Soong, Sobkowicz, and Xie 2022).

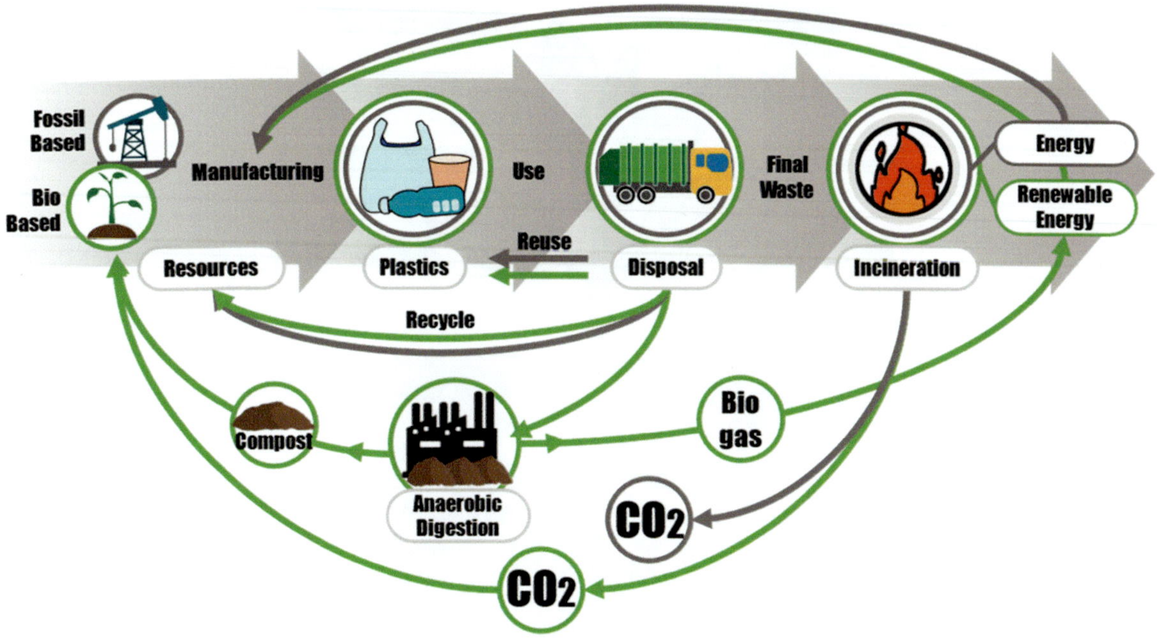

FIGURE 11.2 Linear economy steps (straight arrows) and circular economy additions (green arrows) for bioplastics (Di Bartolo, Infurna, and Dintcheva 2021).

(Naser, Deiab, and Darras 2021). These plastics are produced through the fermentation of renewable feedstocks such as corn starch or sugarcane. The key advantage is their designed end-of-life. A PLA-based product, for example, can be composted in an industrial facility, where microbes break it down into CO_2, water, and biomass, returning the biogenic carbon to the atmosphere for fixation by the next crop of feedstock (Haq et al. 2025). PHAs offer the added benefit of being biodegradable in a wider range of environments, including soil and marine settings. By shifting the manufacturing paradigm from producing a permanent pollutant to producing a transient material designed for a biological life cycle, biomanufacturing provides a powerful preventative tool against future plastic accumulation. This approach embodies the principles of a circular carbon economy, where materials are created, used, and then biologically recycled (Figure 11.2) (Di Bartolo, Infurna, and Dintcheva 2021).

11.4.3 Revitalizing Industrial Brownfields

Industrial sites ("brownfield") are generally left abandoned due to contamination with a complex mixture of pollutants, including heavy metals (e.g., chromium, lead, arsenic) from plating or manufacturing processes, and POPs (e.g., PCBs, solvents) from chemical storage and use. These sites are a source of potential groundwater contamination and are unsuitable for redevelopment.

Integrated Restoration Strategy:

1. **Phase 1: Site characterization and contaminant immobilization**. The first step is to precisely map the type and concentration of pollutants. This could be aided by biomanufactured biosensors deployed in the field for rapid, on-site analysis (Pham, Ling, and Chang 2022). For areas with high concentrations of mobile heavy metals, a targeted intervention using nanobiotechnology would be employed. A solution of biopolymer-stabilized nanoscale zero-valent iron (nZVI) particles could be injected into these hotspots. The nZVI have been reported to chemically reduce toxic and mobile metals, such as hexavalent chromium (Cr(VI)) to the far less toxic and less mobile Cr(III), effectively immobilizing the metal and reducing the immediate risk of leaching (Ye et al. 2021).
2. **Phase 2: Bioaugmented phytoremediation**. The entire site would then be planted with a carefully selected mix of hardy, fast-growing plant species known for their remediation potential, such as poplars, willows, or Indian mustard (*Brassica juncea*) (Suman et al. 2018). These plants would be inoculated with a biomanufactured, multi-functional microbial consortium (Figure 11.3). This designer consortium would contain: (i) endophytic bacteria capable of degrading the organic pollutants (PCBs and solvents) inside the plant's tissues, protecting the plant from phytotoxicity (Karaś, Wdowiak-Wróbel, and Sokołowski 2021), and (ii) PGPR to enhance the plants' growth and stress tolerance in the metal-contaminated soil.

FIGURE 11.3 Illustration of bioaugmented phytoremediation.

3. **Phase 3: Long-term management and value creation**. The plants would be allowed to grow for several seasons. They would perform multiple functions simultaneously: their roots would stabilize the soil to prevent erosion, stimulate the rhizodegradation of organic pollutants, and gradually phytoextract a portion of the heavy metals into their biomass (Aryal 2024). The harvested plant biomass would need to be carefully managed. If heavy metal concentrations are high, the biomass would be harvested and incinerated in a controlled facility to reduce its volume and recover the metals from the ash for disposal or recycling. This process, known as phytomining, permanently removes the metals from the site (Suman et al. 2018). This integrated system not only remediates the soil but also sets the stage for ecological recovery. Over time, as contaminant levels decrease, the site can be transitioned to a state suitable for redevelopment or establishment as a green space, demonstrating how a combination of biomanufactured agents can turn a derelict brownfield into a valuable asset (Karaś, Wdowiak-Wróbel, and Sokołowski 2021).

Table 11.3 provides a comparison between conventional physical/chemical methods and the biomanufacturing-based approaches central to this chapter.

11.5 PATHWAYS FOR ADVANCING BIOMANUFACTURING IN ENVIRONMENTAL RESTORATION

Biomanufacturing for environmental applications faces major hurdles of economic viability and large-scale deployment (Zhou et al. 2024). Conventional methods such as "dig and dump" (Siddiqua et al., 2022), precipitation, adsorption, ion exchange, and electrochemical technologies (Razzak et al. 2022), benefit from established infrastructure and supply chains, making them the default, and often cheaper, option in the short term. In contrast, biomanufacturing requires sterile fermentation, downstream processing, and purification, which remain costly (McNulty et al. 2021). The key transition lies in adapting production models from low-volume pharmaceuticals to high-volume, low-cost environmental products. Process innovations such as continuous fermentation, waste-derived feedstocks, and high-titer microbes (Osuoha, Anyanwu, and Ejileugha 2023).

Logistical and ecological uncertainties add complexity: enzymes may degrade quickly in the field, and engineered consortia may be outcompeted by native microbes (Suman et al. 2018). Regulation and public perception further slow deployment, particularly for GMOs (Rafeeq et al. 2023). Future progress depends on robust safety data, adaptive regulatory frameworks, and clear communication of ecological benefits.

TABLE 11.3
Comparative Analysis of Remediation Strategies

Parameter	Conventional Remediation	Biomanufacturing-Based Remediation	References
Ecological impact	High; often destructive to the local ecosystem, can produce harmful byproducts.	Low; aims to work with and restore biological systems; agents are often biodegradable.	Reichelt-Brushett (2023)
Specificity	Low to moderate; often broad-spectrum chemical reactions that affect non-target components.	High; enzymes and microbes can be selected or engineered to target specific pollutants.	Rafeeq et al. (2023)
Primary mechanism	Physical removal or aggressive chemical transformation.	Biological catalysis and transformation (degradation and sequestration).	Alori et al. (2022)
Cost	Often high initial capital cost (heavy machinery and chemicals) and long-term liability (landfill fees).	Potentially lower operational cost, but Research and Development (R&D) and production scaling can be expensive.	Zhou et al. (2024)
Sustainability	Low; resource-intensive (fossil fuels and chemicals) and often linear (moving waste elsewhere).	High, based on renewable biological processes, enables circular economy approaches (waste to value).	Vandy, Mohanty, and Mohapatra (2025)
Public perception	Generally negative, associated with disruption, traffic, and hazardous materials.	Generally positive, perceived as a "green" or "natural" solution, although Genetically Modified Organisms (GMOs) can be a concern.	Rafeeq et al. (2023)

Instead of treating remediation as a cost, biomanufacturing can anchor value-creating cycles within a circular bioeconomy (Khanna et al. 2024). Restored land can serve not only as a recovered ecosystem but also as a productive landscape. For example, energy crops such as Miscanthus, switchgrass, willow, or poplar can grow on contaminated soils while supporting long-term cleanup and producing biomass for fuels, bioplastics, or high-value chemicals (Abdalla et al. 2024). This approach creates a sustainable revenue cycle, turning a contaminated site into a productive landscape, as illustrated in Figure 11.4 (Ansari et al. 2023).

This logic extends to pollutants themselves: enzymatic PET recycling turns waste into purified TPA and EG, closing the loop and reducing petrochemical dependence (Soong, Sobkowicz, and Xie 2022). By valorizing waste and aligning with carbon circularity, biomanufacturing can link remediation directly to climate and economic goals.

11.6 FUTURE PROSPECTIVE

To realize the full potential of biomanufacturing for environmental stewardship, the scientific and engineering communities must address several critical unanswered questions. The answers will be essential for developing safe, effective, and truly sustainable technologies. Key questions include:

- **What are the long-term ecological impacts?** The introduction of large quantities of biomanufactured agents or engineered microbes could have unintended consequences on native microbial communities, nutrient cycling, and overall ecosystem function (Mahdizade Ari et al. 2024). Long-term monitoring and ecological modeling are needed to assess these risks.
- **How can we design for environmental complexity?** How can we create microbial consortia that are not only effective but also resilient and stable in the face of competition from native microbes and fluctuating environmental conditions? Can artificial intelligence and machine learning be used to predict and design the behavior of complex biological systems in the field (Ibrahim, Raajaraam, and Raman 2021)?
- **How can we optimize synergy?** What are the optimal combinations and application sequences for different biotechnologies (e.g., biosurfactants, then enzymes, then phytoremediation)? Developing a predictive understanding of how these different tools interact is crucial for designing the most efficient and cost-effective strategies for mixed-contaminant sites (Bala et al. 2022).
- **What is the true life-cycle impact?** To claim that a biomanufactured solution is sustainable, a rigorous cradle-to-grave life-cycle assessment (LCA) is essential (Muazu et al. 2024). This analysis must account for the emissions and resource use of the entire process, from growing the feedstock to producing the agent and its ultimate fate in the environment, and compare it honestly against conventional methods.
- **What policies and partnerships are needed?** What economic incentives, policy frameworks, and public-private partnership models are required to de-risk

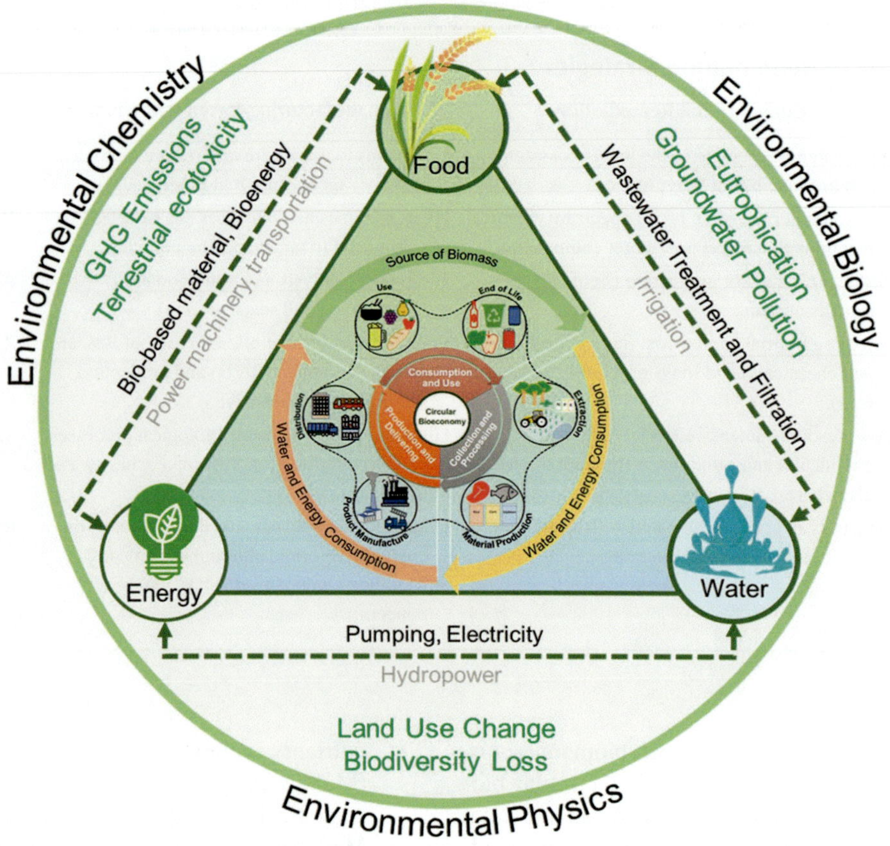

FIGURE 11.4 Circular economy approach for bioremediation and ecological restoration (Ansari et al. 2023).

investment and encourage the adoption of these biotechnologies for restoring public lands and waters (Ghosh and Priyadarshini 2024).

11.7 CONCLUSION

The escalating crises of environmental pollution and ecosystem degradation demand solutions that are not only effective but also inherently sustainable. Biomanufacturing represents a monumental shift in our capacity to address these challenges, moving us beyond mitigation and containment toward active, biologically based restoration. By harnessing the exquisite precision of microbial and enzymatic systems, we can now design and produce a sophisticated toolkit of agents capable of targeting and neutralizing some of the most persistent pollutants in our environment.

As we have explored, this involves manufacturing cell-free enzymes to break down toxic chemicals, biosurfactants to wash contaminants from soil, and novel biopolymers to sequester metals or filter water. More powerfully, it enables the assembly of these components into engineered systems, from rationally designed microbial consortia that can digest complex pollutant mixtures, to bioaugmented phytoremediation strategies that revitalize entire landscapes. While significant challenges related to scale, cost, and regulation remain, framing these efforts within a circular bioeconomy, where restoration generates economic and social value, provides a tangible path forward. The continued convergence of biotechnology, ecology, and materials science will be critical. By investing in this research and fostering the development of these technologies, we can leverage biomanufacturing as a cornerstone of a more sustainable future, harnessing the planet's own biological machinery to heal itself.

REFERENCES

Abdalla, Mohamed, Astley Hastings, Grant Campbell, Heyu Chen, and Pete Smith. 2024. "Assessing the Sustainability of Miscanthus and Willow as Global Bioenergy Crops: Current and Future Climate Conditions (Part 1)." *Agronomy* 14 (12): 3020. doi:10.3390/agronomy14123020

Abhilash, P. C., Vishal Tripathi, Sheikh Adil Edrisi, Rama Kant Dubey, Mansi Bakshi, Pradeep K. Dubey, H. B. Singh, and Stephen D. Ebbs. 2016. "Sustainability of Crop Production from Polluted Lands." *Energy, Ecology and Environment* 1 (1): 54–65. doi:10.1007/s40974-016-0007-x

Alaidaroos, Bothaina A. 2023. "Advancing Eco-Sustainable Bioremediation for Hydrocarbon Contaminants: Challenges and Solutions." *Processes* 11 (10): 3036. doi:10.3390/pr11103036

Alizadeh-Sani, Mahmood, Hamed Hamishehkar, Arezou Khezerlou, Maryam Azizi-Lalabadi, Yaghob Azadi, Elyas

Nattagh-Eshtivani, Mehdi Fasihi, Abed Ghavami, Aydin Aynehchi, and Ali Ehsani. 2018. "Bioemulsifiers Derived from Microorganisms: Applications in the Drug and Food Industry." *Advanced Pharmaceutical Bulletin* 8 (2): 191–199. doi:10.15171/apb.2018.023

Alori, Elizabeth Temitope, Alhasan Idris Gabasawa, Chinyere Edna Elenwo, and Oluwadolapo Ololade Agbeyegbe. 2022. "Bioremediation Techniques as Affected by Limiting Factors in Soil Environment." *Frontiers in Soil Science* 2 (August). doi:10.3389/fsoil.2022.937186

Alvarez, Analía, Beatriz Rodríguez-Garrido, Andrea Cerdeira-Pérez, Alba Tomé-Pérez, Petra Kidd, and Angeles Prieto-Fernández. 2022. "Enhanced Biodegradation of Hexachlorocyclohexane (HCH) Isomers by Sphingobium Sp. Strain D4 in the Presence of Root Exudates or in Co-Culture with HCH-Mobilizing Strains." *Journal of Hazardous Materials* 433 (July): 128764. doi:10.1016/j.jhazmat.2022.128764

Anjum, A., Rajni Garg, Rishav Garg, D. Gupta, and N. O. Eddy. 2024. "Efficient Sequestration of Zinc and Copper from Aqueous Media: Exploring Strategies, Mechanisms, and Challenges." *International Journal of Environmental Science and Technology* 2 (0123456789). Springer Berlin Heidelberg. doi:10.1007/s13762-024-06060-2

Ansari, Andrianto, Shafira Wuryandani, Arin Pranesti, Mareli Telaumbanua, Ngadisih, Muhammad Yusril Hardiansyah, Taufan Alam, Supriyanta, Tri Martini, and Taryono. 2023. "Optimizing Water-Energy-Food Nexus: Achieving Economic Prosperity and Environmental Sustainability in Agriculture." *Frontiers in Sustainable Food Systems* 7 (August). doi:10.3389/fsufs.2023.1207197

Ari, Mahdizade, Leila Dadgar Marzie, Zahra Elahi, Roya Ghanavati, and Behrouz Taheri. 2024. "Genetically Engineered Microorganisms and Their Impact on Human Health." *International Journal of Clinical Practice* 2024 (March): 1–38. doi:10.1155/2024/6638269

Aryal, Mahendra. 2024. "Phytoremediation Strategies for Mitigating Environmental Toxicants." *Heliyon* 10 (19): e38683. doi:10.1016/j.heliyon.2024.e38683

Asin-Garcia, Enrique, James D. Fawcett, Christos Batianis, and Vitor A.P. Martins dos Santos. 2025. "A Snapshot of Biomanufacturing and the Need for Enabling Research Infrastructure." *Trends in Biotechnology* 43 (5): 1000–1014. doi:10.1016/j.tibtech.2024.10.014

Bala, Saroj, Diksha Garg, Banjagere Veerabhadrappa Thirumalesh, Minaxi Sharma, Kandi Sridhar, Baskaran Stephen Inbaraj, and Manikant Tripathi. 2022. "Recent Strategies for Bioremediation of Emerging Pollutants: A Review for a Green and Sustainable Environment." *Toxics* 10 (8): 484. doi:10.3390/toxics10080484

Behrendorff, James B. Y. H. 2021. "Reductive Cytochrome P450 Reactions and Their Potential Role in Bioremediation." *Frontiers in Microbiology* 12 (April). doi:10.3389/fmicb.2021.649273

Bielská, Lucie, Lucia Škulcová, Natália Neuwirthová, Gerard Cornelissen, and Sarah E. Hale. 2018. "Sorption, Bioavailability and Ecotoxic Effects of Hydrophobic Organic Compounds in Biochar Amended Soils." *Science of the Total Environment* 624 (May): 78–86. doi:10.1016/j.scitotenv.2017.12.098

Bilal, Muhammad, Muhammad Asgher, Munawar Iqbal, Hongbo Hu, and Xuehong Zhang. 2016. "Chitosan Beads Immobilized Manganese Peroxidase Catalytic Potential for Detoxification and Decolorization of Textile Effluent." *International Journal of Biological Macromolecules* 89 (August): 181–89. doi:10.1016/j.ijbiomac.2016.04.075

Cecchin, Iziquiel, Krishna R. Reddy, Antônio Thomé, Eloisa Fernanda Tessaro, and Fernando Schnaid. 2017. "Nanobioremediation: Integration of Nanoparticles and Bioremediation for Sustainable Remediation of Chlorinated Organic Contaminants in Soils." *International Biodeterioration & Biodegradation* 119 (April): 419–28. doi:10.1016/j.ibiod.2016.09.027

Chugh, Gaurav, Kadambot H.M. Siddique, and Zakaria M. Solaiman. 2021. "Nanobiotechnology for Agriculture: Smart Technology for Combating Nutrient Deficiencies with Nanotoxicity Challenges." *Sustainability (Switzerland)* 13 (4): 1–20. doi:10.3390/su13041781

Cochard, Bastien, Basile Giroud, Julien Crovadore, Romain Chablais, Lucas Arminjon, and François Lefort. 2022. "Endophytic PGPR from Tomato Roots: Isolation, In Vitro Characterization and In Vivo Evaluation of Treated Tomatoes (Solanum Lycopersicum L.)." *Microorganisms* 10 (4): 765. doi:10.3390/microorganisms10040765

Crittenden, Joshua, Daniel Raudabaugh, and Claudia K. Gunsch. 2025. "Isolation, Characterization, and Mycostimulation of Fungi for the Degradation of Polycyclic Aromatic Hydrocarbons at a Superfund Site." *Biodegradation* 36 (1): 15. doi:10.1007/s10532-024-10106-0

Dadi, Seyma, Cagla Celik, and Ismail Ocsoy. 2020. "Gallic Acid Nanoflower Immobilized Membrane with Peroxidase-like Activity for m-Cresol Detection." *Scientific Reports* 10 (1): 1–9. doi:10.1038/s41598-020-73778-7

Desai, Chirayu, Hilor Pathak, and Datta Madamwar. 2010. "Advances in Molecular and '-Omics' Technologies to Gauge Microbial Communities and Bioremediation at Xenobiotic/Anthropogen Contaminated Sites." *Bioresource Technology* 101 (6): 1558–1569. doi:10.1016/j.biortech.2009.10.080

Di Bartolo, Alberto, Giulia Infurna, and Nadka Tzankova Dintcheva. 2021. "A Review of Bioplastics and Their Adoption in the Circular Economy." *Polymers* 13 (8): 1229. doi:10.3390/polym13081229

Dobler, Leticia, Helen Conceição Ferraz, Livia Vieira Araujo de Castilho, Leandro Stefano Sangenito, Ilson Paranhos Pasqualino, André Luis Souza dos Santos, Bianca Cruz Neves, Ricardo Rodrigues Oliveira, Denise Maria Guimarães Freire, and Rodrigo Volcan Almeida. 2020. "Environmentally Friendly Rhamnolipid Production for Petroleum Remediation." *Chemosphere* 252 (August): 126349. doi:10.1016/j.chemosphere.2020.126349

Dong, Haoran, Long Li, Yue Lu, Yujun Cheng, Yaoyao Wang, Qin Ning, Bin Wang, Lihua Zhang, and Guangming Zeng. 2019. "Integration of Nanoscale Zero-Valent Iron and Functional Anaerobic Bacteria for Groundwater Remediation: A Review." *Environment International* 124 (March): 265–277. doi:10.1016/j.envint.2019.01.030

Elias, Charbel, Stéphane Ranque, and Laure Malleret. 2025. "State of the Art on Fungal Biodegradation of Persistent Organic Pollutants in Soils and Innovative Strategies for Isolating Relevant Candidate Strains." *Environmental Technology & Innovation* 39 (August): 104247. doi:10.1016/j.eti.2025.104247

Feng, Nai-Xian, Jiao Yu, Hai-Ming Zhao, Yu-Ting Cheng, Ce-Hui Mo, Quan-Ying Cai, Yan-Wen Li, Hui Li, and Ming-Hung Wong. 2017. "Efficient Phytoremediation of Organic Contaminants in Soils Using Plant–Endophyte Partnerships." *Science of The Total Environment* 583 (April): 352–368. doi:10.1016/j.scitotenv.2017.01.075

Garg, Rajni, Priya Rani, Rishav Garg, and Nnabuk Okon Eddy. 2021. "Study on Potential Applications and Toxicity Analysis of Green Synthesized Nanoparticles." *Turkish Journal of Chemistry* 45 (6): 1690–1706. doi:10.3906/kim-2106-59

Garg R, Garg R, Anjum A, et al (2025) Waste Management and Classification: Repurposing Garbage to Create Nanoparticles for Water Purification. pp. 111–137

Ghosh, Sahana, and Subhra Priyadarshini. 2024. "Policy Push for India's Bioeconomy." *Nature*. doi:10.1038/d44151-024-00136-1

Halecký, Martin, and Evguenii Kozliak. 2020. "Modern Bioremediation Approaches: Use of Biosurfactants, Emulsifiers, Enzymes, Biopesticides, GMOs." In *Advanced Nano-Bio Technologies for Water and Soil Treatment* (pp. 495–526). Springer Nature. doi:10.1007/978-3-030-29840-1_24

Haq, Fazal, Mehwish Kiran, Iffat Ayesha Khan, Sahid Mehmood, Tariq Aziz, and Muhammad Haroon. 2025. "Exploring the Pathways to Sustainability: A Comprehensive Review of Biodegradable Plastics in the Circular Economy." *Materials Today Sustainability* 29 (March): 101067. doi:10.1016/j.mtsust.2024.101067

Hibbing, Michael E., Clay Fuqua, Matthew R. Parsek, and S. Brook Peterson. 2010. "Bacterial Competition: Surviving and Thriving in the Microbial Jungle." *Nature Reviews Microbiology* 8 (1): 15–25. doi:10.1038/nrmicro2259

Ibrahim, Maziya, Lavanya Raajaraam, and Karthik Raman. 2021. "Modelling Microbial Communities: Harnessing Consortia for Biotechnological Applications." *Computational and Structural Biotechnology Journal* 19: 3892–3907. doi:10.1016/j.csbj.2021.06.048

Jadhav, Jagruti V., Amit P. Pratap, and Sandeep B. Kale. 2019. "Evaluation of Sunflower Oil Refinery Waste as Feedstock for Production of Sophorolipid." *Process Biochemistry* 78: 15–24. doi:10.1016/j.procbio.2019.01.015

Jahan, Israt. 2023. "Advances and Applications of Nanomaterial-Assisted Phytoremediation." *Nanotechnology for Sustainable Agriculture, Food and Environment* (pp. 159–170). doi:10.1201/9781003397861-12

Kalvandi, Sahar, Hamidreza Garousin, Ahmad Ali Pourbabaee, and Hossein Ali Alikhani. 2022. "Formulation of a Culture Medium to Optimize the Production of Lipopeptide Biosurfactant by a New Isolate of Bacillus Sp.: A Soil Heavy Metal Mitigation Approach." *Frontiers in Microbiology* 13 (March). doi:10.3389/fmicb.2022.785985

Karaś, Magdalena Anna, Sylwia Wdowiak-Wróbel, and Wojciech Sokołowski. 2021. "Selection of Endophytic Strains for Enhanced Bacteria-Assisted Phytoremediation of Organic Pollutants Posing a Public Health Hazard." *International Journal of Molecular Sciences* 22 (17): 9557. doi:10.3390/ijms22179557

Khanna, Madhu, David Zilberman, Gal Hochman, and Bruno Basso. 2024. "An Economic Perspective of the Circular Bioeconomy in the Food and Agricultural Sector." *Communications Earth & Environment* 5 (1): 507. doi:10.1038/s43247-024-01663-6

Kim, Yumi, and Yul Roh. 2019. "Environmental Application of Biogenic Magnetite Nanoparticles to Remediate Chromium(III/VI)-Contaminated Water." *Minerals* 9 (5): 260. doi:10.3390/min9050260

Kumar, Ajay, Sandeep Kumar Singh, Chandra Kant, Hariom Verma, Dharmendra Kumar, Prem Pratap Singh, Arpan Modi, et al. 2021. "Microbial Biosurfactant: A New Frontier for Sustainable Agriculture and Pharmaceutical Industries." *Antioxidants* 10 (9): 1472. doi:10.3390/antiox10091472

Li, Junhui, Cuiyu Wu, Shaohua Chen, Qihong Lu, Hojae Shim, Xia Huang, Chongjian Jia, and Shanquan Wang. 2020. "Enriching Indigenous Microbial Consortia as a Promising Strategy for Xenobiotics' Cleanup." *Journal of Cleaner Production* 261 (July): 121234. doi:10.1016/j.jclepro.2020.121234

Lothe, Anjali G., Shashank Singh Kalra, Meng Wang, Elizabeth Erin Mack, Claudia Walecka-Hutchison, Valerie A. Kickhoefer, Leonard H. Rome, and Shaily Mahendra. 2020. "Vault Packaged Enzyme Mediated Degradation of Amino-Aromatic Energetic Compounds." *Chemosphere* 242 (March): 125117. doi:10.1016/j.chemosphere.2019.125117

Marshall, Tatianna, Kristine Lamont, Alejandro G. Marangoni, Loong Tak Lim, Xiuju Wang, and Erica Pensini. 2021. "Trypan Blue Removal from Water with Zein Sorbents and Laccase." *SN Applied Sciences* 3 (1): 1–10. doi:10.1007/s42452-020-04107-w

McNulty, Matthew J., Aaron J. Berliner, Patrick G. Negulescu, Liber McKee, Olivia Hart, Kevin Yates, Adam P. Arkin, Somen Nandi, and Karen A. McDonald. 2021. "Evaluating the Cost of Pharmaceutical Purification for a Long-Duration Space Exploration Medical Foundry." *Frontiers in Microbiology* 12 (October). doi:10.3389/fmicb.2021.700863

Muazu, Rukayya Ibrahim, Polina Yaseneva, Nilay Shah, and Maria-Magdalena Titirici. 2024. "Life Cycle Sustainability Assessment of Bioderived Advanced Materials: A State-of-the-Art Review." *Journal of Environmental Chemical Engineering* 12 (6): 114387. doi:10.1016/j.jece.2024.114387

Mutlu, Baris R., Jonathan K. Sakkos, Sujin Yeom, Lawrence P. Wackett, and Alptekin Aksan. 2016. "Silica Ecosystem for Synergistic Biotransformation." *Scientific Reports* 6 (1): 27404. doi:10.1038/srep27404

Naser, Ahmed Z., I. Deiab, and Basil M. Darras. 2021. "Poly(Lactic Acid) (PLA) and Polyhydroxyalkanoates (PHAs), Green Alternatives to Petroleum-Based Plastics: A Review." *RSC Advances* 11 (28): 17151–17196. doi:10.1039/D1RA02390J

Neal, Andrew L., Shakoor Ahmad, Ruth Gordon-Weeks, and Jurriaan Ton. 2012. "Benzoxazinoids in Root Exudates of Maize Attract Pseudomonas Putida to the Rhizosphere." *PLoS ONE* 7 (4): e35498. doi:10.1371/journal.pone.0035498

Osuntokun, Jejenija, Damian C. Onwudiwe, and Eno E. Ebenso. 2018. "Aqueous Extract of Broccoli Mediated Synthesis of CaO Nanoparticles and Its Application in the Photocatalytic Degradation of Bromocrescol Green." *IET Nanobiotechnology* 12 (7): 888–894. doi:10.1049/iet-nbt.2017.0277

Osuoha, Justice Obinna, Brilliance Onyinyechi Anyanwu, and Chisom Ejileugha. 2023. "Pharmaceuticals and Personal Care Products as Emerging Contaminants: Need for Combined Treatment Strategy." *Journal of Hazardous Materials Advances* 9 (February): 100206. doi:10.1016/j.hazadv.2022.100206

Pascual-García, Alberto, Sebastian Bonhoeffer, and Thomas Bell. 2020. "Metabolically Cohesive Microbial Consortia and Ecosystem Functioning." *Philosophical Transactions of the Royal Society B: Biological Sciences* 375 (1798): 20190245. doi:10.1098/rstb.2019.0245

Pham, Hoang Long, Hua Ling, and Matthew Wook Chang. 2022. "Design and Fabrication of Field-Deployable Microbial Biosensing Devices." *Current Opinion in Biotechnology* 76 (August): 102731. doi:10.1016/j.copbio.2022.102731

Phulpoto, Irfan Ali, Zhisheng Yu, Bowen Hu, Yanfen Wang, Fabrice Ndayisenga, Jinmei Li, Hongxia Liang, and Muneer Ahmed Qazi. 2020. "Production and Characterization of Surfactin-like Biosurfactant Produced by Novel Strain Bacillus Nealsonii S2MT and It's Potential for Oil Contaminated

Soil Remediation." *Microbial Cell Factories* 19 (1): 1–12. doi:10.1186/s12934-020-01402-4

Rafeeq, Hamza, Nadia Afsheen, Sadia Rafique, Arooj Arshad, Maham Intisar, Asim Hussain, Muhammad Bilal, and Hafiz M.N. Iqbal. 2023. "Genetically Engineered Microorganisms for Environmental Remediation." *Chemosphere* 310 (January): 136751. doi:10.1016/j.chemosphere.2022.136751

Rajtor, Monika, and Zofia Piotrowska-Seget. 2016. "Prospects for Arbuscular Mycorrhizal Fungi (AMF) to Assist in Phytoremediation of Soil Hydrocarbon Contaminants." *Chemosphere* 162 (November): 105–116. doi:10.1016/j.chemosphere.2016.07.071

Raji, Zarifeh, Ahasanul Karim, Antoine Karam, and Seddik Khalloufi. 2023. "Adsorption of Heavy Metals: Mechanisms, Kinetics, and Applications of Various Adsorbents in Wastewater Remediation: A Review." *Waste* 1 (3): 775–805. doi:10.3390/waste1030046

Ramírez, Ramsés E., Ricardo E. Buendia-Corona, Ivonne Pérez-Xochipa, and Thomas Scior. 2024. "Computational Binding Study Hints at Ecdysone 20-Mono-Oxygenase as the Hitherto Unknown Target for Ring C-Seco Limonoid-Type Insecticides." *Molecules* 29 (7). doi:10.3390/molecules29071628

Ravelo-Ortega, Gustavo, Javier Raya-González, and José López-Bucio. 2023. "Compounds from Rhizosphere Microbes That Promote Plant Growth." *Current Opinion in Plant Biology* 73 (June): 102336. doi:10.1016/j.pbi.2023.102336

Razzak, Shaikh A., Mohammed O. Faruque, Zeyad Alsheikh, Laila Alsheikhmohamad, Deem Alkuroud, Adah Alfayez, S. M. Zakir Hossain, and Mohammad M. Hossain. 2022. "A Comprehensive Review on Conventional and Biological-Driven Heavy Metals Removal from Industrial Wastewater." *Environmental Advances* 7 (April): 100168. doi:10.1016/j.envadv.2022.100168

Reichelt-Brushett, Amanda. 2023. "Pollution Mitigation and Ecological Restoration." In *Marine Pollution – Monitoring, Management and Mitigation* (pp. 317–337). Springer Nature. doi:10.1007/978-3-031-10127-4_15

Satpute, Surekha K., Grażyna A. Płaza, and Arun G. Banpurkar. 2017. "Biosurfactants' Production from Renewable Natural Resources: Example of Innovativeand Smart Technology in Circular Bioeconomy." *Management Systems in Production Engineering* 25 (1): 46–54. doi:10.1515/mspe-2017-0007

Schirpke, Uta, Erich Tasser, Stefan Borsky, Martin Braun, Josef Eitzinger, Veronika Gaube, Michael Getzner, et al. 2023. "Past and Future Impacts of Land-Use Changes on Ecosystem Services in Austria." *Journal of Environmental Management* 345 (November): 118728. doi:10.1016/j.jenvman.2023.118728

Scown, Corinne D. 2022. "Prospects for Carbon-Negative Biomanufacturing." *Trends in Biotechnology* 40 (12): 1415–24. doi:10.1016/j.tibtech.2022.09.004

Sharma, Mohit, Nihed Tellili, Imen Kacem, and Tarek Rouissi. 2024. "Microbial Biopolymers: From Production to Environmental Applications—A Review." *Applied Sciences* 14 (12): 5081. doi:10.3390/app14125081

Siddiqua, Ayesha, John N. Hahladakis, Wadha Ahmed, K. A. Al-Attiya. 2022. "An Overview of the Environmental Pollution and Health Effects Associated with Waste Landfilling and Open Dumping." *Environmental Science and Pollution Research* 29 (39): 58514–58536. doi:10.1007/s11356-022-21578-z

Soong, Ya-Hue Valerie, Margaret J. Sobkowicz, and Dongming Xie. 2022. "Recent Advances in Biological Recycling of Polyethylene Terephthalate (PET) Plastic Wastes." *Bioengineering* 9 (3): 98. doi:10.3390/bioengineering9030098

Sreekanth, K., P. Anjali, B.S. Akshatha, Riya Alex, and E.K. Radhakrishnan. 2024. "Enhanced Biodegradation of Chlorpyrifos in the Presence of Sub-Inhibitory Concentration of ZnONPs by Pseudomonas Sp. CF7b." *Process Safety and Environmental Protection* 190 (October): 256–263. doi:10.1016/j.psep.2024.07.079

Suman, Jachym, Ondrej Uhlik, Jitka Viktorova, and Tomas Macek. 2018. "Phytoextraction of Heavy Metals: A Promising Tool for Clean-Up of Polluted Environment?" *Frontiers in Plant Science* 9 (October). doi:10.3389/fpls.2018.01476

Trivedi, Pankaj, Chakradhar Mattupalli, Kellye Eversole, and Jan E. Leach. 2021. "Enabling Sustainable Agriculture through Understanding and Enhancement of Microbiomes." *New Phytologist* 230 (6): 2129–2147. doi:10.1111/nph.17319

Thomas, Amy, Laura Bentley, Chris Feeney, Stephen Lofts, Ciaran Robb, Ed C Rowe, Amanda Thomson, Eleanor Warren-Thomas, and Bridget Emmett. 2023. "Land Degradation Neutrality: Testing the Indicator in a Temperate Agricultural Landscape." *Journal of Environmental Management* 346 (November): 118884. doi:10.1016/j.jenvman.2023.118884

Umar, Aiman, Aneeqa Zafar, Hasina Wali, Meh Para Siddique, Muneer Ahmed Qazi, Afshan Hina Naeem, Zulfiqar Ali Malik, and Safia Ahmed. 2021. "Low-Cost Production and Application of Lipopeptide for Bioremediation and Plant Growth by Bacillus Subtilis SNW3." *AMB Express* 11 (1): 165. doi:10.1186/s13568-021-01327-0

Vandy, John Festus, Sasmita Mohanty, and Shibani Mohapatra. 2025. "Biotechnology Industry as a Catalyst for Circular Economy in India." *Discover Sustainability* 6 (1): 728. doi:10.1007/s43621-025-01473-6

Weng, Yilun, Guangze Yang, Yang Li, Letao Xu, Xiaojing Chen, Hao Song, and Chun-Xia Zhao. 2023. "Alginate-Based Materials for Enzyme Encapsulation." *Advances in Colloid and Interface Science* 318 (August): 102957. doi:10.1016/j.cis.2023.102957

Xun, Feifei, Baoming Xie, Shasha Liu, and Changhong Guo. 2015. "Effect of Plant Growth-Promoting Bacteria (PGPR) and Arbuscular Mycorrhizal Fungi (AMF) Inoculation on Oats in Saline-Alkali Soil Contaminated by Petroleum to Enhance Phytoremediation." *Environmental Science and Pollution Research* 22 (1): 598–608. doi:10.1007/s11356-014-3396-4

Ye, Jien, Yating Luo, Jiacong Sun, and Jiyan Shi. 2021. "Nanoscale Zero-Valent Iron Modified by Bentonite with Enhanced Cr(VI) Removal Efficiency, Improved Mobility, and Reduced Toxicity." *Nanomaterials* 11 (10): 2580. doi:10.3390/nano11102580

Yoshida, Shosuke, Kazumi Hiraga, Ikuo Taniguchi, and Kohei Oda. 2021. "Ideonella Sakaiensis, PETase, and MHETase: From Identification of Microbial PET Degradation to Enzyme Characterization." *Methods in Enzymology* 648: 187–205. doi:10.1016/bs.mie.2020.12.007

Zafra, German, Angélica Moreno-Montaño, Ángel E. Absalón, and Diana V. Cortés-Espinosa. 2015. "Degradation of Polycyclic Aromatic Hydrocarbons in Soil by a Tolerant Strain of Trichoderma Asperellum." *Environmental Science and Pollution Research* 22 (2): 1034–1042. doi:10.1007/s11356-014-3357-y

Zhou, Qiaohong, Kaixuan Wu, Lu Yao, Ruonan Chen, Shouzhuang Liu, Hao Xing, Linmei Nie, and Zhenbin Wu. 2024. "Bio-Ecological Remediation of Freshwater Aquaculture Environments: A Systematic Review and Bibliometric Analysis." *Water Biology and Security* 3 (1): 100229. doi:10.1016/j.watbs.2023.100229

12 Global Biomanufacturing Research Trends, Innovations, and Challenges

Prasann Kumar
Department of Agronomy, School of Agriculture, Lovely Professional University, Phagwara, Punjab, India

Joginder Singh
Department of Botany, Nagaland University, Lumami, Nagaland, India

12.1 INTRODUCTION

Agro-based biomanufacturing marks a significant shift in the field. It is no longer confined to traditional uses of biological systems, such as medicinal plant cells, microorganisms, or animal cells, for producing valuable products on a large scale. Instead, it positions the entire agricultural sector as the foundation for creating a new bioeconomy (Adamo et al. 2025; Gao et al. 2025a). In this future, advanced biological techniques convert renewable plant and microbial biomass into a wide variety of products that were once derived from fossil fuels or intensive traditional farming. This shift from the old to the new bioeconomy is driven by technological convergence, starting with ancient practices such as brewing, cheesemaking, and composting, which relied on empirical microbial knowledge before the science was fully understood. It then progresses through 20th-century breakthroughs, including large-scale antibiotic production (e.g., penicillin) and single-cell protein (SCP), which established core principles for scaling bioreactions and industrial microbiology (Duan et al. 2024; Li et al. 2025). Moving into the 21st century, the integration of cutting-edge fields such as synthetic biology, allowing precise microbial design, advanced genomics with Clustered Regularly Interspaced Short Palindromic Repeats (CRISPR) gene editing for optimizing microbes and crops at the DNA level, and artificial intelligence (AI) combined with automation for rapid testing and sophisticated bioreactor management has transformed fermentation into an accurate, efficient, and predictable industrial process. This evolution enables the production of novel biomolecules, functional proteins, bio-based materials, and sustainable agricultural inputs (García Martínez et al. 2024; Gavrilescu 2024).

The strategic importance of this agricultural biomanufacturing revolution is multifaceted. It plays a vital role in global sustainability by replacing petrochemical industries with bio-based products, such as plastics, biofuels, and chemicals, that utilize atmospheric carbon fixed by plants, supporting a circular carbon economy and reducing environmental pollution and dependence on finite fossil resources (Fan et al. 2025; Sonnenberg et al. 2025). It is also closely tied to food security, not by replacing traditional farming but by enhancing it through the production of precision-fermented alternative proteins, such as lab-grown meat and yeast-made dairy proteins. These foods can be nutritious and familiar while requiring far less land, water, and greenhouse gases than conventional animal farming. Additionally, innovations include the production of feed additives to reduce methane emissions from livestock, as well as biostimulants and biopesticides that enhance crop resilience and yields in the face of climate change. These innovations also aim to establish a more robust and efficient global food system that can sustainably feed a growing population (Pandey and Singhal 2022; Wang et al. 2024).

The growth of agricultural biomanufacturing based on integrated bioprocessing platforms offers significant new opportunities. It can create high-value markets and generate revenue for farmers by shifting from bulk commodity production to supplying specialized feedstocks, such as energy crops and agricultural residues. It encourages demand-driven investments in biotech and biomanufacturing hubs that develop new materials and chemicals from biological sources. It promotes strategies for healthy material aging by addressing end-use requirements from the crop's inception. It diversifies domestic supply chains for essential components, thereby enhancing national competitiveness and economic resilience (Love et al. 2013; Asin-Garcia et al. 2025).

This chapter clearly defines its scope and goals. It presents a broad, systematic framework for understanding this transformative field from establishing a precise definition of agricultural biomanufacturing, distinct from traditional agriculture and broader industrial biotech, to tracing its historical and technological evolution. It provides contextual insight into the present and future of the field, supported by empirical case studies highlighting the strategic value of agricultural biomanufacturing (Chen et al. 2022; Tellechea-Luzardo

 DOI: 10.1201/9781003637158-12

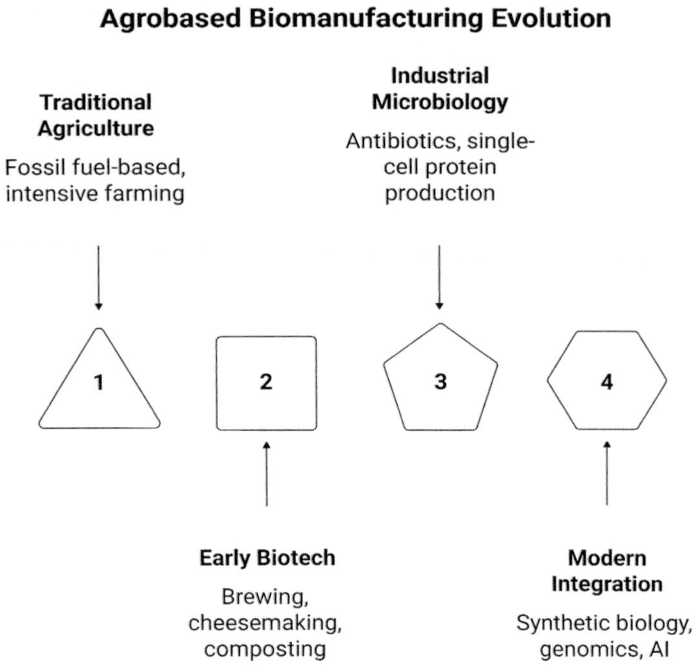

FIGURE 12.1 Agro-based biomanufacturing evolution.

et al. 2022). The chapter emphasizes the necessity of this technology in addressing global sustainability, food security, and economic development, while acknowledging the regulatory, technological, and social challenges that lie ahead. Its ultimate aim is to equip policymakers, industry leaders, researchers, and students with essential knowledge and critical insights to responsibly and sustainably advance this key intersection of biology, agriculture, and industry, guiding us toward a greener, more hopeful future (Figure 12.1).

12.2 GLOBAL RESEARCH TRENDS IN AGRICULTURAL BIOMANUFACTURING

Global research and development trends in agricultural biomanufacturing are primarily driven by a strategic shift away from the first-generation feedstock model, which relies heavily on non-dedicated, edible, and sugar-rich food crops, such as corn and sugarcane. This approach raises ethical concerns about conflicts between food and fuel, as well as limitations to scalability (Gu et al. 2023; Gao et al. 2025b). The new strategy favors a second-generation feedstock system emphasizing non-food, lignocellulosic biomass sources found in today's agricultural residues (e.g., corn stover, wheat straw, rice husks, and sugarcane bagasse), dedicated energy crops such as fast-growing perennial grasses, such as switchgrass and miscanthus, short-rotation woody crops such as poplar and willow, and third-generation aquatic biomass such as macroalgae (seaweed) and microalgae (Hetzler et al. 2023; Lu et al. 2025). These sources do not compete with traditional crops or freshwater supplies, as they benefit from higher photosynthetic efficiencies and faster growth rates. The focus on feedstock innovation aims to overcome the recalcitrance of lignocellulosic biomass, a complex matrix of cellulose, hemicellulose, and lignin resistant to breakdown, by developing advanced, energy-efficient, environmentally friendly pre-treatment technologies (such as steam explosion, ammonia-assisted steam explosion, and ionic liquids) along with discovering and engineering novel microbial consortia and enzyme mixes to lower sugar extraction costs. At the same time, algal biotechnology efforts focus on genetically modifying strains to increase their lipid content for biofuel production or to produce high-value compounds, as well as optimizing photobioreactor and cultivation systems for scalable, cost-effective yields (Harcum and Kiss 2022; Germain and Winn 2025). The goal is to turn waste streams and undervalued resources into standardized, sustainable, and scalable feedstock building blocks that support the entire biomanufacturing economy. This shift in feedstock strategy is closely linked to another major trend: a worldwide adoption of waste valorization and circular bioeconomy principles. This approach goes beyond simple waste-to-energy ideas, favoring integrated biorefineries designed to extract maximum value from each molecular component, mimicking the zero-waste philosophy of petroleum refineries. Research in this area focuses on designing fully integrated processes that can separate heterogeneous biomass into components, such as C5 and C6 sugars, lignin, proteins, and nutrients, and convert each into reusable products. Examples include fermenting sugars into bioethanol, biobutanol, and platform chemicals such as succinic and lactic acids for bioplastics like PLA; transforming lignin, once considered waste, into renewable aromatic chemicals, bio-based plastics, carbon fibers, or binders; concentrating nutrients and proteins from process streams for animal feed or fertilizers; and using methane from anaerobic digestion of wastewaters for energy (Sørensen et al. 2022; Zhao et al. 2025). These closed-loop systems, where waste from one process becomes input for another, significantly boost lifecycle sustainability, reduce environmental impacts, and improve economic viability by generating multiple revenue streams. However, regional differences influence global research programs, shaped by specific geopolitical factors, resource availability, and policy environments. In North America, particularly the United States and Canada, research is driven by large-scale, market-oriented efforts rooted in a strong tradition of agri-biochemical engineering, utilizing extensive feedstocks from the corn and wheat belts and forest residues to produce biofuels (in response to mandates like the Renewable Fuel Standard [RFS]) and bio-based chemicals, with major contributions from the US Department of Energy's National Laboratories and significant investments by leading agricultural companies. In the European Union (EU), the research focus is mainly policy-driven, emphasizing lifecycle assessment (LCA), sustainability certification, and advanced waste valorization technologies, often within cross-border Horizon Europe consortia and collaborations with higher education institutions (Hetzler et al. 2023; Del Hierro et al. 2024; Asin-Garcia et al. 2025; Germain and Winn 2025). In the Asia-Pacific region, efforts balance necessity with ambition: China invests heavily in developing advanced infrastructure and new technologies to reduce air pollution caused by agricultural waste burning and to cut dependence on chemical and energy imports, using rice straw and other residues; Japan and South Korea lead in high-value biotechnologies, such as genetically modifying microalgae and cyanobacteria to produce nutraceuticals, cosmetics, and fine chemicals; Australia concentrates on drought-tolerant crops and seaweed cultivation (Harcum and Kiss 2022; Gu et al. 2023; Hetzler et al. 2023; Del Hierro et al. 2024; Asin-Garcia et al. 2025). Underlying these efforts are significant and evolving investments from both the domestic public and private sectors. Government funding in the United States, the EU, and increasingly, China support high-risk, early stage research in synthetic biology, metabolic engineering, and process improvement through large grants and national initiatives. Venture capital and corporate investments continue to fund startups progressing toward pilot and commercial scales, shifting from solely biofuel companies to those developing alternative proteins, innovative biomaterials, and specialized enzymes. Social entrepreneurs seek models that transform waste streams into revenue or reduce inputs for biofuel production, particularly those that are scalable and bio-based, such as aviation fuels (sustainable aviation fuel [SAF]). This interconnected landscape of feedstock innovation, circular economy principles, regional priorities, and investment trends collectively shapes the global biomanufacturing sector. Moving toward sustainability, economic viability, and long-term resilience, the ultimate goal remains to decouple human industry from fossil resources and minimize environmental harm (Figure 12.2)

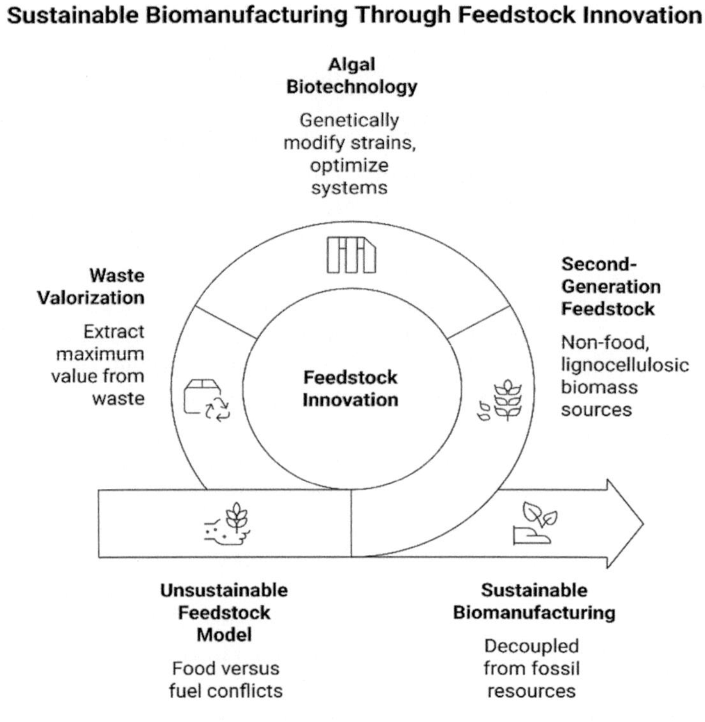

FIGURE 12.2 Sustainable biomanufacturing through feedstock innovation.

12.3 KEY TECHNOLOGICAL INNOVATIONS DRIVING THE SECTOR

Several innovative technologies are transforming the way biological systems function, making them more efficient, scalable, and sustainable. This has spurred rapid growth in biotechnology for agriculture and environmental science. Synthetic biology and metabolic engineering drive this revolution, enabling the design of microbial cell factories that produce high-value compounds, such as biofuels, bioplastics, plant growth enhancers, and medicines, in agricultural settings. Scientists can modify microorganisms, such as *Bacillus* spp., Pseudomonas, and engineered *Escherichia coli*, to utilize their resources more efficiently and produce desired products while reducing energy use and byproduct formation (Love et al. 2013; Harcum and Kiss 2022; Sørensen et al. 2022; Tellechea-Luzardo et al. 2022; Gao et al. 2025b). These microbial hosts are becoming increasingly capable of surviving in natural conditions, such as soil with limited nutrients or heavy metal stress, making them ideal for *in situ* bioremediation and biofertilizer applications. The Clustered Regularly Interspaced Short Palindromic Repeats and CRISPR-associated protein (CRISPR-Cas) system is a crucial tool that facilitates these advances by enabling precise genome editing, knocking out multiple genes simultaneously, and fine-tuning biosynthetic pathways. CRISPR-based tools have accelerated strain development by optimizing enzyme levels, eliminating competing pathways, and introducing genes from extremophiles or other resilient organisms to enhance stress tolerance and catalytic performance. For example, CRISPR has been used to create bacterial strains with higher phytase activity, increasing phosphorus solubility and improving nitrogen fixation, which helps sustainable farming reduce chemical fertilizer use. Combining synthetic biology with microbial community design further enhances functionality; for example, one species might produce signaling molecules that activate secondary metabolite synthesis in another, thereby mimicking natural ecological interactions and making agro-biotechnological applications more stable and resilient (Love et al. 2013; Sørensen et al. 2022; Tellechea-Luzardo et al. 2022; Hetzler et al. 2023; Zhao et al. 2025).

Along with advances in genetics, advanced bioprocess engineering and process intensification have substantially improved the scalability and cost-effectiveness of bio-based production systems. High-throughput fermentation platforms with real-time monitoring of pH, dissolved oxygen, temperature, and biomass are increasingly replacing traditional batch fermentation methods. These platforms enable you to adjust culture conditions to maximize yield and productivity dynamically. They utilize small bioreactors and automated liquid-handling systems to test hundreds of microbial strains simultaneously under various environmental conditions, thereby accelerating the process of identifying the optimal strains and conditions. The shift from batch to continuous bioprocessing is even more impactful because it boosts volumetric productivity, ensures consistent product quality, and reduces operational downtime. In continuous mode, a new medium is added to the bioreactor, and products are collected in real time, thereby maintaining the cells in their most productive physiological state for extended periods (Harcum and Kiss 2022; Gu et al. 2023; Hetzler et al. 2023; Castillo-Saldarriaga et al. 2024; Asin-Garcia et al. 2025; Zhao et al. 2025). This approach not only optimizes resource use but also aligns well with decentralized, on-farm biomanufacturing models. Small-scale, modular bioreactors can produce biofertilizers or biostimulants locally, thereby reducing transportation costs and lowering carbon footprints. Downstream processing has advanced significantly through techniques such as membrane filtration, aqueous two-phase extraction, and affinity-based purification strategies. These methods consume less energy and solvents while recovering a higher percentage of target biomolecules. *In situ* product removal (ISPR) is a new approach that integrates separation units directly into the bioreactor, preventing product inhibition and enabling longer cultivation times. Using immobilized enzymes or whole-cell biocatalysts on porous supports also enhances their stability and reusability, making large-scale enzymatic conversions more cost-effective (Tellechea-Luzardo et al. 2022; Wang et al. 2024; Asin-Garcia et al. 2025; Fan et al. 2025; Gao et al. 2025b).

The use of AI and machine learning (ML) in strain development and bioprocess optimization is also expanding. This complements the biological and engineering breakthroughs already underway. AI-driven algorithms analyze large amounts of data from genomics, transcriptomics, proteomics, and phenomics to predict gene functions, identify regulatory elements, and model metabolic fluxes with remarkable accuracy. ML models trained on experimental data can predict optimal genetic modifications or media compositions without the need for numerous lab experiments, significantly accelerating development (Pandey and Singhal 2022; Tellechea-Luzardo et al. 2022; García Martínez et al. 2024; Asin-Garcia et al. 2025; Li et al. 2025). In fermentation processes, ML models supported by sensor networks help detect contamination, forecast optimal harvest times, and adjust feeding strategies in real time to sustain peak performance. Researchers are exploring reinforcement learning frameworks to automatically control bioreactor parameters in real time, creating systems that can optimize themselves and adapt to changing conditions. Federated learning, a privacy-protecting form of distributed ML, is also becoming a key tool in the agricultural sector. It allows multiple farms or research institutions to collaboratively train predictive models using local microbial performance data without sharing raw data. This maintains data sovereignty while making models more applicable across various agroecosystems. This approach is beneficial for developing inoculants tailored for specific regions or microbial formulations effective in diverse climates, considering soil type, crop type, and local weather patterns (Gu et al. 2023; Hetzler et al. 2023;

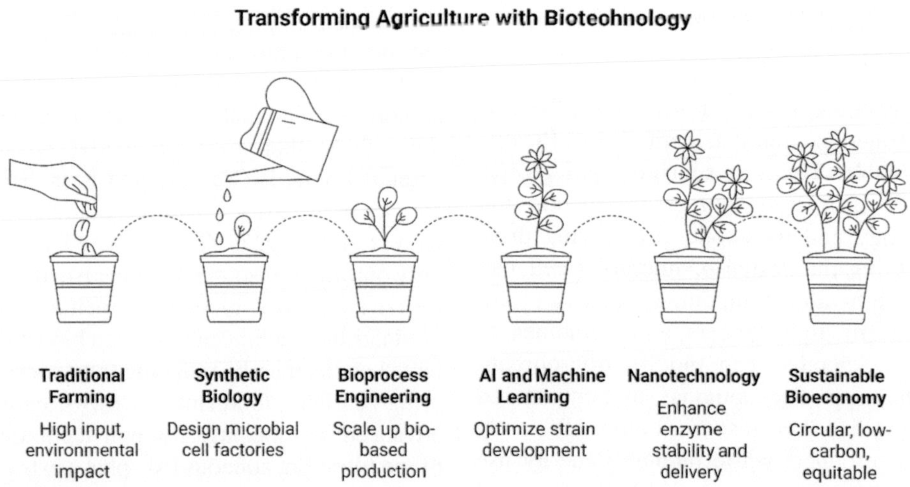

FIGURE 12.3 Transforming agriculture with biotechnology.

Wang et al. 2024; Asin-Garcia et al. 2025; Germain and Winn 2025). The integration of nanotechnology into biocatalysis and product delivery systems is equally important, offering unprecedented levels of precision and efficiency. Nanomaterials, such as mesoporous silica nanoparticles, carbon nanotubes, and magnetic nanocomposites, are highly effective in transporting enzymes. They enhance enzyme stability, recyclability, and catalytic activity, even under demanding industrial or environmental conditions. When enzymes are immobilized on nanostructured surfaces, they exhibit a higher surface-to-volume ratio and improved mass transfer kinetics, thereby accelerating reactions and enhancing yields. Nanobiohybrids are under development for agricultural purposes to deliver microbial inoculants or bioactive compounds directly to the rhizosphere (Tellechea-Luzardo et al. 2022; Hetzler et al. 2023; Germain and Winn 2025; Song et al. 2025; Zhao et al. 2025). This shields sensitive microbes from drying out, UV radiation, or being eaten during application. Nanoencapsulation techniques allow plant-growth-promoting rhizobacteria (PGPR) to release growth-promoting hormones, siderophores, and volatile organic compounds (VOCs) in a controlled manner, thereby extending their effectiveness in the soil. Quantum dots and plasmonic nanoparticles are also used as biosensors to detect pathogens, nutrient deficiencies, and pollution levels in real time, enabling early intervention. Organic semiconductor-based bioelectronics are paving the way for smart implants and bioelectronic interfaces that monitor root health, modify microbial activity, or deliver electrical stimulation to improve nutrient uptake and stress resistance. These soft, biocompatible devices work seamlessly with living tissues, enabling closed-loop systems that adapt to environmental changes (Love et al. 2013; Harcum and Kiss 2022; Asin-Garcia et al. 2025; Gao et al. 2025b).

These technological fields, synthetic biology, advanced bioprocessing, AI/ML, and nanotechnology, are converging to foster an innovative, sustainable bioeconomy rooted in agriculture. They support the shift from traditional farming, which relies heavily on inputs, to precision microbiome management, where biological solutions are developed, refined, and applied with scientific rigor and environmental foresight. For example, combining CRISPR-edited microbial communities with AI-guided fermentation protocols and nano-enabled delivery systems enables the large-scale production of next-generation biofertilizers. These biofertilizers outperform single-strain products in colonization efficiency, durability, and functional output. Likewise, integrating real-time sensing with edge computing allows for adaptive decision-making on farms, supporting practices such as variable-rate inoculant application or microbial activation in response to drought conditions (Harcum and Kiss 2022; Sørensen et al. 2022; Gu et al. 2023; Del Hierro et al. 2024; Zhao et al. 2025). These innovations also contribute to larger goals, such as climate-smart agriculture, including carbon sequestration through enhanced root exudation, reduction of greenhouse gases via improved nitrogen cycling, and land restoration through microbial-assisted phytostabilization. Recent research indicates that synthetic microbial communities are more resilient and stable in the rhizosphere than individual strains, suggesting that they can be utilized to modify ecosystems over time. The combination of these technologies is not merely additive; it's synergistic, creating feedback loops in which field data inform laboratory-based strain redesign, which in turn improves process performance and product effectiveness. Looking ahead, future focus will likely be on cross-layer integration of digital twins that simulate entire production pipelines from gene circuit design to final formulation, allowing the creation of virtual prototypes before deployment. Furthermore, portable bioreactors and blockchain-tracked supply chains could make cutting-edge agri-biotech solutions accessible to everyone, especially in resource-limited areas. However, challenges remain, such as ensuring biosafety and regulatory compliance, shifting public perception, and bridging the gap between academic innovation and commercial application (Love et al. 2013;

Sørensen et al. 2022; Tellechea-Luzardo et al. 2022; Hetzler et al. 2023; Gao et al. 2025b; Song et al. 2025; Zhao et al. 2025). Addressing these issues will require collaboration among molecular biologists, engineers, data scientists, agronomists, and policymakers across disciplines. The integration of synthetic biology, bioprocess intensification, AI, and nanotechnology signifies a profound transformation in how we harness living systems for sustainable growth. It promises to increase crop productivity, bolster environmental resilience, and fundamentally alter agricultural innovation. Ultimately, it aims to establish biology as the foundation of a global food system that is circular, low-carbon, and equitable (Figure 12.3).

12.4 EMERGING PRODUCT PORTFOLIO FROM AGRICULTURAL BIOMANUFACTURING

The convergence of biotechnology, synthetic biology, and advanced manufacturing is driving a significant transformation in the agricultural sector. It is no longer just a place to grow food; it is now a dynamic platform for agricultural biomanufacturing that can make a wide range of high-value, long-lasting products. Bio-based materials and polymers, such as polylactic acid (PLA) and polyhydroxyalkanoates (PHA), are at the forefront of this shift. Increasingly, these materials are being produced through microbial fermentation using agricultural feedstocks such as sugarcane, corn stover, and lignocellulosic biomass (Gu et al. 2023; Gavrilescu 2024; Wang et al. 2024; Asin-Garcia et al. 2025; Fan et al. 2025; Lu et al. 2025). These biopolymers are more environmentally friendly than petroleum-based plastics. PLA is used in biodegradable packaging, mulch films, and controlled-release agrochemical delivery systems. PHAs, conversely, are produced by bacteria such as *Cupriavidus necator* and have better mechanical properties, allowing them to be composted entirely under natural conditions. This makes them ideal for precision agriculture tools, seed coatings, and temporary soil sensors. Combining engineered microbial cell factories with circular economy principles enables us to convert crop waste and by-products into valuable biomaterials. This reduces environmental damage and enhances supply chain efficiency. Technologies such as precision fermentation and cellular agriculture are transforming the way we approach nutritional security by utilizing microorganisms, such as yeast (*Pichia pastoris*) or filamentous fungi, that have been genetically engineered to produce animal-free proteins, dairy components (like whey and casein), and alternative fats that mimic the taste and texture of traditional animal products. These fermentation-derived ingredients are now being scaled for inclusion in plant-based meats, aquaculture feeds, and livestock supplements, offering scalable solutions to reduce deforestation, methane emissions, and land-use competition associated with traditional animal farming. At the same time, SCPs from bacteria, algae, or yeasts grown on inexpensive substrates such as CO_2, methane, or industrial flue gases are gaining popularity as protein-rich feed additives (Chen et al. 2022; Tellechea-Luzardo et al. 2022; Wang et al. 2024; Asin-Garcia et al. 2025; Li et al. 2025; Sonnenberg et al. 2025). These can be produced locally, even in dry or degraded areas, which makes feed more sustainable and resilient in agricultural regions vulnerable to climate change.

At the same time, the field of bioenergy and biofuels has evolved from first-generation ethanol to more advanced biofuels, including biobutanol, renewable diesel, green aviation fuel (SAF), and biogas-to-energy systems. These new bioprocessing methods have made this possible. Second- and third-generation biofuels utilize algal feedstocks and non-food biomass sources, such as switchgrass, miscanthus, and agricultural residues. This means less competition with food production and more potential for carbon sequestration. Consolidated bioprocessing (CBP) strategies utilize engineered microbes to integrate enzyme production, saccharification, and fermentation into a single step. This significantly reduces production costs and energy consumption. Additionally, syngas fermentation and microbial electrosynthesis platforms enable acetogenic bacteria, such as *Clostridium ljungdahlii*, to convert gaseous carbon sources (CO and CO_2) into liquid fuels. This creates new methods for capturing and utilizing carbon (CCU) in agricultural areas. When used in conjunction with anaerobic digestion systems on farms, these new ideas enable on-site energy generation from manure and organic waste. This helps with energy self-sufficiency and rural electrification. Along with these large-scale results, a growing supply of high-value biochemicals, enzymes, and biostimulants is being developed for sustainable crop management. Engineered microbes are now used to produce specialty chemicals such as 1,3-propanediol, succinic acid, and muconic acid (Love et al. 2013; Chen et al. 2022; Pandey and Singhal 2022; Duan et al. 2024; García Martínez et al. 2024; Gao et al. 2025a; Gao et al. 2025b). These chemicals are used to make bioplastics, resins, and agrochemical intermediates. Phytases, cellulases, and laccases are examples of industrial enzymes produced by recombinant expression in microbial hosts. They are used to make animal feed, improve soil, and break down pesticides, thereby making nutrients more available and reducing the need for chemicals. The development of microbial biostimulants, comprising PGPR, fungal endophytes, and synthetic microbial consortia (SynComs), is probably the most significant. These biostimulants improve root structure, nitrogen fixation, phosphorus solubilization, and stress tolerance. Recent research suggests that inoculating crops with optimized bacterial communities, specifically those comprising *Bacillus cereus*, *Bacillus thuringiensis*, and *Herbaspirillum huttiense*, can significantly enhance crop performance under conditions of heavy metal stress, drought, and salinity, thereby establishing biostimulants as essential facilitators of regenerative agriculture and soil health restoration (Love et al. 2013; Chen et al. 2022; Pandey and Singhal 2022; Gu et al. 2023; Li et al. 2025; Sonnenberg et al. 2025).

The production of nutraceuticals, cosmeceuticals, and biopharmaceuticals, high-margin, bioactive compounds derived from engineered crops and microbial systems, further expands the scope of agricultural biomanufacturing. Fermentation tanks and transgenic plants are now being used to biosynthesize nutraceuticals, including omega-3 fatty acids (eicosapentaenoic acid [EPA]/docosahexaenoic acid [DHA]), resveratrol, curcuminoids, and vitamin B12 analogs. These are sustainable and traceable sources for dietary supplements and functional foods. Algae and yeast platforms have become proficient in producing astaxanthin, lutein, and squalene. These are chemicals that are often used in anti-aging skin care products and immune-boosting formulas. They are also cruelty-free and eco-friendly alternatives to ingredients that come from animals or petrochemicals (Love et al. 2013; Tellechea-Luzardo et al. 2022; Hetzler et al. 2023; Song et al. 2025; Zhao et al. 2025). Molecular pharming, which utilizes genetically modified plants or plant cell cultures to produce therapeutic proteins, is gaining popularity in the pharmaceutical industry. For example, monoclonal antibodies, vaccines (such as oral vaccines in edible plants), and recombinant enzymes have all been successfully expressed in species such as tobacco, lettuce, and duckweed. These "biopharming" platforms promise lower production costs, a reduced risk of mammalian pathogens contaminating the products, and the ability to manufacture products in multiple locations, which is beneficial for regions with limited resources. The combination of organic semiconductor-based bioelectronics with biological systems also opens up new avenues for monitoring the real-time production of metabolites during fermentation or field cultivation. This enables closed-up control of bioprocesses and the use of innovative release mechanisms for bioactive payloads. Advancements in continuous bioprocessing, AI-driven strain optimization, and nanotechnology-enabled delivery systems have made these new product lines even more scalable. For example, nanoparticles protect sensitive molecules during application and ensure they are released in the right place, such as the rhizosphere or gastrointestinal tract (Chen et al. 2022; Gu et al. 2023; Hetzler et al. 2023; Gavrilescu 2024; Wang et al. 2024; Fan et al. 2025). The transition from lab-scale proof of concept to commercial deployment is contingent upon addressing challenges associated with regulatory approval, public acceptance, and LCA. Recent frameworks for precision microbiome management demonstrate that interdisciplinary collaboration among agronomists, microbiologists, engineers, and policymakers is fostering innovation ecosystems that combine scientific rigor with practical applications. Federated learning models and edge-computing architectures are being investigated to enhance fermentation parameters across dispersed farms while safeguarding data privacy, thereby guaranteeing fair access to advanced biomanufacturing tools. Additionally, these technologies align with global sustainability goals, including Sustainable Development Goal 2 (SDG 2) (Zero Hunger), SDG 12 (Responsible Consumption and Production), and SDG 13 (Climate Action). This demonstrates the importance of these initiatives in creating robust, low-carbon agricultural economies. As climate change puts more stress on natural resources, the ability to convert biomass, CO_2, and unused waste into valuable products makes agricultural biomanufacturing a key component of the bioeconomy. Future directions encompass the advancement of self-replicating biofactories, living materials with adaptive functionalities, and autonomous microbial consortia capable of detecting environmental stimuli and adjusting their output accordingly (Love et al. 2013; Chen et al. 2022; Pandey and Singhal 2022; Gu et al. 2023; Li et al. 2025; Sonnenberg et al. 2025). With ongoing funding for research and development, infrastructure, and policy support, the new products emerging from agricultural biomanufacturing have the potential to transform agriculture into a multi-output, circular, and innovation-driven industry. This industry will not only feed the world but also power its industries, heal its people, and restore its ecosystems (Figure 12.4).

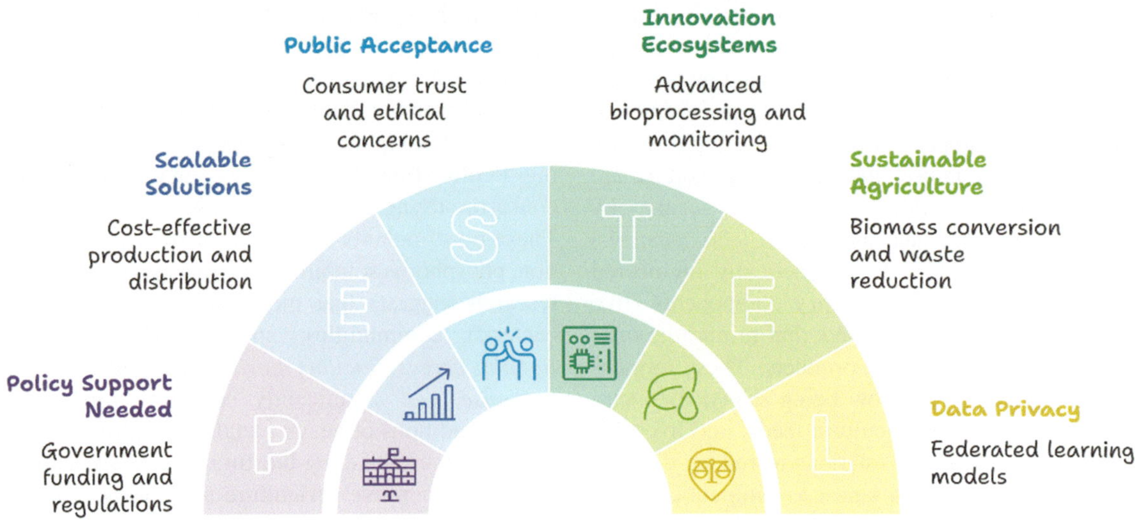

FIGURE 12.4 Agricultural biomanufacturing transformation.

12.5 CRITICAL CHALLENGES AND BARRIERS TO ADOPTION

The progression of disruptive biotechnological breakthroughs, especially those based on bioproduction platforms such as microbial cell factories and synthetic biology-enabled pathways, faces a complex set of critical challenges. These include technological, economic, logistical, regulatory, social, and environmental issues that collectively slow down progress and limit the potential for scaling. Central to these obstacles is the inherent difficulty of transitioning processes from controlled lab settings to full industrial production (Chen et al. 2022; Gu et al. 2023; Gavrilescu 2024; Wang et al. 2024; Fan et al. 2025; Sonnenberg et al. 2025). Challenges such as mass transfer limitations, inefficient heat dissipation, product inhibition, and microbial instability under industrial fermentation conditions often result in performance below laboratory expectations. Additionally, the large investments needed for bioreactor infrastructure and processing equipment, including sterilization, introduce significant upfront risks without guaranteed returns. These challenges are exacerbated by process integration issues such as the need for continuous versus batch operations and real-time adaptive monitoring that increase economic risks, particularly in emerging markets that lack established engineering standards or modular frameworks (Love et al. 2013; Chen et al. 2022; Pandey and Singhal 2022; Gu et al. 2023; Gavrilescu 2024; Gao et al. 2025a; Li et al. 2025; Sonnenberg et al. 2025). Moreover, reliable, cost-effective, and environmentally friendly access to lignocellulosic biomass depends heavily on local availability, seasonal supply, transportation costs, and competition with land-use systems, creating uncertainty and price fluctuations. The pre-treatment of tough, heterogeneous feedstocks to release fermentable sugars requires substantial resources and often generates toxic by-products, such as furans, phenolics, and organic acids, which harm cell viability and reduce yields, necessitating detoxification or strain engineering. Economics remain a major obstacle, as bio-based products, including bulk chemicals, biofuels, pharmaceuticals, and bioplastics, must compete with well-established petrochemical industries that benefit from centuries of infrastructure and economies of scale. Historically low fossil fuel prices further widen the cost gap, which narrows only occasionally through policy support, emissions pricing, or breakthroughs in titers, rates, and yields (TRY). Even with improved bioprocessing, total ownership costs encompassing feedstock, utilities, labor, waste disposal, and depreciation often exceed market tolerances, particularly in commodity markets characterized by price sensitivity and narrow margins (Pandey and Singhal 2022; Gavrilescu 2024; Asin-Garcia et al. 2025; Li et al. 2025). Consequently, private venture capital is cautious, limiting commercial efforts mostly to niche markets without strong public-private partnerships or incentives. Regulatory barriers also hinder progress; pathways for approving novel bioproducts, especially those from Genetically Modified Organisms (GMOs), engineered pathways, or synthetic biology, are fragmented, uncertain, and often lag behind technological advances. This results in lengthy approval times, high compliance costs, and legal uncertainties, for example, the EU's strict GMO regulations that enforce containment and labeling, which can make certain applications unfeasible. Environmental safety assessments covering ecotoxicology, gene transfer models, and ecosystem impacts are costly and time-consuming, discouraging smaller companies. Social acceptance and public perception are equally important; despite scientific backing for their safety, GMO-derived products face mistrust rooted in past controversies, misinformation, and ethical concerns. This influences market adoption, especially where "natural" and "non-GMO" branding commands higher prices, leading to a paradox in which advanced, sustainable alternatives are often rejected more on political grounds than on evidence. Lack of transparency and corporate opacity further erode public trust, encouraging activist-driven policies and investment hurdles. To manage these issues and develop effective policies, comprehensive LCA is vital for accurately assessing sustainability beyond basic carbon metrics. Bio-based systems can mask other environmental impacts, such as land-use change emissions, water use, nutrient runoff, biodiversity loss, and end-of-life biodegradability, which may offset greenhouse gas savings. Current LCA methods struggle to quantify regional effects, such as water scarcity, which can mislead sustainability claims and provoke accusations of greenwashing (Tellechea-Luzardo et al. 2022; Wang et al. 2024; Gao et al. 2025b; Sonnenberg et al. 2025). Furthermore, there are no standardized metrics across circularity, labor conditions, or energy policies for future modeling, limiting stakeholders' ability to make evidence-based decisions. Moving forward, systems-level approaches are crucial to overcome these interconnected challenges. These include improving strain resilience through adaptive evolution, creating modular and continuous bioprocesses to reduce capital costs, employing digital twin technology for predictive scale-up, and utilizing blockchain for enhanced traceability. Harmonizing global regulations can make oversight more proportional based on risk and impact. At the same time, engagement with the public, participatory assessments, and next-generation LCAs that consider planetary boundaries and social equity are essential. Only through multidisciplinary efforts can biomanufacturing meet its potential as a key driver of a circular bioeconomy, reducing reliance on fossil fuels and providing significant environmental and societal benefits. Achieving this will depend on sustained public funding for translational research aligned with energy, agriculture, and industry policies, as well as a cultural shift that prioritizes long-term sustainability over short-term gains, transforming today's pressing challenges into opportunities for resilient, equitable, and decarbonized industrial systems (Figure 12.5)

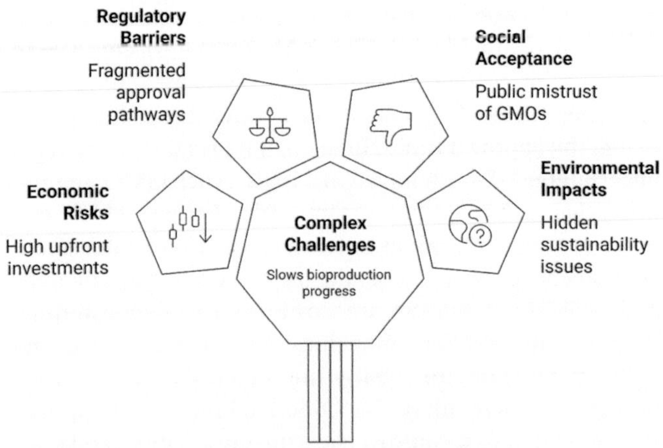

FIGURE 12.5 Challenges hinder bioproduction scaling.

12.6 FUTURE PERSPECTIVES AND CONCLUDING REMARKS

The future of biomanufacturing as a transformative pillar of the global bioeconomy depends on its strategic integration with existing agricultural systems, the establishment of coherent and adaptable policy frameworks, the cultivation of interdisciplinary research ecosystems, and a long-term vision that reimagines agriculture not merely as a source of food but as a dynamic, multifunctional platform for sustainable production. To become commercially viable on a large scale, next-generation biomanufacturing must move beyond isolated biorefinery models and integrate into the agricultural value chain. This involves utilizing crop residues, cover crops, underutilized marginal lands, and integrated agro-industrial waste streams as feedstocks, while also enhancing soil health, biodiversity, and rural livelihoods. This integration requires the co-design of cropping systems and bioprocessing infrastructure. Precision agriculture, remote sensing, and digital agronomy can help with feedstock quality and logistics. Decentralized, modular bioreactors located near biomass sources can reduce transportation costs and emissions, enhance the supply chain's resilience, and foster localized circular economies. Additionally, dual-use cropping strategies, such as growing oilseed crops for both food and oleochemical feedstocks, or creating nitrogen-fixing cover crops that enhance soil fertility and yield fermentable carbohydrates simultaneously, can balance the need for food security with bio-based production goals. This reduces competition for land use and follows the principles of sustainable intensification. To make this vision a reality, however, we need strong, forward-thinking policy frameworks that reduce the risk of private investment, account for environmental costs, and level the playing field between bio-based and fossil-derived products. Governments need to put in place carbon pricing systems that are the same across the board, give non-energy bioproducts production tax credits or feed-in tariffs, pay for pre-commercial demonstration facilities, and make it easier for new biomanufactured goods, especially those made from engineered biology, to get through the regulatory process by using science-based, tiered risk assessment protocols that separate contained industrial fermentation from environmental release scenarios. Policies that encourage circularity, such as extended producer responsibility programs for biodegradable polymers or requirements for bio-based content in public procurement, are equally important. These can create early market demand and help businesses grow. International coordination is necessary to prevent regulatory fragmentation and trade barriers, particularly as global supply chains for bio-based materials grow. Multilateral agreements on sustainability certification, GMO handling, and carbon accounting standards will be crucial for facilitating cross-border collaboration and market access. The pressing necessity for interdisciplinary research and education that surpasses conventional academic silos is fundamental to both technological progress and the effectiveness of policy. Biomanufacturing systems are very complicated. They include metabolic engineering, process scale-up, supply chain logistics, environmental impact assessment, and consumer behavior. Due to this, synthetic biologists, chemical engineers, agronomists, economists, social scientists, ethicists, and policy analysts must collaborate. To encourage systems thinking, colleges and universities need to change their curricula. This could include joint degree programs, collaborative laboratories involving students from different departments, and hands-on learning opportunities that demonstrate the real-world problems that arise when bio-industrial deployment occurs. Public research funding agencies should prioritize convergence grants that require consortia from multiple institutions and disciplines, as well as stakeholder involvement from the outset. This will ensure that new ideas are not only scientifically sound but also socially responsible, economically viable, and environmentally sustainable. Additionally, job-finding programs must address the growing skills gaps in bioprocessing operations, bioinformatics, and sustainable supply chain management. This is especially important in rural and developing economies, where biomanufacturing could drive inclusive growth. The long-term goal is a regenerative, biomanufacturing-driven agricultural future in which farms function as integrated biorefineries, producing food, feed, fiber, fuel, and high-value biochemicals in a closed-loop, low-waste system that actively sequesters carbon, enhances ecosystem services, and makes communities more resilient. In this model, microbial cell factories collaborate with agroecological practices to convert crop waste into bioplastics, manure into biofertilizers or renewable natural gas, and saline or wastewater into non-food biomass. This separates economic productivity from environmental damage. Digital technologies, such as AI-driven strain design, Internet of Things (IoT)-enabled field monitoring, and blockchain-based traceability, will provide the data we need to enhance resource flows, verify sustainability claims, and earn the consumer trust. This transition must be fair, ensuring that smallholder farmers, indigenous communities, and developing countries are not only suppliers of raw materials but also active participants and beneficiaries in the bioeconomy through equitable licensing models, effective technology transfer mechanisms, and partnerships that foster capacity building. This vision would not come true just because of technology. It will require careful governance, inclusive innovation, and a new definition of value that prioritizes planetary boundaries and fairness between generations over profit. In conclusion, although substantial scientific, economic, and societal obstacles persist, the convergence of innovations in synthetic biology, systems engineering, and sustainability science presents an unparalleled opportunity to reconceptualize the interface between agriculture and industry. Embedding biomanufacturing in resilient agricultural landscapes, supported by policies that foster interdisciplinary collaboration, enables society to transition toward a circular bioeconomy that meets human needs without compromising ecological integrity. This future is not only technologically achievable but also ethically essential in response to climate disruption, resource scarcity, and increasing global inequality. To move forward, we need to make significant investments, reform our governance, and collaborate to develop innovative ideas that benefit both people and the planet.

ACKNOWLEDGMENT

The authors acknowledge the use of AI tools solely for the purpose of improving the clarity, grammar, and fluency of the language in this manuscript. All scientific content, conceptual development, data interpretation, and substantive revisions were performed exclusively by the authors. The AI-assisted refinements were limited to linguistic enhancement and did not influence the intellectual direction, analysis, or conclusions of the work. The final responsibility for the accuracy, integrity, and originality of the manuscript rests entirely with the authors.

REFERENCES

Adamo G, Amarantos I, Arzimanoglou I, Bergese P, Bussolati B, Bongiovanni A, Cauda V, Chiari M, Garinis GA, Giebel B, Guazzi P, Jurga M, Neri C, Ohlsson P, Rohde E, Sales G, Trapani I, Vader P, Vainio S, Wauben MHM, Yáñez-Mó M, Zarovni N (2025) Spotlight on Innovation: Key Insights from the Extracellular Vesicle Cluster Meeting at the European Innovation Council. *Cytotherapy*. https://doi.org/10.1016/j.jcyt.2025.08.008

Asin-Garcia E, Fawcett JD, Batianis C, Martins dos Santos VAP (2025) A snapshot of biomanufacturing and the need for enabling research infrastructure. *Trends Biotechnol* 43(5):1000–1014. https://doi.org/10.1016/j.tibtech.2024.10.014

Castillo-Saldarriaga C, Sarria S, Santos CNS, Ajikumar PK, Takors R (2024) Semi-continuous biomanufacturing for maximizing the production of complex chemicals and fuels: a case study of amorpha-4,11-diene. *Trends Biotechnol* 42(12):1777–1794. https://doi.org/10.1016/j.tibtech.2024.08.010

Chen L, Guttieres D, Koenigsberg A, Barone PW, Sinskey AJ, Springs SL (2022) Large-scale cultured meat production: Trends, challenges and promising biomanufacturing technologies. *Biomaterials* 280:121274. https://doi.org/10.1016/j.biomaterials.2021.121274

Del Hierro AG, Moreno-Cid JA, Casey E (2024) Continuous biomanufacturing for sustainable bioeconomy applications. *EFB Bioeconomy J* 4:100071. https://doi.org/10.1016/j.bioeco.2024.100071

Duan Y, Chen L, Ma L, Amin FR, Zhai Y, Chen G, Li D (2024) From lignocellulosic biomass to single cell oil for sustainable biomanufacturing: Current advances and prospects. *Biotechnol Adv* 77:108460. https://doi.org/10.1016/j.biotechadv.2024.108460

Fan Z, Gao Z, Fu Y, Liao Y, Xiao W, Ding Y, Yang C, Li X, Li G, Su D, Shan Y, Guo J (2025) Unlocking the potential of fruit and vegetable waste: Sustainable biomanufacturing strategies for synergistic nutrient recovery and high-value bioproduct synthesis. *Trends Food Sci Technol* 165:105288. https://doi.org/10.1016/j.tifs.2025.105288

Gao F, Shi S, Zhao Y, Yang D, Liao X (2025a) Intersecting precision fermentation for global cell-based food production innovation: Challenges and opportunities. *Biotechnol Adv* 85:108712. https://doi.org/10.1016/j.biotechadv.2025.108712

Gao L, Khoo SC, Zhang Z, Wu X (2025b) Trends in sustainable single-cell protein from non-grain feedstocks. *Trends Biotechnol*. https://doi.org/10.1016/j.tibtech.2025.04.018

García Martínez JB, Behr J, Denkenberger DC (2024) Food without agriculture: Food from CO2, biomass and hydrocarbons to secure humanity's food supply against global catastrophe. *Trends Food Sci Technol* 150:104609. https://doi.org/10.1016/j.tifs.2024.104609

Gavrilescu M (2024) From pollutants to products: Microbial cell factories driving sustainable biomanufacturing and environmental conservation. *Chem Eng J* 500:157152. https://doi.org/10.1016/j.cej.2024.157152

Germain LS, Winn LM (2025) Commercialization of cell and gene therapy in Canada: Current landscape, challenges and opportunities. *Drug Discov Today* 30(10):104453. https://doi.org/10.1016/j.drudis.2025.104453

Gu X, Huang L, Lian J (2023) Biomanufacturing of γ-linolenic acid-enriched galactosyldiacylglycerols: Challenges in microalgae and potential in oleaginous yeasts. *Synth Syst Biotechnol* 8(3):469–478. https://doi.org/10.1016/j.synbio.2023.06.007

Harcum SW, Kiss RD (2022) Editorial overview: Pharmaceutical biotechnology: the pandemic years — new modality development and biomanufacturing innovations in a maturing field. *Curr Opin Biotechnol* 78:102846. https://doi.org/10.1016/j.copbio.2022.102846

Hetzler Z, Lott N, Dey Poonam A, Dalgan S, Wei Q (2023) Single-use biosensors for biomanufacturing: Perspective on the state-of-the-art. *Curr Opin Biomed Eng* 28:100512. https://doi.org/10.1016/j.cobme.2023.100512

Li W, Mao Z, Xiao Z, Liao X, Koffas M, Chen Y, Ma H, Tang YJ (2025) Large language model for knowledge synthesis and AI-enhanced biomanufacturing. *Trends Biotechnol* 43(8):1864–1875. https://doi.org/10.1016/j.tibtech.2025.02.008

Love JC, Love KR, Barone PW (2013) Enabling global access to high-quality biopharmaceuticals. *Curr Opin Chem Eng* 2(4):383–390. https://doi.org/10.1016/j.coche.2013.09.002

Lu M, Li N, Rao D, Wu J, Xia W (2025) High-valued hexoses bioproduction from natural sugar resources: Status, trends, challenges and perspectives. *Bioresour Technol* 435:132930. https://doi.org/10.1016/j.biortech.2025.132930

Pandey M, Singhal B (2022) Chapter 8 - Blockchain technology in biomanufacturing: Current perspective and future challenges. In: Islam SKH, Pal AK, Samanta D, Bhattacharyya SBT (eds) *Hybrid Computational Intelligence for Pattern Analysis*. Academic Press, pp 207–237.

Song F, Zhang H, Qin Z, Zhou J (2025) Intelligent biomanufacturing of water-soluble vitamins. *Trends Biotechnol*. https://doi.org/10.1016/j.tibtech.2025.04.007

Sonnenberg M, Miehe R, Kiemel S, Bauernhansl T (2025) Pre-structuring product development challenges in the context of advanced biomanufacturing. *Procedia CIRP* 136:296–301. https://doi.org/10.1016/j.procir.2025.08.052

Sørensen M, Andersen-Ranberg J, Hankamer B, Møller BL (2022) Circular biomanufacturing through harvesting solar energy and CO2. *Trends Plant Sci* 27(7):655–673. https://doi.org/10.1016/j.tplants.2022.03.001

Tellechea-Luzardo J, Otero-Muras I, Goñi-Moreno A, Carbonell P (2022) Fast biofoundries: coping with the challenges of biomanufacturing. *Trends Biotechnol* 40(7):831–842. https://doi.org/10.1016/j.tibtech.2021.12.006

Wang G, Mohsin A, Chu J, Zhuang Y, Zhang S (2024) Chapter 1 - Advances and prospects for advanced biomanufacturing. In: Wei Y, Ji X-J, Cao MBT-S (eds) *Developments Microbial Cell Factories: From Design to Commercial Production*. Academic Press, pp 1–16

Zhao R, He T, Wang M, Guo J, Chen N, Yuan Q (2025) Challenges and pathways for enhanced CO2-derived biosynthesis in cell factories. *Cell Reports Phys Sci* 6(7):102677. https://doi.org/10.1016/j.xcrp.2025.102677

13 Artificial Intelligence and Machine Learning Approach in Sustainable Biomanufacturing

Vishal Sharma
Department of Seafood Science, National Kaohsiung University of Science and Technology, Kaohsiung City, Taiwan
and
School of Biotechnology, University of Jammu, Jammu, India
Department of Marine Environmental Engineering, National Kaohsiung University of Science and Technology, Kaohsiung City, Taiwan

Parushi Nargotra
Department of Seafood Science, National Kaohsiung University of Science and Technology, Kaohsiung City, Taiwan
School of Biotechnology, University of Jammu, Jammu, India

Bijender Kumar Bajaj
School of Biotechnology, University of Jammu, Jammu, India

Diksha Sharma
Department of Seafood Science, National Kaohsiung University of Science and Technology, Kaohsiung City, Taiwan

13.1 INTRODUCTION

In response to the recent pandemic and climate-related crises, global focus has shifted toward cleaner, self-sustaining, and circular economic models, especially in bioprocessing, where the growing demand for sustainability is propelling the transition from traditional manufacturing to biologically driven biomanufacturing. Biomanufacturing, inherently aligned with green chemistry and renewable resources, presents a promising pathway toward a circular bioeconomy producing materials, biochemicals, fuels, enzymes, and pharmaceuticals to meet the growing demand for sustainable solutions (Nargotra et al. 2023; Sharma et al., 2023a). However, despite its potential, the complexity and variability of biological systems pose significant challenges to process optimization, scalability, and economic viability. To effectively harness different bioproducts, it is essential to understand the entire biological process, from raw material conversion and bioproduct synthesis to purification and bio-waste valorization, enabling the production of value-added products at an industrial scale (Singh and Singhal 2022). Moreover, meeting the future energy demands requires developing biorefinery processes based on renewable resources, such as lignocellulosic biomass (LCB), which is abundant in agricultural and forestry wastes (Sharma et al., 2023c; Yadav et al. 2025; Sharma et al. 2024).

Although the industrial sector has gained significant momentum, there is an urgent need to adopt digitalization along with advanced process analytics and computational biology tools to accelerate progress in bioprocess engineering. The digital transformation of traditional bioprocesses into smart ones involves automation, real-time online monitoring, artificial intelligence (AI), and machine learning (ML) (Khanal et al. 2023). Bioprocesses are inherently difficult to harness and stabilize due to their high dimensionality and complexity. Their performance is further influenced by numerous process variables and intricate interconnections, making it challenging to achieve consistent and optimal bioproduct yields. Integrating AI and ML can optimize these complex bioprocesses, enhancing efficiency, scalability, and sustainability in bio-based production systems (Duong-Trung et al. 2023) (Figure 13.1).

DOI: 10.1201/9781003637158-13

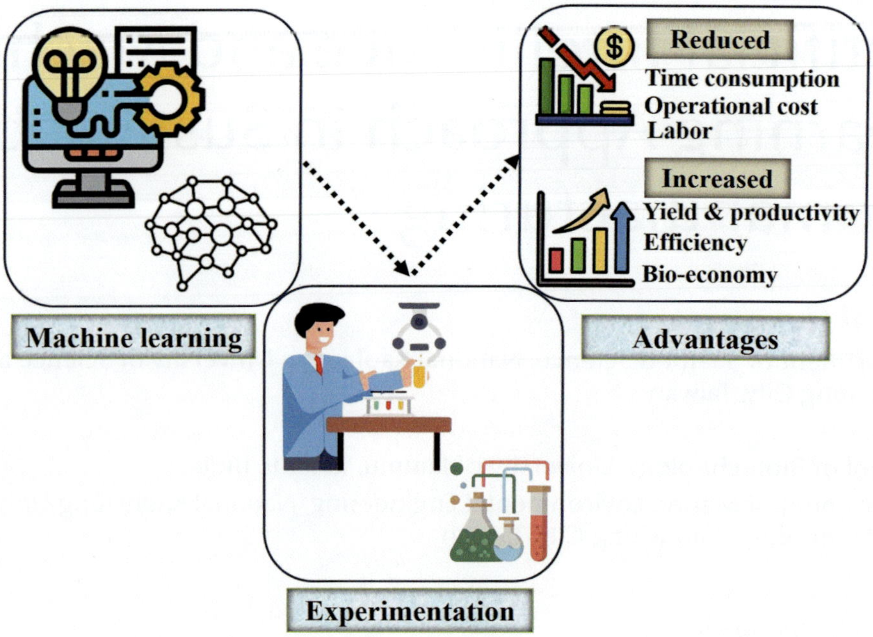

FIGURE 13.1 Role of machine learning in the biomanufacturing process.

From selecting the right microbial strains to optimizing fermentation parameters and scaling up with minimal loss of efficiency, these technologies are gradually becoming embedded in the heart of industrial biotechnology. Their applications are no longer confined to academic experimentation but are driving real-time monitoring, adaptive control systems, and intelligent automation across different industries and biorefineries.

Recent years have witnessed the transition from Biomanufacturing 4.0 to Biomanufacturing 5.0, reflecting a broader shift in industrial paradigms. Biomanufacturing 4.0, aligned with the principles of Industry 4.0, primarily emphasizes digitalization, automation, data integration, and advanced analytics for process optimization. It focuses on enhancing efficiency, precision, and productivity through smart sensors, IoT-based monitoring, and ML-driven control systems. In contrast, Biomanufacturing 5.0 represents the next evolutionary stage, rooted in the principles of Industry 5.0 and centered on human-AI collaboration, personalization, and sustainability. Rather than replacing human intelligence, it aims to create synergistic interactions between humans and intelligent systems for more adaptive, ethical, and resilient biomanufacturing practices. The fourth industrial revolution is largely driven by the rapid expansion of ML, which enables computer systems to learn and improve without explicit programming. By adapting to external inputs, ML has advanced numerous industrial applications, including data mining, automation, predictive maintenance, tomography, process monitoring, fault detection, and optimization (Velidandi et al. 2023). The development of modern bioprocesses generates large amounts of varied experimental data that contain crucial information about the process. The integration of new analytical methods, digital technologies, and automation can enhance this development by streamlining and accelerating experiments (Yang et al. 2023). ML enhances cost-effectiveness, efficiency, reliability, safety, and data-driven decisions through powerful optimization algorithms. Its workflow typically involves three stages: training, validation, and testing. Various ML tools, such as linear regression, artificial neural networks (ANNs), decision trees (DTs), support vector machines (SVMs), genetic algorithms (GAs), and random forest (RF), are most commonly applied in the sustainable biomanufacturing applications (Sharma et al., 2023a).

The shift toward renewable energy is driven by environmental concerns and fossil fuel depletion, with LCB emerging as a key feedstock for bioenergy (Nargotra et al. 2025). However, its natural variability poses challenges in standardizing biorefinery processes. ML techniques, such as ANN, RF, and DTs, are increasingly used to predict biomass properties, enhance feedstock consistency, and optimize steps such as enzymatic hydrolysis and solid-state fermentation for efficient biofuel production. Additionally, the application of ML design in optimization, bioreactor management in fermentations and monitoring, treatment pipeline fault detection, and integrated processes is guided by research investigations. ML is also playing a growing role in enzyme purification, protein engineering, and downstream processing by enabling accurate prediction of membrane fouling, optimizing ultrafiltration parameters, and reducing reliance on trial-and-error methods (Lu et al. 2025). It also aids in designing advanced membrane materials and improving process efficiency through models such as RF, neural networks, and convolutional neural networks.

Considering these advancements, the current chapter focuses on the integration of AI and ML approaches in sustainable biomanufacturing. It explores how ML tools are being utilized across the entire bioprocessing value chain, from biomass characterization and feedstock optimization to fermentation, bioreactor control, enzyme purification, and downstream processing. Emphasis is placed on data-driven strategies that improve process efficiency, scalability, and sustainability while minimizing costs and resource usage. This chapter aims to provide a comprehensive overview of state-of-the-art ML applications, highlight key case studies, and discuss future directions for leveraging AI/ML to accelerate the transition toward a smart, circular bioeconomy.

13.2 WHY ARTIFICIAL INTELLIGENCE OR MACHINE LEARNING IS REQUIRED IN BIOMANUFACTURING?

In the modern industrial setup, biomanufacturing, which is the industrial production of biological products such as biofuels, bioplastics, pharmaceuticals, enzymes, and food additives, is inherently complex and influenced by a multitude of biological, chemical, and physical parameters. As the industry moves toward precision, efficiency, and sustainability, traditional trial-and-error approaches to process optimization are proving to be inadequate. This is where AI/ML have emerged as transformative tools, offering data-driven insights, real-time process control, and predictive modeling capabilities that are revolutionizing the biomanufacturing landscape (Mondal et al. 2023). Data from several sources, including process control systems, bioreactor conditions, sensors, applied analytical techniques, omics data, material input, and process output, are used in the manufacturing process. Large bioprocessing companies are now investing on data management through cloud computing, ML, and the Internet of Things (IoT). Since AI/ML uses statistics and probability to assess whether a digital algorithm is effective in solving or optimizing the problem for which it was designed, it has a direct relationship with mathematics. Its foundation is the examination of the level of suitability between the values derived from the model and the actual ones, for instance, using linear regression (Arias et al. 2023).

One of the key reasons AI and ML are indispensable in biomanufacturing is their ability to handle large and complex datasets. Bioprocesses generate vast amounts of data from sensors, bioreactors, omics technologies (genomics, proteomics, and metabolomics), and laboratory experiments. ML algorithms, such as SVM, neural networks, and DTs, can mine this data to identify patterns and relationships among variables that are not evident through traditional statistical methods. This facilitates the development of predictive models that can optimize conditions such as pH, temperature, substrate concentration, and fermentation time to maximize product yield and quality.

AI and ML also enable real-time monitoring and control of bioprocesses. Using AI-enabled control systems, operators can detect deviations or abnormalities in the process early and automatically make necessary adjustments. This leads to increased consistency, reduced batch failures, and improved compliance with regulatory standards, which are particularly important in pharmaceutical and food-grade biomanufacturing. The design of bioreactors has advanced significantly, enabling continuous online data to be generated for reinforcement learning (RL)-based bioprocess optimization and optimal control. The drawback of this approach was that it was based on fixed models, which meant that it needed to be updated and improved continuously to generate excess data in automated fermentation systems (Singh and Singhal 2022). Thus, learning techniques for regulating final ethanol titers during yeast fermentations have been created and successfully implemented to improve model-free reinforcement.

Another crucial area where AI and ML are required is in strain design and metabolic engineering (Kim et al. 2020). Modern synthetic biology relies heavily on genome-scale modeling to predict the best genetic modifications for enhanced production. Metabolic engineering strives to enhance microbial production of valuable compounds by manipulating genetics and cellular processes, supported by vast multi-omics data. However, integrating these complex datasets is challenging using traditional methods. AI and ML offer powerful tools to analyze, predict, and optimize metabolic pathways, driving innovation in industrial biotechnology (Helmy et al. 2020). ML models trained on experimental datasets can predict gene targets, simulate metabolic fluxes, and reduce the number of laboratory trials required to develop a high-performing microbial cell factory. This shortens development cycles and significantly cuts costs. In the context of process scale-up, ML models help bridge the gap between lab-scale experiments and industrial-scale operations. Moreover, mechanistic modeling in metabolic engineering depends on enzyme parameters, such as k_{cat} and K_m, which were historically scarce. ML models, including RF and deep neural networks, now enable accurate prediction of these parameters from structured datasets (Lu et al. 2024). These ML-derived predictions enhance model accuracy and improve insights into protein abundance and enzyme kinetics. However, current models are often organism-specific, such as in *Escherichia coli*, and may not generalize across species.

Changes in mass transfer, hydrodynamics, and climatic conditions frequently lead to unforeseen difficulties when scaling up a bioprocess. ML tools can reduce the chance of failure during scale-up by simulating various scenarios and suggesting reliable operating conditions. Moreover, AI's contribution to techno-economic and life cycle analysis of bioprocesses is also increasing. ML algorithms can assess different process setups and raw material selections for their environmental impact and economic feasibility. This aids the decision-making process for economical, environmentally friendly manufacturing.

13.3 OVERVIEW OF ARTIFICIAL INTELLIGENCE AND MACHINE LEARNING

13.3.1 Key Concepts of AI/ML

AI refers to the broad field of developing systems capable of performing tasks that normally require human intelligence, such as problem-solving, language comprehension, pattern recognition, and adaptive learning. Within this domain, ML is a specialized subset focused on enabling systems to learn and improve from data. Rather than being explicitly programmed for every task, ML algorithms use statistical methods to identify patterns and refine their performance over time based on experience (Okolie 2024). ML enables computers to independently learn, analyze data, and make predictions after initial programming, without the need for constant manual coding. In the bioprocessing industry, ML has found wide-ranging applications, offering deep process insights, driving innovation, and reducing the reliance on manual operations and traditional predictive methods (Mondal et al. 2023).

ML techniques are broadly classified into three main categories, namely, supervised learning, unsupervised learning, and RL (Khanal et al. 2023). In supervised learning, the model is trained on a labeled dataset, enabling it to make accurate predictions on new, unseen data. This categorization depends on how datasets are structured according to specific problem types, allowing ML algorithms to establish mathematical relationships and create models to solve those problems. With rapid technological progress, computational tools now produce vast amounts of biologically derived data (Velidandi et al. 2023). As a result, the past decade has witnessed significant expansion in the application of various ML algorithms in the development and production of bioproducts. This approach encompasses various subfields, including DT learning, association rule learning, inductive logic programming, and SVM. In supervised learning, a computer is trained using datasets that contain both inputs and their corresponding labeled outputs, allowing it to learn the relationship between the two and develop a predictive model (Oyetunde et al. 2018). Among the most commonly used supervised learning algorithms are linear regression, logistic regression, SVM, RF, ANNs, DTs, naive Bayes, and K-Nearest Neighbors (KNN). These tools are employed to predict unmeasured variables. For instance, linear discriminant analysis (LDA) is used to interpret spectroscopic data and analyze bio-algal metabolic processes, while partial least squares (PLS) and principal component analysis (PCA) are commonly applied in bioreactor monitoring (Mondal et al. 2023). In classification tasks, categorical variables help detect design flaws and process faults.

Unsupervised learning, such as cluster analysis, is used to discover hidden patterns and underlying structures within data without the need for labeled outputs. In this method, clustering techniques are employed to identify underlying data structures, often using methods such as PCA and hierarchical clustering. For association tasks, the k-means algorithm is commonly used to uncover relationships within the data. The unsupervised learning does not require labeled data, and rather than using classification or regression, the system uses clustering to find patterns in the data (Oyetunde et al. 2018). Clustering, GAs, similarity, and metric learning are more subcategories of this kind of learning. Without the assistance of a network trainer or supervisor, RL teaches a system by rewarding or punishing it according to how well its output is received. This kind of learning can be further divided into deep learning, manifold learning, and Bayesian networks. Data mining, process automation, process-aware assaults on industrial control systems, predictive maintenance, optimization, process monitoring and fault detection, and industrial tomography are just a few of the industrial processes that have benefited from the application of ML. ML has been shown to be a useful tool across a range of domains, including business models, biological sciences, physical sciences, chemical sciences, medical sciences, and geoscience, due to its capacity to recognize patterns in parameters and generate predictions based on novel inputs. Cost effectiveness, process dependability, efficiency, and safety, as well as data-driven decision-making, are all greatly aided by the optimization algorithms employed in ML. Training, validation, and testing are the three phases of ML.

13.3.2 ML Tools and Platforms in Bioprocess Engineering

Bioprocess engineering integrates multiple disciplines to enable the transition of bioproducts from laboratory research to commercial-scale production. Due to the complex nature and variability of biological systems, the processes involved are often highly heterogeneous. Scaling up requires managing numerous process parameters and understanding their interactions. The multidimensional nature of biological datasets makes it difficult to apply conventional mathematical modeling and simulation approaches effectively. In response, ML techniques have shown strong potential in overcoming these challenges and enhancing the efficiency of large-scale bioproduction systems (Singh and Singhal 2022). Incorporating decision support tools during the initial stages of developing new biorefinery production models can greatly enhance efficiency by enabling the integration and analysis of all relevant variables to identify optimal solutions. AI or ML-based tools facilitate the processing of large datasets through advanced algorithms, allowing for accurate prediction of complex outcomes, improved implementation of logistics and automation, and early detection of errors or anomalies, thereby supporting timely corrective measures. In biochemical engineering, commonly used ML algorithms include multivariate statistical analysis (MSA), SVM, ANN, and RL. Effectively applying these techniques requires a solid grasp of their underlying mathematical principles.

13.3.3 Multivariate Statistical Analysis

MSA, enables the simultaneous analysis of multiple variables, using complex mathematical computations to derive accurate insights from diverse datasets. MSA offers a flexible approach for analyzing complex scenarios in which numerous factors influence the variability of dependent variables, resulting in more accurate and realistic predictions. Meanwhile, multiple discriminant analysis classifies data into predefined groups using predictor variables, aiming to reduce classification errors (Saha et al. 2024). MSA techniques, such as PCA, PLS, and polynomial regression, remain widely used in the bioprocess industry due to their reliability and effectiveness. PCA is a dimensionality reduction technique that transforms original variables into new, uncorrelated ones known as latent variables, by maximizing variance (Table 13.1).

PCA is based on the principle of preserving maximum variance when projecting data from a high-dimensional space to a lower dimensional one. The process begins with standardizing the dataset to achieve a mean of zero and a standard deviation of one, ensuring uniform scaling across all variables. It has been widely applied in various areas such as interpreting spectroscopic data to identify chemical substances, distinguishing cell types, uncovering relationships among bioreactor operating conditions in biogas and biofuel production, and detecting anomalies in large-scale cell culture processes (Mowbray et al. 2021). PCA transforms an n-dimensional dataset into an m-dimensional feature space with mutually orthogonal axes. PCA has been applied to Raman spectroscopy data to reveal underlying correlations among reactor operating parameters in biogas and biofuel production, and to detect anomalies in large-scale cell culture processes (Saha et al. 2024).

In practice, PLS is primarily utilized as a multiple-input single-output (MISO) model, although it is a multiple-input multiple-output (MIMO) model. Consequently, it is common to build multiple independent PLS models in parallel when multiple output variables need to be correlated with input factors (Mowbray et al. 2021). PLS has been used in bioprocess monitoring, reactor control, and linking amino acid sequences to protein function. While PCA and PLS are useful for high-dimensional linear data, they often fall short in modeling the non-linear, low-dimensional nature of biological systems. PLS regression projects both input (X) and output (Y) variables into a new space, making it a bilinear factor model. However, the accuracy of PCA and PLS is frequently found to be inadequate when simulating biological reaction systems because most of these systems are low dimensional and highly non-linear. Therefore, in practice, they have often been complemented or outperformed by ANN in non-linear contexts. Neural network-based methods, such as autoencoders, are emerging as powerful alternatives to PCA and PLS in protein engineering or bioprocess applications, potentially reducing their future relevance.

TABLE 13.1
Principal Component Analysis versus Partial Least Squares

Aspect	Principal Component Analysis	Partial Least Squares
Primary purpose	Reduces data dimensionality by extracting latent variables that capture maximum variance in predictors (X).	Models relationships between predictors (X) and responses (Y) while reducing dimensionality.
Type of model	Unsupervised (depends only on X variables).	Supervised (relates X to Y).
Mathematical basis	Eigenvalue decomposition of the covariance matrix of X.	Simultaneous projection of X and Y to maximize covariance.
Input/output relationship	Focuses on variance in input data only.	Focuses on the covariance between input and output data.
Common applications	Spectroscopic data interpretation, fault detection, bioreactor monitoring, and anomaly detection.	Process modeling, prediction of product yield or quality, bioprocess control, and correlation of sequence–function relationships.
Strengths	Simple, fast, and effective for exploratory analysis and visualization.	Strong predictive capability; handles multicollinearity and noisy data.
Limitations	Assumes linearity; cannot model non-linear biological interactions.	Primarily linear; may require non-linear extensions or hybrid models.
Typical role in bioprocessing	Data exploration and feature extraction.	Process prediction, optimization, and control.
Alternatives/advancements	Kernel PCA, Autoencoders.	Non-linear PLS, neural networks, hybrid mechanistic–ML models.

13.3.4 Artificial Neural Network (ANN)

The most well-known ML technique is probably the ANNs, which were inspired by how the human brain works. ANNs are composed of layers of interconnected nodes, or neurons, each with a distinct purpose. An ANN is trained by varying the weights and biases of the neurons that make up the network until the network is able to correctly categorize or predict the training data (Frontistis et al. 2023). ANNs are biologically inspired computer programs that simulate

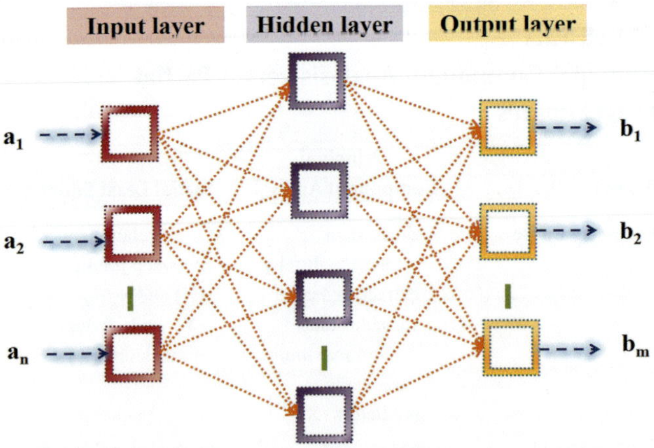

FIGURE 13.2 Schematic representation of a single hidden-layer back-propagation neural network. Here, a and b denote the input and output vectors of the perceptron, respectively; a_n and b_m represent the number of neurons in the input and output layers of the artificial neural network (ANN), respectively.

the way the human brain processes information. Instead of using programming, ANNs learn (or are taught) by finding patterns and relationships in data. The layers of the neural structure were made up of hundreds of individual units called processing elements (PE) or artificial neurons that were connected by coefficients (weights) (Yang et al. 2023) (Figure 13.2).

Neural computations are powered by the network of connections between neurons. ANNs can also integrate and use data from literature and experiments to solve problems. The most popular method for estimating the ANN's error, the discrepancy between the simulated and actual values is backpropagation, which also updates the weights and biases of the neurons to reduce the estimated error (Frontistis et al. 2023).

Optimization of process parameters in a bioprocess or biomanufacturing is essential for boosting the industry relevant products, including enzyme, antibiotics, biofuels, etc., besides reducing their costs. While the traditional one factor at a time (OFAT) method is simple, it is inefficient and overlooks interactions between variables. Advanced methods, such as response surface methodology (RSM), although offer more accurate and efficient optimization, yet with sophistication of the experiments and product-oriented research, AI-based tools such as ANNs have proven their mettle in delivering robust and predictive optimization solutions (Sharma et al., 2023b).

As a supervised learning technique, an ANN utilizes a dataset containing measurable inputs, such as pH levels or substrate composition, to predict target outputs such as biomass concentration or specific material properties. Due to their vast number of tunable parameters, neural networks possess the theoretical capability to approximate any function, granting them exceptional versatility (Mowbray et al. 2021). By employing a defined model structure and training the network with paired input-output data, these parameters are iteratively optimized. This process gradually refines the predictions, aligning them more closely with the expected outcomes based on the training data. In biorefinery development, sustainability assessment must also consider agricultural and industrial background activities, given their significant environmental impacts. Studies using digitization tools, such as ANNs, have effectively evaluated the environmental performance of such systems, including inputs and labor. Since ANNs can represent non-linear interactions in dynamic systems, they are the most popular modeling technique in bioprocessing. ANN (as Model-1) and PLS-PCA (as Model-2) are the two models that yield the most successful results when compared, but ANN is thought to be replacing Model 2 soon. As technology advances, ANN learns from historical plant data to produce more precise results (Mondal et al. 2023).

ANNs have been applied in various areas of biochemical engineering, including the prediction of fermentation parameters, analysis of amino acid sequences, bioprocess simulation, biomass concentration estimation, and modeling of groundwater bioremediation. In LCB processing into biofuels, fermentable sugar release is a crucial step in the fermentation. The initial screening tools used to improve the efficiency of fermentable sugar extraction from LCB are pre-experimental data from prediction models. The combined severity factor index and fermentable sugar content have been known to be better predicted by the ANN models than by earlier intelligent models. ANN and adaptive neuro-fuzzy inference systems (ANFIS) have frequently been used to model the lignocellulosic biomass value-adding process (Pradhan et al. 2022). Various forms of ANNs, including convolutional neural networks (CNNs), deep neural networks (DNNs), and recurrent neural networks (RNNs), are widely utilized in biochemical engineering. CNNs have demonstrated high accuracy in tasks such as enzyme structure classification, algal biofuel process optimization, bioreactor performance enhancement, and protein structure prediction (Saha et al. 2024). Meanwhile, RNNs are employed in genetic engineering to identify promoter sequences, in protein engineering to analyze amino acid sequences, and in biosensor systems for soft sensor calibration. However, one of the major challenges in neural network implementation is selecting the right hyperparameters, such as the number of hidden layers, neurons per layer, and the activation function. This is especially important when using networks for feature extraction and dimensionality reduction in complex biological datasets.

13.3.5 Support Vector Machine

SVMs are widely used in biological research for identifying complex, non-linear relationships within data. Unlike traditional polynomial regression, which minimizes the squared error between predicted and actual values, SVMs focus on minimizing the generalization error bound to enhance

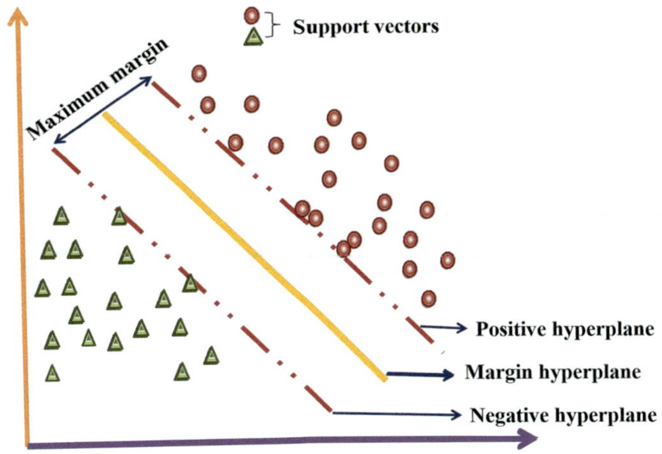

FIGURE 13.3 An illustration of a support vector machine (SVM) classification for two linearly separable classes, showing the optimal separating hyperplane and the support vectors that define the margin.

predictive performance (Sharma et al., 2023b). SVM does not need explicit equations for class boundaries, in contrast to many other ML techniques. Rather, it uses the margin, the distance between the hyperplane and the closest data points from each class, to choose the best hyperplane. Support vectors, or the data points nearest to the decision border, are used to do this. SVM works by converting a low-dimensional input vector into a high-dimensional feature space and then looking for the best separating hyperplane to classify the data points in such a way that the space's misclassification error is as small as possible (Ji et al. 2023) (Figure 13.3).

Through the kernel method, SVM can effectively handle both linear and non-linear relationships without explicitly requiring equations for decision boundaries, which is what gives it its mathematical elegance (Saha et al. 2024). SVMs have shown outstanding predictive capabilities in areas such as compost quality assessment, anaerobic digestion (AD), and classification of organic waste. When applied to regression tasks, this method is referred to as support vector regression (SVR) (Chen et al. 2024). Studies have shown that SVM can outperform ANN in predicting outcomes in AD processes, offering greater accuracy in certain applications. Recent research has also addressed several early limitations of SVMs, particularly by improving their effectiveness on medium-sized datasets through innovative optimization techniques. As a result, SVMs are often used as benchmark models across various applications. However, their application to large datasets remains challenging due to the high computational cost and memory demands associated with kernel functions (Mowbray et al. 2021).

13.3.6 Decision Trees

DTs are powerful ML tools with valuable applications in biomanufacturing, enabling the prediction and analysis of outcomes based on multiple process parameters. Widely used for both classification and regression tasks, DTs segment the feature space by identifying optimal split points. However, recursive splitting can lead to overly complex trees that are prone to overfitting. To address this, pruning methods, often guided by cost–complexity criteria, are applied to balance model accuracy with generalization, ensuring reliable performance on unseen bioprocess data (Akinpelu et al. 2023). Modeling bioprocesses, such as biofuel production, poses challenges due to the need to identify key operational factors, such as extraction rate and process temperature, and understand their interdependencies. To tackle these complexities, researchers have increasingly turned to DT-based classification methods in recent years (Chen et al. 2022; Sharma et al., 2023b). The tree-structure computation classifies the data inputs and extracts decisions from branches based on the hierarchical if/else questions (Figure 13.4).

Two popular ensemble techniques are gradient boosting (GB) and RF. Fundamentally, RF averages the outcomes of DTs to get around each constraint, while GB uses weak learners to enhance prediction as a sequential technique. Although RF contains fewer hyperparameters and can produce good predictions even with suboptimal tuning, GB models typically perform better when it comes to hyperparameter tweaking (Tsui et al., 2023). Tree-based models can be trained quickly and at lower computational cost than other ML algorithms, but they are also more prone to overfitting.

13.4 INTEGRATION OF AI/ML IN BIOMANUFACTURING PROCESSES

13.4.1 Upstream Processes: Raw Material Selection and Optimization

AI/ML tools enhance upstream biomanufacturing by enabling data-driven selection of optimal raw materials and feedstocks. They predict bioprocess outcomes based on substrate properties, reducing reliance on trial-and-error. ML models help optimize conditions, such as nutrient ratios and pH, for improved yields. This integration accelerates process development, lowers costs, and improves scalability. If we talk about the biorefinery setup and its growth, the biomass feedstock supply chain, including cultivation, harvesting, pretreatment, storage, transportation, and other upstream processes, is highly complex due to feedstock variability, geographical dispersion, and seasonal fluctuations. These factors significantly affect efficiency, cost, and overall process sustainability (Zhao et al. 2024). Specifically, pretreatment is an inherent step in the fractionation of LCB into value-added products and a cost-determining factor.

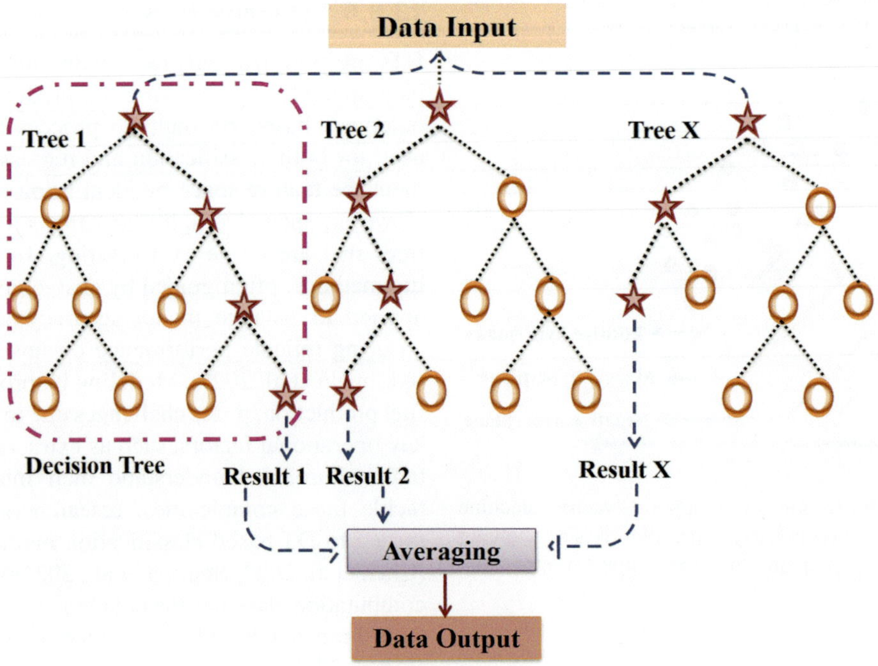

FIGURE 13.4 Overview of the decision tree (DT) and random forest (RF) models. The figure depicts multiple individual trees that collectively form the RF ensemble.

Among different biomaterial extraction methods, green solvents such as deep eutectic solvents and low-cost physical techniques, such as ultrasound and microwaves, have attracted significant research attention (Nargotra et al. 2024; Ortizo et al. 2024; Sharma et al. 2025). ML offers powerful tools to manage and optimize these complexities by enabling accurate forecasting, dynamic decision-making, and efficient resource allocation across the biomass supply chain, ultimately enhancing both environmental and economic outcomes of bio-based production systems. Moreover, ANN, SVM, DTs, RFs, and other models support the biomass energy sector by optimizing raw material identification, evaluating feedstock characteristics, and improving storage and transportation efficiency. It helps analyze biomass origin, component composition, and post-production logistics using diverse algorithms (Peng and Sadaghiani 2023).

The seasonal availability and limited harvest windows make biomass feedstock supply inconsistent, requiring efficient storage systems to ensure year-round availability for biofuel or biomaterial production. Various storage methods, from low-cost open-air to climate-controlled silos, differ in cost, moisture control, and material loss. The logistic regression (LR), DTs, RFs, multilayer perceptron (MLP), and SVM play a critical role in optimizing storage strategies (Goettsch et al. 2020). These models help in selecting ideal depot locations, predicting moisture content, minimizing losses, and improving overall efficiency and sustainability of biomass storage and upstream logistics. Additionally, transportation is a key component of the biomass supply chain, ensuring the reliable movement of feedstock from farms to biorefineries. Due to varying distances, modes, and operational conditions, optimizing transportation is critical for cost and efficiency. The AI/ML models enhance biomass transportation by enabling demand forecasting, route optimization, fuel consumption prediction, truck traffic analysis, and infrastructure planning. Algorithms, such as ANN, SVM, Long Short-Term Memory (LSTM), Gaussian Processes (GP), and DTs, are used to predict shipment volumes, delivery times, and asset performance, supporting smarter, data-driven logistics for sustainable biomass supply (Kim 2017; Peng and Sadaghiani 2023).

13.4.2 Fermentation and Bioconversion Monitoring

Fermentation and bioconversion processes often face challenges, including process variability, non-linear dynamics, sensitivity to environmental conditions, and difficulties in real-time monitoring of microbial activity and metabolite concentrations. Traditional monitoring methods can be time-consuming, labor intensive, and insufficient for capturing the complexity of these biological systems. ML/AI tools can accurately forecast key process parameters, detect anomalies early, and optimize conditions such as pH, temperature, and nutrient levels. This leads to improved process control, higher product yields, and more efficient, scalable bioprocessing operations. Integrating ML into fermentation processes aims to determine the optimal parameter set that enhances overall performance. Unlike conventional methods that often optimize a single metric such as product yield or titer, ML enables multi-objective optimization

(MOO), where multiple goals, such as product formation rate, substrate utilization, and controlled cell growth, can be simultaneously balanced based on their relative importance (Wang et al. 2025). In fed-batch fermentation, for example, the focus might be on maximizing product output while minimizing feed input to reduce material waste. Popular algorithms used in MOO include GAs and Non-dominated Sorting Genetic Algorithm II (NSGA-II), which help in efficiently navigating complex parameter spaces to improve both productivity and cost-effectiveness (Li et al. 2024). Improper control can lead to substrate waste, by-product build-up, or reduced yields. Models, such as ANN and Least Squares Support Vector Machines (LS-SVM), have been used to predict future metabolite concentrations and automate feeding strategies, enhancing production efficiency (Wang et al. 2020). ANN-based systems have successfully optimized glycerol feeding and airflow, increasing yields of 1,3-propanediol and ethanol. RL has also been applied to design adaptive control strategies in fed-batch processes by learning optimal feeding patterns over time (Mesquita et al. 2021).

It has been demonstrated that ML methods, such as SVMs, neural networks, and Gaussian process regression, are useful for simulating intricate correlations between performance and experimental data (Vinestock et al. 2024). To identify the most effective experimental parameters for maximizing biomass concentration in a fermentation of *Yarrowia lipolytica*, Pensupa et al. (2023) used data mining to create ML models to screen for experimental designs. Gaussian processes were found to have the best predictive power. The potential of these methods was also shown by Packiam et al. (2022), who created PERISCOPE-Opt for forecasting ideal fermentation conditions utilizing XGBoost, RF, and SVMs. Interest in low-carbon feedstocks and waste valorization to establish circular bioprocesses has surged in tandem with the growing significance of renewable feedstocks. The intricacy of the biological system and the complexity and diversity of the feedstock composition present a major modeling problem (Vinestock et al. 2024). Since old procedures are frequently erroneous or inapplicable, new methods, instruments, and strategies are needed to solve this. Applying synthetic biology technologies, enhancing experimental design through Bayesian optimization, and using data-driven modeling techniques were proposed as the keys to unlocking greater production from such systems.

13.4.3 Downstream Processing and Product Purification

The main bioprocess issues that need to be resolved to make market-ready bioproducts, such as biofuels competitively priced, are feedstock prices, technological challenges, logistics, moving biomass from its field of origin, low biofuel yields as a result of inefficient raw material conversion, and costly downstream processing (Shelare et al. 2023). ML techniques also hold great promise for end-to-end biomanufacturing and downstream unit operations. Downstream purification of bioproducts in biorefineries, food and pharmaceutical industries primarily involves filtration techniques, such as diafiltration, membrane ultrafiltration, and dead-end filtration, and various chromatography methods, including hydrophobic interaction, cation and anion exchange, affinity, size exclusion, high-performance liquid chromatography (HPLC), multimodal chromatography, and others (Rathore et al. 2023). ANNs are increasingly used to monitor and optimize end-to-end biopharmaceutical manufacturing. A previous study used ANNs with mechanistic models to optimize a three-step chromatography process, avoiding manual testing of 400 process flowsheets (Pirrung et al. 2017). This approach was later expanded to include ultrafiltration and diafiltration, enabling identification of optimal, cost-effective manufacturing workflows.

In the microalgal biorefineries, the collection, extraction, and purification of important macromolecules, biofuels, nutraceuticals, and other value-added products are required. The variation in microalgae species, separation/extraction conditions, and product yields, however, makes the optimization of these processes challenging. By introducing innovations that combine digitalization, large-scale data, and automation to boost effectiveness, AI/ML models have emerged as useful tools for overcoming bottlenecks in the extraction and quantification of bioactive compounds (Chong et al. 2023). Enhancing microalgal harvesting and drastically lowering related expenses are two areas where AI/ML models can be especially useful. Furthermore, there is a growing need to reduce the extraction process's input costs. Due to their advantages, such as faster turnaround times, reduced use of hazardous solvents, acceptable accuracy, strong predictive capabilities, cost-effectiveness, and environmentally friendly operation, ANFIS, ANN, and SVMs are increasingly used for identifying, quantifying, and predicting bioactive compounds (Imamoglu 2024).

13.4.4 Bioreactor Design and Scale-Up Using ML

To produce high-quality, high-value biomaterials for a more sophisticated industrial sector, there is a significant scientific need and commercial potential for the development of bioreactors for biomanufacturing processes based on smart manufacturing systems. Advanced ML models and algorithms can minimize experimental efforts and reduce the overall cost of bioprocess development, while also enhancing the economic efficiency of large-scale bioreactor control. ML is increasingly influencing bioprocess engineering by enabling data-driven solutions; however, its integration across various subdomains remains limited, constraining its full potential for automating and optimizing bioprocesses. In bioreactor engineering, ML models have been instrumental in modeling the complex interactions among variables such as contaminant levels, pH, temperature, mixing rates, and nutrient availability, and their effects on cellular metabolism. ML has facilitated critical tasks such as forecasting state variables (e.g., biomass concentration), monitoring dynamic parameters that affect reactor performance, and enabling automated, safe process control. Moreover, in the domain of biodevices and biosensors, ML applications span the optimization and control of microbial fuel cells, the design of soft and microfluidic sensor systems, and the analysis of complex datasets from real-time monitoring. These developments demonstrate how ML is transforming conventional approaches by improving precision, adaptability, and responsiveness in bioprocess environments.

13.5 MACHINE LEARNING FOR PROCESS OPTIMIZATION AND CONTROL

Data-driven decision-making and real-time process adaptability are made possible by various ML tools, which have revolutionized bioprocess optimization and control. ML models may learn from historical and real-time data to find complicated, non-linear correlations among process variables, while traditional process optimization frequently depends on laborious experimental designs and manual modifications (Sharma et al., 2023b). As a result, important characteristics, such as product output, biomass concentration, and substrate consumption, can be accurately predicted. ML methods, including SVM, RL, and ANNs, can dynamically modify operating conditions in process control to preserve system stability and optimal performance.

ML-based process optimization can be broadly categorized into two types. The first involves improving product quality without altering production parameters during operation. This includes tasks such as identifying root causes of quality issues, predicting outcomes early to eliminate unnecessary steps, and diagnosing faults in products or equipment. These approaches enhance quality indirectly. The second type involves real-time adjustment of production parameters during manufacturing. By analyzing ongoing process data, ML models fine-tune their parameters to meet specific optimization goals, thereby improving product quality. This is achieved through ML algorithms integrated with optimization modules or self-optimizing control systems built on analytical models (Weichert et al. 2019). It is observed that the hyperparameters of ML techniques affect their performance. To get an optimum model, all hyperparameters must be adjusted. By combining ML techniques with optimization methods, a hybrid super learner model can be produced by automatically adjusting the hyperparameters (Mondal et al. 2023). Using microalgae oil as a feedstock, Sultana et al. (2022) created a hybrid model using the Bayesian optimization algorithm (BOA) with ANN and SVR to forecast the rate of biodiesel production. The optimal hyperparameters (nano-catalyst dose, process temperature, reaction time, and algal oil-to-methanol ratio) that yielded the lowest validation error were selected using BOA. Several performance measures (such as residual analysis, root mean square error [RMSE], Residual Error [RE], R^2, and Mean Absolute Error [MAE]) were used to evaluate the models' performance, and the BOA-SVR model outperformed the BOA-ANN model (Cheng et al. 2023).

Bioprocess control and optimization have advanced significantly in the last few decades. To find the ideal parameters to attain a good performance in bioprocesses, for instance, orthogonal experimental design (OED) and RSM are widely used (Sharma et al. 2021). Additionally, the non-linear interactions between variables in bioprocesses have been studied through the progressive use of ML methods. For instance, the fermentation-based generation of poly(3-hydroxybutyrateco-3-hydroxyvalerate) was modeled and optimized using ANN (Zafar et al. 2012). To improve bioprocess performance, a growing variety of AI technologies have recently been used in bioprocess optimization and control. To maximize the biomass and production yield of microorganisms for industrial production, for example, a hybrid multi-objective method was employed (Saini et al. 2021). Additionally, deep learning techniques were used to forecast key process parameters (Critical Process Parameters [CPPs]) in AD based on online sensor data. However, there are potential challenges in using ML models to optimize bioprocesses. The first significant issue is how to efficiently acquire the accessible data. Many studies on bioprocesses have been carried out recently; however, because the experimental conditions used in these studies differ greatly, it might be challenging to combine their data for additional study (Cheng et al. 2023). More thorough experimental data messages should be given to address this issue. Furthermore, the comprehensive message can be used to create a database for gathering and organizing bioprocess data.

13.6 BIOPROCESS APPLICATIONS OF MACHINE LEARNING

13.6.1 Enzyme Engineering and Production

Enzyme engineering involves altering amino acid sequences in enzymes to enhance their functional properties. Notwithstanding the vast array of potential changes, this process has already produced impressive outcomes in the form of novel enzyme designs and optimizations for food production, waste biodegradation, chemical and pharmaceutical biosynthesis, regenerative medicine, and biosensing. Typically, enzymes are engineered to improve catalytic activity, substrate specificity, tolerance to cosolvents, thermodynamic stability, solubility, and expression efficiency (Mazurenko et al. 2020). ML is increasingly being used in enzyme engineering to predict structure-function relationships, including enzyme structure, catalytic activity, and substrate specificity. It also aids in predicting solubility and stability changes resulting from amino acid substitutions. Additionally, ML-guided directed evolution is emerging as a powerful tool, supported by a growing history of ML-based enzyme prediction models

(Yang et al. 2024). In contrast to the model-driven rational designs, the ML method is data-driven since it forecasts previously unobserved properties of similar inputs by using established data trends. Unlike directed evolution, which selects existing mutants iteratively, ML-based design can create new and promising variants that have never been seen before (Ndochinwa et al. 2024). This is accomplished by looking for patterns in the collected data. Like how traditional directed evolution and rational design interact, ML is being used in tandem with both strategies.

The ability of ML to generalize is what makes it so appealing in enzyme engineering. ML algorithms can quickly produce predictions for novel variants after being trained on a known input, sometimes known as a training set. However, both the quality of the training data and the performance of the underlying algorithm have a substantial impact on how well an ML prediction performs when applied to previously untested data. One major challenge in using ML for biocatalyst design is the wide range of enzyme mechanisms, reactions, and experimental setups. The lack of large, consistent datasets for model training, the difficulty of standardizing data formats, the requirement for strict quality control in data collection and reporting, and the sluggishness of gathering new data for model testing are the main causes of this. Annotating known protein sequences could reveal numerous functional enzymes, but creating completely new sequences that evolution has not examined could also be beneficial because it may reveal previously undiscovered combinations of characteristics and possibly non-natural activity (Voigt et al. 2002). Deep learning techniques to put together compatible structural domains in enzymes have been influenced by chimeragenesis, a method of producing energetically advantageous proteins by recombining functional homologous proteins (Yang et al. 2024).

To enhance the effectiveness of ML in enzyme engineering, it is crucial to improve supervised learning methods for more accurate identification of high-fitness variants. While increasing data through high-throughput screening helps, computational strategies, such as better protein representations and evaluation metrics, are key to improving ML predictions (Detlefsen et al. 2022). Current encoding methods, like one-hot vectors and fixed descriptors, are being supplemented by learned embeddings from protein language models (PLMs), although their advantage in supervised fitness prediction remains limited (Michael et al. 2023). Techniques, such as fine-tuning and semi-supervised learning, show promise, especially with limited labeled data, but require further validation through robust benchmarking.

However, there are certain challenges in applying ML models in protein or enzyme engineering. The inherent multidisciplinarity of the approach is one of the primary obstacles in using ML for enzyme engineering. To define objectives, conduct thorough analysis and training, and steer clear of typical mistakes, improper method usage, and misunderstandings, biochemists, molecular biologists, mathematicians, and computer scientists must establish a common vocabulary (Mazurenko et al. 2020). While ready-to-use software packages undoubtedly aid in standardizing the training of ML algorithms for non-specialists, it may not be the best course of action to pile all the available data and run a variety of ML algorithms to choose the best predictor. The necessity for reliable comparison of different predictors is becoming increasingly crucial as ML applications in enzyme engineering continue to expand. The absence of new datasets for testing, as well as defined processes for comparison, are the primary barriers to this comparison. Among the most urgent problems are the absence of benchmark datasets, disparities in the performance metrics employed, incomplete or erroneous disclosure in publications, and the challenge of identifying reviewers with a wide enough range of experience (Walsh et al. 2016). While ML in enzyme engineering shows great promise, many studies prioritize predictive accuracy over understanding the underlying biological mechanisms. There is a lack of interpretability-focused analyses, such as feature importance, learning curves, or error diagnostics, which limits scientific insight. Emphasizing explainable ML models is essential for uncovering molecular principles and improving the utility of predictions.

13.6.2 ML in Biofuel and Bioplastic Manufacturing

13.6.2.1 Biofuel Production System

Bio-based polymers, derived from renewable biomass, are gaining attention as sustainable alternatives to petroleum-based materials. Their biodegradable nature helps reduce environmental pollution. With favorable properties, such as strength, lightness, and flexibility, they support innovations in biofuels and biopolymers (Sofian et al., 2024b). These materials play a key role in driving sustainability across energy, fuel, and biochemical industries. There are numerous uses for ML in various lignocellulosic biorefinery sectors, assisting in "cradle to grave" notion in the circular economy. The identification of critical operational parameters, such as the operating temperature and the rate of biofuel extraction, and the interdependence between those factors often leads to modeling issues in the context of biofuel production. In recent research investigations, ML-based algorithm techniques have been widely used and compared to address these problems (Sharma et al., 2023a).

There are many factors involved in the production of biofuels and biopolymers, from the collection of lignocellulosic/microalgal waste from the source till the product is ready to be delivered to the market. ML applications in biofuel technology are divided into three categories that include the discovery and screening of the feedstock, process optimization, and decision-making. ML techniques assist in predicting catalyst behavior and performance for thermochemical conversion processes. They also help uncover correlations between catalyst properties and their efficiency. Recent research has shown the effective use of

ML in selecting catalysts for biofuel production. For example, a study on guaiacol hydrodeoxygenation demonstrated how ML can optimize catalyst composition and link surface characteristics with conversion outcomes (Okolie 2024). ANNs specifically have shown potential in enhancing microalgae selection for biofuel production. ANN models, such as MLPs, can predict traits, such as chlorophyll content, which is key to assessing biofuel (biodiesel) potential. Tang et al. (2023) applied an ANN to predict microalgae chlorophyll content from color model data using 700 experimental data points. The ANN achieved an R^2 of 0.66 and RMSE of 2.36, outperforming a linear regression model ($R^2 = 0.58$, RMSE = 3.16), thereby demonstrating its potential for species identification and chlorophyll estimation. This approach supports efficient screening of microalgae strains with desirable bioactive properties. Integrating ANN with image-based tools can further streamline the biofuel-oriented selection process.

ML is also proving effective in addressing challenges associated with designing sustainable systems and multiscale process engineering. Traditional modeling approaches for such systems are often non-linear and computationally intensive. ML provides efficient surrogate models that replicate system behavior without detailed mechanistic inputs (Okolie 2024). These models reduce computational demands while maintaining accuracy. In the biofuel sector, ML supports early evaluation of biomass conversion technologies. It aids in assessing both environmental and economic impacts using lifecycle assessment (LCA) and techno-economic analysis (TEA) (Akinpelu et al. 2023). Due to inconsistent or limited data, conventional TEA and LCA are difficult. ML-based approaches help fill these gaps by generating reliable, high-throughput estimates for informed decision-making.

13.6.2.2 Bioplastics Production

Bioplastics are environmentally friendly alternatives to petroleum-based plastics, emitting scarcer greenhouse gases and containing no harmful pollutants. Their renewable nature makes them suitable for long-term use, largely in packaging, with potential for expansion into electronics and automotive sectors. However, their carbon footprint depends on whether the captured carbon remains stored in the final product (Senosha et al. 2025). In particular, ML algorithms such as ANNs, SVMs, and DTs are being used to model and predict the production yield of biodegradable bioplastics such as, polyhydroxyalkanoates (PHAs) (Sofian et al., 2024a). These models help identify optimal nutrient concentrations, temperature, pH, and substrate types to maximize PHA synthesis by microorganisms. Similarly, ML aids in improving the production efficiency of other bioplastics, such as polylactic acid (PLA) and starch-based polymers, by analyzing large datasets from experimental runs and bioreactor monitoring systems. Additionally, ML can assist in predicting the mechanical, thermal, and biodegradation properties of different bioplastic formulations, helping researchers design materials suited for specific applications such as packaging, agriculture, or medical use (Senosha et al. 2025).

Most of the bioplastics could be potentially derived from microalgal species. The development of algae-based bioplastics is aided by the increased productivity and sustainability of seaweed and microalgae production, which has been demonstrated to be possible through the integration of IoT and AI/ML (Esposito et al. 2017; Lim et al. 2022). Due to these advancements, smart farming techniques have been developed, including remote control of cultivation operations, real-time monitoring, and predictive modeling for decision-making. Despite advancements in using ML for microalgae identification, challenges, such as model complexity, high costs, and limited data, persist. Issues such as species overlap, variability, and dependence on preprocessing affect classification accuracy. To improve bioplastics production from algae, more robust algorithms, better image processing, and scalable datasets are needed (Sofian et al., 2024a). Enhancing data sharing and model evaluation will support the development of more reliable and widely applicable ML systems. More research is needed to develop reliable automation for large-scale cultivation. Enhancements in sensor technology and algae strain selection are crucial for future progress (Tham et al. 2022). Through strong linear correlations and low normalized root mean square error (NRMSE) values, (Saboe et al. (2021) successfully correlated microbial photoelectrosynthesis (MPS) signals with high algae concentrations in an algal cultivation pond, proving the feasibility of using AI/ML in the monitoring and control of microalgae cultivation processes. Potential disruptions in microalgae cultivation can be quickly detected and fixed by integrating ML with real-time monitoring systems, guaranteeing a consistent and high-quality production of bioplastic.

13.7 MACHINE LEARNING-BASED LIFECYCLE ASSESSMENT OF BIOPROCESSES

The growing concern over environmental deterioration has led policymakers and industries to adopt proactive strategies, such as pollution prevention and environmental management systems, to minimize their ecological impact. LCA has emerged as a key tool in this context, enabling the evaluation of the environmental impacts of products or processes from "cradle to grave" (Yadav et al. 2023). In biorefineries, LCA plays a vital role in assessing sustainability and guiding policy decisions. LCA is a widely adopted and robust method for evaluating the environmental impacts of a product, process, or value chain across its entire life cycle, from raw material extraction to disposal. Guided by ISO 14040 standards, it provides a structured framework for such analysis (Arias et al. 2023). Its flexibility in handling diverse impact categories makes it highly applicable across sectors. The standardized methodology ensures consistency and comparability. As a result, LCA remains the most preferred tool for environmental assessment. While biomass use can mitigate climate change, it may also contribute to other environmental issues such as eutrophication, acidification, and land-use emissions. Therefore, comprehensive LCA is essential to verify the true environmental benefits of biorefineries.

ML has the potential to improve the precision and thoroughness of LCA results by predicting missing data and filling data gaps. Moreover, the study of more intricate models is made possible by ML's capacity to analyze vast volumes of data precisely and swiftly. As a result, attempts to increase the sustainability of lignocellulosic bioethanol production may be better informed by more thorough and trustworthy insights into the environmental effects of this process. In coupling LCA with ML, the process begins with defining the goal and scope, as in conventional LCA. A key enhancement is the creation of a digital data architecture that enables rapid analysis of large datasets, clarifying the process objectives, data interconnections, and expected outcomes. After data collection, a life cycle inventory is constructed, followed by an impact assessment to quantify environmental burdens (Arias et al. 2023). To enhance interpretation, dimensionality reduction techniques are applied to develop a matrix that clusters data–score relationships. This approach provides clearer insights for decision-making by highlighting the most significant variables influencing environmental performance.

In the past ten years, ML has been applied in several LCA application domains to forecast life cycle inventory and close data gaps. A kinetic-based process simulation and a high-accuracy ANN model were previously created to forecast key energy consumption and greenhouse gas emissions across various woody biomass (Liao et al. 2020). Likewise, for slow pyrolysis of numerous feedstocks under specific processing circumstances, the RF algorithm was combined with the LCA model to forecast net positive energy production and net negative global warming potential emissions (Cheng et al. 2020). Based on a chemical's physicochemical and structural characteristics, the ANN algorithm was used to estimate its environmental and toxicological characteristics (hazardous concentration values [HC50]) (Hou et al. 2020). These previous studies demonstrate how ML approaches can be used in LCA to improve data availability and forecast missing variables, increasing the confidence and accuracy of LCA results.

13.8 CHALLENGES AND FUTURE PERSPECTIVES

As industries embrace digitalization, biomanufacturing is gradually exploring advanced technologies such as AI or ML, although its adoption remains limited. The AI or ML-based technology is undoubtedly one of the most promising additions in optimizing and providing innovative bioprocess engineering. Modern bioprocess development generates vast amounts of heterogeneous experimental data containing vital process insights. For this, the engineers select appropriate ML models by benchmarking their performance on relevant datasets. ML, combined with mathematical models, enables exploration across diverse design spaces. Applications such as modeling, prediction, classification, and pattern recognition, especially using ANNs, enhance understanding of the process. Advancements in computer modeling and biomedical image analysis further support bioprocess optimization. Hybrid ML models that integratephysical and data-driven approaches offer improved predictive power. The rise of high-throughput technologies has amplified big data availability, fueling growth in biotechnology. However, concerns over the risks of faulty products and the absence of robust, representative datasets hinder ML's integration. The complex, multi-parametric nature of biological systems adds to the unpredictability of ML models, while their complexity challenges interpretability in bioprocess contexts. Despite these barriers, rapid innovations in ML and computational tools are expected to address these issues (Saha et al. 2024). The future holds strong potential for ML in enhancing and scaling bioproduct development through advanced bioprocess engineering.

ML is typically applied to systems with large datasets, but biochemical engineering often involves limited experimental data. To address this, data augmentation strategies, such as adding random noise within realistic bounds, are used to expand datasets. This approach leverages the natural variability in bioprocesses and experimental measurements to generate synthetic yet valid data. These expanded datasets enable effective model construction for bioprocess modeling and soft-sensing. The need for large datasets stems from the numerous parameters in ML models, especially neural networks. Proper augmentation helps prevent overfitting and supports robust model training. Moreover, in fields such as metabolic and protein engineering, extensive databases have been developed to support ML model construction. However, a major challenge lies in the lack of standardization and compatibility across these databases. Inconsistent formats and missing records hinder effective data integration and model training. This issue is common across scientific areas such as chemical synthesis, molecular design, and process upscaling, limiting the full potential of data-driven approaches (Mowbray et al. 2021).

To effectively advance Industry 4.0, integrated policies are needed to guide the integration of digital technologies into biomanufacturing. These should address challenges such as workforce displacement, process adaptation, feedstock supply chain, and minimum job requirements. Since digitized systems often focus on specific linear processes, care must be taken to ensure they support circular production models. Without proper guidance, the benefits of AI/ML and digitization could be undermined. Establishing standardized benchmarks in bioprocess engineering is crucial for its advancement and for encouraging cross-disciplinary collaboration. Drawing from lessons in the ML field, a good benchmark should offer publicly accessible datasets with defined tasks, such as classification, regression, and prediction, and be flexible and general enough to fit various bioprocess research workflows (Duong-Trung et al. 2023). Such standards will support consistent evaluation, promote data sharing, and streamline ML integration into bioprocessing. Moreover, beyond advancing ML methods, there is a pressing need to grow a culture of research reproducibility. Reproducible models enhance clarity, reliability, and adaptability, making them easier to build upon and encouraging collaborative, transparent science. Supporting replication efforts helps validate findings and contributes to developing more complex, multi-scale models. Rather than dismissing studies with non-replicable results, the researchers should focus on verifying and improving research integrity. Funding agencies and journals should promote reproducibility as a core scientific value, even creating dedicated criteria to evaluate it.

13.9 SUMMARY AND CONCLUSIONS

ML has emerged as a transformative tool in biomanufacturing and bioprocess engineering, empowered by advances in automation, high-throughput analytics, and data-driven modeling. However, to move from potential to practice, a structured roadmap is needed to guide AI/ML adoption in sustainable biomanufacturing. In the short term, efforts should focus on improving data quality, standardization, and accessibility across bioprocess operations. Establishing robust data pipelines and integrating automation with real-time monitoring systems will ensure reliable inputs for ML models. Training the workforce in data literacy and AI fundamentals is equally critical to facilitate meaningful adoption. In the medium term, the emphasis should shift toward developing hybrid and mechanistic ML models that combine the interpretability of classical process knowledge with the predictive strength of data-driven algorithms. These integrated approaches can accelerate process optimization, scale-up design, and biochemical analysis, especially for valorizing waste biomass and developing sustainable bioproducts. In the long term, the vision should be to achieve self-optimizing, closed-loop biomanufacturing systems driven by autonomous ML frameworks. With the support of advanced computation, these systems can dynamically learn from process data, adjust control parameters in real time, and continuously enhance efficiency and sustainability. The collaboration between academia, industry, and policymakers should be initiated to ensure ethical, transparent, and responsible AI deployment. In short, AI or ML's success in sustainable biomanufacturing will depend not on model complexity but on intelligent integration, linking data, domain expertise, and innovation to enable a more resilient and circular bioeconomy.

REFERENCES

Akinpelu, David Akorede, Oluwaseun A. Adekoya, Peter Olusakin Oladoye, Chukwuma C. Ogbaga, and Jude A. Okolie. 2023. Machine Learning Applications in Biomass Pyrolysis: From Biorefinery to End-of-Life Product Management. *Digital Chemical Engineering* 8 (September): 100103. https://doi.org/10.1016/j.dche.2023.100103

Arias, Ana, Gumersindo Feijoo, and Maria Teresa Moreira. 2023. How Could Artificial Intelligence Be Used to Increase the Potential of Biorefineries in the near Future? *A Review. Environmental Technology & Innovation* 32 (November): 103277. https://doi.org/10.1016/j.eti.2023.103277

Chen, Long, Pinjing He, Hua Zhang, Wei Peng, Junjie Qiu, and Fan Lü. 2024. Applications of Machine Learning Tools for Biological Treatment of Organic Wastes: Perspectives and Challenges. *Circular Economy* 3 (2): 100088. https://doi.org/10.1016/j.cec.2024.100088

Chen, Wei-Hsin, Hsiu-Ju Lo, Ria Aniza, et al. 2022. Forecast of Glucose Production from Biomass Wet Torrefaction Using Statistical Approach along with Multivariate Adaptive Regression Splines, Neural Network and Decision Tree. *Applied Energy* 324 (October): 119775. https://doi.org/10.1016/j.apenergy.2022.119775

Cheng, Fangwei, Hongxi Luo, and Lisa M. Colosi. 2020. Slow Pyrolysis as a Platform for Negative Emissions Technology: An Integration of Machine Learning Models, Life Cycle Assessment, and Economic Analysis. *Energy Conversion and Management* 223 (November): 113258. https://doi.org/10.1016/j.enconman.2020.113258

Cheng, Yang, Xinyu Bi, Yameng Xu, et al. 2023. Artificial Intelligence Technologies in Bioprocess: Opportunities and Challenges. *Bioresource Technology* 369 (February): 128451. https://doi.org/10.1016/j.biortech.2022.128451

Chong, Jun Wei Doris Ying Roy, Ying Tang, Hui Yi Leong, Kuan Shiong Khoo, Pau Loke Show, and Kit Wayne Chew. 2023. Bridging Artificial Intelligence and Fucoxanthin for the Recovery and Quantification from Microalgae. *Bioengineered* 14 (1): 2244232. https://doi.org/10.1080/21655979.2023.2244232

Detlefsen, Nicki Skafte, Søren Hauberg, and Wouter Boomsma. 2022. What Is a Meaningful Representation of Protein Sequences? *Nature Communications* 13 (1): 1914. https://doi.org/10.1038/s41467-022-29443-w

Duong-Trung, Nghia, Stefan Born, Jong Woo Kim, et al. 2023. When Bioprocess Engineering Meets Machine Learning: A Survey from the Perspective of Automated Bioprocess Development. *Biochemical Engineering Journal* 190 (January): 108764. https://doi.org/10.1016/j.bej.2022.108764

Esposito, Serena, Antonio Cafiero, Francesco Giannino, Stefano Mazzoleni, and Marcello Maria Diano. 2017. A Monitoring, Modeling and Decision Support System (DSS) for a Microalgae Production Plant Based on Internet of Things Structure. *Procedia Computer Science, The 8th International Conference on Emerging Ubiquitous Systems and Pervasive Networks (EUSPN 2017)/The 7th International Conference on Current and Future Trends of Information and Communication Technologies in Healthcare (ICTH-2017)/ Affiliated Workshops*, vol. 113 (January): 519–24. https://doi.org/10.1016/j.procs.2017.08.316

Frontistis, Zacharias, Grigoris Lykogiannis, and Anastasios Sarmpanis. 2023. Machine Learning Implementation in Membrane Bioreactor Systems: Progress, Challenges, and Future Perspectives: A Review. *Environments* 10 (7): 7. https://doi.org/10.3390/environments10070127

Goettsch, Diana, Krystel K. Castillo-Villar, and Maria Aranguren. 2020. Machine-Learning Methods to Select Potential Depot Locations for the Supply Chain of Biomass Co-Firing. *Energies* 13 (24): 24. https://doi.org/10.3390/en13246554

Helmy, Mohamed, Derek Smith, and Kumar Selvarajoo. 2020. Systems Biology Approaches Integrated with Artificial Intelligence for Optimized Metabolic Engineering.

Metabolic Engineering Communications 11 (December): e00149. https://doi.org/10.1016/j.mec.2020.e00149

Hou, Ping, Bu Zhao, Olivier Jolliet, Ji Zhu, Peng Wang, and Ming Xu. 2020. Rapid Prediction of Chemical Ecotoxicity Through Genetic Algorithm Optimized Neural Network Models. *ACS Sustainable Chemistry & Engineering* 8 (32): 12168–76. https://doi.org/10.1021/acssuschemeng.0c03660

Imamoglu, Esra. 2024. Artificial Intelligence and/or Machine Learning Algorithms in Microalgae Bioprocesses. *Bioengineering* 11 (11): 11. https://doi.org/10.3390/bioengineering11111143

Ji, Huizhuo, Dandan Pu, Wenjing Yan, Qingchuan Zhang, Min Zuo, and Yuyu Zhang. 2023. Recent Advances and Application of Machine Learning in Food Flavor Prediction and Regulation. *Trends in Food Science & Technology* 138 (August): 738–51. https://doi.org/10.1016/j.tifs.2023.07.012

Khanal, Samir Kumar, Ayon Tarafdar, and Siming You. 2023. Artificial Intelligence and Machine Learning for Smart Bioprocesses. *Bioresource Technology* 375 (May): 128826. https://doi.org/10.1016/j.biortech.2023.128826

Kim, Gi Bae, Won Jun Kim, Hyun Uk Kim, and Sang Yup Lee. 2020. Machine Learning Applications in Systems Metabolic Engineering. *Current Opinion in Biotechnology, Analytical Biotechnology*, vol. 64 (August): 1–9. https://doi.org/10.1016/j.copbio.2019.08.010

Kim, Joo-Sung. 2017. Vessel Target Prediction Method and Dead Reckoning Position Based on SVR Seaway Model. *International Journal of Fuzzy Logic and Intelligent Systems* 17 (4): 279–88. https://doi.org/10.5391/IJFIS.2017.17.4.279

Li, Hui, Jiajun Chen, Xingyan Li, et al. 2024. Artificial Neural Network and Genetic Algorithm Coupled Fermentation Kinetics to Regulate L-Lysine Fermentation. *Bioresource Technology* 393 (February): 130151. https://doi.org/10.1016/j.biortech.2023.130151

Liao, Mochen, Stephen Kelley, and Yuan Yao. 2020. Generating Energy and Greenhouse Gas Inventory Data of Activated Carbon Production Using Machine Learning and Kinetic Based Process Simulation. *ACS Sustainable Chemistry & Engineering* 8 (2): 1252–61. https://doi.org/10.1021/acssuschemeng.9b06522

Lim, Hooi Ren, Kuan Shiong Khoo, Wen Yi Chia, Kit Wayne Chew, Shih-Hsin Ho, and Pau Loke Show. 2022. Smart Microalgae Farming with Internet-of-Things for Sustainable Agriculture. *Biotechnology Advances* 57 (July): 107931. https://doi.org/10.1016/j.biotechadv.2022.107931

Lu, Hongzhong, Luchi Xiao, Wenbin Liao, Xuefeng Yan, and Jens Nielsen. 2024. Cell Factory Design with Advanced Metabolic Modelling Empowered by Artificial Intelligence. *Metabolic Engineering* 85 (September): 61–72. https://doi.org/10.1016/j.ymben.2024.07.003

Lu, Qinglin, Hao Zhang, Rong Fan, Yinhua Wan, and Jianquan Luo. 2025. Machine Learning-Based Bayesian Optimization Facilitates Ultrafiltration Process Design for Efficient Protein Purification. *Separation and Purification Technology* 363 (August): 132122. https://doi.org/10.1016/j.seppur.2025.132122

Mazurenko, Stanislav, Zbynek Prokop, and Jiri Damborsky. 2020. Machine Learning in Enzyme Engineering. *ACS Catalysis* 10 (2): 1210–23. https://doi.org/10.1021/acscatal.9b04321

Mesquita, Thiago J. B., Gilson Campani, Roberto C. Giordano, Teresa C. Zangirolami, and Antonio C. L. Horta. 2021. Machine Learning Applied for Metabolic Flux-Based Control of Micro-Aerated Fermentations in Bioreactors. *Biotechnology and Bioengineering* 118 (5): 2076–91. https://doi.org/10.1002/bit.27721

Michael, Richard, Jacob Kæstel-Hansen, Peter Mørch Groth, et al. 2023. Assessing the Performance of Protein Regression Models. *Preprint, bioRxiv*, September 2. https://doi.org/10.1101/2023.06.18.545472

Mondal, Partha Pratim, Abhinav Galodha, Vishal Kumar Verma, et al. 2023. Review on Machine Learning-Based Bioprocess Optimization, Monitoring, and Control Systems. *Bioresource Technology* 370 (February): 128523. https://doi.org/10.1016/j.biortech.2022.128523

Mowbray, Max, Thomas Savage, Chufan Wu, et al. 2021. Machine Learning for Biochemical Engineering: A Review. *Biochemical Engineering Journal* 172 (August): 108054. https://doi.org/10.1016/j.bej.2021.108054

Nargotra, Parushi, Bijender Kumar Bajaj, Reeta Rani Singhania, Diksha Sharma, and Vishal Sharma. 2024. Transition from Linear to Circular Bioeconomy for the Sustainable Production of Green Chemicals. In *Bioeconomy for Sustainability*, edited by Vinod Kumar Garg and Navish Kataria. Springer Nature. https://doi.org/10.1007/978-981-97-1837-5_3

Nargotra, Parushi, Vishal Sharma, Yi-Chen Lee, et al. 2023. Microbial Lignocellulolytic Enzymes for the Effective Valorization of Lignocellulosic Biomass: A Review. *Catalysts* 13 (1): 1. https://doi.org/10.3390/catal13010083

Nargotra, Parushi, Vishal Sharma, Hui-Min David Wang, Chwen-Jen Shieh, Yung-Chuan Liu, and Chia-Hung Kuo. 2025. Biocatalysis for Lignin Conversion and Valorization: Driving Sustainability in the Circular Economy. *Catalysts* 15 (1): 1. https://doi.org/10.3390/catal15010091

Ndochinwa, Obinna Giles, Qing-Yan Wang, Oyetugo Chioma Amadi, et al. 2024. Current status and emerging Frontiers in Enzyme Engineering: An Industrial Perspective. *Heliyon* 10 (11). https://doi.org/10.1016/j.heliyon.2024.e32673

Okolie, Jude A. 2024. Introduction of Machine Learning and Artificial Intelligence in Biofuel Technology. *Current Opinion in Green and Sustainable Chemistry* 47 (June): 100928. https://doi.org/10.1016/j.cogsc.2024.100928

Ortizo, Rhessa Grace Vishal Sharma Guanga, Mei-Ling Tsai, et al. 2024. Exploring the Potential of Magnetic Deep Eutectic Solvents and DES-Functionalized Nanomaterials for Food Analysis: Advancements and Current Trends. *Food Bioscience* 61 (October): 104764. https://doi.org/10.1016/j.fbio.2024.104764

Oyetunde, Tolutola, Forrest Sheng Bao, Jiung-Wen Chen, Hector Garcia Martin, and Yinjie J. Tang. 2018. Leveraging Knowledge Engineering and Machine Learning for Microbial Bio-Manufacturing. *Biotechnology Advances* 36 (4): 1308–15. https://doi.org/10.1016/j.biotechadv.2018.04.008

Packiam, Kulandai Arockia Chien Wei Rajesh, Fuyi Li Ooi, et al. 2022. PERISCOPE-Opt: Machine Learning-Based Prediction of Optimal Fermentation Conditions and Yields of Recombinant Periplasmic Protein Expressed in *Escherichia coli*. *Computational and Structural Biotechnology Journal* 20 (January): 2909–20. https://doi.org/10.1016/j.csbj.2022.06.006

Peng, Wei, and Omid Karimi Sadaghiani. 2023. An Analytical Review on the Utilization of Machine Learning in the Biomass Raw Materials, Their Evaluation, Storage, and Transportation. *Archives of Computational Methods in Engineering* 30 (8): 4711–32. https://doi.org/10.1007/s11831-023-09950-9

Pensupa, Nattha, Treesukon Treebuppachartsakul, and Suejit Pechprasarn. 2023. Machine Learning Models Using Data Mining for Biomass Production from Yarrowia Lipolytica Fermentation. *Fermentation* 9 (3): 3. https://doi.org/10.3390/fermentation9030239

Pirrung, Silvia M., Luuk A. M. van der Wielen, Ruud F. W. C. van Beckhoven, Emile J. A. X. van de Sandt, Michel H. M. Eppink, and Marcel Ottens. 2017. Optimization of Biopharmaceutical Downstream Processes Supported by Mechanistic Models and Artificial Neural Networks. *Biotechnology Progress* 33 (3): 696–707. https://doi.org/10.1002/btpr.2435

Pradhan, Dileswar, Swarna Jaiswal, and Amit K. Jaiswal. 2022. "Artificial Neural Networks in Valorization Process Modeling of Lignocellulosic Biomass." *Biofuels, Bioproducts and Biorefining* 16 (6): 1849–68. https://doi.org/10.1002/bbb.2417

Rathore, Anurag S., Saxena Nikita, Garima Thakur, and Somesh Mishra. 2023. Artificial Intelligence and Machine Learning Applications in Biopharmaceutical Manufacturing. *Trends in Biotechnology* 41 (4): 497–510. https://doi.org/10.1016/j.tibtech.2022.08.007

Saboe, Daniel, Hamidreza Ghasemi, Ming Ming Gao, et al. 2021. Real-Time Monitoring and Prediction of Water Quality Parameters and Algae Concentrations Using Microbial Potentiometric Sensor Signals and Machine Learning Tools. *Science of The Total Environment* 764 (April): 142876. https://doi.org/10.1016/j.scitotenv.2020.142876

Saha, Ritika, Ashutosh Chauhan, and Smita Rastogi Verma. 2024. Machine Learning: An Advancement in Biochemical Engineering. *Biotechnology Letters* 46 (4): 497–519. https://doi.org/10.1007/s10529-024-03499-8

Saini, Dinesh Kumar, Amit Rai, Alka Devi, et al. 2021. A Multi-Objective Hybrid Machine Learning Approach-Based Optimization for Enhanced Biomass and Bioactive Phycobiliproteins Production in *Nostoc* Sp. CCC-403. *Bioresource Technology* 329 (June): 124908. https://doi.org/10.1016/j.biortech.2021.124908

Senosha, Maphuti Pinkie, Anthony Njuguna Matheri, and Belaid Mohamed. 2025. Review on Application of Sustainability, Circular and Digital Economy on Bioplastics Production. *Circular Economy and Sustainability* 5 (2): 1269–89. https://doi.org/10.1007/s43615-024-00453-8

Sharma, Dayanand, Ashutosh Kumar Pandey, Kunwar Durg Yadav, and Sunil Kumar. 2021. Response Surface Methodology and Artificial Neural Network Modelling for Enhancing Maturity Parameters during Vermicomposting of Floral Waste. *Bioresource Technology* 324 (March): 124672. https://doi.org/10.1016/j.biortech.2021.124672

Sharma, Diksha, Vishal Sharma, Mei-Ling Tsai, et al. 2025. Recent Advances in Ultrasound-Assisted Technologies for Improved Bioproduct Recovery in Lignocellulosic and Microalgal Biorefineries. *Journal of Industrial and Engineering Chemistry*, January 2. https://doi.org/10.1016/j.jiec.2024.12.082

Sharma, Diksha, Mei-Ling Tsai, Cheng-Di Dong, et al. 2024. An Innovative Recyclable Deep Eutectic Solvent-Ultrasound Pretreatment of Pineapple Leaf Waste Biomass for Enhanced Bioethanol Production. *Biomass Conversion and Biorefinery*, ahead of print, October 8. https://doi.org/10.1007/s13399-024-06223-8

Sharma, Surbhi, Mei-Ling Tsai, Vishal Sharma, et al. 2023a. Environment Friendly Pretreatment Approaches for the Bioconversion of Lignocellulosic Biomass into Biofuels and Value-Added Products. *Environments* 10 (1): 1. https://doi.org/10.3390/environments10010006

Sharma, Vishal, Mei-Ling Tsai, Chiu-Wen Chen, Pei-Pei Sun, Parushi Nargotra, and Cheng-Di Dong. 2023b. Advances in Machine Learning Technology for Sustainable Biofuel Production Systems in Lignocellulosic Biorefineries. *Science of The Total Environment* 886 (August): 163972. https://doi.org/10.1016/j.scitotenv.2023.163972

Sharma, Vishal, Mei-Ling Tsai, Parushi Nargotra, et al. 2023c. Journey of Lignin from a Roadblock to Bridge for Lignocellulose Biorefineries: A Comprehensive Review. *Science of The Total Environment* 861 (February): 160560. https://doi.org/10.1016/j.scitotenv.2022.160560

Shelare, Sagar D., Pramod N. Belkhode, Keval Chandrakant Nikam, et al. 2023. Biofuels for a Sustainable Future: Examining the Role of Nano-Additives, Economics, Policy, Internet of Things, Artificial Intelligence and Machine Learning Technology in Biodiesel Production. *Energy* 282 (November): 128874. https://doi.org/10.1016/j.energy.2023.128874

Singh, Ashutosh, and Barkha Singhal. 2022. Role of Machine Learning in Bioprocess Engineering: Current Perspectives and Future Directions. In *Design and Applications of Nature Inspired Optimization: Contribution of Women Leaders in the Field*, edited by Dipti Singh, Vanita Garg, and Kusum Deep. Springer International Publishing. https://doi.org/10.1007/978-3-031-17929-7_3

Sofian, Bin Abu, Abu Danish Aiman, Hooi Ren Lim, Sivakumar Manickam, Wei Lun Ang, and Pau Loke Show. 2024a. Towards a Sustainable Circular Economy: Algae-Based Bioplastics and the Role of Internet-of-Things and Machine Learning. *ChemBioEng Reviews* 11 (1): 39–59. https://doi.org/10.1002/cben.202300028

Sofian, Bin Abu, Abu Danish Aiman, Xun Sun, Vijai Kumar Gupta, et al. 2024b. Advances, Synergy, and Perspectives of Machine Learning and Biobased Polymers for Energy, Fuels, and Biochemicals for a Sustainable Future. *Energy & Fuels* 38 (3): 1593–617. https://doi.org/10.1021/acs.energyfuels.3c03842

Sultana, Nahid, S. M. Zakir Hossain, M. Abusaad, N. Alanbar, Y. Senan, and S. A. Razzak. 2022. Prediction of Biodiesel Production from Microalgal Oil Using Bayesian Optimization Algorithm-Based Machine Learning Approaches. *Fuel* 309 (February): 122184. https://doi.org/10.1016/j.fuel.2021.122184

Tang, Ying Ying, Kit Wayne Doris Huong-Yong Ting Chew, et al. 2023. Application of Regression and Artificial Neural Network Analysis of Red-Green-Blue Image Components in Prediction of Chlorophyll Content in Microalgae. *Bioresource Technology* 370 (February): 128503. https://doi.org/10.1016/j.biortech.2022.128503

Tham, Pei En, Yan Jer Ng, Navintran Vadivelu, et al. 2022. Sustainable Smart Photobioreactor for Continuous Cultivation of Microalgae Embedded with Internet of things. *Bioresource Technology* 346 (February): 126558. https://doi.org/10.1016/j.biortech.2021.126558

Tsui, T.-H., Mark C. M. van Loosdrecht, Yanjun Dai, and Yen Wah Tong. 2023. Machine Learning and Circular Bioeconomy: Building New Resource Efficiency from Diverse Waste Streams. *Bioresource Technology* 369 (February): 128445. https://doi.org/10.1016/j.biortech.2022.128445

Velidandi, Aditya, Pradeep Kumar Gandam, Madhavi Latha Chinta, et al. 2023. State-of-the-Art and Future Directions of Machine Learning for Biomass Characterization and for Sustainable Biorefinery. *Journal of Energy Chemistry* 81 (June): 42–63. https://doi.org/10.1016/j.jechem.2023.02.020

Vinestock, Tom, Michael Short, Keeran Ward, and Miao Guo. 2024. Computer-Aided Chemical Engineering Research Advances in Precision Fermentation. *Current Opinion in Food Science* 58 (August): 101196. https://doi.org/10.1016/j.cofs.2024.101196

Voigt, Christopher A., Carlos Martinez, Zhen-Gang Wang, Stephen L. Mayo, and Frances H. Arnold. 2002. Protein Building Blocks Preserved by Recombination. *Nature Structural Biology* 9 (7): 553–58. https://doi.org/10.1038/nsb805

Walsh, Ian, Gianluca Pollastri, and Silvio C. E. Tosatto. 2016. Correct Machine Learning on Protein Sequences: A Peer-Reviewing Perspective. *Briefings in Bioinformatics* 17 (5): 831–40. https://doi.org/10.1093/bib/bbv082

Wang, Bo, Muhammad Shahzad, Xianglin Zhu, Khalil Ur Rehman, and Saad Uddin. 2020. A Non-Linear Model Predictive Control Based on Grey-Wolf Optimization Using Least-Square Support Vector Machine for Product Concentration Control in l-Lysine Fermentation. *Sensors* 20 (11): 11. https://doi.org/10.3390/s20113335

Wang, Zhen-Zhi, Du-Wen Zeng, Yi-Fan Zhu, et al. 2025. Fermentation Design and Process Optimization Strategy Based on Machine Learning. *BioDesign Research* 7 (1): 100002. https://doi.org/10.1016/j.bidere.2025.100002

Weichert, Dorina, Patrick Link, Anke Stoll, Stefan Rüping, Steffen Ihlenfeldt, and Stefan Wrobel. 2019. A Review of Machine Learning for the Optimization of Production Processes. *The International Journal of Advanced Manufacturing Technology* 104 (5): 1889–902. https://doi.org/10.1007/s00170-019-03988-5

Yadav, Aditya, Cheng-Di Dong, Diksha Sharma, et al. 2025. Integrated Choline Chloride/Citric Acid-Microwave Pretreatment for Efficient Nanolignin Extraction and Bioethanol Production from Cocoa Pod Husk Waste. *Energy & Environment* 36 (5): 2213–30. https://doi.org/10.1177/0958305X241270269

Yadav, Aditya, Vishal Sharma, Mei-Ling Tsai, et al. 2023. Development of Lignocellulosic Biorefineries for the Sustainable Production of Biofuels: Towards Circular Bioeconomy. *Bioresource Technology, May* 9, 129145. https://doi.org/10.1016/j.biortech.2023.129145

Yang, Chao-Tung, Endah Kristiani, Yoong Kit Leong, and Jo-Shu Chang. 2023. Big Data and Machine Learning Driven Bioprocessing – Recent Trends and Critical Analysis. *Bioresource Technology* 372 (March): 128625. https://doi.org/10.1016/j.biortech.2023.128625

Yang, Jason, Francesca-Zhoufan Li, and Frances H. Arnold. 2024. Opportunities and Challenges for Machine Learning-Assisted Enzyme Engineering. *ACS Central Science* 10 (2): 226–41. https://doi.org/10.1021/acscentsci.3c01275

Zafar, Mohd., Shashi Kumar, Surendra Kumar, and Amit K. Dhiman. 2012. Artificial Intelligence Based Modeling and Optimization of Poly(3-Hydroxybutyrate-Co-3-Hydroxyvalerate) Production Process by Using *Azohydromonas lata* MTCC 2311 from Cane Molasses Supplemented with Volatile Fatty Acids: A Genetic Algorithm Paradigm. *Bioresource Technology* 104 (January): 631–41. https://doi.org/10.1016/j.biortech.2011.10.024

Zhao, Jinghan, Jingxin Wang, and Nathaniel Anderson. 2024. Machine Learning Applications in Forest and Biomass Supply Chain Management: A Review. *International Journal of Forest Engineering* 35 (3): 371–80. https://doi.org/10.1080/14942119.2024.2380230

14 Legal Regime and Governance Policies in Sustainable Production

Mahesh Jaiwar
Department of Applied Agriculture, Central University of Punjab, Bathinda, Punjab, India

Yashi Srivastava
Centre of Food Technology, University of Allahabad, Prayagraj, Uttar Pradesh, India

Tejasvita Sirohi
Department of Law, Kurukshetra University, Kurukshetra, India

14.1 INTRODUCTION

In recent years, global attention has focused on the need to shift toward more sustainable ways of producing and consuming. This shift requires certain policies that promote production systems that use resources efficiently and reduce environmental harm throughout the entire product lifecycle (Fadly, 2020). This paradigm shift requires a comprehensive understanding of how legal regimes and governance policies shape the behavior of producers, consumers, and governments, steering them toward more sustainable practices (Ruggerio 2021; Bhamra and Hernández 2021). Incorporating sustainability into legal and institutional frameworks is essential for tackling global challenges and protecting the environment (Castro 2012). Sustainable production also means recognizing that environmental damage can happen at any given stage of the product's lifecycle, whether it be from resource extraction or disposal (Bhamra and Hernández 2021). Public involvement is critical for achieving sustainability, and increasing awareness can help promote more responsible consumption and production habits (Mensah 2019).

The development of sustainable design has been shaped by growing awareness of global social, environmental, and economic issues. International initiatives like the United Nations' Millennium Development Goals (MDGs) and Sustainable Development Goals (SDGs) have played a major role in focusing attention on key areas (Bhamra and Hernández 2021). The initial focus was primarily on addressing environmental concerns through end-of-pipe solutions, involving corrective measures to mitigate pollution created during the manufacturing process (Bhamra and Hernández 2021). However, since the late 1980s, the emphasis has shifted toward proactively preventing pollution through process and product design and promoting cleaner production methods. Policy instruments, including environmental regulations and incentives, have been instrumental in influencing the behavior of producers and consumers, encouraging them to adopt more sustainable practices. The circular economy (CE), which aims to operate in closed cycles without generating waste, strongly aligns with the systemic changes needed for sustainable development (Bhamra and Hernández 2021). The Rio de Janeiro meeting highlighted the importance of discussing environmental and developmental issues outside the political arena, focusing instead on more practical applications. The United Nations (UN) has progressively advanced the field of sustainability since at least 1972 through national meetings and conferences (Garren and Brinkmann 2018). The 1992 UN Conference in Rio de Janeiro marked a pivotal moment by establishing the need to limit greenhouse gas (GHG) emissions worldwide, which later led to the 1997 Kyoto Protocol with set emission-reduction targets (Bhamra and Hernández 2021). Despite early scientific discussions on sustainable development, environmental terms began appearing in the International Center for Environment Audit and Sustainable Development (ICED) conferences in 1990, marking a transition period in design for sustainability (Figure 14.1).

The term "Sustainable Development" became widely known after the Brundtland Report was published in 1987. It defined sustainable development as meeting current needs without preventing future generations from meeting theirs (Jayawardhana et al., 2022; Benevene et al., 2021; Shi et al., 2019). The Hannover principles, which were developed in 2000, offered a foundation for sustainable design, with the eco-design manual laying out ways to integrate these concerns into product development (Bhamra and Hernández 2021). SDG has grown significantly since the 1972 UN Conference on the Human Environment (Barral, 2010). This chapter focuses on the legal and governance frameworks that support sustainable production globally.

DOI: 10.1201/9781003637158-14

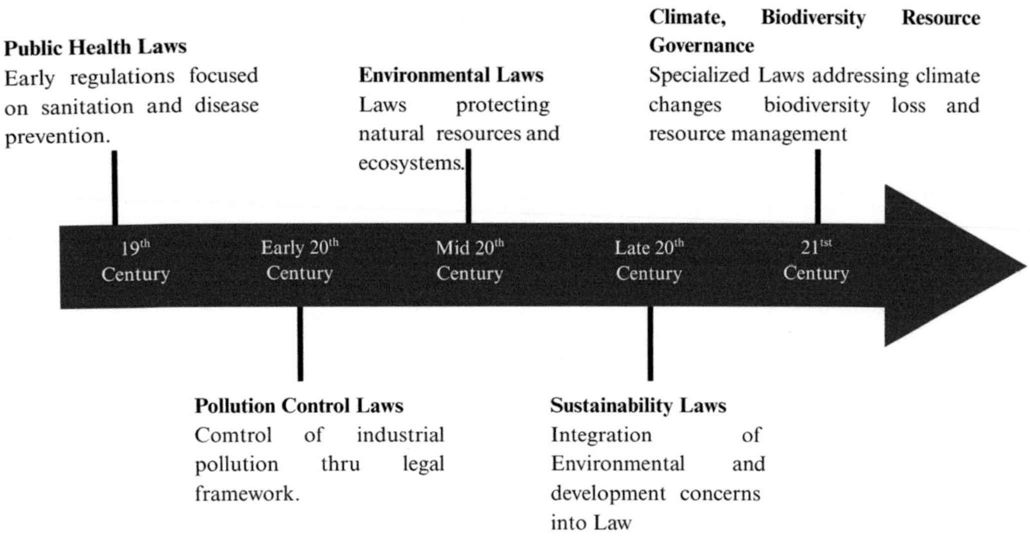

FIGURE 14.1 Evolution of sustainability laws.

Corporate governance contributes through Environmental, Social, and Governance (ESG) standards, Corporate Social Responsibility (CSR), and sustainability reporting. Public–private partnerships (PPPs) and green finance, including sovereign bonds and incentives, promote sustainable investments. Technologies like AI and blockchain enhance transparency in supply chains. Regulatory tools include environmental assessments, carbon pricing, and eco-certifications, while voluntary standards, such as Certified B Corporation (B-Corp), promote self-regulation.

14.2 LEGAL FRAMEWORKS GOVERNING SUSTAINABLE PRODUCTION

Legal frameworks play a crucial role in promoting sustainable production and consumption by establishing standards, regulations, and incentives that guide the behavior of businesses and consumers (Arı and Yikmaz 2019). These frameworks can range from environmental regulations that set limits on pollution and resource extraction to consumer protection laws that promote transparency and accountability in product labeling and advertising. Extended Producer Responsibility (EPR) schemes, for example, hold producers accountable for the end-of-life management of their products, incentivizing them to design products that are more durable, repairable, and recyclable. Governments worldwide have implemented environmental laws and administrative frameworks to address environmental challenges (Lashari et al. 2022). Implementing environmental policy instruments, such as taxes, subsidies, and regulations, can encourage the adoption of sustainable production practices and discourage environmentally harmful activities (Villeneuve et al. 2017). Also, governments are able to promote more sustainable products and services through public procurement policies that prioritize products with lower environmental impacts and greater social benefits. Furthermore, international agreements and conventions, such as the Paris Agreement and the Convention on Biological Diversity, provide a framework for international cooperation on sustainable development issues, setting global targets and standards for environmental protection and resource management (Culiao, Tae, and Kim 2018). Governments can influence both consumer and producer behavior to encourage sustainability by linking sustainable consumption initiatives to policies that aim to improve production sustainability within the private sector at the national and international levels (Stevens 2010).

14.2.1 International Legal Instruments

14.2.1.1 United Nations Sustainable Development Goals (SDGs)

The SDGs, adopted by all 193 UN member states in 2015, represent a universal call for action to end poverty, protect the planet, and ensure prosperity for all by 2030. SDGs are the successors to the MDGs and consist of 17 goals and 169 targets, encompassing areas such as health, education, gender equality, clean energy, climate action, and reduced inequalities (Figure 14.2).

What makes the SDGs groundbreaking is their integrated and indivisible nature—each goal is interconnected, and progress in one area often supports others. For example, advancing clean water and sanitation (Goal 6) improves health outcomes (Goal 3) and gender equality (Goal 5), particularly in developing countries.

The SDGs also emphasize "leaving no one behind," ensuring that development benefits the most vulnerable populations. Countries are required to integrate these goals into their national development strategies, report their progress through Voluntary National Reviews (VNRs), and collaborate through international partnerships (Goal 17) (Figure 14.3).

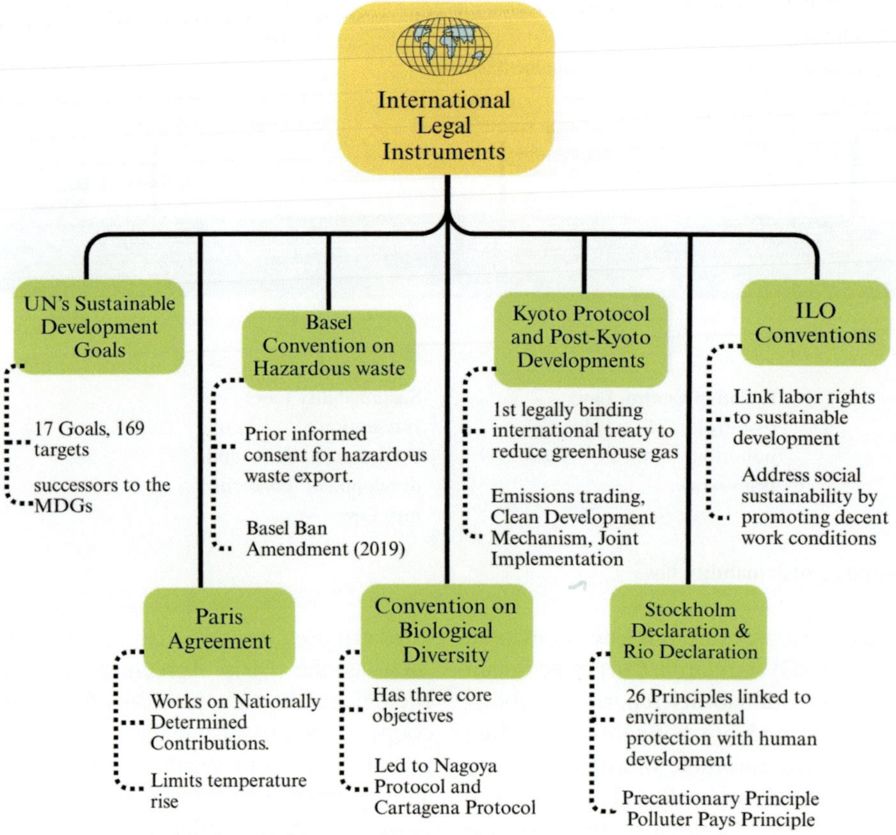

FIGURE 14.2 International Legal Instruments for Sustainable Development.

14.2.1.2 Paris Agreement on Climate Change

The Paris Agreement, adopted in December 2015 under the United Nations Framework Convention on Climate Change (UNFCCC), is a landmark treaty in the fight against climate change. Its primary objective is to limit global temperature rise to well below 2°C above pre-industrial levels, with efforts to restrict it to 1.5°C. The agreement legally binds parties to take climate action but allows flexibility in how nations meet their goals through Nationally Determined Contributions (NDCs).

Unlike the Kyoto Protocol, which imposed binding targets only on developed countries, the Paris Agreement adopts a bottom-up approach, requiring all parties—both developed and developing—to submit emission-reduction plans. Every five years, countries are expected to update their NDCs and increase their ambition, a process known as the "ratchet mechanism." The agreement also includes provisions for climate finance, adaptation measures, loss and damage, and technology transfer, aiming to support developing nations in building climate resilience.

14.2.1.3 Basel Convention on Hazardous Waste

The Basel Convention, signed in 1989 and entering into force in 1992, was created in response to growing concerns about the transboundary movement of hazardous waste, especially from developed to less developed countries. It prohibits the export of hazardous waste without prior informed consent from the receiving country, and aims to minimize waste generation at its source. This treaty was especially important following scandals in the 1980s, where toxic waste from European countries was illegally dumped in Africa and Latin America. The Basel Convention includes detailed lists of hazardous wastes and outlines procedures for tracking, labeling, packaging, and transporting waste to ensure environmental and human health protection. The treaty has been further strengthened by amendments such as the Basel Ban Amendment (2019), which bans the export of hazardous waste from developed to developing countries for any reason.

14.2.1.4 Kyoto Protocol and Post-Kyoto Developments

The Kyoto Protocol, adopted in 1997 and enforced in 2005, was the first legally binding international treaty requiring industrialized countries to reduce GHG emissions. It established clear targets and introduced market-based mechanisms, such as Emissions Trading, the Clean Development Mechanism (CDM), and Joint Implementation (JI), to help countries meet their targets cost-effectively. However, the Protocol faced several challenges. Some major emitters, such as the United States, refused to ratify it, while others, such as Canada, later withdrew. The lack of obligations for developing countries and the rise of emissions from emerging economies also limited its effectiveness.

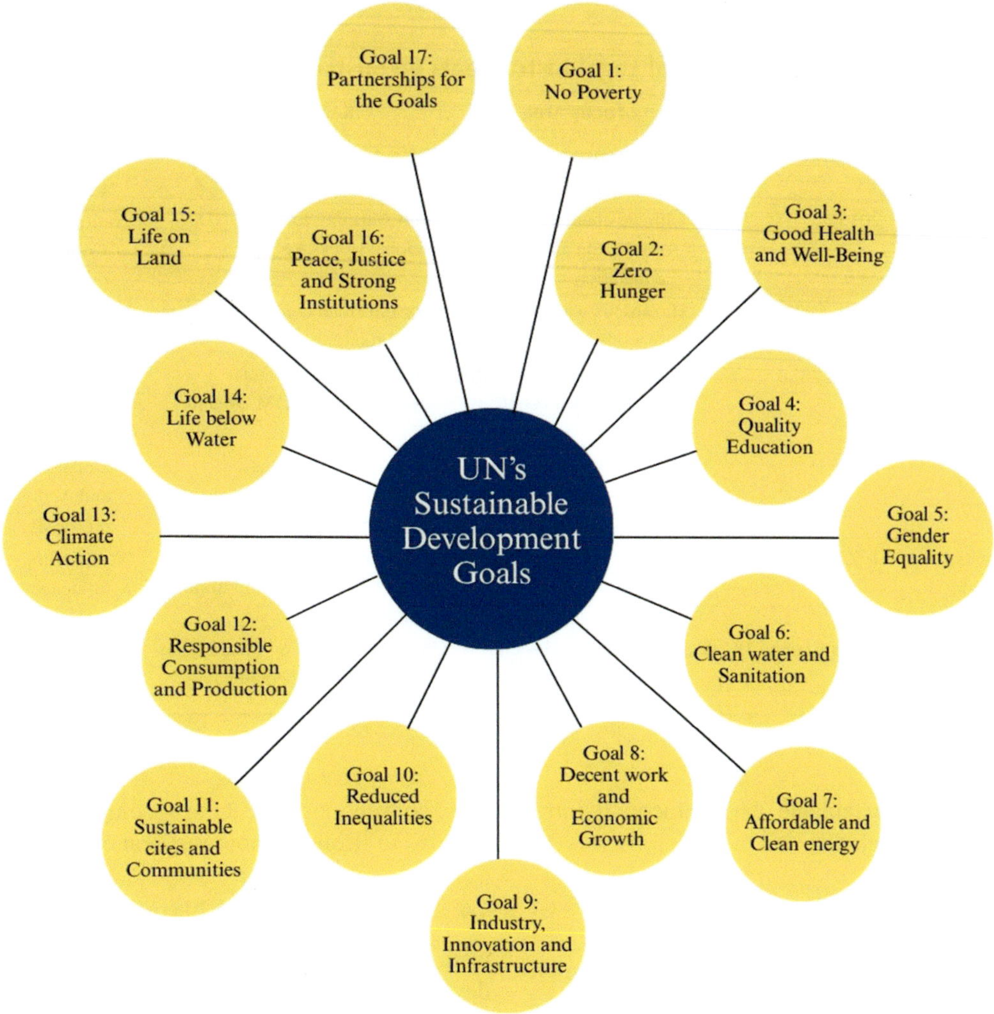

FIGURE 14.3 United Nations' 17 Sustainable Development Goals (SDGs).

The Doha Amendment (2012) extended the Kyoto commitments to 2020, but participation declined. Ultimately, the Paris Agreement replaced the Kyoto Protocol with a more inclusive, flexible, and ambitious framework involving all nations.

14.2.1.5 Convention on Biological Diversity (CBD)

Adopted at the 1992 Rio Earth Summit, the Convention on Biological Diversity (CBD) is one of the most important global agreements for the conservation and sustainable use of biological diversity. The CBD has three core objectives:

1. The conservation of biodiversity;
2. The sustainable use of its components;
3. The fair and equitable sharing of benefits arising from genetic resources.

The Convention has led to important developments such as the Nagoya Protocol (2010) on access to genetic resources and the Cartagena Protocol (2003) on biosafety. In 2022, the CBD parties adopted the Kunming-Montreal Global Biodiversity Framework, which aims to reverse biodiversity loss by 2030.

CBD implementation includes National Biodiversity Strategies and Action Plans (NBSAPs), supported by scientific research, indigenous knowledge, and ecosystem-based approaches.

14.2.1.6 Stockholm Declaration (1972) and Rio Declaration (1992)

The Stockholm Declaration, adopted during the 1972 United Nations Conference on the Human Environment, is considered the foundation of modern environmental law. It marked the first major international effort to link environmental protection with human development and produced 26 principles, including the famous Principle 1: "Man has the fundamental right to freedom, equality and adequate conditions of life, in an environment of a quality that permits a life of dignity and well-being."

Twenty years later, the Rio Declaration on Environment and Development (1992) built on this legacy, establishing

TABLE 14.1
Comparative Overview of Key International Legal Instruments on Sustainability

S. No.	Instrument	Year	Focus Area	Key Features	Implementation Mechanism
1.	UN SDGs	2015	Holistic sustainable development	17 Goals, 169 Targets	Voluntary National Reviews (VNRs), National Integration
2.	Paris Agreement	2015	Climate change mitigation and adaptation	Limit warming to <2°C, NDCs, global stocktake	Nationally Determined Contributions (NDCs), 5-year cycle
3.	Basel Convention	1989 (1992)	Hazardous waste control	Prior informed consent for waste exports	National legislation, Basel Ban Amendment
4.	Kyoto Protocol	1997 (2005)	GHG emission reductions in developed countries	Binding targets, market mechanisms (CDM, JI, and ETS)	Compliance through emission targets and trading
5.	CBD	1992	Biodiversity conservation	3 objectives: conservation, sustainable use, fair benefit sharing	National Biodiversity Strategies and Action Plans (NBSAPs), protocols
6.	Stockholm and Rio Declarations	1972 and 1992	Environmental principles and rights	Precautionary and polluter pays principles, human–environment relationship	Non-binding, normative influence on treaties
7.	ILO Conventions	Various (since 1919)	Labor rights and social sustainability	Core conventions on forced labor, child labor, and equal pay	National labor laws, reporting to the ILO supervisory bodies

27 principles to guide sustainable development. It introduced critical concepts such as:

- Precautionary Principle (Principle 15): Take preventive action in the face of uncertainty.
- Polluter Pays Principle (Principle 16): Those responsible for pollution must bear the costs.
- Common but Differentiated Responsibilities (Principle 7): Acknowledge different capabilities and responsibilities among countries.

These principles remain foundational in contemporary environmental policy and treaty-making.

14.2.1.7 ILO Conventions on Labor Rights and Sustainability

The International Labour Organization (ILO) has developed a set of core conventions that link labor rights to sustainable development. These conventions address social sustainability by promoting decent work, fair wages, occupational safety, and the elimination of forced and child labor.

Key conventions include:

- Convention No. 87 on Freedom of Association (1948)
- Convention No. 98 on Collective Bargaining (1949)
- Convention No. 100 on Equal Remuneration (1951)
- Convention No. 182 on the Worst Forms of Child Labour (1999)

These standards are essential for achieving SDG 8: Decent Work and Economic Growth. Countries that implement these conventions contribute not only to workers' well-being but also to stronger, more resilient economies (Table 14.1).

14.2.2 Regional and National Legal Frameworks

National policies play a key role in achieving sustainable development by implementing measures to reduce pollution, promote renewable energy, and conserve resources (Liyanage, Netswera, and Motsumi 2020). Governments across the globe are enacting laws and regulations to promote sustainable production practices and reduce environmental impacts. The creation of sustainable development strategies necessitates discussion at the regional level, peer learning, and the exchange of best practices to accomplish the designated objectives (Haq, Välisuo, and Niemi 2021). These strategies can be used to direct national policy, guarantee resource efficiency, and encourage regional cooperation. Subnational governments have become increasingly involved in sustainability management and reporting, with many regions and states developing their own initiatives to address environmental issues (Garren and Brinkmann 2018). They can create regional environmental policies, support sustainable projects, and collaborate to address issues that cross boundaries (Mensah 2019).

Many factors impede the adoption of sustainable practices in developing countries, including limited resources, infrastructure, and technical expertise. Additionally, weak governance structures and corruption can undermine the implementation of environmental regulations and policies. Improving environmental outcomes and fostering long-term sustainability require effective environmental governance that encompasses regulatory frameworks, monitoring, and enforcement. It is crucial to resolve this issue immediately, and manufacturers should be aware of their corporate social responsibilities, which include employees, the environment, the community, and customers (Ma et al. 2023). It is essential to guarantee fair business practices, reduce waste production, and preserve natural resources to promote sustainable development. Firms must build capabilities to comply with national and international standards to maintain their competitive advantage in both local and international markets (Mushangai 2023). Eco-innovation is spreading more slowly in Africa than in other regions, such as Europe and Asia (Nnaji and Igbuku 2018). Eco-innovation, including practices such as waste management and renewable energy use, is crucial for economic growth and environmental sustainability, but its slow adoption in Africa could hinder the continent's development and environmental protection (Enuoh and Bisong 2014).

14.3 GOVERNANCE POLICIES FOR SUSTAINABLE PRODUCTION

14.3.1 Corporate Governance and Sustainability

Corporate governance and sustainability are now closely linked and no longer limited to theory or activism. Companies are increasingly expected to consider environmental and social impacts alongside profits, reflecting growing demands from regulators, investors, and the public. Sustainability is now a key part of how companies make decisions, manage risks, and define their role in society.

14.3.1.1 Environmental, Social, and Governance (ESG) Integration in Governance

The use of ESG criteria is a growing requirement for ensuring companies operate responsibly. Businesses are assessed on their environmental impact and stakeholder relations. More and more countries require businesses to disclose sustainability risks and performance metrics (Khamisu & Paluri, 2024). ESG has shifted from a voluntary framework to being integral to corporate strategies as it affects financing opportunities and market perception (Xin et al., 2025). Disclosures Global Reporting Initiative (GRI), Task Force on Climate-related Financial Disclosures (TCFD), International Sustainability Standards Board (ISSB) have published reporting standards which enable disclosures to be made with accuracy and uniformity across borders (Barker, 2025).

ESG standards have emerged as key frameworks to guide companies toward sustainable operations. ESG captures how corporations manage their environmental impact, treat their stakeholders, and uphold transparency and ethical management. According to Khamisu and Paluri (2024), ESG disclosure is increasingly mandated by regulatory bodies worldwide, with many jurisdictions requiring public firms to report on sustainability-related risks and performance.

From a governance perspective, ESG has evolved from a voluntary public relations tool into a structured framework shaping strategic decisions. As Xin et al. (2025) highlight in their study of Chinese manufacturers, ESG practices now significantly influence a company's access to capital, its market competitiveness, and even its legitimacy in supply chains.

This trend is reinforced by sustainability reporting initiatives such as the GRI, the TCFD, and the ISSB. These frameworks aim to ensure that corporate disclosures are not only comprehensive but also comparable and credible across markets (Barker 2025).

14.3.1.2 Corporate Foundations and Strategic CSR

One way corporations institutionalize social responsibility is through corporate foundations, which often serve as arms-length entities focused on addressing social issues. These foundations can act as "sensors" for emerging social needs and as channels for building reputational trust (Westhues and Einwiller 2006). When properly aligned with the parent firm's values and operations, they enhance transparency and stakeholder engagement. However, challenges arise when the lines between corporate philanthropy and branding blur, raising ethical questions about the authenticity of these efforts.

14.3.1.3 Greenwashing and Governance Failures

Despite the growing adoption of ESG and CSR language, not all sustainability claims reflect genuine change. The danger of greenwashing, where companies exaggerate or misrepresent their environmental credentials, remains a serious governance challenge.

Greenwashing can take many forms: vague environmental messaging, selective reporting, or misleading branding. For example, some fossil fuel companies have publicly promoted minor investments in renewables while continuing to expand oil and gas extraction. Similarly, fashion brands may launch "eco-friendly" clothing lines while maintaining exploitative labor practices and unsustainable mass production models (Wassénius et al. 2024).

Shaikh and Tripathy (2025) emphasize that inconsistency in sustainability reporting, even among companies that follow recognized frameworks, can undermine stakeholder trust. Disclosures often lack depth, omit negative impacts, or cherry-pick data to present a favorable image.

Governance structures that enable or ignore such practices risk reputational damage, legal liabilities, and loss of stakeholder confidence. Robust sustainability governance, on the other hand, demands the following:

- Third-party assurance of sustainability reports;
- Clear metrics and targets;
- Accountability mechanisms for unmet ESG goals;
- Inclusive stakeholder engagement in decision-making processes.

As Barker (2025) notes, a well-designed corporate sustainability report should not simply highlight achievements; it should openly address trade-offs, disclose challenges, and reflect a credible path toward continuous improvement.

14.3.2 Public–Private Partnerships (PPP) for Sustainability

PPPs are effective for advancing sustainable production by combining government policies with private-sector innovation. Governments offer support through tax benefits, subsidies, and regulations, while businesses bring in capital, expertise, and technologies. Achieving sustainability also depends on public awareness, responsible consumption, and active civil society engagement (Mensah, 2019).

14.3.2.1 Role of Governments and Corporations

For sustainability to become a reality, public involvement needs to grow. People must understand that responsible consumption, environmental care, and socially inclusive values are essential for long-term survival and well-being (Mensah 2019). Although sustainability is a shared responsibility, it depends on the integration of environmental, economic, and social goals by national, regional, and international institutions. Despite a clear definition from global bodies, sustainability is often interpreted differently across organizations, leading them to align it with their own agendas. While current global systems may promote inequality and overconsumption, it is not justified to delay sustainability efforts until large-scale structural reforms are made (Garren and Brinkmann 2018). Policymakers can use integrated modeling tools to understand the interconnections between environmental, economic, and social factors and make better decisions.

14.3.2.2 Incentives and Policy Mechanisms for Green Investments

Investment strategies and policy frameworks play a critical role in driving sustainable business practices. Governments can use legal measures, fiscal tools, and incentive programs to support long-term investment in green technologies and environmental initiatives. By setting standards and environmental regulations, they can reduce ecological harm and support the shift toward sustainability. Addressing climate change effectively requires comprehensive efforts that include both government action and market mechanisms (Iqbal et al. 2025). Support for green growth and climate finance policies is essential to protect ecosystems and promote efficient resource use. Procurement policies can also incorporate sustainability principles to encourage eco-friendly production and consumption.

Sustainable finance, which guides investment toward green initiatives, is a powerful tool for promoting responsible manufacturing. It can shape capital flows and stimulate market demand for sustainable goods and services (Iqbal et al. 2025). Governments contribute by enforcing laws, defining environmental quality standards, and creating financial and regulatory structures that reward sustainable practices.

14.3.2.3 Case Studies of PPP in Sustainable Production

PPPs are widely used, especially in developing countries, to improve infrastructure. These partnerships allow the private sector to participate in the design, funding, construction, operation, and maintenance of infrastructure projects (Shen et al. 2016). When aligned with sustainability goals, PPP projects can greatly improve economic, social, and environmental outcomes. The success of PPPs depends on how responsibilities are divided between the public and private sectors. As infrastructure has lasting impacts, it is vital that PPPs follow sustainable development principles.

In many regions, value-for-money evaluations are used to determine the most efficient and beneficial PPP model. While PPPs offer advantages such as speed, innovation, and access to private capital, their main goal is to make infrastructure more cost-effective and sustainable. Governance mechanisms are essential to align private-sector involvement with sustainability objectives. These hybrid models balance efficiency and public interest, and are becoming more common worldwide.

The public sector benefits from private-sector expertise and technology, which improve implementation and service delivery. Additionally, sharing project risks between sectors helps ensure efficient management and execution. In sectors such as agriculture, PPPs are used not only to improve crop productivity and rural incomes but also to support sustainable practices and the adoption of high-value or experimental crops. Successful PPP implementation depends on strong planning, coordination, and institutional frameworks. It also requires careful consideration of the entire value chain, from Research and Development (R&D) to marketing and financing (Cheng et al. 2020).

Project success is influenced by several factors, including effective communication, risk awareness, political backing, clear responsibilities, expert guidance, and public support. PPPs have been used in both agricultural and transportation infrastructure to improve food security, increase productivity, and boost rural economies.

14.3.2.4 Government-Led Green Finance Initiatives

Governments are increasingly adopting PPPs to implement green policies and sustainability projects. These partnerships can help meet long-term goals by combining public oversight with private resources. Clearly defining roles, responsibilities, and timelines is critical for managing PPPs effectively. Multi-objective decision-making tools can support sustainability assessments and help in choosing appropriate PPP models.

Strong regulations, efficient resource use, and the integration of knowledge and skills all contribute to successful project delivery. In countries with limited private funding, PPPs are especially important for advancing infrastructure development and addressing financial risks. In agriculture, PPPs not only help fund projects but also increase crop productivity and support innovative approaches. A clear institutional setup and strategic planning are key to achieving intended sustainability outcomes. In addition to infrastructure development, PPPs in agriculture aim to revitalize rural economies and support sustainable food systems.

14.3.2.5 Tech Industry's Role in Sustainability Governance

Technology plays a vital role in enhancing thc performance and governance of PPPs. AI-based systems can support real-time decision-making in sectors like agriculture by providing weather updates and advising on planting and irrigation. Blockchain technology improves supply chain transparency by allowing consumers to verify product origins and sustainability claims, promoting ethical sourcing. However, to ensure responsible use, regulatory frameworks must address concerns, such as data privacy, algorithm bias, and accountability (Zhang et al. 2022).

AI can also streamline project monitoring, automate bidding and contract processes, and provide data-driven insights for all stakeholders (Prasad et al. 2024). These tools can support the long-term development of IT infrastructure by leveraging private-sector innovation and public funding. When used effectively, technology can promote sustainable supply chains, reduce environmental harm, and support fair labor practices.

Frameworks like the Four-Level AI-based PPP model help optimize collaboration between sectors in large infrastructure projects. Automated tools such as machine learning and natural language processing can evaluate contractor qualifications and improve procurement efficiency. AI-integrated Build–Operate–Transfer (BOT) models can also enhance productivity and reduce costs (Prasad et al. 2024). The Hybrid Annuity Model is increasingly preferred due to its risk-sharing structure, better financial outcomes, and long-term viability.

Other PPP formats include leasing agreements, where private companies manage public assets, and concession models, where firms build and operate facilities under long-term agreements before handing them over to the government (Prasad et al. 2024). These arrangements combine private-sector innovation and efficiency with the public sector's regulatory guidance and sustainability goals.

Ultimately, AI, blockchain, and digital transparency are critical tools in strengthening governance and sustainability across PPP projects. Hybrid PPP models are now a growing global trend, helping governments meet development targets more efficiently.

14.3.3 Governance of Supply Chain and Sustainable Standards

Managing supply chains in a sustainable way means following clear rules that support ethical behavior and protect the environment. As PPPs evolve, their structure becomes more complex due to various economic and social factors. This complexity can lead to confusion and make it harder for partnerships to stay focused on their original goals (Agarwal et al. 2023). In agriculture, PPPs often struggle because of slow processes, poor transparency, and a lack of trust between different parties. Simply filling the gap where government capacity is weak isn't enough—strong public institutions and legal systems are essential.

Making documents related to PPPs accessible to the public helps build trust and reduce corruption. Unfortunately, this transparency is often limited due to legal or ethical concerns. A key improvement in newer PPP frameworks is the focus on fairness through standards like those from the International Federation of Consulting Engineers (FIDIC), which aim to make agreements balanced for all sides. Private companies also offer performance guarantees that give governments more confidence that the project will be completed properly. Community participation and clear communication are essential in this process.

Supply chain transparency helps customers make better choices and ensures that businesses are held responsible. It also makes it easier to trace products to their origin, which supports ethical sourcing. Rules that promote democracy, fight corruption, and hold institutions accountable are vital to this transparency. Technologies like blockchain help by offering reliable records that can't be changed. When private companies are rewarded for meeting clear performance goals, it helps align their actions with the overall mission of the project (Prasad et al. 2024). In many PPPs, insurance firms handle risks, while both governments and private firms provide funding and resources. Laws and institutional norms determine how PPPs operate by defining who owns what and who makes the decisions.

14.3.3.1 Circular Economy and Extended Producer Responsibility (EPR)

The CE focuses on reducing waste and using resources efficiently. EPR means that companies are responsible for what happens to their products even after they've been sold. This includes setting up programs to take products back for reuse or recycling, which reduces landfill waste and encourages better product design.

Governments can support this by creating policies and incentives that promote circular practices. EPR also pushes manufacturers to create products that are easier to repair, recycle, and reuse, thereby reducing environmental damage. Being sustainable also means being open, honest, and actively involved with communities—not just following rules. Businesses can build trust by participating in audits, engaging with local groups, and being clear about their operations (Jang, McSparren, and Rashchupkina 2016; Aschehoug and Schulte 2017).

14.3.3.2 Fair-Trade and Ethical Sourcing Policies

Fair-trade practices and moral sourcing policies are crucial for guaranteeing equitable labor conditions, environmental protection, and community development in production processes, and fair-trade standards ensure that farmers and workers receive fair prices and safe working conditions, promoting economic empowerment and reducing poverty in developing nations. Supply chains can promote ethical sourcing by setting and upholding stringent environmental and social standards for suppliers, thereby protecting ecosystems and human rights. Businesses are able to assess and address human rights issues in their supply chains through human rights due diligence, which is essential for moral sourcing, and to conserve the environment and promote human values, education is essential because it can influence society's attitudes and behaviors. Also, ethical business practices and sustainable procurement are becoming more and more important for businesses around the world, as it is controlling the entire supply chain to meet the demands of various stakeholders for goods, services, and processes. (Islam and Alharthi 2020). In addition, businesses are using technology and data analytics more and more to track and improve sustainability throughout their supply chains.

14.3.3.3 Green Procurement and Sustainable Supply Chain Initiatives

Green procurement means including environmental criteria in purchasing decisions. This helps promote products and services that are eco-friendly, which encourages innovation and shifts the market toward sustainability. Companies can reduce emissions and resource use through sustainable sourcing and distribution (Hariyani et al. 2024).

Blockchain technology improves accountability by tracking raw materials and confirming their sustainability. Internet of Things (IoT) tools help monitor materials in real-time, promoting energy efficiency and reducing waste. CE ideas, such as product reuse, eco-design, and waste reduction, can be applied at every stage of the supply chain (Kharat et al. 2024). These changes not only meet customer expectations but also ensure that businesses comply with environmental regulations.

Businesses are also rethinking how they design products—using fewer materials, extending product life, and improving recycling options. These changes support the SDGs, especially the goal of responsible production and consumption. A company's ability to adopt green supply chain practices often depends on having the right technologies, skilled workers, and processes in place (Hariyani et al. 2024). Lifecycle assessments help companies understand and reduce the environmental impact of their products throughout their lifecycles.

14.3.3.4 Waste Management and Circular Economy in Supply Chains

A circular supply chain reduces waste by designing systems where materials and products are reused, recycled, or remanufactured. Companies following this model design products for durability, easy repair, and disassembly (Hariyani et al. 2024). Integrating these systems into traditional supply chains supports sustainable production and consumption but can be challenging due to financial, technical, social, and regulatory barriers (Kharat et al. 2024).

Many organizations are adopting upcycling, which transforms waste materials into higher value products. They also invest in renewable resources and carbon offsetting to reduce emissions. Waste reduction strategies, such as audits, process improvements, and switching to better materials, are essential. Green logistics, including electric vehicles and route optimization, help reduce emissions and boost energy efficiency.

Industries like pharmaceuticals face additional complexity due to high R&D costs, regulatory requirements, and reverse logistics. Companies in these sectors must find ways to control waste and lower operational costs while maintaining high safety standards (Kharat et al. 2024). Environmental sustainability is becoming critical in supply chain planning, offering advantages such as improved reputation, innovation, and stakeholder trust.

Green supply chain management (GSCM) embeds sustainability into procurement, logistics, and reverse logistics processes. Supplier engagement is essential for enforcing environmental standards and encouraging material innovation. Reverse logistics, where products are collected for reuse or recycling, increases distribution efficiency and supports long-term sustainability (Hariyani et al. 2024).

14.3.3.5 Transparency and Traceability in Global Supply Chains

As consumers become more aware of the social and environmental impacts of their purchases, demand for transparent, traceable supply chains is growing. Traceability involves following a product's journey from origin to final user, which is especially important in sectors such as food, clothing, and electronics.

Blockchain technology offers a secure way to track compliance with labor, environmental, and safety standards. Increased data visibility leads to better decision-making, while lifecycle assessments help companies reduce their environmental impact. Collaboration across supply chain actors—like suppliers, regulators, and consumers—is essential for improving accountability and sustainability.

Digital tools such as cloud computing, IoT, AI, machine learning, and big data analytics improve operational efficiency and supply chain monitoring. Mobile apps, e-commerce platforms, and social media empower consumers to make more informed buying decisions (Khan et al. 2022).

Traceability includes three parts: tracking a product's journey, recording its history, and monitoring follow-up actions. These processes are vital for managing recalls and identifying quality issues. Digital traceability tools offer data trails and help businesses use resources more efficiently. As technology advances, it plays a key role in building more transparent and sustainable global supply chains (Hariyani et al. 2024; Khan et al. 2022).

14.4 REGULATORY APPROACHES TO SUSTAINABILITY COMPLIANCE

Regulations and policy tools are essential to getting companies to adopt more sustainable production methods. Governments around the world are increasingly putting environmental rules in place that set clear standards for waste management, emissions control, and the efficient use of natural resources (Hariyani et al. 2024). Many countries have introduced EPR programs that require manufacturers to take responsibility for their products throughout their lifecycles.

To meet environmental and social standards, businesses are also working closely with suppliers to earn third-party certifications like International Organization for Standardization (ISO) 14001 and fair-trade labels. These certifications help improve supply chain transparency and accountability. Additionally, companies are now adopting green logistics, using energy-efficient warehouses and eco-friendly packaging and transport to reduce environmental impacts. Governments further support these efforts by offering incentives, such as tax breaks and subsidies, to businesses that invest in clean technologies and sustainable operations.

14.4.1 Command-and-Control Regulations

14.4.1.1 Environmental Impact Assessments (EIA)

Environmental Impact Assessments (EIAs) are structured evaluations used to predict the potential environmental effects of a new project, policy, or development plan. EIAs ensure that environmental issues are considered during early decision-making stages by identifying risks, proposing solutions, and informing both policymakers and the public. This process plays a crucial role in promoting sustainable development by guiding actions that reduce harm and deliver long-term environmental benefits.

The 1992 Rio Declaration—specifically Principle 17, established EIAs as the preferred tool for sustainable development (Roos et al. 2020). EIAs help protect ecosystems, encourage conservation, and support responsible development. In South Africa, for example, regulators view EIAs as a way to protect biodiversity, ensure public participation, and support legal compliance (Roos et al. 2020). Still, there are concerns about how much of EIA's potential is realized in practice, especially regarding follow-up and impact mitigation.

To improve their effectiveness, experts suggest changes such as making pre-application consultations mandatory, adjusting timeframes, increasing objectivity, including cumulative impact analysis, and placing greater focus on sustainability outcomes. Some authorities even argue that EIAs help balance economic development with environmental and social responsibilities (Roos et al. 2020).

14.4.1.2 Licensing and Permits for Sustainable Production

Licensing and permit procedures are essential tools for governments to oversee and control industrial operations, guaranteeing adherence to environmental laws and encouraging sustainable production techniques (Hariyani et al. 2024). These procedures stipulate that companies need to get permission before starting operations that could have a substantial environmental impact. The terms of licenses and permits frequently include restrictions on emissions, waste management protocols, and resource usage, which pushes businesses to put environmentally friendly practices into place (Roos et al. 2020). To guarantee compliance and reduce the possibility of environmental harm, regulatory agencies conduct routine inspections and audits. Furthermore, certain licensing schemes provide incentives for businesses that go above and beyond the minimal standards for environmental performance by, for example, expediting the permitting procedure for sustainable projects.

14.4.1.3 Role of Environmental Courts and Special Sustainability Tribunals

Environmental courts and special tribunals are vital for resolving disputes, interpreting environmental law, and ensuring justice in sustainability-related cases. These specialized courts hold companies and individuals accountable, promoting compliance and offering fair solutions to environmental conflicts (Angstadt and Schink 2023).

They are increasingly recognized as important institutions for addressing the environmental challenges of the Anthropocene era. Scholars in ESG fields are especially interested in how these courts function and contribute to broader environmental justice and legal reform (Angstadt and Schink 2023). For example, Thailand's Supreme Court has established a dedicated "Green Bench" to handle environmental cases.

Environmental courts are becoming more common around the world in countries such as Austria, Sweden, and India. They are praised for their specialized expertise, efficiency, and ability to enforce environmental law more effectively. These courts help ensure environmental concerns are not overlooked and that violations are handled seriously (Angstadt and Schink 2023). However, defining what qualifies as an "environmental" issue remains a legal challenge in many systems.

14.4.1.4 Case Study: The Impact of Stricter EU Carbon Border Tax on Global Trade

A good example of how regulations can affect global trade is the European Union's (EU's) proposed Carbon Border Adjustment Mechanism (CBAM). This policy would impose import taxes on goods from countries with weaker climate laws to reduce "carbon leakage" and promote global decarbonization (Böhringer et al. 2022). The goal is to encourage other nations to strengthen their environmental standards without putting EU industries at a disadvantage.

However, this policy must align with World Trade Organization (WTO) rules, especially regarding fair treatment of foreign producers. If import taxes differ across countries, they could violate WTO obligations—unless justified under Article XX, which allows environmental exceptions (Böhringer et al. 2022). The most WTO-compliant approach would be to match import rules to the same obligations faced by domestic producers, such as requiring foreign companies to surrender emissions allowances or credits.

Opinions on CBAM's effectiveness vary. Some predict that it will reduce emissions and strengthen the competitiveness of EU industries. Others warn it could hurt developing countries and shift global trade patterns unfairly. CBAM's design—such as setting average-emissions benchmarks—might not reflect differences between firms and could discourage emission reduction efforts from foreign exporters (Böhringer et al. 2022).

Deciding which products to tax is another challenge. For example, it is unclear whether only steel should be taxed, or if steel tubes and higher value-added products made from steel should be included. Narrowing CBAM's scope may reduce its environmental impact but help maintain WTO compliance. Excluding export rebates and offering fewer protections for domestic producers may reduce its fairness and effectiveness (Böhringer et al. 2022).

Despite the challenges, most agree that strong, coordinated climate policies are necessary to achieve sustainable production. Complementary policies—such as promoting green technologies, revising trade agreements, and supporting sustainable finance can help ease the transition. Redirecting investment from polluting industries to environmentally friendly ones can lower funding costs for green businesses and encourage sustainable innovation (Iqbal et al. 2025). Rather than relying solely on carbon border taxes, countries might also consider other approaches, such as output-based rebates or targeted subsidies for emissions reductions. These alternatives could be less controversial and more effective under WTO rules (Böhringer et al. 2022).

14.4.2 Market-Based Mechanisms

Market-based regulatory tools play a crucial role in promoting sustainability by using economic incentives to encourage businesses to reduce environmental harm. Rather than relying solely on prescriptive laws, these mechanisms allow companies to remain flexible while guiding them toward greener choices. This section covers a range of strategies—carbon pricing systems such as cap-and-trade and carbon taxes, green subsidies, financial incentives, sustainability certifications, ESG-integrated banking practices, and sustainable stock indices. Each of these tools is designed to embed environmental costs into market decisions, stimulate innovation, and promote long-term investment in sustainable development.

14.4.2.1 Carbon Pricing: Cap-and-Trade and Carbon Taxes

Carbon pricing is a key market-driven strategy aimed at lowering greenhouse gas emissions and promoting sustainable production. This approach typically takes two forms: carbon taxes and emissions trading systems, also known as cap-and-trade programs. A carbon tax imposes a direct fee on each unit of carbon dioxide released, offering a predictable cost for emissions (Chen et al. 2020). In comparison, cap-and-trade systems establish a maximum allowable level of emissions and distribute permits that businesses can trade, allowing flexibility in how reductions are achieved (Chen et al. 2020).

Cap-and-trade programs are considered more effective in encouraging technological innovation and achieving significant emission reductions. However, their success depends heavily on how the emissions cap is set and on the price of emission permits (Chen et al. 2020). If the cap is too lenient or permit prices are too low, companies may lack motivation to reduce emissions meaningfully.

A challenge commonly associated with carbon pricing is the risk of carbon leakage. This occurs when companies move production to countries with weaker environmental regulations to avoid the costs associated with emissions control. Such actions can undermine the environmental benefits of carbon pricing while harming domestic industries' competitiveness (Böhringer et al. 2022). One way to address this issue is through border carbon adjustments, which place an equivalent carbon cost on imported goods. This helps ensure that local industries are not unfairly disadvantaged and that environmental goals are maintained (Böhringer et al. 2022).

The structure of carbon pricing policies must be designed to balance environmental effectiveness with economic feasibility. Output-based allocation can be used to protect industries vulnerable to competition from less-regulated markets by allocating emission allowances based on production levels. Additionally, consumption-based taxation on high-emission goods can transfer some of the environmental costs to consumers and simulate the effect of comprehensive border adjustments without discriminating between domestic and foreign producers (Böhringer et al. 2022).

Both carbon taxes and trading schemes can lead to innovation and cleaner production. Cap-and-trade tends to be more effective in sectors with low barriers to market entry, while carbon taxes can become more impactful when paired with financial support for cleaner technologies (Chen et al. 2020). Although both tools are widely used, more research is needed to understand how each affects innovation over the long term.

Properly designed carbon pricing policies can stimulate investment in low-carbon technologies. If they are poorly implemented, however, they may increase operational costs without delivering meaningful environmental outcomes. Combining carbon pricing with subsidies for green technology can address this inefficiency and improve overall effectiveness (Jin, Ding, and Yang 2021).

When production benchmarks are already established, as in the European Union Emissions Trading System, introducing carbon consumption taxes can be relatively straightforward. These types of taxes expand existing frameworks and help prevent emissions from simply being shifted to other markets (Böhringer et al. 2022).

Even though such environmental regulations may affect global trade flows, they are necessary for reducing pollution and fostering sustainable development. Broader strategies that include investment in clean technologies, environmentally responsible finance, and integration of sustainability principles into international trade policy are critical to transitioning toward an economy with minimal environmental impact (Böhringer et al. 2022).

14.4.2.2 Green Subsidies and Financial Incentives

Governments use green subsidies and financial incentives to promote environmental sustainability and accelerate the shift toward cleaner production. These tools help encourage businesses to invest in green technologies, support low-carbon economic growth, and adopt CE practices (Jin, Ding, and Yang 2021). By offering financial support to companies undertaking environmentally beneficial projects, governments can reduce the costs and risks associated with sustainability transitions.

Subsidies play an important role in motivating companies to take environmental actions that they may otherwise delay due to cost concerns. Financial tools such as green loan guarantees and interest subsidies are widely used to support such initiatives. Loan guarantees reduce the lending risks for financial institutions, while interest subsidies lower the borrowing costs for businesses, making it easier for them to fund green projects (Jin, Ding, and Yang 2021).

For example, China has promoted green lending by providing tax benefits and subsidized interest rates to banks that finance environmental projects. Shenzhen city, for instance, offered a 50 percent interest subsidy to businesses borrowing funds for carbon reduction initiatives (Jin, Ding, and Yang 2021). Similarly, Malaysia's Green Technology Financing Scheme offers borrowers a 2 percent interest rebate, along with a 60 percent government guarantee on the approved loan amount. These types of programs have proven effective in expanding production and improving economic outcomes.

Choosing the right incentive depends on the efficiency of green investment. Loan guarantees are particularly useful when investment efficiency is low, while interest subsidies may be more appropriate as green technologies become more cost-effective and accessible (Jin, Ding, and Yang 2021).

Financial incentives are increasingly linked to broader sustainability strategies. Tools such as sustainable finance, clean energy investments, and inclusive green growth have become vital components in the fight against environmental degradation (Iqbal et al. 2025). Sustainable finance encourages companies to use raw materials in ways that minimize waste and maximize ecological value, ensuring that production remains environmentally conscious.

Instruments such as green bonds and ESG-focused investments support these objectives by directing capital toward businesses that follow responsible environmental practices. Technological innovation also plays a crucial role in reducing emissions and improving environmental quality. Research shows that increased investment in innovation can significantly reduce greenhouse gas emissions while supporting long-term well-being and economic stability (Iqbal et al. 2025).

By combining these financial strategies with regulatory measures, governments can create more supportive environments for sustainable industrial growth. Green subsidies, when well-designed, not only help businesses manage costs but also guide national economies toward cleaner and more resilient futures.

14.4.2.3 Sustainable Certifications (ISO 14001, LEED, and FSC)

Sustainable certifications and eco-labels serve as important tools that help consumers identify products and services that meet specific environmental and social standards. These certifications are typically awarded by independent third parties based on pre-defined sustainability criteria. They offer consumers a reliable way consumers to make informed decisions and encourage companies to adopt more environmentally responsible practices (Iqbal et al. 2025).

Ecolabels strengthen consumer trust and can increase demand for certified products. When consumers trust these labels, businesses are more motivated to reduce their environmental footprint. Certifications also promote transparency by requiring companies to meet measurable and verifiable sustainability benchmarks. However, the impact of these programs depends heavily on the clarity, consistency, and credibility of their certification processes.

Certification programs often lead to innovation by pushing firms to adopt cleaner technologies, improve resource efficiency, and reduce waste. Many industries use these frameworks not just for compliance, but to enhance performance and build competitive advantages. Stakeholders tend to view certifications positively because they support improvements in ESG performance (Kaboor and Ameena Febin 2024).

These programs provide structured pathways for companies to evaluate and improve their sustainability practices. Businesses that earn certifications can distinguish themselves in competitive markets, especially as consumers and investors increasingly prioritize sustainability. At the same time, concerns have been raised about the potential misuse of certifications as marketing tools when genuine environmental commitment is lacking (Kaboor and Ameena Febin 2024).

One of the most widely used frameworks is ISO 14001, which helps organizations establish and maintain environmental management systems. It guides companies in monitoring their environmental impacts, improving performance, and using resources more efficiently. Adoption of ISO 14001 supports goal-setting, data tracking, and continuous improvement in environmental operations (Kaboor and Ameena Febin 2024). Many stakeholders now expect companies to offer clear, verifiable evidence of their sustainability performance, making third-party audits an essential feature of robust certification systems.

In the construction sector, Leadership in Energy and Environmental Design (LEED) certification is a key standard. It focuses on promoting green building practices by encouraging energy and water efficiency, indoor air quality, and waste reduction. LEED-certified buildings serve as models for sustainable urban development by minimizing environmental impacts and creating healthier living environments (Garren and Brinkmann 2018).

Forest Stewardship Council (FSC) certification applies to the forestry sector. It ensures that forest products are sourced in ways that protect biodiversity, respect indigenous rights, and support long-term economic viability. The FSC system is based on ten core principles that address environmental, social, and economic dimensions of sustainable forestry (Alvarez, 2010). This holistic approach strengthens supply chain sustainability by ensuring that raw materials come from responsibly managed sources.

Certification also plays a role in supply chain evaluation. Companies can assess suppliers based on their adherence to sustainability standards like ISO 14001 to ensure responsible sourcing (Hariyani et al. 2024). This helps align procurement decisions with broader corporate sustainability goals.

Despite the benefits, there are challenges. A lack of standardization across different certification schemes can confuse consumers and create inconsistency in enforcement. Some programs may have weak monitoring, which undermines their credibility (Kaboor and Ameena Febin 2024). Small and medium-sized enterprises may also struggle to meet certification requirements due to limited resources, leading to unequal access and market disadvantages.

To overcome these challenges, it is important to establish uniform reporting guidelines and improve monitoring practices. Strong certification systems can influence ESG performance, support transparency, and promote meaningful sustainability improvements across industries.

14.4.2.4 Green Banking and Sustainable Finance: Banks Adopting ESG Investment Policies

Green banking and sustainable finance have become essential in supporting the shift toward environmentally responsible economic practices. By incorporating ESG principles into investment decisions, financial institutions are directing capital toward companies and projects that demonstrate a commitment to sustainability (Cunha et al. 2019). This approach aligns long-term financial performance with environmental stewardship and social responsibility.

Banks and other financial institutions increasingly adopt ESG-focused investment policies to encourage responsible corporate behavior. These policies aim to reduce the environmental impact of economic activity by financing projects that contribute to cleaner production, renewable energy, climate resilience, and inclusive development (Hariyani et al. 2024). Despite progress, there is a growing need to evaluate the real-world outcomes of sustainability investments—at the company level and across broader systems (Alvarez, 2010).

Sustainable investment strategies, which have existed for decades, are gaining momentum, particularly with the global push to achieve the UN SDGs by 2030 (Cunha et al. 2019). These goals encourage governments, civil society, and businesses to work together toward common sustainability objectives. As a result, investors are increasingly expected to help address climate change, resource depletion, and inequality by allocating funds to initiatives that support environmental and social progress.

Sustainable finance involves a broad range of investment instruments and asset classes. These include ethical investing, socially responsible investing, green bonds, impact investing, and ESG integration. Each method emphasizes the importance of evaluating non-financial risks and opportunities alongside traditional financial performance indicators (Jain, Sharma, and Srivastava 2019).

The main goal of sustainable finance is to deliver financial returns while also creating positive social or environmental outcomes. This approach reflects the ethical, ecological, and economic concerns of a growing number of investors. ESG-focused investing seeks to balance profitability with accountability by ensuring that companies are managing risks related to climate, resource use, labor practices, and governance structures (Cunha et al. 2019).

Sustainable investment practices also contribute to the development of more transparent and accountable financial systems. By integrating ESG criteria into lending and investment decisions, financial institutions encourage better corporate behavior and more responsible business models. As interest in sustainability continues to grow, green finance will remain a powerful lever for promoting environmentally sound economic development.

14.4.2.5 Sustainable Stock Indices

Sustainable stock indices are tools developed by financial markets to promote responsible investing by evaluating companies based on ESG criteria. These indices help investors track the performance of businesses that meet specific sustainability standards and serve as benchmarks for comparing companies' commitments to sustainable development (Cunha et al. 2019).

Many global stock exchanges have introduced sustainability indices to support ethical investing. Popular examples include the Dow Jones Sustainability Index, FTSE4Good, E.Capital, and Vigeo. These indices assess companies based on their ESG performance and include only those that lead their industry in sustainability (Jain, Sharma, and Srivastava 2019). By setting high performance standards across social and environmental dimensions, these benchmarks help investors identify companies that prioritize long-term value creation over short-term profit.

The first known sustainability index was the Domini 400 Social Index, launched in 1990. It was followed by the Dow Jones Sustainability Index in 1999, which became the first global benchmark for corporate sustainability performance (Cunha et al. 2019). Since then, many regional and sector-specific indices have emerged. The popularity of these tools reflects the rising importance of ESG criteria in shaping investment strategies and company valuations.

These indices serve multiple purposes. They guide investors who want to align their portfolios with sustainability values, offer a reference for evaluating corporate ESG efforts, and create pressure on firms to improve their performance to be included. Only companies that perform well relative to their peers in environmental protection, social responsibility, and governance practices are selected (Jain, Sharma, and Srivastava 2019).

Sustainable indices are particularly useful for portfolio diversification and risk management. Studies show that the performance of sustainable indices is influenced by several external factors, including interest rates, currency fluctuations, gold prices, and market sentiment (Cunha et al. 2019). As a result, financial professionals often use a combination of traditional and ESG indices to balance their portfolios and achieve both financial and ethical goals.

There is also growing academic interest in comparing the performance of sustainable indices with that of conventional ones. Some studies report competitive or superior returns for ESG indices, while others note mixed results, suggesting that sustainability-focused investing can be as financially rewarding as traditional strategies, depending on context and market conditions (Jain, Sharma, and Srivastava 2019).

The field of sustainable investing continues to evolve. Future research may further explore how ESG criteria affect other asset classes, such as fixed income, real estate, and private equity. More analysis is also needed to assess how actively managed sustainable portfolios perform relative to passive ones (Cunha et al. 2019).

As investor interest grows, sustainable stock indices will remain vital tools for promoting transparency, guiding ethical investment, and holding companies accountable for their environmental and social impacts.

14.4.3 Voluntary Compliance and Self-Regulation

In addition to legal mandates, voluntary sustainability measures and self-regulation are becoming essential strategies for promoting responsible production. Many businesses now adopt industry certifications, stakeholder engagement models, and internal governance systems to meet environmental and social expectations without direct government enforcement. This section explores key components of voluntary sustainability efforts, including certifications, stakeholder-driven initiatives, and the growing influence of self-regulatory mechanisms such as B-Corp standards and internal sustainability audits.

14.4.3.1 Role of Industry Standards and Certifications

Industry-specific standards and sustainability certifications are widely used to promote environmentally and socially responsible production. These programs establish measurable benchmarks for performance and give both consumers and investors greater confidence in a company's environmental claims (Kaboor and Ameena Febin 2024). Certifications confirm that a product or process adheres to defined sustainability criteria, signaling corporate accountability.

Despite their usefulness, some challenges persist. Smaller businesses may be excluded due to the high costs and complexity of certification requirements. Additionally, the growing number of certification schemes can create confusion and skepticism in the market (Kaboor and Ameena Febin 2024). To address these issues, harmonizing standards, increasing transparency, and offering support to small and medium-sized enterprises are necessary to ensure broader adoption and effectiveness.

Research comparing different certification programs can reveal variations in criteria, implementation strategies, and impacts on ESG outcomes. Long-term studies could help determine how certifications influence a firm's financial results, reputation, and relationships with stakeholders. It is also important to explore how certification affects small and large firms differently, particularly in terms of resources and operational adjustments (Kaboor and Ameena Febin 2024).

14.4.3.2 Stakeholder-Led Sustainability Initiatives

Stakeholder-led sustainability initiatives are crucial in driving sustainable production by fostering collaboration, sharing best practices, and promoting transparency across industries (Cunha et al. 2019). These initiatives often involve partnerships between businesses, non-governmental organizations (NGOs), governments, and communities to address specific sustainability challenges and achieve common goals (Kaboor and Ameena Febin 2024). By engaging a diverse range of stakeholders, these initiatives can leverage collective knowledge, resources, and influence to create meaningful change and drive systemic improvements in environmental and social performance. Such partnerships can promote innovation, build trust, and enhance accountability in supply chains, leading to more resilient and sustainable outcomes. Increased transparency and improved stakeholder engagement are governance changes noted in sustainability initiatives (Kaboor and Ameena Febin 2024).

Companies with robust governance structures often demonstrate higher levels of transparency in reporting and decision-making processes, fostering trust among stakeholders (Kaboor and Ameena Febin 2024). Effective stakeholder engagement enables companies to better understand and respond to environmental and social concerns, aligning business practices with societal expectations (Kaboor and Ameena Febin 2024). However, the effectiveness of stakeholder-led initiatives can vary depending on factors such as the level of commitment from participants, the clarity of objectives, and the mechanisms for monitoring and enforcement. To maximize the impact of these initiatives, it is essential to establish clear governance frameworks, define measurable targets, and ensure inclusive participation from all relevant stakeholders. Social auditing techniques integrated into stakeholder dialogues could build greater trust between researchers and stakeholders, while also promoting cooperation (Venkataraman 2019). Also, studies on corporate motivations and stakeholder involvement in sustainability programs could reduce associated risks.

14.4.3.3 Case Study: How B-Corps Are Leading the Self-Regulation Movement – Corporate Ethics Boards and Internal Sustainability Audits

Certified B Corporations, or B-Corps, represent a leading model of voluntary self-regulation. These companies commit to high standards of environmental and social responsibility and are evaluated across several ESG dimensions. B-Corps integrate sustainability directly into their core business strategies, showing that ethical practices can support long-term financial and societal value.

Other forms of self-regulation, such as corporate ethics boards and internal sustainability audits, also play a vital role. These mechanisms ensure accountability by helping companies identify risks, improve internal controls, and align with legal and ethical standards. Ethics boards help manage complex ESG issues, while audits verify compliance with environmental, safety, and labor requirements (Rakipi and D'Onza 2023).

The internal audit function is becoming a central part of sustainability assurance. It improves the quality of ESG disclosures, supports regulatory compliance, and prevents unethical or illegal behavior. Internal audits also allow managers to detect operational risks and adjust strategies accordingly. The extent of audit involvement often reflects the perceived importance of various stakeholders, such as legal advisors or ethics officers. Where these actors are seen as highly influential, audits tend to be more rigorous and better integrated into the company's governance system (Rakipi and D'Onza 2023).

These practices help organizations build trust with investors, employees, and the public. They also provide mechanisms to monitor ESG progress, assess the effectiveness of responses, and improve long-term governance quality. Ultimately, B-Corps and companies with strong ethics and auditing programs demonstrate how voluntary systems can complement legal requirements and lead to responsible, forward-looking business models.

The success of these models highlights the importance of aligning legal systems, governance frameworks, and voluntary action to promote sustainability. Businesses that implement these practices reduce their environmental footprint, increase social responsibility, and create long-term benefits for all stakeholders (Alvarez, 2010; Rakipi and D'Onza 2023).

Achieving sustainability goals requires the combined efforts of researchers, consumers, and policymakers. Scientific understanding, market choices, and effective regulation must all come together to shape sustainable behavior and production at every level (Venkataraman 2019).

14.5 CHALLENGES IN IMPLEMENTING LEGAL AND GOVERNANCE FRAMEWORKS

Despite significant global momentum around sustainability and the establishment of various legal frameworks and governance structures, the actual implementation of sustainable production policies remains deeply challenging. These difficulties are not merely technical or administrative; they reflect systemic political, economic, and institutional barriers that hinder the operationalization of sustainability principles on the ground.

This section explores six interrelated challenges that continue to obstruct effective implementation: regulatory gaps and weak enforcement, political inertia and fragmented governance, regulatory arbitrage, conflicting priorities between economic growth and environmental protection, lack of harmonized policy standards, and the often underutilized role of civil society (Figure 14.4).

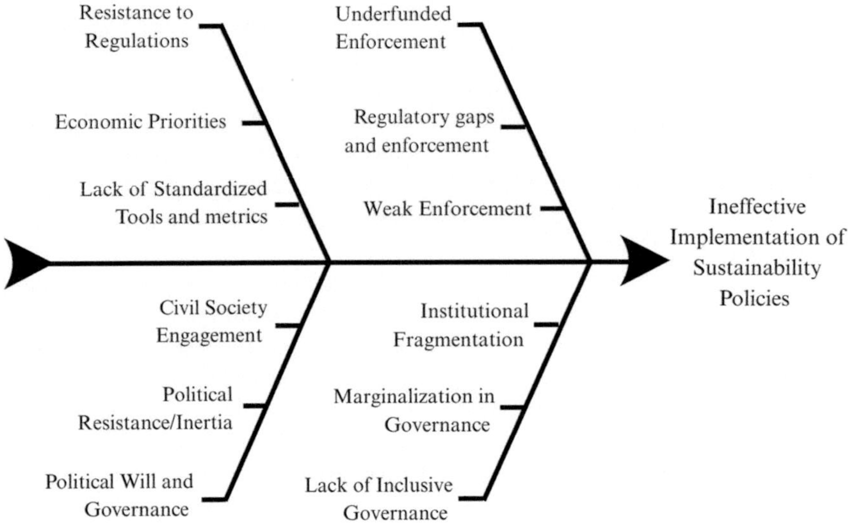

FIGURE 14.4 Challenges in the implementation of sustainable development policies.

14.5.1 Weak Enforcement and Regulatory Gaps

One of the most consistent issues across jurisdictions is the disconnect between regulation and enforcement. Even when sustainability policies are enshrined in law, enforcement mechanisms are often underfunded, politically deprioritized, or institutionally weak. Many countries adopt progressive environmental or sustainability legislation under global pressure, only to struggle to follow through.

This implementation gap is especially evident in sectors such as municipal solid waste management. In Arctic cities such as Anchorage (United States), Murmansk (Russia), and Tromsø (Norway), sustainable waste practices remain limited due to logistical, climatic, and infrastructural constraints. According to Filimonova and Birchall (2024), even when municipalities are legally responsible for sustainability, they are often constrained by budgetary limitations, lack of technical expertise, and conflicting national or regional priorities.

Furthermore, enforcement efforts are often undermined by regulatory ambiguity. As Aureli et al. (2020) note, sustainability reporting, particularly in the form of ESG disclosures, may legally be required—but the lack of clarity on what constitutes "meaningful" reporting leads to superficial compliance. This results in "tick-box" exercises, where companies do the bare minimum to appear compliant without making substantial internal changes.

14.5.2 Political Will and Institutional Fragmentation

The implementation of sustainability laws is deeply political. While the 2030 Agenda and the SDGs provide a global framework, how these goals are pursued depends heavily on national and local political contexts. Beisheim et al. (2025) argue that governments often selectively adopt SDG targets that align with their existing political agendas while ignoring or downplaying others. The flexibility built into the SDGs—which allows states to tailor them to national priorities—can, paradoxically, become a structural limitation on ambitious implementation.

Political will is not simply a matter of declaring support for sustainability—it requires sustained investment, bureaucratic reform, and the courage to confront vested interests. Fragmentation within government is also a major issue. Different ministries or agencies may operate in silos, leading to conflicting policies. For example, a ministry of industry may support the expansion of fossil fuels for economic reasons, while a ministry of environment advocates decarbonization. Without mechanisms for inter-ministerial coordination, such conflicts become implementation bottlenecks.

Moreover, sustainability goals often fall victim to electoral cycles. Politicians tend to focus on short-term achievements that can generate quick political gains, whereas sustainability policies usually require long-term planning and benefits that materialize over decades. This temporal mismatch discourages deep commitment to systemic sustainability reforms.

14.5.3 Regulatory Arbitrage and Transboundary Loopholes

In today's globalized economy, companies can—and often do—exploit differences between national legal systems to avoid compliance with stringent sustainability rules. This phenomenon, known as regulatory arbitrage, allows firms to relocate their operations or financial transactions to jurisdictions with weaker or less enforced regulations.

As Heikki Marjosola (2021) explains, regulatory arbitrage is not accidental, it is often a strategic choice made possible by the elasticity and mobility of capital. For instance, multinational corporations might base manufacturing in countries with lower environmental standards while maintaining headquarters in jurisdictions with stronger reputations. This undermines the integrity of sustainability governance and exacerbates environmental inequality.

The challenge here lies not only in regulating corporate behavior, but in reforming the architecture of international law to close legal gaps between countries. Without some degree of global coordination, sustainability efforts in one country can be easily undone by unsustainable practices in another.

14.5.4 Conflict Between Growth and Environmental Goals

Sustainability is often framed as a balancing act between economic, environmental, and social priorities. However, in practice, economic growth frequently takes precedence. Many governments, especially in developing and resource-dependent economies, see stringent sustainability regulations as a threat to competitiveness, investment, and job creation.

This growth-first mentality creates constant tension in policy circles. For example, in the EU negotiations over the Non-Financial Disclosure (NFD) Directive, many governments—most notably Germany—opposed strong reporting obligations due to the expected adjustment costs for businesses (Kinderman 2020). Even France, which supported the directive, had to navigate domestic opposition from industry associations. The final directive was thus significantly watered down compared to the initial proposals.

This pattern illustrates a wider trend: sustainability is accepted rhetorically but resisted practically when it conflicts with dominant economic interests. Unless this structural contradiction is addressed, implementation will remain shallow and selective.

14.5.5 Lack of Harmonized Standards and Comparative Benchmarks

Even where sustainability reporting is mandatory, a major issue remains: the lack of standardized, comparable metrics. Multiple frameworks exist—such as the GRI, the Sustainability Accounting Standards Board (SASB), and the TCFD—but they vary in scope, format, and rigor.

As Shaikh and Tripathy (2025) point out, this lack of uniformity leads to inconsistent disclosures, making it difficult for stakeholders to compare performance across firms and sectors. This inconsistency undermines trust and allows companies to selectively disclose information that casts them in a favorable light.

The EU's push for upward regulatory harmonization through the 2014/95/EU directive was meant to address this issue. However, as Kinderman (2020) documents, the final law allowed significant national discretion in implementation. As a result, differences persist, and true harmonization remains an aspirational goal rather than a present reality.

14.5.6 The Underused Role of Civil Society and Grassroots Engagement

Finally, while civil society organizations, youth movements, and Indigenous groups have played a vital role in pushing for environmental justice and sustainable practices, they are often marginalized in formal governance processes. Their input is solicited in consultations, but rarely integrated into binding policy decisions.

Ives et al. (2018) argue that reconnecting with nature—on emotional, philosophical, and social levels—is essential for transforming sustainability from an external policy obligation into a shared societal value. However, such "reconnection strategies" require inclusive governance models that go beyond technocratic decision-making and foster participatory dialogue across all segments of society.

The absence of this inclusive ethos not only limits the richness of sustainability governance but also reduces its legitimacy and public support. Empowering civil society can thus be seen as both a normative and practical necessity for effective implementation.

14.6 FUTURE TRENDS AND RECOMMENDATIONS

14.6.1 Emerging Legal Trends in Sustainability

Legal frameworks addressing sustainability are evolving rapidly in response to environmental and societal demands. Notable developments include the broader implementation of EPR programs, mandatory ESG disclosures, and comprehensive legal frameworks to address climate change and biodiversity preservation. EPR policies require manufacturers to manage the full lifecycle of their products, encouraging the design of more durable, recyclable, and reusable goods. ESG disclosure rules, increasingly mandated across jurisdictions, compel companies to report on environmental and social impacts, offering stakeholders greater transparency (Rakipi and D'Onza 2023).

Many industries are now embedding sustainability principles directly into their operations by adopting cleaner production technologies, sourcing eco-friendly inputs, and managing supply chains with reduced environmental impact (Kharat et al. 2024). Strategies such as upcycling are also gaining momentum, allowing companies to repurpose waste materials and reduce dependence on finite resources (Hariyani et al. 2024). For example, in the fashion sector, used textiles are being transformed into new products to minimize waste and promote resource efficiency.

GSCM has become integral, encompassing material selection, responsible sourcing, packaging reduction, optimized logistics, and emissions reduction. Companies increasingly favor materials certified by bodies such as the FSC or those meeting the Global Organic Textile Standard (GOTS) to ensure sustainable practices (Hariyani et al. 2024). Lifecycle assessments help identify opportunities to reduce environmental impacts at each production stage. Integrating CE principles allows firms to optimize supply chain performance through waste reduction, reuse, and process innovation. Advanced technologies such as IoT, big data, and cloud-based systems also support reverse logistics and product lifecycle assessments for sustainable outcomes (Kharat et al. 2024).

14.6.2 Technology and AI in Governance Policies

Technological innovations—particularly artificial intelligence (AI)—are transforming sustainability governance by enabling smarter monitoring, compliance, and optimization strategies. AI systems are now capable of analyzing large-scale environmental data from satellites, sensors, and supply chain platforms to detect irregularities, monitor resource use, and measure the impact of sustainability initiatives. IoT devices, AI algorithms, and big data analytics are widely used to improve procurement, predict supply disruptions, and evaluate supplier compliance (Hariyani et al. 2024).

Machine learning enhances forecasting of consumer behavior and allows companies to better manage inventories, reducing overproduction and minimizing waste (Khan et al. 2022). Integrating blockchain with AI significantly boosts supply chain transparency by verifying product origins and supporting ethical, traceable sourcing practices.

Embedding sustainability into strategic planning has proven beneficial in creating long-term value for both investors and society (Kharat et al. 2024). GSCM now encompasses all operational stages, from sourcing to distribution, ensuring sustainability considerations are embedded in everyday decisions. Blockchain also improves data security, transparency, and traceability, reinforcing good governance and responsible consumption patterns (Hariyani, Hariyani, and Mishra 2025).

14.6.3 Strengthening Multi-Stakeholder Engagement in Sustainability

Effective governance in sustainable production relies heavily on multi-stakeholder collaboration. Bringing together governments, private-sector actors, civil society, and local communities ensures that sustainability policies reflect diverse needs and values. Such partnerships facilitate inclusive decision-making and enable knowledge exchange and shared accountability across industries (Hariyani et al. 2024).

Through collective engagement, stakeholders can co-develop sustainability targets, align resources, and build trust. Businesses that involve their supply chain partners can jointly reduce emissions, improve labor conditions, and increase resilience. This shared approach strengthens compliance and drives broader participation in sustainable initiatives (Hariyani et al. 2024).

14.6.4 Policy Recommendations for More Effective Sustainable Production

To foster the adoption of GSCM, it is important to offer incentives, such as tax breaks or subsidies, to encourage businesses to invest in sustainable practices (Hariyani et al. 2024). Additionally, governments can promote sustainable consumption by raising public awareness and encouraging responsible purchasing decisions. Collaboration between supply chain partners is essential for fostering innovative solutions and effectively implementing GSCM practices (Kharat et al. 2024). To support sustainable production, governments should encourage the development and adoption of standardized metrics for assessing environmental and social performance, enabling better tracking and comparison of sustainability efforts. Governments can facilitate the adoption of sustainable practices by providing training and technical assistance to businesses, helping them overcome barriers to implementation. To promote sustainability effectively, it is imperative to increase public participation by raising awareness of responsible consumption and production.

Organizations should actively engage with policymakers and industry groups to advocate for more consistent and supportive regulations, developing a deep understanding of varying regional and international standards to manage compliance complexities and reduce operational costs (Hariyani et al. 2024). In addition, organizations should perform comprehensive risk assessments to identify potential environmental and social vulnerabilities within their supply chains, implementing robust monitoring and auditing systems to ensure adherence to sustainability standards. Finally, companies can utilize digital platforms and data analytics to improve supply chain visibility, identifying opportunities for efficiency gains and waste reduction (Hariyani et al. 2024).

To overcome high implementation costs, governments can offer subsidies, encourage PPPs, and promote green loans (Kharat et al. 2024). Addressing regulatory and policy constraints is crucial for promoting sustainable practices, harmonizing environmental standards across different regions, and providing clear guidelines for sustainable packaging (Hariyani et al. 2024).

14.6.5 Climate Litigation as a Legal Trend: Rising Lawsuits against Corporations and Governments

Climate litigation is an increasingly important legal mechanism for enforcing environmental accountability. Lawsuits initiated by individuals, NGOs, or public agencies aim to hold corporations and governments accountable for climate-related damage and for failing to act. These legal actions often rely on emerging human rights frameworks, asserting that environmental harm violates basic rights such as access to clean air, health, and life (Ge, Wang, and Epede 2022; Peel and Osofsky 2015).

The scope of climate litigation has expanded globally, with significant activity in North America, Europe, Asia, and the Global South. These legal proceedings serve not only to obtain remedies but also to push for regulatory reform and raise awareness (Lin 2014; Taylor 2021). Tort law is also being explored as a way to address both individual and collective harm caused by climate change (Semmelmayer 2021). Strategic litigation has emerged as a tool to influence social and political change, particularly in cases related to business and human rights.

Early climate litigation began in the 1980s, focusing on procedural deficiencies in environmental assessments. Today, lawsuits are more sophisticated, employing international legal standards and demanding enforceable commitments to reduce emissions (Peel and Osofsky 2015; Peel and Lin 2019).

14.6.6 Decentralized Governance and Community-Led Sustainability Initiatives

Decentralized governance, which empowers local communities and stakeholders, can foster more effective and context-specific approaches to sustainable production. By shifting decision-making authority to the local level, governance systems can be more responsive to the unique environmental, social, and economic conditions of a particular region (Garren and Brinkmann 2018). This approach encourages innovation and experimentation, as communities are free to develop and implement solutions that best fit their needs and priorities. Community-led initiatives also promote greater accountability and transparency, as local stakeholders are more directly involved in the decision-making process and can hold leaders accountable for their actions.

A shift toward localized governance structures encourages active community participation, guaranteeing that sustainability projects are in accordance with local values and needs, hence promoting greater responsibility and support. These initiatives frequently depend on locally available resources, decreasing reliance on outside supply chains, and fostering local economic development. Decentralized governance makes it easier to include indigenous knowledge and traditional practices into sustainability plans, acknowledging the important contributions of local cultures to sustainable resource management. Local communities are more likely to accept and support sustainability initiatives when they are actively involved in the planning and execution phases, which promotes long-term success and resilience.

Grassroots movements and local climate protection networks play a critical role in pushing elected officials to take action, form partnerships, and gather public support, resulting in the creation of new climate policies and standards that significantly reduce GHG emissions (Table 14.2).

TABLE 14.2
Trends and Recommendations in Sustainable Production Governance

S. No	Key Area	Description	Examples/Tools	Anticipated Impact	Reference
1.	Emerging Legal Trends in Sustainability	Legal frameworks are evolving toward mandatory ESG disclosures, Extended Producer Responsibility (EPR), and laws focused on climate change and biodiversity protection.	EPR laws, ESG disclosure mandates	Drives transparency, product lifecycle accountability, and climate action.	Rakipi and D'Onza (2023); Kharat et al. (2024); Hariyani et al. (2024)
2.	Circular economy (CE) integration	Industries adopting CE principles: waste reduction, reuse, upcycling, and sustainable supply chains, supported by certifications (FSC and GOTS) and lifecycle assessments.	FSC and GOTS certifications, CE models	Promotes resource efficiency and sustainable innovation across industries.	Hariyani et al. (2024); Kharat et al. (2024)
3.	Green supply chain management (GSCM)	Lifecycle responsibility from sourcing to logistics; incorporating eco-friendly materials, packaging reduction, and emissions minimization.	Lifecycle assessment tools, eco-certifications	Reduces environmental footprint and enhances supply chain sustainability.	Hariyani et al. (2024); Kharat et al. (2024)
4.	Technology and AI in governance	AI and IoT tools enhance supply chain monitoring, predictive analytics, and resource efficiency. Blockchain ensures data security and traceability for ethical sourcing.	IoT, AI, blockchain, big data	Enables smarter, more transparent, and traceable supply chains.	Hariyani et al. (2025); Khan et al. (2022); Prasad et al. (2024)
5.	AI for compliance and optimization	AI enables real-time compliance monitoring, environmental impact tracking, and waste reduction through smarter procurement and forecasting.	AI algorithms, machine learning models	Enhances governance efficiency and reduces operational waste and emissions.	Hariyani et al. (2025); Khan et al. (2022); Prasad et al. (2024)
6.	Multi-stakeholder engagement	Collaborative governance involving governments, private sectors, NGOs, and communities fosters shared responsibility, inclusivity, and practical sustainability solutions.	Public–private partnerships (PPPs), community programs	Strengthens legitimacy, accountability, and practical implementation of policies.	Kaboor and Ameena Febin (2024); Cunha et al. (2019)
7.	Supply chain collaboration	Engagement throughout supply chains to jointly reduce emissions, enhance labor standards, and build resilience.	Green procurement platforms	Encourages systemic change through collaboration rather than isolation.	Hariyani et al. (2024)
8.	Policy recommendations	Incentives (tax breaks, subsidies), public awareness, and standard metrics to promote green supply chain practices. PPPs and training to remove barriers to implementation.	Green loans, subsidies, eco-label policies	Supports widespread adoption of sustainability through practical, policy-backed support.	Hariyani et al. (2024); Kharat et al. (2024)
9.	Harmonizing standards	Advocating for consistent global standards in ESG reporting and sustainable packaging to ease compliance and improve effectiveness.	ISO standards, EU taxonomy,	Reduces complexity, strengthens comparability, and drives international alignment.	Kinderman (2020); Shaikh and Tripathy (2025)
10.	Climate litigation	Rising lawsuits targeting corporations and governments for climate accountability; strategic litigation advancing rights-based and environmental legal standards.	Climate litigation strategies, Environmental Impact Assessment (EIA)	Spurs regulatory reforms and set precedents for environmental accountability.	Peel and Osofsky (2015); Peel and Lin (2019); Lin (2014); Taylor (2021); Semmelmayer (2021)
11.	Decentralized governance and community initiatives	Localized governance empowers communities to innovate, ensures alignment with cultural and regional needs, and enhances participation and accountability in sustainability efforts.	Local climate networks, grassroots movements	Increases policy relevance, community engagement, and long-term resilience.	Garren and Brinkmann (2018)
12.	Integration of indigenous knowledge	Incorporating traditional ecological knowledge into modern sustainability strategies, enhancing cultural alignment and environmental stewardship.	Co-management agreements, local governance models	Enriches sustainability solutions with context-sensitive, time-tested knowledge.	Garren and Brinkmann (2018)

14.7 CONCLUSION

The transition to sustainable production is not just an environmental necessity; it is a complex and multidimensional shift that calls for an integrated legal and governance response. Throughout this chapter, we explored how sustainability principles are being embedded into international legal instruments, national regulations, and corporate governance structures. Despite clear progress in policy development, implementation remains inconsistent due to a range of institutional, economic, and political barriers.

From the foundational role of international frameworks such as the SDGs and the Paris Agreement to regional and national efforts, such as EIAs and PPPs, it is evident that legal mechanisms play a vital role in driving sustainable practices. These tools not only establish boundaries for industrial behavior but also provide incentives for innovation, investment, and collaboration.

However, the path forward is riddled with challenges. Weak enforcement, fragmented institutions, and regulatory arbitrage continue to hinder meaningful change. Companies may still engage in greenwashing, and governments often struggle to harmonize economic growth with ecological protection. As this chapter has demonstrated, effective governance must go beyond symbolic commitments; it requires accountability, transparency, and inclusive decision-making.

Equally crucial is the role of corporate actors and financial systems. The rise of ESG standards, green finance, sustainable procurement, and voluntary certifications marks a promising shift in how businesses operate. Hitherto, to truly transform production systems, these efforts must be underpinned by robust legal frameworks and actively monitored governance processes. Voluntary compliance, while valuable, cannot replace enforceable obligations.

Looking ahead, the integration of digital technologies such as AI, blockchain, and big data analytics offers new opportunities to enhance transparency, traceability, and efficiency in sustainable production. When aligned with sound policy and legal guidance, these innovations can support more ethical, resource-efficient supply chains and drive systemic transformation.

Ultimately, sustainable production is not the responsibility of a single actor or sector; it is a collective endeavor. Legal regimes and governance policies are essential scaffolding, but real progress depends on the commitment of all stakeholders: governments, businesses, civil society, and consumers. Building resilient, fair, and environmentally responsible production systems is not only possible but also imperative for the well-being of current and future generations.

REFERENCES

Agarwal, Vernika, Snigdha Malhotra, and Vishal Dagar. 2023. "Coping with Public–Private Partnership Issues: A Path Forward to Sustainable Agriculture." *Socio-Economic Planning Sciences* 89: 101703.

Alvarez, Gabriela. 2010. "Fair Trade and Beyond: Voluntary Standards and Sustainable Supply Chains." In *Delivering Performance in Food Supply Chains*, 478–510. Woodhead Publishing.

Angstadt, J. Michael, and Maddison S. Schink. 2023. "Specialist Environmental Courts and Tribunals: A Systematic Literature Review and Case for Earth System Governance Analysis." *Earth System Governance* 18 (September): 100192.

Arı, İzzet, and Riza Fikret Yikmaz. 2019. "Greening of Industry in a Resource- and Environment-Constrained World." In *Elsevier eBooks*, 53. Elsevier BV.

Aschehoug, Silje Helene, and Kjersti Øverbø Schulte. 2017. "Sustainability Strategies in Industrial Practice." In *IFIP Advances in Information and Communication Technology*, 358. Springer Science+Business Media.

Aureli, Selena, et al. 2020. "Nonfinancial Reporting Regulation and Challenges in Sustainability Disclosure and Corporate Governance Practices." *Business Strategy and the Environment* 29: 2392–2403.

Barker, Richard. 2025. "Corporate Sustainability Reporting." *Journal of Accounting and Public Policy* 49: 107280.

Barral, Virginie. 2010. "Sustainable Development." *Yearbook of International Environmental Law* 21 (1): 322.

Beisheim, Marianne, et al. 2025. "Politics Matters! Political Will as a Critical Condition for Implementing the Sustainable Development Goals." *Earth System Governance* 24: 100244.

Benevene, Paula, Ilaria Buonomo, Eric Kong, Martina Pansini, and Maria Luisa Farnese. 2021. "Management of Green Intellectual Capital: Evidence-Based Literature Review and Future Directions." *Sustainability* 13 (15): 8349.

Bhamra, Tracy, and Ricardo J. Hernández. 2021. "Thirty Years of Design for Sustainability: An Evolution of Research, Policy and Practice." *Design Science* 7 (January).

Böhringer, Christoph, Carolyn Fischer, Knut Einar Rosendahl, and Thomas Fox Rutherford. 2022. "Potential Impacts and Challenges of Border Carbon Adjustments." *Nature Climate Change* 12 (1): 22–29.

Castro, Paula. 2012. "*Legal Innovation for Social Change: Exploring Change and Resistance to Different Types of Sustainability*."

Chen, You-hua, Chan Wang, Pu-yan Nie, and Zirui Chen. 2020. "A Clean Innovation Comparison between Carbon Tax and Cap-and-Trade System." *Energy Strategy Reviews* 29 (March): 100483.

Cheng, Zhe, Huanming Wang, Wei Xiong, Dajian Zhu, and Le Cheng. 2020. "Public–Private Partnership as a Driver of Sustainable Development: Toward a Conceptual Framework of Sustainability-Oriented PPP." *Environment, Development and Sustainability* 23 (1): 1043.

Culiao, Rudiliza, Sungho Tae, and Rakhyun Kim. 2018. "A Review of the Philippine Green Building Rating System, BERDE, in Comparison with G-SEED and LEED." *International Journal of Sustainable Building Technology and Urban Development*.

Cunha, Felipe Arias Fogliano de Souza, Erick Meira, Renato J. Orsato, Marcelo Cabús Klötzle, Fernando Luiz Cyrino Oliveira, and Rodrigo Goyannes Gusmão Caiado. 2019. "Can Sustainable Investments Outperform Traditional Benchmarks? Evidence from Global Stock Markets." *Business Strategy and the Environment* 29 (2): 682.

Enuoh, Oliver O. O., and Francis E. Bisong. 2014. "Global Sustainable Development Agenda: An Implication for Conservation Challenges in Cross River State, Nigeria." *Journal of Sustainable Development* 7 (4).

Fadly, Dalia. 2020. "Greening Industry in Vietnam: Environmental Management Standards and Resource Efficiency in SMEs." *Sustainability* 12 (18): 7455.

Filimonova, N., and S. J. Birchall. 2024. "Sustainable Municipal Solid Waste Management: A Comparative Analysis of Enablers and Barriers." *Journal of Environmental Management* 371: 123111.

Garren, Sandra J., and Robert Brinkmann. 2018. "Sustainability Definitions, Historical Context, and Frameworks." In *Springer eBooks*, 1. Springer Nature.

Ge, Genhasi, Daoping Wang, and Mesumbe Bianca Epede. 2022. "Pricing Policies of Green Dual-Channel Supply Chain with Fairness Concerns and Altruistic Preferences Based on Consumers' Environmental Awareness and Channel Preference." *International Journal of Environmental Research and Public Health* 19 (20): 13564.

Haq, Hafiz, Petri Välisuo, and Seppo Niemi. 2021. "Modelling Sustainable Industrial Symbiosis." *Energies* 14 (4): 1172.

Hariyani, Dharmendra, Poonam Hariyani, and Sanjeev Mishra. 2025. "Digital Technologies for the Sustainable Development Goals." *Green Technologies and Sustainability, March*: 100202.

Hariyani, Dharmendra, Poonam Hariyani, Sanjeev Kumar Mishra, and Milind Kumar Sharma. 2024. "A Literature Review on Green Supply Chain Management for Sustainable Sourcing and Distribution." *Waste Management Bulletin*. Elsevier BV.

Iqbal, Muhammad Ahsan, Wasim Abbass Shaheen, Suheela Shabir, Usman Ullah, Ionel-Alin Ienciu, Marius-Ioan Mihut, António Raposo, and Heesup Han. 2025. "Towards a Green Economy: Investigating the Impact of Sustainable Finance, Green Technologies, and Environmental Policies on Environmental Degradation." *Journal of Environmental Management* 374 (January): 124047.

Islam, Md. Mazharul, and Majed Alharthi. 2020. "Relationships among Ethical Commitment, Ethical Climate, Sustainable Procurement Practices, and SME Performance: A PLS-SEM Analysis." *Sustainability* 12 (23): 10168.

Ives, Christopher D., et al. 2018. "Reconnecting with Nature for Sustainability." *Sustainability Science* 13: 1389–1397.

Jain, Mansi, Gagan Deep Sharma, and Mrinalini Srivastava. 2019. "Can Sustainable Investment Yield Better Financial Returns: A Comparative Study of ESG Indices and MSCI Indices." *Risks*, 7 (1), 15.

Jang, Jinseop, Jason J. McSparren, and Yuliya Rashchupkina. 2016. "Global Governance: Present and Future." *Palgrave Communications* 2 (1).

Jayawardhana, Kumudu, Imali Fernando, and Janaka Siyambalapitiya. 2022. "Sustainability in Social Enterprise Research: A Systematic Literature Review." *SAGE Open* 12 (3).

Jin, Wei, Wen Ding, and Jun Yang. 2021. "*Impact of Financial Incentives on Green Manufacturing: Loan Guarantee vs. Interest Subsidy*."

Kaboor, A., and T. Ameena Febin. 2024. "Evaluating the Effectiveness of Sustainable Business Certification Programs: A Critical Assessment of Environmental, Social, and Governance Impacts." *In 9th International Conference on Economic Growth and Sustainable Development-Emerging Trends*. Mysuru, India: Shri Dharmasthala Manjunatheshwara Institute for Management Development.

Khamisu, Muhammad Sani, and Ratna Achuta Paluri. 2024. "Emerging Trends of Environmental Social and Governance (ESG) Disclosure Research." *Cleaner Production Letters* 7: 100079.

Khan, Muhammad, Gohar Saleem Parvaiz, Alisher Tohirovich Dedahanov, Odiljon Sobirovich Abdurazzakov, and Dilshodjon Alidjonovich Rakhmonov. 2022. "The Impact of Technologies of Traceability and Transparency in Supply Chains." *Sustainability* 14 (24): 16336.

Kharat, Manoj Govind, Samridhi Kapoor, Shreyanshu Parhi, Mukesh Govind Kharat, and Shatrudhan Pandey. 2024. "Operationalizing Sustainability in Pharmaceuticals: Green Supply Chain Metrics for Circular Economy." *Sustainable Futures* 9 (December): 100413.

Kinderman, Daniel. 2020. "The Challenges of Upward Regulatory Harmonization: The Case of Sustainability Reporting in the EU." *Regulation & Governance* 14: 674–697.

Lashari, Ishfaque Ahmed, Qiyuan Li, Qamaruddin Maitlo, Faraz Ali Bughio, Ashique Ali Jhatial, and Obed Rashidi Syed. 2022. "Environmental Sustainability through Green HRM: Measuring the Perception of University Managers." *Frontiers in Psychology* 13 (November).

Lin, Jolene. 2014. "Litigating Climate Change in Asia." *Climate Law* 4 (July): 140.

Liyanage, Shantha, Fulu Godfrey Netswera, and Abel Motsumi. 2020. "Insights from EU Policy Framework in Aligning Sustainable Finance for Sustainable Development in Africa and Asia." *International Journal of Energy Economics and Policy* 11 (1): 459.

Ma, Chao, Mazhar Farid Chishti, Muhammad Kashif Durrani, Rizwana Bashir, Sofia Safdar, and Rana Tanveer Hussain. 2023. "The Corporate Social Responsibility and Its Impact on Financial Performance: A Case of Developing Countries." *Sustainability* 15 (4): 3724.

Marjosola, Heikki. 2021. "The Problem of Regulatory Arbitrage: A Transaction Cost Economics Perspective." *Regulation & Governance* 15: 388–407.

Mensah, Justice. 2019. "Sustainable Development: Meaning, History, Principles, Pillars, and Implications for Human Action." *Cogent Social Sciences* 5 (1).

Mushangai, Dandira. 2023. "Dynamic Capabilities: Axiomatic Formation of Firms' Competitive Competencies." *Social Sciences & Humanities Open* 8 (1): 100654.

Nnaji, J. C., and Augustine Igbuku. 2018. "Improving the Eco-Innovation Status of Africa Using Sustainable Chemistry Principles." *European Journal of Sustainable Development Research* 3 (1).

Peel, Jacqueline, and Jolene Lin. 2019. "Transnational Climate Litigation: The Contribution of the Global South." *American Journal of International Law* 113 (4): 679.

Peel, Jacqueline, and Hari M. Osofsky. 2015. *Climate Change Litigation*.

Prasad, Kanishk, Santoshachandra Rao Karanam, D. Ganesh, Kazi Kutubuddin Sayyad Liyakat, Vamsidhar Talasila, and P. Purushotham. 2024. "AI in Public–Private Partnership for IT Infrastructure Development." *Journal of High Technology Management Research* 35 (1): 100496.

Rakipi, Romina, and Giuseppe D'Onza. 2023. "The Involvement of Internal Audit in Environmental, Social, and Governance Practices and Risks: Stakeholders' Salience and Insights from Audit Committees and Chief Executive Officers." *International Journal of Auditing* 28 (3): 522.

Roos, Claudine, Dirk Cilliers, François Retief, Reece Alberts, and Alan Bond. 2020. "Regulators' Perceptions of Environmental Impact Assessment (EIA) Benefits in a Sustainable Development Context." *Environmental Impact Assessment Review* 81 (January): 106360.

Ruggerio, Carlos Alberto. 2021. "Sustainability and Sustainable Development: A Review of Principles and Definitions." *Science of the Total Environment.*

Semmelmayer, Philipp. 2021. "Climate Change and the German Law of Torts." *German Law Journal* 22 (8): 1569.

Shaikh, Imlak, and Arun K. Tripathy. 2025. "Variability in Sustainability Reporting." *Journal of Environmental Management* 373: 123851.

Shen, Liyin, Vivian W. Y. Tam, Lin Gan, Kunhui Ye, and Zongnan Zhao. 2016. "Improving Sustainability Performance for Public–Private–Partnership (PPP) Projects." *Sustainability* 8 (3): 289.

Shi, Longyu, Linwei Han, Fengmei Yang, and Lijie Gao. 2019. "*The Evolution of Sustainable Development Theory: Types, Goals, and Research Prospects.*"

Stevens, Candice. 2010. "Linking Sustainable Consumption and Production: The Government Role." *Natural Resources Forum* 34 (1): 16.

Taylor, Mark B. 2021. "Counter Corporate Litigation: Remedy, Regulation, and Repression in the Struggle for a Just Transition." *Sustainability* 13 (19): 10742.

Venkataraman, S. 2019. *Stakeholder Approach to Corporate Sustainability: A Review.* RePEc: Research Papers in Economics. Federal Reserve Bank of St. Louis.

Villeneuve, Claude, David Tremblay, Olivier Riffon, Georges Lanmafankpotin, and Sylvie Bouchard. 2017. "A Systemic Tool and Process for Sustainability Assessment." *Sustainability* 9 (10): 1909.

Wassénius, Emmy, Beatrice Crona, and Sasha Quahe. 2024. "Essential Environmental Impact Variables." *One Earth* 7: 211–228.

Westhues, Martina, and Sabine Einwiller. 2006. "Corporate Foundations and CSR." *Corporate Reputation Review* 9 (2): 144–153.

Xin, Chunhua, Ran Zhu, and Fengqi Guo. 2025. "Corporate Sustainability and ESG Consistency." *Journal of Environmental Management* 377: 124688.

Zhang, Wushuai, Yuan Qiao, Prakash Lakshmanan, Liuzheng Yuan, Jiayou Liu, Chenghu Zhong, and Xinping Chen. 2022. "Combing Public–Private Partnership and Large-Scale Farming Increased Net Ecosystem Carbon Budget and Reduced Carbon Footprint of Maize Production." *Resources, Conservation and Recycling* 184 (May): 106411.

15 Life Cycle Assessment Approaches for Manufacturing

Bharti Singh, Sandeep Kumar, and Rachna Bhateria
Department of Environmental Sciences, Maharshi Dayanand University, Rohtak, Haryana, India

15.1 INTRODUCTION

Sustainability and sustainable manufacturing (SM) have emerged as critical priorities for governments and industries worldwide. In response, numerous conceptual frameworks and assessment methodologies have been developed to support sustainable practices. Despite these advances, precisely assessing sustainability outcomes at the product level remains a difficult and unsolved challenge. One of the well-known models is the triple bottom line (TBL) method, which emphasizes the inclusion of social, economic, and environmental variables in sustainability evaluations (Remmen, 2007, Elkington and Rowlands, 1999). Furthermore, when evaluating the assessment of sustainability in products and processes through life cycle thinking (LCT)—which accounts for the entire life cycle from "cradle to grave"—are becoming increasingly prevalent. These methodologies allow for a more comprehensive review by capturing environmental, economic, and social aspects throughout a product's life cycle. By adopting LCT, potential trade-offs and the shifting of impacts between different life cycle stages can be identified and mitigated, thereby supporting more informed and balanced decision-making (Finkbeiner et al. (2010). The Life Cycle Sustainability Assessment (LCSA) framework, which combines the TBL theory with LCT techniques, has been presented as a comprehensive method of analyzing product sustainability performance. LCSA covers the environmental, economic, and social aspects of sustainability through the use of three complementary methodologies: LCA, LCC, and SLCA. This integrated method allows for a comprehensive evaluation of products throughout their life cycles, supporting more informed, sustainable decision-making.

A key factor that has contributed to the widespread adoption and methodological rigor of LCA is the establishment of internationally recognized standards by the International Organization for Standardization (ISO) (2006). The ISO 14040 series, particularly ISO 14040: 2006a and ISO 14044:2006b, provides the fundamental principles, framework, requirements, and guidelines for conducting scientifically robust and comparable LCA studies. These standards serve as the "constitution" for LCA practice, ensuring consistency and credibility across applications in various sectors. Building upon this foundation, additional standards such as ISO 14046 (Water Footprint), ISO 14067 (Carbon Footprint [CFP]), and ISO 14072 (Organizational LCA) have expanded the scope of life cycle methodologies to address specific environmental concerns and organizational practices. These standardized frameworks not only support methodological harmonization but also facilitate the practical application of life cycle approaches in both policy and industry contexts.

Despite these developments, significant scientific challenges remain. The reliability and comprehensiveness of the LCSA framework are limited, primarily due to the uneven maturity of its underlying methodologies (Valdivia et al. (2013), Neugebauer et al. (2015). While LCA is well-established and widely applied, SLCA and LCC face methodological limitations, including the lack of standardized guidance for indicator selection, unclear links between indicators and impact areas, and inconsistent practical implementation. As a result, sustainability assessments may yield incomplete, inconsistent, or non-comparable results, reducing their effectiveness for guiding SM decisions (Lehmann (2013), Neugebauer et al. (2016). Addressing these gaps and strengthening the robustness of LCSA is therefore a critical scientific problem in the field.

In response to these challenges, various novel strategies have been introduced to strengthen the methodological rigor and practical application of LCSA. One such approach is the Tiered Approach, which introduces a hierarchical categorization system to support the structured application of LCSA, particularly within the manufacturing sector. Additionally, efforts have been made to define social impact pathways—such as fair wage considerations—to strengthen the foundation of Social Life Cycle Assessment (SLCA). In parallel, a revised Life Cycle Costing (LCC) method, known as the Economic Life Cycle Assessment (ELCA) framework, has been introduced to resolve some of the key methodological limitations associated with traditional LCC, thereby contributing to a more integrated and comprehensive sustainability assessment (Chang et al. (2017).

The subsequent sections provide an in-depth examination of the three core methodologies underpinning LCSA—LCA, LCC, and SLCA—while highlighting current progress, unresolved research challenges, and future directions. This chapter makes a significant contribution to the field by

DOI: 10.1201/9781003637158-15

systematically identifying methodological gaps in SLCA and LCC, discussing strategies such as the Tiered Approach and ELCA to strengthen LCSA, and providing practical insights for its application in the manufacturing sector. By integrating recent research, international standards, and emerging approaches, the chapter also includes existing knowledge that offers guidance for researchers and practitioners to conduct more robust, reliable, and comprehensive sustainability assessments, thereby advancing both the scientific understanding and practical implementation of SM practices.

15.2 BACKGROUND AND RELATED WORK

This section critically examines existing review literature on the sustainability of manufacturing processes and identifies the research gaps that this study aims to address. Most prior studies have primarily focused on energy consumption as the key indicator of environmental performance, while other important resources and sustainability factors have received comparatively less attention. Liu et al. (2018) conducted a comprehensive review of energy consumption in metal manufacturing processes across various levels, including the machine and process levels, and under different operational modes such as idle, standby, and active operation. Their study also examined the inherent energy of raw materials used in manufacturing and suggested strategies for minimizing energy use. Similarly, Garcia et al. (2018) assessed the sustainability of manufacturing processes, revealing that energy usage was the dominant focus, featured in 87% of the reviewed literature. However, only 25% of the studies incorporated LCA to assess the environmental impacts of manufacturing categories. The same pattern was noted by Agrawal and Vinodh (2019), who analyzed advancements in sustainable additive manufacturing (AM) and observed a predominant focus on energy consumption. Collectively, these studies underscore a critical research gap: the limited consideration of other resource consumptions and process emissions. Addressing these factors, alongside energy use, is essential for a more comprehensive assessment of manufacturing sustainability.

Kellens et al. (2017) investigated the environmental aspects of manufacturing procedures and observed that, despite the increasing focus on sustainability, comprehensive LCA studies comparing the environmental impacts of manufacturing processes with other conventional manufacturing (CM) processes remain limited. In a related study, Saade et al. (2020) analyzed previously published research on LCA manufacturing processes, focusing specifically on comparisons of global warming potential (GWP) and inherent energy between modern manufacturing and CM technologies. Their findings revealed that GWP is primarily influenced by the energy use of manufacturing processes, whereas raw material use is the dominant contributor in CM processes. The chapter emphasized the importance of improving energy efficiency in manufacturing processes to enhance their environmental competitiveness. These studies collectively indicate that energy usage in manufacturing processes has gained significantly greater focus than other inventory components. These include raw material usage, inert gas consumption, material waste, support structure requirements, and material removal during post-processing operations. Consequently, there is a critical need for analysis of diverse life cycle inventory (LCI) data across various manufacturing process conditions. Such an effort would provide a multidimensional understanding of the environmental performance of manufacturing technologies.

15.3 RESEARCH CONCEPT

15.3.1 Environmental Life Cycle Assessment

Environmental LCA is an internationally standardized methodology, as defined in ISO 14044:2006b, for quantifying the environmental impacts of a product or process throughout all stages of its life cycle, implemented through a structured sequence of steps (Figure 15.1):

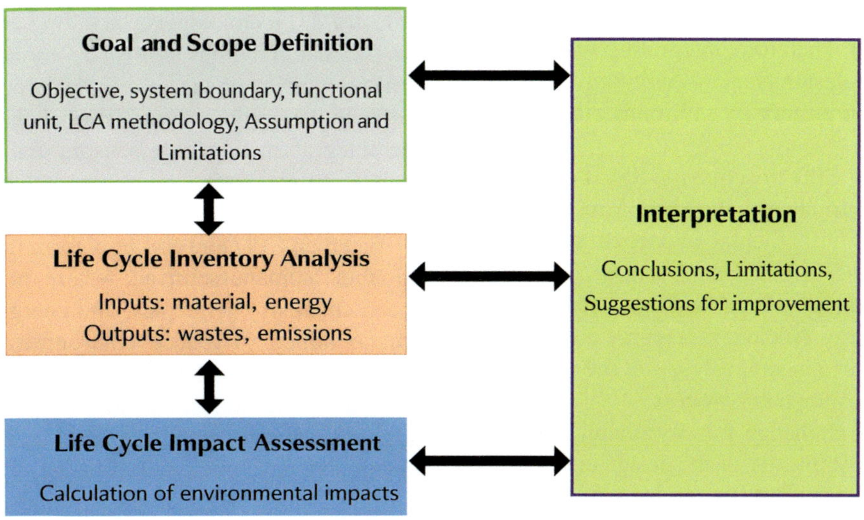

FIGURE 15.1 Key Steps in Environmental Life Cycle Assessment (LCA) as per ISO 14044:2006b (Kokare et al. (2023).

- **Goal and Scope Definition:** This phase entails establishing the study's objectives, defining the life cycle stages under evaluation, specifying the functional unit to standardize all inputs and outputs, selecting a suitable impact assessment methodology, and identifying any underlying assumptions and limitations.
- **Life Cycle Inventory Analysis:** This step involves the quantitative compilation of all input and output flows related to every operational stage inside the system boundaries of the study. Inputs typically comprise primary resources such as materials, energy, water, and auxiliary substances, whereas outputs include releases to air, water bodies, and land, as well as solid waste. Data for this inventory analysis are derived from experimental data, established LCI databases, technical reports, and peer-reviewed literature.
- **Life Cycle Impact Assessment:** In this phase, the inventory data are translated into potential environmental impacts by applying characterization models. The assessment typically includes categories such as GWP, acidification potential, ozone depletion potential, along with other applicable environmental indicators as specified by the selected assessment method.
- **Interpretation:** This step involves interpreting the assessment findings used to determine major contributors and critical environmental hotspots across the life cycle. It also includes an evaluation of limitations, the exploration of improvement opportunities, and the formulation of strategies to reduce environmental impacts (Kokare et al., 2023).

15.3.1.1 Sustainable Product Design (SPD) and Sustainable Manufacturing (SM) Process

Product design plays a pivotal role in determining the environmental, economic, and functional performance of a product across its life cycle. As highlighted by several studies, decisions made during the design phase have long-lasting impacts on subsequent stages such as material selection, manufacturing, usage, and end-of-life disposal (Rebitzer et al. (2004). Therefore, integrating sustainability principles early in the design phase—commonly referred to as SPD—is crucial for promoting environmentally responsible innovation.

The significance of SPD in achieving SM is well-recognized. According to estimates by the European Union, up to 80% of the overall environmental impact of a product is determined during the initial design stages. This underscores the importance of embedding sustainability criteria—such as energy efficiency, resource conservation, recyclability, and environmental safety—at the very beginning of the product development process.

However, a major challenge for sustainability-oriented designers lies in the complexity of the design environment. Design decisions are often influenced by a wide array of conflicting factors, including cost, aesthetics, functionality, user preferences, and market demands. These constraints may hinder the incorporation of environmental considerations, even when sustainability is a stated objective (Bogue (2014). Despite these challenges, recent literature suggests that Sustainable Material Design (SMD) can play a transformative role in reducing environmental harm. This involves strategies aimed at eliminating hazardous substances, reducing material waste and scrap, and selecting materials that are renewable, recyclable, or less energy-intensive to process (Zsidisin & Siferd (2001).

On the manufacturing side, traditional production processes have been identified as major contributors to environmental degradation. Manufacturing consumes a large amount of energy and generates waste across all three phases—solid, liquid, and gaseous (Duflou et al. (2012). In response, the concept of SM has emerged, emphasizing the need to minimize the ecological footprint of production activities. Various measures taken in this direction include improving energy efficiency, reducing CO_2 and other gaseous emissions, recovering valuable by-products, and adopting cleaner production technologies (Despeisse et al. (2012).

Both SPD and SM form a unified approach in achieving sustainability across the product life cycle. When effectively integrated, they enable the development of products that are not only functionally superior and economically viable but also environmentally responsible.

Recent developments in SPD emphasize the use of digital design tools, life cycle data integration, and simulation-based assessment to predict environmental and economic performance during the early design stages. Tools such as Computer-Aided Design (CAD) integrated with LCA enable designers to visualize material flows, energy use, and potential emissions even before manufacturing begins. Furthermore, the incorporation of eco-design principles—such as modularity, dematerialization, and product-service systems (PSS)—encourages circular economy practices where products are designed for reuse, remanufacturing, and recycling.

SM, conversely, has evolved through the adoption of Industry 4.0 technologies, including automation, AM, and the Internet of Things (IoT). These innovations enable real-time monitoring of energy and resource efficiency, reducing waste and improving production flexibility. For instance, the integration of smart sensors and digital twins helps manufacturers assess environmental impacts dynamically and optimize production for minimal emissions.

Together, SPD and SM now form the foundation of sustainable biomanufacturing, where renewable feedstocks, green chemistry principles, and energy-efficient processes are combined to reduce environmental footprints while maintaining product performance. This integration supports the shift from linear to circular production systems, aligning industrial practices with the broader objectives of sustainable development and LCSA. Table 15.1. highlights how SPD and SM, individually and in combination, use specific strategies and tools to enhance environmental,

TABLE 15.1
Key Aspects of Sustainable Design and Manufacturing

Aspect	Focus	Key Strategies/Tools	Impact/Outcome
Sustainable Product Design (SPD)	Early stage design decisions shaping environmental, economic and functional performance	Sustainable Material Design (SMD): reduce waste, use renewable/recyclable materials; Digital tools: CAD + LCA, simulation; Eco-design principles: modularity, dematerialization, PSS	Predict environmental and economic performance; Promote circular economy: reuse, recycling, remanufacturing; Reduce material-related environmental harm
Sustainable Manufacturing (SM)	Minimizing ecological footprint during production	Cleaner production technologies; Energy efficiency and emission reduction; Industry 4.0: automation, additive manufacturing (AM), IoT, digital twins	Real-time monitoring of energy/resources; Minimized emissions and waste; Flexible and optimized production
Integration of SPD and SM	Unified approach across the product life cycle	Renewable feedstocks and green chemistry; Energy-efficient processes; LCSA	Environmentally responsible, functionally superior, and economically viable products; Supports circular production and sustainable development

economic, and functional performance, leading to reduced waste, energy efficiency, and circular production.

15.3.2 Life Cycle Costing (LCC)

The LCC encompasses all costs incurred during its entire lifespan, including acquisition, operation, maintenance, and end-of-life disposal. LCC is a structured, widely used methodology that enables comprehensive evaluation of these costs. It supports decision-making by providing a complete economic perspective over the product's or system's life cycle. LCC serves as a valuable decision-making tool for manufacturers by enabling the identification of cost-saving opportunities throughout the life cycle of a product. Its application during the initial design and development stages is particularly impactful, with studies indicating potential cost savings of up to 70–85% of the total product cost (Chatti et al. (2019). Over time, numerous LCC models and techniques have been proposed. However, no single approach has been universally adopted as a standard due to differing user preferences, varying problem contexts, inconsistencies in cost data collection systems, and the diversity of equipment and systems under analysis (Dhillon 2017). While standards such as ISO 15686–5 (ISO (2017) offer general frameworks for implementing LCC, Green and Shaw (1990) have introduced a simple yet impactful step-by-step approach for conducting LCC analyses. An LCC analysis begins with the clear definition of its objective, which may include comparative cost evaluation between alternative systems, life cycle cost estimation for budgeting purposes, or cost-benefit analysis, among others. Once the objective is established, the purpose of the analysis is defined by reviewing key parameters such as system performance, technical requirements, and scheduling constraints. This step also involves the selection of relevant systems and subsystems to be included in the study. Based on the objective and scope, the analyst selects the most appropriate cost estimation method and LCC model. Subsequently, relevant data are gathered from various sources, including historical records, expert input, published literature, and technical databases. Using this information, input parameters for the LCC model are formulated. The collected input and output data are then subjected to coherence, precision, reliability, and thoroughness checks. Where applicable, uncertainty and risk evaluations are performed to evaluate how changes in input variables influence the cost outcomes, allowing for the presentation of cost ranges instead of single-point estimates. Once the data are validated, the results of the LCC analysis are interpreted to identify major cost drivers, evaluate cost trade-offs, and determine the most economically viable alternatives. The selected option is justified based on the cost findings. The process concludes with comprehensive documentation of the methodology and results, followed by an effective presentation to the intended audience, which may include decision-makers or stakeholders. Importantly, the LCC analysis should be revisited periodically to reflect changes in financial conditions, system performance, technical requirements, or scheduling constraints, thereby ensuring its continued relevance and usefulness in decision-making.

15.3.2.1 Social-Life Cycle Assessment (SLCA)

SLCA is a systematic approach designed to assess social and socio-economic effects (both positive and negative) of goods, operations, or services across their full life span. Unlike environmental LCA, which focuses on environmental emissions and resource use, SLCA addresses how a product influences people and society at various stages of the product life cycle, from initial raw material sourcing to final disposal. It evaluates the influence on a wide range of stakeholders, from the workforce and consumers to communities, society, and global supply chain participants.

The conceptual foundation and methodological guidance for SLCA are provided jointly by the United Nations Environment Programme (UNEP) and Society of Environmental Toxicology and Chemistry (SETAC). Their 2009 Guidelines and the 2020 Methodological Sheets have become the primary references for practitioners and researchers conducting SLCA (Benoit et al., 2013). SLCA is increasingly seen as a crucial tool for supporting sustainable development by making social aspects more visible in product and policy decision-making.

While it shares the structural framework of environmental LCA (as defined in ISO 14040 and ISO 14044 (ISO, 2006)), SLCA includes unique challenges related to the qualitative nature of social data, regional and cultural variability, and stakeholder perceptions. Despite these challenges, SLCA can be conducted either independently or integrated with environmental and economic assessments to support a comprehensive LCSA.

The four main phases of SLCA, as aligned with the LCA framework, are described below:

- **Goal and Scope Definition:** This phase establishes the purpose of the SLCA study, identifying the key questions to be addressed, the intended audience, and the decision-making context. The system boundaries are defined to determine which life cycle stages and stakeholder groups will be assessed. The functional unit or reference flow is also specified, ensuring that results can be meaningfully interpreted and compared. Similar to environmental LCA, the scope must consider data availability, time frame, and geographical relevance.
- **Life Cycle Inventory (LCI) Analysis:** The inventory phase involves the systematic collection of information concerning the social dimensions of the product system. Data may be both quantitative (e.g., number of working hours, injury rates) and qualitative (e.g., perceptions of well-being, labor conditions). Sources include literature reviews, audits, company reports, interviews, expert consultations, stakeholder engagement, and surveys. Tools such as structured and semi-structured questionnaires are commonly employed. The complexity of data collection in SLCA is often higher due to cultural diversity, sensitivity of social issues, and reliance on qualitative judgments.
- **Life Cycle Impact Assessment (LCIA):** At this stage, collected social data are sorted and assessed within pertinent social impact categories (e.g., fair wages, child labor, health and safety, and access to services). These categories are further grouped into subcategories and indicators as defined by UNEP/SETAC. Indicator scores are assigned to evaluate the extent of the impact, often using rating scales, stakeholder feedback, or expert scoring. The output can include the identification of social hotspots—critical areas where the most significant social risks or opportunities occur along the life cycle.
- **Interpretation:** This final step involves analyzing the results in light of the original goal and scope. Key social issues and hotspots are identified, and uncertainties, limitations, and data gaps are discussed. The interpretation also includes recommendations for improving the social performance of the product, such as supplier engagement, worker protection measures, or community development initiatives. Interpretation should be transparent, reproducible, and aligned with the ethical context of the study.

As summarized in Table 15.2, the three methodologies differ in scope, strengths, and maturity levels.

15.3.3 ISO 14040 Series: International Standards Guiding LCA

The advancement and standardization of LCA practices have been strongly guided by the ISO, particularly through the ISO 14040 series. These standards serve as the "constitution" for LCA, providing universally recognized frameworks that ensure consistency, transparency, and scientific rigor in environmental assessments (Finkbeiner (2014).

TABLE 15.2
Comparison of LCSA Methodologies

Methodology	Scope	Strengths	Limitations	Maturity Level
Life Cycle Assessment (LCA)	Environmental impacts	Well-established, standardized, widely applied	Methodological challenges remain in some impact categories	High
Life Cycle Costing (LCC)	Economic evaluation	Integrates cost over the life cycle	Lack of standard indicators, unclear metrics, weak linkages	Medium
Social Life Cycle Assessment (SLCA)	Social impacts	Captures social aspects often neglected	No standardized guidance, limited comparability	Low

15.3.3.1 The ISO 14040 Series: A Framework for Environmental Soundness

The ISO 14040 series consists primarily of the following:

15.3.3.1.1 ISO 14040:2006 – Principles and Framework

This foundational standard defines the fundamental principles and structure of LCA, encompassing goal and scope definition, inventory analysis, impact assessment, and interpretation. It emphasizes the iterative nature of LCA and promotes a systems-based approach to environmental decision-making (ISO 14040 2006a).

15.3.3.1.2 ISO 14044:2006 – Requirements and Guidelines

This standard supplements ISO 14040 by detailing methodological standards and protocols for performing a complete and consistent LCA study. It outlines data quality requirements, allocation procedures, and how to handle multi-functionality in complex systems. Together, these standards form the backbone of modern LCA, ensuring that studies conducted across industries and geographies maintain a high degree of comparability and credibility (ISO 14044 2006b).

15.3.3.1.3 The Offspring Standards: Specialized Guidance and Expansions

Several derivative standards have evolved from the ISO 14040 series, targeting specific needs within environmental management and LCT. These include:

ISO 14046:2014a – Water Footprint: Focuses on water-related impacts in LCA, offering a framework to assess and report water use in terms of quantity and quality (ISO 14046 2014a).

ISO 14067:2018 – Carbon Footprint (CFP) of Products: Outlines principles, criteria, and protocols for measuring and reporting a product's CFP, grounded in LCA methodology (ISO 14067 2018).

ISO 14072:2014b – LCA for Organizations: Extends LCA methodology to organizational-level assessments, enabling businesses to evaluate their environmental impacts comprehensively (ISO 14072 2014b).

These offspring standards allow for deeper, more focused environmental assessments while staying aligned with the fundamental ISO 14040 framework.

15.3.3.1.4 Global Relevance and Impact

The ISO 14040 series has played a crucial role in the global adoption of LCA, making it an essential tool in environmental policy, product development, sustainability reporting, and corporate environmental strategies. Governments, industries, and researchers widely rely on these standards to support environmentally informed decision-making and meet sustainability goals.

15.4 TIERED APPROACH FOR APPLYING LCSA IN MANUFACTURING SYSTEMS

To address this gap, the Tiered Approach has been proposed as a structured, step-by-step methodology that facilitates the gradual implementation of LCSA. It enables practitioners to start with a simplified LCSA and progressively move toward a more detailed and comprehensive assessment (Neugebauer et al. (2015).

15.4.1 Framework of the Tiered Approach

The Tiered Approach serves as a practical roadmap for applying LCSA in real-world settings. It introduces a hierarchical structure of impact and cost categories, guiding users in selecting appropriate indicators at different levels of assessment maturity. This structured method not only assists practitioners in indicator selection but also suggests potential areas for future methodological development in LCSA.

The selection of categories in this framework is informed by authoritative references such as:

- ILCD Guidebook for Life Cycle Assessment (JRC, 2011),
- Guidelines for Conducting Social Life Cycle Assessment of Products (UNEP/SETAC 2009), and
- Standards for Life Cycle Costing (Swarr et al. (2011).

Core criteria—importance, methodological integrity, and implementation viability are used to determine the inclusion of each category. In LCA, impact categories are chosen at the intermediate impact, as these are typically associated with more consensus-based characterization techniques with reduced statistical uncertainties compared to final outcomes (Bare et al., 2000).

The Tiered Approach is organized into three progressively detailed levels:

15.4.1.1 Tier 1: Sustainability Footprint (Entry-Level LCSA)

This tier represents a basic, low-threshold entry point for LCSA implementation. It considers a minimal yet essential set of sustainability categories—typically limited to climate impact, production expenditures, and worker remuneration. Tier 1 allows for an initial screening across all three pillars of sustainability, making it accessible even to non-experts and industrial practitioners with limited resources or data. It aligns the differing levels of maturity among LCA, LCC, and SLCA and helps initiate sustainability dialogues within manufacturing organizations.

15.4.1.2 Tier 2: Best Practice LCSA

Tier 2 reflects a more advanced and commonly practiced level of LCSA. It incorporates additional categories that are widely recognized and used in existing guidelines, such as the ILCD Handbook. For instance:

- LCA categories: Ozone depletion, eutrophication, acidification, and photochemical oxidant formation.
- LCC categories: Consumer-related costs such as purchase price, energy usage, and maintenance expenses.
- SLCA categories: Social indicators beyond the worker level, including health impacts on consumers and local communities, and working conditions.

This tier offers a deeper insight into environmental and economic impacts and begins to broaden the social dimension beyond basic workforce-related aspects.

15.4.1.3 Tier 3: Comprehensive LCSA

The third tier offers the most detailed and inclusive level of assessment. It integrates a wide range of additional and emerging sustainability indicators:

- For LCA: New and emerging impact categories such as water consumption and land occupation.
- For LCC: Expanded amounts including operational risks, accident-related costs, and environmental damage costs (when not already covered in LCA or SLCA).
- For SLCA: Advanced social aspects such as education, human rights, and preservation of cultural heritage.

Tier 3 is intended for in-depth analyses, particularly in contexts involving strategic planning, regulatory compliance, or stakeholder reporting. To provide a concise comparison of the three levels of the Tiered Approach, Table 15.3 summarizes the key features, indicators, and typical applications for each tier.

15.4.2 Implementation of the Tiered Approach in Manufacturing

The Tiered Approach has been practically validated through several case studies involving a range of manufacturing technologies and products. Notable examples include the analysis of modular machine tool frames, wireless microsystems, turning technologies, conventional bikes, and e-bikes (Peukert et al. (2015), Benecke et al. (2015), Neugebauer et al. (2013), Buchert et al. (2015). These studies primarily implemented Tier 1 of the approach, focusing on simplified indicators such as climate impact, operational costs, and fair labor practices. Selected examples of LCSA applications across manufacturing sectors are summarized in Table 15.4, highlighting the focus areas and outcomes of each case study.

The application of Tier 1 allowed for the identification of the following:

- Environmental hotspots across product life cycles;
- Key social issues (e.g., fair wages);
- Economic performance concerns (e.g., production costs);
- Opportunities for sustainability improvements in the evaluated technologies and processes.

TABLE 15.3
Tiered Approach to LCSA in Manufacturing

Tier	Focus/Purpose	Key Indicators	Applications
Tier 1: Entry Level	Basic sustainability screening	LCA: Climate impact LCC: Production costs SLCA: Worker remuneration	Initial assessments; easy implementation for non-experts
Tier 2: Best Practice	Widely adopted, deeper insight	LCA: Ozone depletion, eutrophication, acidification LCC: Purchase, energy, maintenance costs SLCA: Health and working conditions	Broader environmental, economic and social analysis
Tier 3: Comprehensive	In-depth, advanced assessment	LCA: Water, land use LCC: Operational and environmental costs SLCA: Education, human rights, cultural heritage	Strategic planning, regulatory compliance, stakeholder reporting

TABLE 15.4
Applications of LCSA in Manufacturing

Sector/Product	LCSA Focus	Outcomes/Benefits
Automotive	LCA, LCC, and SLCA	Identified environmental hotspots, optimized production costs, and improved social compliance
Electronics	LCA and LCC	Reduced energy use, cost-effective life cycle management
Consumer goods	SLCA	Highlighted social impact areas, improved labor conditions
Food and beverage	LCA and SLCA	Minimized environmental footprint, enhanced supply chain transparency

In addition to the above, these case studies provided insights into the real-world challenges of implementing LCSA across environmental, economic, and social dimensions. For example, for advanced modular machine tool frames manufactured across Germany, Brazil, and China, country-specific social indicators such as fair wages and worker health were assessed. These comparisons enabled actionable recommendations regarding sustainable material selection, supplier management, and design improvements.

Moreover, the case studies highlighted potential trade-offs among environmental, economic, and social aspects. For instance, replacing conventional coolant-based turning processes with inner-cooled alternatives offered environmental profits but also introduced increased social risks. These risks were associated with the involvement of low-wage labor in certain regions, thereby demonstrating the need for balanced decision-making that accounts for such trade-offs.

15.4.3 Research Gaps and Future Directions

While the Tiered Approach marks a significant step toward practical LCSA implementation, several research gaps and methodological limitations persist:

1. Incomplete Indicator Sets: Comprehensive and standardized sets of impact indicators for SLCA and LCC are still underdeveloped. This limits the consistency and comparability of assessments, particularly beyond Tier 1.
2. Lack of Defined Impact Pathways: Unlike environmental LCA, which has established impact categories and pathways, SLCA and LCC lack clearly defined frameworks for evaluating social and economic outcomes.
3. Trade-off Interpretation Challenges: A major issue during the interpretation phase is addressing the trade-off between the environmental, economic, and social dimensions. The Tiered Approach allows for transparent presentation of these trade-offs, but does not yet offer mechanisms to resolve or prioritize conflicting outcomes.
4. Absence of Weighting Systems: The current framework does not support the weighting or ranking of sustainability dimensions, which limits its utility in decision-making contexts where trade-offs are unavoidable (Zamagni et al. 2013; Arcese et al. 2013).

To address these gaps, future research should aim to the following:

- Develop standardized and widely accepted indicators for SLCA and LCC;
- Define robust and actionable social and economic impact pathways;
- Integrate decision-support tools like multi-criteria decision analysis (MCDA);
- Expand case study applications across a broader range of manufacturing contexts;
- Improve methods for stakeholder engagement and context-specific weighting.

By overcoming these limitations, the Tiered Approach can be refined into a comprehensive, practical, and scalable tool for conducting LCSA in the manufacturing sector, facilitating more sustainable strategic planning from design to disposal.

15.5 CONCLUSION

The transition toward SM demands comprehensive methodologies that assess environmental, economic, and social impacts throughout a product's life cycle. This chapter has explored the evolution, implementation, and challenges of LCSA within the manufacturing sector by systematically analyzing its core components—LCA, LCC, and SLCA and their integration through the Tiered Approach.

The chapter began by contextualizing LCSA within the broader sustainability discourse, emphasizing the relevance of LCT and the TBL approach. It was shown that while LCA is a methodologically mature and widely applied tool for evaluating environmental performance, LCC and SLCA still face considerable challenges regarding standardization, data collection, and the development of consistent impact assessment frameworks.

A critical review of existing literature revealed a predominant focus on energy consumption in sustainability studies, with limited attention paid to other essential factors such as material waste, social equity, and full economic costing. This research gap highlights the urgent need for more comprehensive life cycle inventories and the expansion of sustainability indicators beyond environmental parameters alone.

Through a detailed discussion of the Tiered Approach, the chapter offered a structured and scalable methodology for implementing LCSA in manufacturing. By categorizing assessment activities into three levels of complexity—Tier 1 (entry-level), Tier 2 (best practice), and Tier 3 (comprehensive)—the approach enables industries with varying resource capacities to progressively adopt LCSA principles. Case studies showcased the practical applicability of Tier 1 in identifying environmental hotspots, economic inefficiencies, and social concerns, while also revealing complex trade-offs among the three sustainability pillars.

Despite its promise, the Tiered Approach—and LCSA more broadly—remains limited by methodological gaps. These include the absence of standardized impact pathways for LCC and SLCA, lack of comprehensive and validated indicator sets, and the challenge of transparently managing trade-offs between conflicting sustainability . Addressing these challenges will require future research to focus on harmonizing frameworks, developing integrated databases, incorporating multi-criteria decision-making tools, and fostering cross-sectoral collaboration.

Importantly, the continued advancement and adoption of LCA—both as a standalone method and as part of the LCSA framework—has been significantly supported by the advancement of internationally recognized standards, particularly the ISO 14040 series. The ISO 14040 and ISO 14044 standards have established the foundational principles, terminology, and methodological guidance necessary for ensuring consistency, scientific robustness, and comparability in LCA applications across sectors. Furthermore, derivative Standards such as ISO 14046 (Water Footprint Assessment) and ISO 14067 (Carbon Footprint of Products), and ISO 14072 (Organizational LCA) have expanded the applicability of life cycle approaches in addressing specific sustainability goals. These standards function as the methodological backbone of LCA, and they represent a critical step toward the harmonization and institutionalization of sustainability assessments on a global scale.

In conclusion, while LCSA represents a powerful conceptual and methodological tool for enabling SM, its effective application depends on continued methodological refinement and broader industry adoption. The Tiered Approach offers a promising pathway for advancing LCSA implementation, but its success ultimately hinges on collaborative efforts to standardize methods, improve data availability, and align stakeholder priorities. Only then can LCSA fulfill its potential as a cornerstone of sustainable production systems and policy-making in the industrial era.

REFERENCES

Agrawal, R., and Vinodh, S. 2019. State of art review on sustainable additive manufacturing. *Rapid Prototyping Journal*, 25(6), 1045–1060. https://doi.org/10.1108/RPJ-04-2018-0085

Arcese, G., Merli, R., and Lucchetti, M. C. 2013. Life cycle approach: A critical review in the tourism sector. In *The 3rd World Sustainability Forum* (1–10).

Bare, J. C., Hofstetter, P., Pennington, D. W., and De Haes, H. A. U. 2000. Midpoints versus endpoints: The sacrifices and benefits. *The International Journal of Life Cycle Assessment*, 5: 319–326.

Benecke, Stephan, Neugebauer, Sabrina, Peukert, Bernd, et al. 2015. Sustainability assessment of wireless micro systems in smart manufacturing environments. *EcoDesign 2015*, Tokyo, Japan. (Siehe Anhang).

Benoit C., Mazijn B., United Nations Environment Programme, et al. 2013 *Guidelines for social life cycle assessment of products*. United Nations Environment Programme, Paris, France.

Bogue, R. 2014. Sustainable manufacturing: a critical discipline for the twenty-first century. *Assembly Automation*, 34(2):117–122.

Buchert, T., Steingrímsson, J. G., Neugebauer, S., Nguyen, T. D., Galeitzke, M., Oertwig, N., Seidel, J., McFarland, R., Lindow, K., Hayka, H., and Stark, R. 2015. Design and manufacturing of a sustainable pedelec. *Procedia CIRP*, 29: 579–584.

Chang, Y. J., Neugebauer, S., Lehmann, A., Scheumann, R., and Finkbeiner, M. 2017. Life cycle sustainability assessment approaches for manufacturing. *Sustainable Manufacturing* 221.

Chatti, S., Laperrière, L., Reinhart, G., and Tolio, T. 2019. *CIRP encyclopedia of production engineering*. Springer.

Despeisse, M., Mbaye, F., Ball, P. D., and Levers, A. 2012. The emergence of sustainable manufacturing practices. *Production Planning & Control*, 23(5): 354–376.

Dhillon, B.S. 2017. *Life cycle costing for engineers, First issued in paperback*. Boca Raton; London; New York: CRC Press.

Duflou, J. R., Sutherland, J. W., Dornfeld, D., Herrmann, C., Jeswiet, J., Kara, S., Hauschild, M., and Kellens, K. 2012. Towards energy and resource efficient manufacturing: A processes and systems approach. *CIRP Annals*, 61(2): 587–609.

Elkington, J., and Rowlands, I. H. 1999. Cannibals with forks: The triple bottom line of 21st century business. *Alternatives Journal*, 25(4): 42.

Finkbeiner, M. 2014. The International Standards as the Constitution of Life Cycle Assessment: The ISO 14040 Series and its Offspring. In: Klöpffer, W. (eds) *Background and Future Prospects in Life Cycle Assessment. LCA Compendium – The Complete World of Life Cycle Assessment*. Springer, Dordrecht. https://doi.org/10.1007/978-94-017-8697-3_3

Finkbeiner, M., Schau, E. M., Lehmann, A., and Traverso, M. 2010. towards life cycle sustainability assessment. *Sustainability* 2: 3309–3322.

Garcia, F. L., Moris, V. A. D. S., Nunes, A. O., and Silva, D. A. L. 2018. Environmental performance of additive manufacturing process–an overview. *Rapid Prototyping Journal*, 24(7): 1166–1177. https://doi.org/10.1108/RPJ-05-2017-0108

Greene, L. E., and Shaw, B. L. 1990. The steps for successful life cycle cost analysis. In *IEEE Conference on Aerospace and Electronics* 1209–1216.

International Organization for Standardization. 2006a. *ISO 14040: Environmental management — Life cycle assessment — Principles and framework*. Geneva, Switzerland: ISO.

International Organization for Standardization. 2006b. *ISO 14044: Environmental management — Life cycle assessment — Principles and framework*. Geneva, Switzerland: ISO.

International Organization for Standardization. 2014a. *ISO 14046: Environmental management — Life cycle assessment — Principles and framework*. Geneva, Switzerland: ISO.

International Organization for Standardization. 2014b. *ISO 14072: Environmental management — Life cycle assessment — Principles and framework*. Geneva, Switzerland: ISO.

International Organization for Standardization. 2017 *ISO 15686–5: Buildings and constructed assets*. Service life planning. Life-cycle costing.

International Organization for Standardization. 2018. *ISO 14067: Environmental management — Life cycle assessment — Principles and framework*. Geneva, Switzerland: ISO.

International Standard Organization. 2006. *Environmental management-Life cycle assessment-Requirements and guidelines*. London: ISO.

JRC. 2011. *ILCD handbook—recommendations for life cycle impact assessment in the European context*. First. Ispra: European Commission.

Kellens, K., Baumers, M., Gutowski, T. G., Flanagan, W., Lifset, R., and Duflou, J. R. 2017. Environmental dimensions of additive manufacturing: mapping application domains and their environmental implications *Research and Analysis*, 21(S1), S49–S68. https://doi.org/10.1111/jiec.12629

Kokare, S., Oliveira, J. P., and Godina, R. 2023. Life cycle assessment of additive manufacturing processes: A review. *Journal of Manufacturing Systems*, 68: 536–559.

Lehmann, A., Zschieschang, E., Traverso, M., Finkbeiner, M., and Schebek, L. 2013. Social aspects for sustainability assessment of technologies—challenges for social life cycle assessment (SLCA). *The International Journal of Life Cycle Assessment*, 18: 1581–1592.

Liu, Z. Y., Li, C., Fang, X. Y., and Guo, Y. B. 2018. Energy consumption in additive manufacturing of metal parts. *Procedia Manufacturing*, 26: 834–845.

Neugebauer, S., Chang, Y. J., Maliszewski, M., Lindow, K., Stark, R., and Finkbeiner, M. 2013. Life cycle sustainability assessment & sustainable product development: A case study on pedal electric cycles (Pedelec). *In Proceedings of the 11th global conference on sustainable manufacturing—innovative solutions* 549–554. Berlin: Universitätsverlag der TU Berlin.

Neugebauer, S., Forin, S., and Finkbeiner, M. 2016. From life cycle costing to economic life cycle assessment—introducing an economic impact pathway. *Sustainability*, 8(5): 428.

Neugebauer, S., Martinez-Blanco, J., Scheumann, R., and Finkbeiner, M. 2015. Enhancing the practical implementation of life cycle sustainability assessment–proposal of a Tiered approach. *Journal of Cleaner Production*, 102: 165–176.

Peukert, B., Benecke, S., Clavell, J., Neugebauer, S., Nissen, N. F., Uhlmann, E., Lang, K. D. and Finkbeiner, M. 2015. Addressing sustainability and flexibility in manufacturing via smart modular machine tool frames to support sustainable value creation. *Procedia CIRP*, 29: 514–519.

Rebitzer, G., Ekvall, T., Frischknecht, R., Hunkeler, D., Norris, G., Rydberg, T., Schmidt, W. P., Suh, S., Weidema, B. P. and Pennington, D. W. 2004. Life cycle assessment: Part 1: Framework, goal and scope definition, inventory analysis, and applications. *Environment International*, 30(5): 701–720.

Remmen, A. 2007. *Life cycle management: a business guide to sustainability*. UNEP/Earthprint.

Saade, M. R. M., Yahia, A., and Amor, B. 2020. How has LCA been applied to 3D printing? A systematic literature review and recommendations for future studies. *Journal of Cleaner Production*, 244, 118803. https://doi.org/10.1016/j.jclepro.2019.118803

Swarr, T., D. Hunkeler, W. Kloepffer, H.L. Pesonen, A. Ciroth, C. Brent, and R. Pagan. 2011. *Environmental life cycle costing: A code of practice*.

UNEP/SETAC. 2009. *Guidelines for social life cycle assessment of products*. Belgium: In de Weer.

Valdivia, S., Ugaya, C. M., Hildenbrand, J., Traverso, M., Mazijn, B., and Sonnemann, G. 2013. A UNEP/SETAC approach towards a life cycle sustainability assessment—our contribution to Rio+ 20. *The International Journal of Life Cycle Assessment*, 18: 1673–1685.

Zamagni, A., Pesonen, H. L., and Swarr, T. 2013. From LCA to Life Cycle Sustainability Assessment: concept, practice and future directions. *The International Journal of Life Cycle Assessment*, 18: 1637–1641.

Zsidisin, G. A., and Siferd, S. P. 2001. Environmental purchasing: a framework for theory development. *European Journal of Purchasing & Supply Management*, 7(1): 61–73.

Index

Pages in *italics* refer to figures and pages in **bold** refer to tables.

For Product Safety Concerns and Information please contact our EU representative GPSR@taylorandfrancis.com Taylor & Francis Verlag GmbH, Kaufingerstraße 24, 80331 München, Germany